Selective Glycosylations: Synthetic Methods and Catalysts

Selective Glycosylations

Synthetic Methods and Catalysts

Edited by Clay S. Bennett

WILEY-VCH
Verlag GmbH & Co. KGaA

Editor

Prof. Clay S. Bennett
Tufts University
Department of Chemistry
62 Talbot Ave.
Medford, MA 02155
United States

Cover

The cover image was kindly provided by the Editor

Library of Congress Card No.: applied for

British Library Cataloguing-in-Publication Data
A catalogue record for this book is available from the British Library.

Bibliographic information published by the Deutsche Nationalbibliothek
The Deutsche Nationalbibliothek lists this publication in the Deutsche Nationalbibliografie; detailed bibliographic data are available on the Internet at http://dnb.d-nb.de.

Print ISBN: 978-3-527-33987-7
ePDF ISBN: 978-3-527-69622-2
ePub ISBN: 978-3-527-69624-6
Mobi ISBN: 978-3-527-69625-3
oBook ISBN: 978-3-527-69623-9

Cover Design Schulz Grafik-Design, Fußgönheim, Germany
Typesetting SPi Global, Chennai, India
Printing and Binding Markono Print Media Pte Ltd, Singapore

Printed on acid-free paper

Contents

List of Contributors

Mithila D. Bandara
University of Missouri - St. Louis
Department of Chemistry and
Biochemistry
One University Blvd.
St. Louis, MO 63121
USA

David Benito-Alifonso
University of Bristol
School of Chemistry
Cantock's Close
Bristol BS8 1TS
UK

Clay S. Bennett
Tufts University
Department of Chemistry
62 Talbot Ave.
Medford, MA 02155
USA

Luis Bohé
Institut de Chimie des Substances
Naturelles
CNRS-ICSN UPR2301
Université Paris-Sud
Avenue de la Terrasse
91198 Gif-sur-Yvette
France

Robin Brabham
York Structural Biology Laboratory
Department of Chemistry
University of York
Heslington, York
YO10 5DD
UK

Jeroen D.C. Codée
Leiden University
Leiden Institute of Chemistry
PO Box 9502
2300 RA Leiden
The Netherlands

David Crich
Wayne State University
Department of Chemistry
5101 Cass Ave.
Detroit, MI 48202
USA

Alexei V. Demchenko
University of Missouri - St. Louis
Department of Chemistry and
Biochemistry
One University Blvd.
St. Louis, MO 63121
USA

Alisa E. R. Fairweather
University of Iowa
Department of Chemistry
E331 Chemistry Building
Iowa City, IA 52245
USA

Martin A. Fascione
York Structural Biology Laboratory
Department of Chemistry
University of York
Heslington, York
YO10 5DD
UK

M. Carmen Galan
University of Bristol
School of Chemistry
Cantock's Close
Bristol BS8 1TS
UK

Carola Gallo-Rodriguez
Universidad de Buenos Aires
CIHIDECAR Departamento de Química Orgánica
Ciudad Universitaria, Pabellón II
1428 Buenos Aires
Argentina

Bas Hagen
Leiden University
Leiden Institute of Chemistry
PO Box 9502
2300 RA Leiden
The Netherlands

Thomas Hansen
Leiden University
Leiden Institute of Chemistry
PO Box 9502
2300 RA Leiden
The Netherlands

Kim Le Mai Hoang
Nanyang Technological University
Division of Chemistry and Biological Chemistry
School of Physical and Mathematical Sciences
SPMS-CBC-02-01
21 Nanyang Link, 637371
Singapore

Akihiro Ishiwata
RIKEN
Synthetic Cellular Chemistry Laboratory
2-1 Hirosawa Wako
Saitama 351-0198
Japan

Yukishige Ito
RIKEN
Synthetic Cellular Chemistry Laboratory
2-1 Hirosawa Wako
Saitama 351-0198
Japan

Gustavo A. Kashiwagi
Universidad de Buenos Aires
CIHIDECAR Departamento de Química Orgánica
Ciudad Universitaria, Pabellón II
1428 Buenos Aires
Argentina

Suvarn S. Kulkarni
Indian Institute of Technology Bombay
Department of Chemistry
Powai
Mumbai 400076
India

Wei-Lin Leng
Nanyang Technological University
Division of Chemistry and Biological Chemistry
School of Physical and Mathematical Sciences
SPMS-CBC-02-01
21 Nanyang Link, 637371
Singapore

Yu-Hsuan Lih
Genomics Research Center, Academia Sinica
128 Academia Road, Section 2, Nankang
Taipei 115
Taiwan

Xue-Wei Liu
Nanyang Technological University
Division of Chemistry and Biological Chemistry
School of Physical and Mathematical Sciences
SPMS-CBC-02-01
21 Nanyang Link, 637371
Singapore

Ravi S. Loka
University of Iowa
Department of Chemistry
E331 Chemistry Building
Iowa City, IA 52245
USA

Gijs van der Marel
Leiden University
Leiden Institute of Chemistry
PO Box 9502
2300 RA Leiden
The Netherlands

KwoK-Kong Tony Mong
National Chiao Tung University
Applied Chemistry Department
1001 Ta Hsueh Road
Hsinchu 300
Taiwan ROC

Hien M. Nguyen
University of Iowa
Department of Chemistry
E331 Chemistry Building
Iowa City, IA 52245
USA

Pham Be Nhi
National Chiao Tung University
Applied Chemistry Department
1001 Ta Hsueh Road
Hsinchu 300
Taiwan ROC

Toshiki Nokami
Tottori University
Department of Chemistry and Biotechnology
4-101 Koyama-Minami
Tottori city
680-8552 Tottori
Japan

George A. O'Doherty
Northeastern University
Department of Chemistry and Chemical Biology
360 Huntington Ave.
Boston, MA 02115
USA

Justin Ragains
Louisiana State University
Department of Chemistry
232 Choppin Hall
Baton Rouge, LA 70803
USA

Pei Shi
Corden Pharma
1-B Gill Street
Woburn, MA 01801
USA

Eric T. Sletten
University of Iowa
Department of Chemistry
E331 Chemistry Building
Iowa City, IA 52245
USA

Yu-Jia Tan
Nanyang Technological University
Division of Chemistry and Biological Chemistry
School of Physical and Mathematical Sciences
SPMS-CBC-02-01
21 Nanyang Link, 637371
Singapore

Mark S. Taylor
University of Toronto
Department of Chemistry
80 St. George Street
Toronto, ON M5S 3H6
Canada

Nhut Thi Thanh Tran
National Chiao Tung University
Applied Chemistry Department
1001 Ta Hsueh Road
Hsinchu 300
Taiwan ROC

Stefan van der Vorm
University of Leiden
Leiden Institute of Chemistry
PO Box 9502
2300 RA Leiden
The Netherlands

Chung-Yi Wu
Genomics Research Center, Academia Sinica
128 Academia Road, Section 2, Nankang
Taipei 115
Taiwan

Jagodige P. Yasomanee
University of Missouri - St. Louis
Department of Chemistry and Biochemistry
One University Blvd.
St. Louis, MO 63121
USA

Preface

The past decade has seen increased recognition of the important roles oligosaccharides play in an array of biological process including (but not limited to) protein folding, pathogen invasion, cell adhesion, and immune response. As a consequence, the field of glycoscience is undergoing rapid growth, with an ever-increasing number of investigators turning their attention to it. Despite all of this, the field is still in its infancy, especially compared to other areas of biology such as genomics and proteomics. There are several reasons for this; chief among them is the fact that glycoscientists do not enjoy ready access to homogeneous material for study, an advantage that was critical for the advances in other areas of biomedical research. This is because, unlike the other major classes of biopolymers, cells produce carbohydrates as heterogeneous mixtures, which are often intractable. As a consequence, organic synthesis (including chemoenzymatic synthesis) remains the only avenue for the production of pure oligosaccharides for biomedical evaluation. The synthesis of most oligosaccharides is a nontrivial undertaking, however, owing to issues of regiochemistry and stereochemistry. Thus, while chemical glycosylation has been known for over a century, the construction of a new oligosaccharide can still be a research project in and of itself.

Among the challenges facing the chemist who wishes to synthesize oligosaccharides, one of the most significant is controlling selectivity in the glycosylation reaction. Typical glycosylation reactions proceed through a mechanism somewhere along the S_N1–S_N2 continuum, which renders controlling selectivity in the reaction immensely difficult. While several elegant solutions to this problem have been devised, a general approach to controlling selectivity in glycosylation reactions with a broad range of substrates remains to be developed. This has prompted calls for the development of new approaches to glycosylation from numerous sectors. Before embarking on developing a new approach to glycosylation, however, it is first necessary to understand the advances that currently constitute the state of the art. The purpose of this volume is to describe the principles of chemical glycosylation. Rather than break down the text into chapters focusing on activating different classes of leaving groups, the focus is instead largely on mechanistic aspects that are responsible for selectivity. Furthermore, technologies for automated and one-pot synthesis have been extensively reviewed elsewhere and will only be covered when relevant.

This volume is organized into five parts. The first part deals with an introduction to the basic principles or carbohydrate synthesis. In Chapter 1, Codeé *et al.* outline the factors responsible for controlling additions to oxocarbenium cations. Next, Demchenko and coworkers describe the roles protecting groups play in both attenuating glycan reactivity and controlling stereoselectivity in glycosylations in Chapter 2. This part concludes with Chapter 3 from Mong, Nokami, and coworkers, which details the roles solvents play in controlling the stereochemical outcome of glycosylation.

Part II describes ways in which electrophilic glycosyl donors can be modified to undergo selective reactions. In Chapter 4, Ishiwata and Ito provide a detailed introduction to the use of Intramolecular Aglycone Delivery (IAD) for the stereoselective synthesis of *cis*-1,2-glycans. This is followed by a discussion of the use of chiral auxiliaries in oligosaccharide synthesis by Brabham and Fascione in Chapter 5. Finally, Bohé and Crich describe how glycosyl sulfonates permit the construction of the so-called difficult linkages through S_N2-like glycosylations in Chapter 6.

The development of methods for catalytic activation of glycosyl donors is the focus of Part III. This part begins with a description of methods for the construction *C*-glycans, often through transition-metal-mediated processes, by Liu *et al.* in Chapter 7. This is followed by a comprehensive overview of recent approaches for catalytic activation of donors for O-glycosylation by Benito-Alfonso and Galan in Chapter 8. In Chapter 9, Nguyen and coworkers provide a case study in catalytic activation, focusing on their Ni-catalyzed 1,2-*cis*-glycoside synthesis. This part concludes with an introduction to the increasingly popular field of photochemical glycosylation by Ragains in Chapter 10.

In addition to the challenges in controlling the stereochemical outcome of glycosylation reactions, regioselectivity is a problem the synthetic chemist must attend to when dealing with glycosides. Current state-of-the-art approaches to addressing this issue are outlined in Part IV of the volume. In Chapter 11, Taylor provides us with a discussion of methods for regioselectively glycosylating unprotected glycosyl acceptors. This is followed by a discussion of methods for one-pot protection and functionalization of unprotected glycans by Kulkarni in Chapter 12.

The final part of the volume provides the reader with examples of classes of glycans where standard approaches to glycosylation do not always apply. This begins with an overview of recent advances in 2-deoxy-sugar synthesis by Bennett in Chapter 13. Gallo-Rodriguez and Kashiwagi follow this up with an introduction to the challenging issue of controlling selectivity in glycosylations with furanoside donors (Chapter 14). Next, Shi and O'Doherty provide us with a description of how the *de novo* synthesis can permit the construction of a number of carbohydrate natural products in Chapter 15. Finally, Lih and Wu provide an overview of the state of the art in the synthesis of sialic acids in Chapter 16.

The goal of this volume is to try to provide a holistic view of chemical glycosylation. Our target audience is not limited to individuals who are currently engaged in carbohydrate chemistry but extends to the larger synthetic community, many of whom may be new to the field. Our hope is that this volume will inspire investigators to make new, and ideally unforeseen, contributions to the field. We do this because we believe that it will be necessary to engage as many investigators as possible if we are to achieve the long-term goal of developing technologies that will permit the routine and rapid construction of oligosaccharide libraries that are desperately needed for the study of glycobiology.

Medford, MA, USA

Clay S. Bennett
Tufts University

Part I

Introduction

1

Stereoselective Glycosylations – Additions to Oxocarbenium Ions

Bas Hagen, Stefan van der Vorm, Thomas Hansen, Gijs A. van der Marel, and Jeroen D.C. Codée

1.1 Introduction

Tremendous progress has been made in the construction of oligosaccharides, and many impressive examples of large and complex oligosaccharide total syntheses have appeared over the years [1]. At the same time, the exact mechanism underlying the union of two carbohydrate building blocks often remains obscure, and optimization of a glycosylation reaction can be a time- and labor-intensive process [2, 3]. This can be explained by the many variables that affect the outcome of a glycosylation reaction: the nature of both the donor and acceptor building blocks, solvent, activator and activation protocol, temperature, concentration, and even the presence and the type of molecular sieves. The large structural variety of carbohydrates leads to building blocks that differ significantly in reactivity, with respect to both the nucleophilicity of the acceptor molecule and the reactivity of the donor species. The reactivity of a donor is generally related to the capacity of the donor to accommodate developing positive charge at the anomeric center, upon expulsion of the anomeric leaving group. This also determines the amount of carbocation character in the transition state leading to the products. Most glycosylation reactions will feature characteristics of both S_N1- and S_N2-type pathways in the transition states leading to the products. It is now commonly accepted that the exact mechanism through which a glycosidic linkage is formed can be found somewhere in the continuum of reaction mechanisms that spans from a completely dissociative S_N1 mechanism on one side to an associative S_N2 pathway on the other side (Figure 1.1) [4–6]. On the S_N1-side of the spectrum, glycosyl oxocarbenium ions are found as product-forming intermediates. On this outer limit of the reaction pathway continuum, the oxocarbenium ions will be separated from their counterions by solvent molecules (solvent-separated ion pairs, SSIPs), and there will be no influence of the counterion on the selectivity of the reaction. Moving toward the S_N2 side of the spectrum contact (or close) ion pairs (CIPs) are encountered, and in reactions of these species, the counterion will have a role to play. Because glycosylation reactions generally occur in apolar solvents (dichloromethane is by far the most used one), ionic intermediates have very limited lifetimes, and activated donor species will primarily be present as a pool of

Selective Glycosylations: Synthetic Methods and Catalysts, First Edition. Edited by Clay S. Bennett.

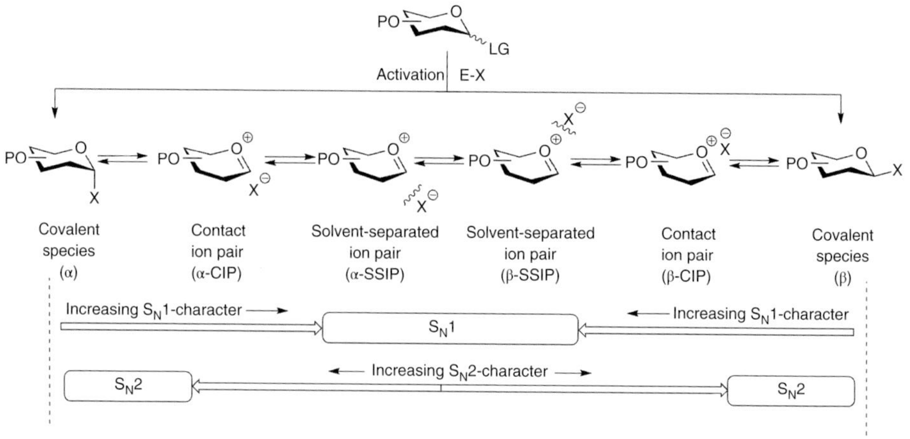

Figure 1.1 Continuum of mechanisms to explain the stereochemical course of glycosylation reactions.

covalent intermediates. The stability, lifetime, and reactivity of an oxocarbenium ion depend – besides the nature of the counterion – on the nature and orientation of the functional groups present on the carbohydrate ring. This chapter explores the role of oxocarbenium ions (and CIPs, featuring a glycosyl cation) in chemical glycosylation reactions. While it was previously often assumed that glycosylations, proceeding via an oxocarbenium ion intermediate, show poor stereoselectivity, it is now clear that oxocarbenium ions can be at the basis of stereoselective glycosylation events. The first part of this chapter deals with the stability, reactivity, and conformational behavior of glycosyl oxocarbenium ions, whereas the second part describes their intermediacy in the assembly of (complex) oligosaccharides.

1.2 Stability, Reactivity, and Conformational Behavior of Glycosyl Oxocarbenium Ions

Amyes and Jencks have argued that glycosyl oxocarbenium ions have a short but significant lifetime in aqueous solution [7]. They further argued that in the presence of properly positioned counterions (such as those derived of expulsion of an aglycon), CIPs will rapidly collapse back to provide the covalent species and that the "first stable intermediate for a significant fraction of the reaction" should be the solvent-separated oxocarbenium ion. By extrapolation of these observations to apolar organic solvents, Sinnott reached the conclusion that intimate ion pairs have no real existence in an apolar environment, such as used for glycosylation reactions [8]. Hosoya *et al.* have studied CIPs by quantum mechanical calculations in dichloromethane as a solvent [9]. In these calculations, they have included four solvent molecules to accurately mimic the real-life situation. In many of the studied cases, CIPs turned out to be less stable than the corresponding solvent-separated ions, as will be described next [10]. Yoshida and coworkers have described that activation of thioglucoside **1** with a sulfonium salt activator, featuring the bulky nonnucleophilic tetrakis(pentafluorophenyl) borate counterion, in a continuous-flow microreactor, provides a reactive species (**2**) that has a lifetime on the

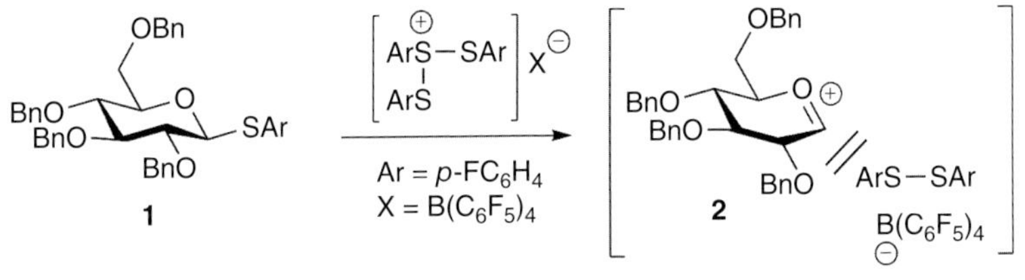

Scheme 1.1 Generation of glucosyl oxocarbenium ions in a continuous-flow microreactor.

order of a second (Scheme 1.1) [11]. They argued that this species was a glucosyl oxocarbenium ion, "somewhat stabilized" by the disulfide generated from the donor aglycon and the activator.

The stability of a glycosyl oxocarbenium ion is largely influenced by the substituents on the carbohydrate ring. The electronegative substituents (primarily oxygen, but also nitrogen-based) have an overall destabilizing effect on the carbocation, and the destabilizing effect can be further enhanced by the presence of electron-withdrawing protecting groups, such as acyl functions. The exact position of the substituent on the ring and its orientation influence the stability of the anomeric cation. The combined influence of all substituents on the ring determines the reactivity of a glycosyl donor, and the extensive relative reactivity value (RRV) charts, drawn up by the Ley and Wong groups for a large panel of thioglycosides, clearly illustrate these functional group effects [12–14]. From these RRV tables, it is clear that the donor reactivity spectrum spans at least eight orders of magnitude. To investigate the influence of the carbohydrate ring substituents on the stereochemical outcome of a glycosylation reaction, Woerpel and coworkers have systematically studied C-glycosylation reactions of a set of furanosides and pyranosides, featuring a limited amount of ring substituents [15–20]. Their studies in the furanose series are summarized in Scheme 1.2a [15, 17]. As can be seen, the alkoxy groups at C2 and C3 have a strong influence on the stereochemical outcome of the reaction, where the alkoxy group at C5 appears to have less effect on the reaction. The presence of an alkoxy or alkyl group at C3 leads to the formation of the allylglycosides **11** and **12** with opposite stereoselectivity. Woerpel and coworkers have devised a model to account for these stereodirecting substituent effects that takes into account the equilibrium between two possible envelope oxocarbenium ion conformers (**13** and **14**, Scheme 1.2b) [17]. Attack on these oxocarbenium ion conformers by the nucleophile occurs from the "inside" of the envelopes, because this trajectory avoids unfavorable eclipsing interactions with the substituent at C2, and it leads, upon rehybridization of the anomeric carbon, to a fully staggered product (**15** and **16**), where attack on the "outside" would provide the furanose ring with an eclipsed C1—C2 constellation. The spatial orientation of the alkoxy groups influences the stability of the oxocarbenium ions. An alkoxy group at C3 can provide some stabilization of the carbocation when it takes up a *pseudo*-axial position. Stabilization of the oxocarbenium ion featuring a C2-alkoxy group is best achieved by placing the electronegative substituent in a *pseudo*-equatorial position to allow for the hyperconjugative stabilization by the properly oriented C2—H2 bond. Alkyl substituents at C3 prefer to adopt a *pseudo*-equatorial position because of steric reasons. With these spatial substituent preferences, the stereochemical outcome of the C-allylation reactions in Scheme 1.2 can be explained. Activation of the C3-benzyloxyfuranosyl acetate with $SnBr_4$ can provide an

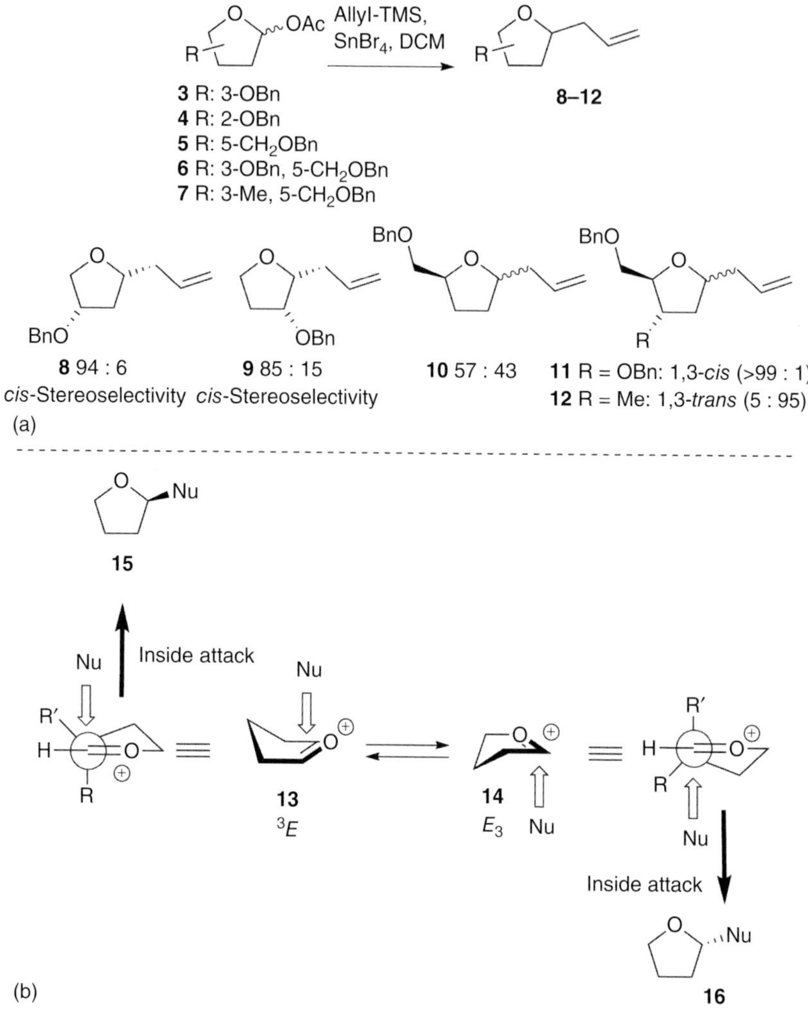

Scheme 1.2 (a) Diastereoselective C-allylations of furanosyl acetates. (b) "Inside" attack model.

oxocarbenium ion intermediate that preferentially adopts an E_3 conformation, as in **14**. Nucleophilic attack on this conformer takes place from the diastereotopic face that leads to the 1,3-*cis* product. In a similar vein, inside nucleophilic attack on the C2-benzyloxy furanosyl oxocarbenium ion E_3 conformer, derived from furanosyl acetate **4**, accounts for the stereochemical outcome of the C-allylation leading to product **9**.

To accurately gauge the combined effect of multiple substituents on a furanosyl ring, van Rijssel *et al.* [21, 22] used a quantum mechanical calculation method, originally developed by Rhoad and coworkers [23], to map the energy of furanosyl oxocarbenium ions related to the complete conformational space they can occupy. Energy maps for all four possible diastereoisomeric, fully decorated furanosyl oxocarbenium ions were generated revealing the lowest energy conformers for the ribo-, arabino-, xylo-, and lyxo-configured furanosyl oxocarbenium ions **17–21** (Scheme 1.3). It became apparent that the orientation of the C5-substituent, having a *gg*, *gt*, or *tg* relation to the substituents at

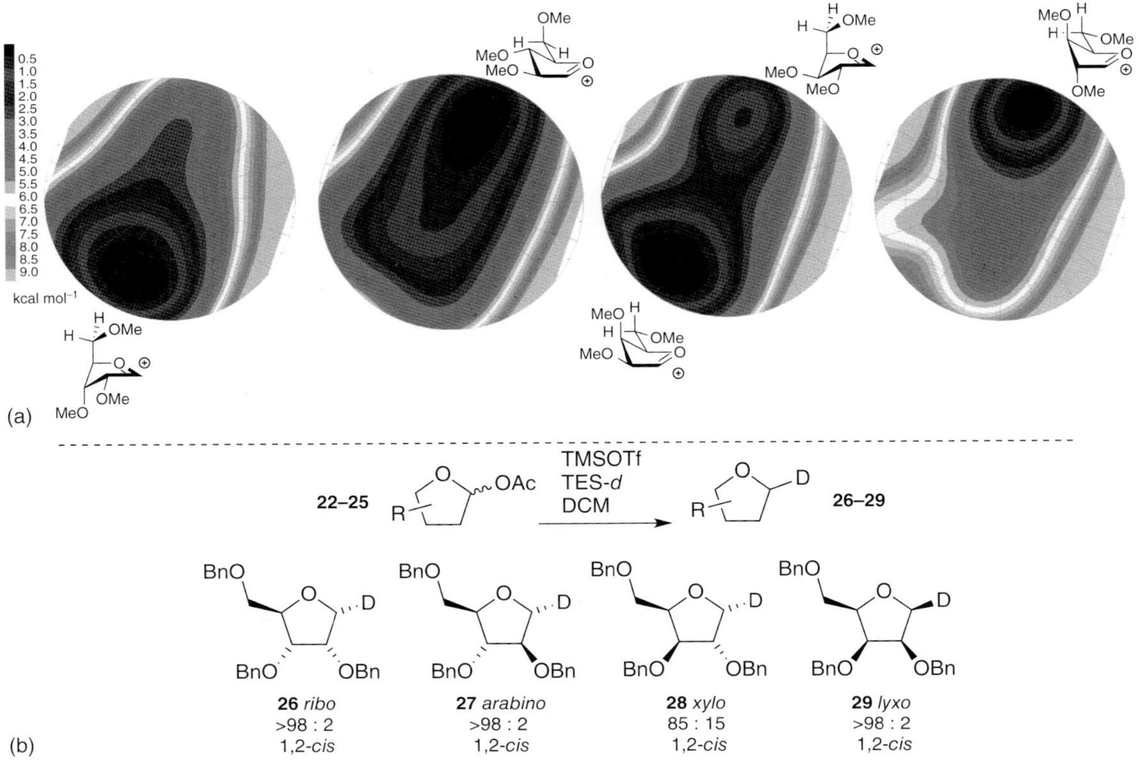

Scheme 1.3 Free energy surface maps of fully decorated furanosyl oxocarbenium ions and diastereoselective reductions of furanosyl acetates.

C4, was of profound influence on the stability of the oxocarbenium ions, and differences up to 4 kcal mol^{-1} were observed for structures only differing in their C4—C5 rotation. These stereoelectronic effects have also been described in the pyranose series, where a C4—C6 acetal can restrict the C6-oxygen in a *tg* position, for *manno-* and *gluco*-configured systems, or in a *gg* position for *galacto*-configured constellations [24–26]. The *tg* orientation represents the most destabilizing orientation because in this situation, the O6 atom is farthest away from the electron-depleted anomeric center, not allowing for any electron density donation for stabilization. With the lowest energy furanosyl oxocarbenium ion conformers found by the free energy surface (FES) mapping method, the stereochemical outcome of reduction reactions at the anomeric center of the four diastereoisomeric furanosyl acetates **22–25** could be explained (Scheme 1.3). Interestingly, all four furanosides reacted in a 1,2-*cis* selective manner with the incoming nucleophile (tri-ethylsilane-*d*). Only xylofuranosyl acetate **24** provided some of the 1,2-*trans* addition products, which could be related to the stability of the ^{3}E *gt* oxocarbenium ion intermediate **20**.

The stereoelectronic substituent effects found in the furanose series are paralleled in the pyranose system, where the following substituent effects have been delineated: the stability of pyranosyl oxocarbenium ions benefits from an equatorial orientation of the C2-alkoxy groups (allowing for hyperconjugative stabilization by the σ_{C2-H2} bond) and an axial orientation of the C3 and C4 alkoxy groups [20]. The C5-alkoxymethylene group has a slight preference for an equatorial position because of steric reasons [18]. These substituent preferences have been used to explain the stereochemical outcome of a series of C-allylations, using a two-conformer model. Woerpel and coworkers reasoned that six-membered oxocarbenium ions preferentially adopt a half-chair structure to accommodate the flat $[C1{=}O5]^+$ oxocarbenium ion moiety (Scheme 1.4a) [20]. These half-chair intermediates are attacked by incoming nucleophiles following a trajectory that leads to a chair-like transition state. Thus, attack of a $^{3}H_4$ half chair **30** preferentially occurs form the β-face (in the case of a D-pyranoside), where attack on the opposite half chair **31** (the $^{4}H_3$) leads to the α-product. With the described spatial substituent preferences and mode of nucleophilic attack, the stereoselectivities in the C-allylation reactions shown in Scheme 1.4b can be accounted for: the C4—OBn is trans-directing, where the C3 and C2—O—Bn promote the formation of the cis-product. In the lyxopyranosyl oxocarbenium ion, these three substituent preferences can be united, and the allylation of 2,3,4-tri-*O*-benzyl lyxopyranosyl acetate **35** proceeds in a highly stereoselective manner to provide the 1,2-*cis* product **40**.

When a C5 benzyloxymethyl group is added to this system, as in a mannosyl cation, it can be reasoned that the $^{3}H_4$ oxocarbenium ion is more stable than its $^{4}H_3$ counterpart (see Scheme 1.5): the C2, C3, and C4 groups are all positioned properly to provide maximal stabilization of the electron-depleted anomeric center, and only the C5 substituent, in itself not a powerful stereodirecting group, is not positioned favorably [18]. However, the axial orientation of this group does lead to a significant 1,3-diaxial interaction with the axially positioned C3-alkoxy group. The allylation of mannose proceeds with α-selectivity, indicating that nucleophilic attack on the β-face of the $^{3}H_4$ oxocarbenium is not a favorable reaction pathway. To account for this stereochemical outcome, Woerpel and coworkers have suggested a Curtin–Hammett kinetic scenario, in which the two half chairs **42** and **43** are in rapid equilibrium. Attack on the $^{3}H_4$ conformer suffers from unfavorable steric interactions between the incoming nucleophile and the

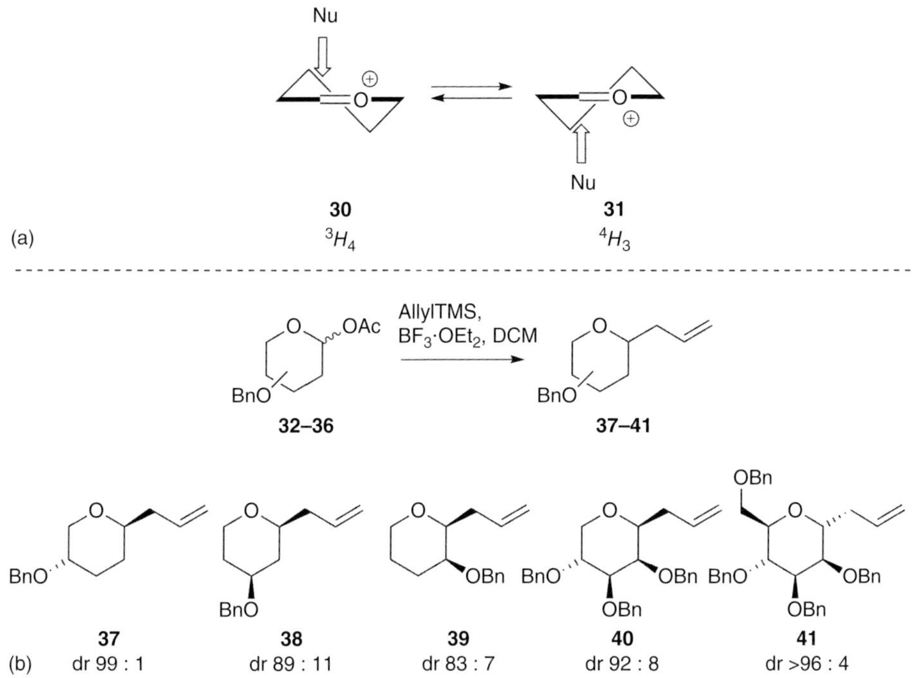

Scheme 1.4 (a) Two-conformer model to explain the stereoselectivity in pyranosyl C-allylations. (b) Observed diastereoselectivity in reactions of (partially) substituted pyranosyl acetates (major products are shown).

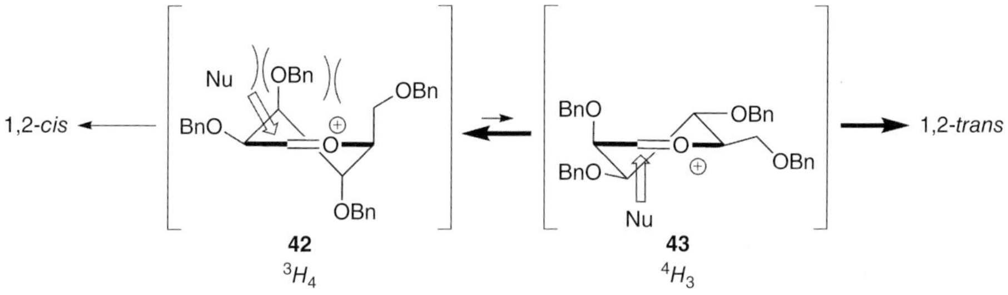

Scheme 1.5 The $^3H_4$ and $^4H_3$ mannosyl oxocarbenium ions and the trajectories of incoming nucleophiles.

substituents at C3 and C5, in addition to the destabilizing C3—C5 interaction, already present in the system. Attack on the α-face of the $^4H_3$ oxocarbenium ion, on the other hand, is devoid of these unfavorable steric interactions, making this transition state overall more favorable.

With strong nucleophiles, the two-conformer oxocarbenium ion model falls short, and S_N2-type pathways come into play [27, 28]. In a continuation of their efforts to understand the stereoselectivities of C-glycosylation reactions of (partially) substituted pyranosyl donors, the Woerpel laboratory studied the addition reactions of a series of *C*-nucleophiles, ranging from weak nucleophiles (such as allyl trimethylsilane) to relatively strong nucleophiles (such as silyl ketene acetals) [27, 28]. Table 1.1 summarizes the stereochemical outcome of the reactions of 2-deoxy glucopyranosyl acetate donor **44** with these nucleophiles under the agency of TMSOTf as a Lewis acid catalyst [28], together with their relative nucleophilicity, as established by Mayr and coworkers [29]. The α-selectivity in the reaction with allyl trimethylsilane can be accounted for by invoking the $^{4}H_{3}$ oxocarbenium ion (**55**, Scheme 1.6a) as most likely product-forming intermediate. Nucleophilic attack on the alternative $^{3}H_{4}$ half chair **54** again suffers from prohibitively large steric interactions to be a reasonable pathway. With reactive nucleophiles, such as silyl ketene acetals **52** and **53** (Table 1.1, entries 4 and 5), the most likely product-forming pathway proceeds with significant S_N2-character taking place on the α-triflate intermediate **56** (Scheme 1.6b) [28]. Of note, no attempts were undertaken to characterize this triflate[30].

1.3 Computational Studies

To better understand the conformational behavior, reactivity, and stability of glycosyl oxocarbenium ions, several quantum mechanical studies have been undertaken (see Table 1.2) [9, 10, 31–35]. Whitfield and coworkers have reported many computational

Table 1.1 Changing diastereoselectivity in the addition of *C*-nucleophiles of increasing reactivity.

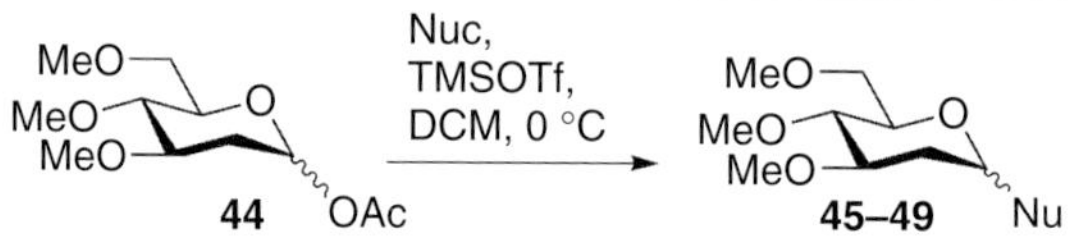

Entry	Nucleophile	N^{α}	product	α/β (yield)
1		1.8	**45**	89:11 (57%)
2	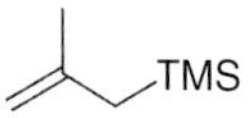 **50**	4.4	**46**	50:50 (73%)
3	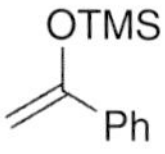 **51**	6.2	**47**	68:32 (94%)
4	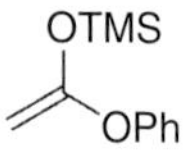 **52**	8.2	**48**	27:73 (78%)
5	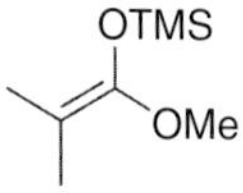 **53**	9.0	**49**	19:81 (68%)

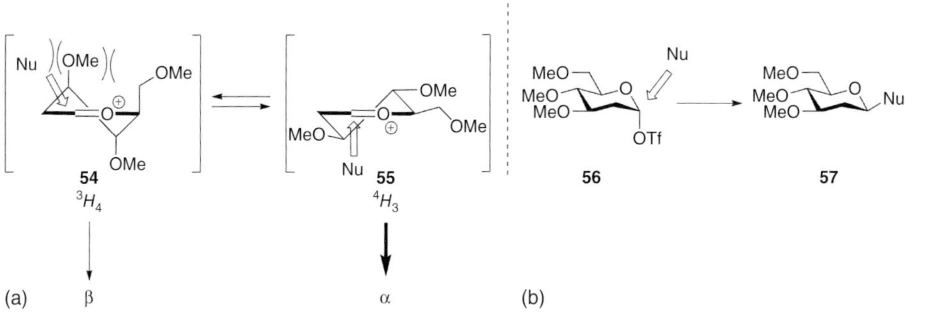

Scheme 1.6 Reactive intermediates in S_N1-type (a) and S_N2-type (b) pathways.

studies in which they investigated the conformational behavior of, among others, tetra-*O*-methyl gluco- and mannopyranosyl triflates as well as their 4,6-*O*-benzylidene congeners upon ionization (i.e., expulsion of the triflate leaving group) and the conformational behavior of the resulting oxocarbenium ions [32]. To prevent collapse of the initially formed ion pair, they used lithium cations to stabilize the departing anionic leaving

Table 1.2 A selection of oxocarbenium ions and their calculated energies (determined by DFT calculations).

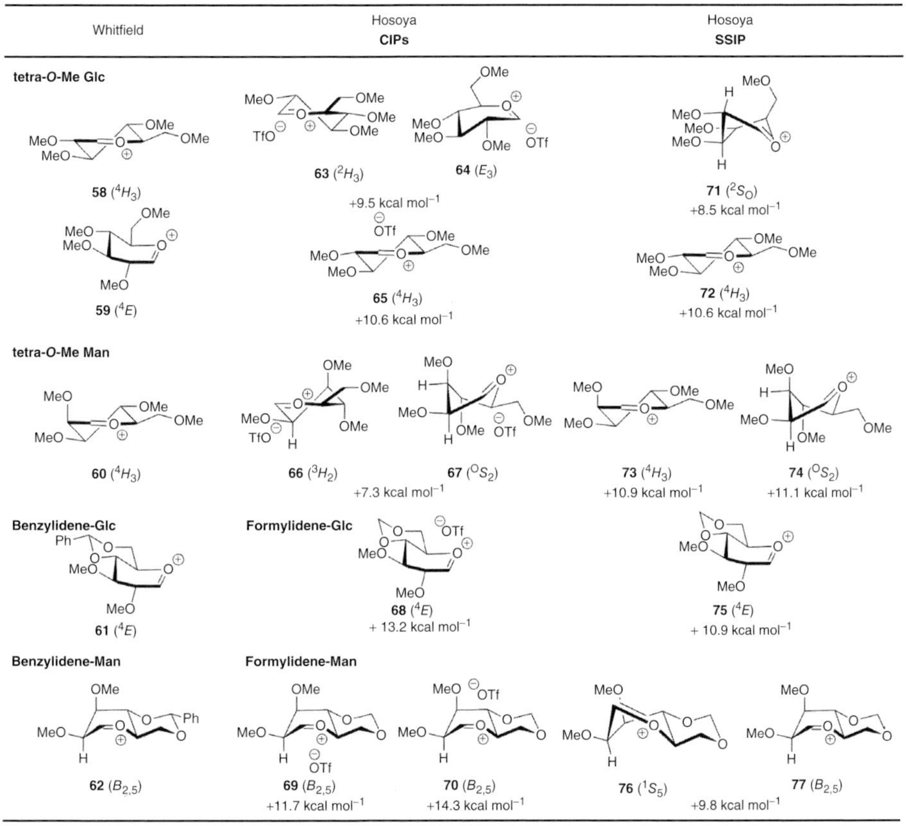
Whitfield
Hosoya CIPs
Hosoya SSIP
tetra-O-Me Glc
58 ($^{4}H_{3}$)
59 (^{4}E)
63 ($^{2}H_{3}$)
64 (E_{3})
+9.5 kcal mol^{-1}
65 ($^{4}H_{3}$)
+10.6 kcal mol^{-1}
71 ($^{2}S_{O}$)
+8.5 kcal mol^{-1}
72 ($^{4}H_{3}$)
+10.6 kcal mol^{-1}
tetra-O-Me Man
60 ($^{4}H_{3}$)
66 ($^{3}H_{2}$)
67 ($^{O}S_{2}$)
+7.3 kcal mol^{-1}
73 ($^{4}H_{3}$)
+10.9 kcal mol^{-1}
74 ($^{O}S_{2}$)
+11.1 kcal mol^{-1}
Benzylidene-Glc
61 (^{4}E)
Formylidene-Glc
68 (^{4}E)
+ 13.2 kcal mol^{-1}
75 (^{4}E)
+ 10.9 kcal mol^{-1}
Benzylidene-Man
62 ($B_{2,5}$)
Formylidene-Man
69 ($B_{2,5}$)
+11.7 kcal mol^{-1}
70 ($B_{2,5}$)
+14.3 kcal mol^{-1}
76 ($^{1}S_{5}$)
77 ($B_{2,5}$)
+9.8 kcal mol^{-1}

group. These calculations revealed that ionization of the tetra-*O*-methyl gluco- and mannopyranosyl α-triflates initially provides $^{4}H_3$ (**58** and **60**, respectively) or closely related ^{4}E-like oxocarbenium ions **59** (see Stoddart's hemisphere representation [36] for *pseudo*-rotational itineraries shown in Figure 1.2a). Expulsion of the anomeric triflate from the β-isomers requires a conformational change, where the glucose and mannose pyranosyl rings distort to an $^{1}S_3$-like structure [32]. In this constellation, the anomeric leaving group can be expelled by assistance of one of the ring oxygen lone pairs leading to an ^{4}E (for the glucose) or $^{4}H_3$ half-chair (for the mannose) oxocarbenium ion. The stability of these ions is primarily governed by sterics, since they lack the electronic stabilization described earlier. Interestingly, similar itineraries have been established to be operational in glycosyl hydrolases. Rovira and coworkers have determined that the hydrolysis of β-glucosides by retaining glucosyl hydrolases, belonging to the GH5, GH7, and GH16 families, proceeds via a trajectory, in which the substrate is first placed in a conformation that allows expulsion of the aglycon (Figure 1.2b) [37, 38]. Then passing through $^{4}H_3$ transition state **80**, which is close in conformational space to the starting $^{1}S_3$ geometry **79**, the $^{4}C_1$ product **81** (the covalent enzyme–glucose adduct) is obtained. This catalytic itinerary was visualized using a combination of X-ray crystallography, free-energy landscape mapping (to determine the intrinsically favorable ground-state conformations), and quantum mechanics/molecular mechanics reaction simulations. Further calculations of the Whitfield group showed that the 4,6-*O*-benzylidene glucose

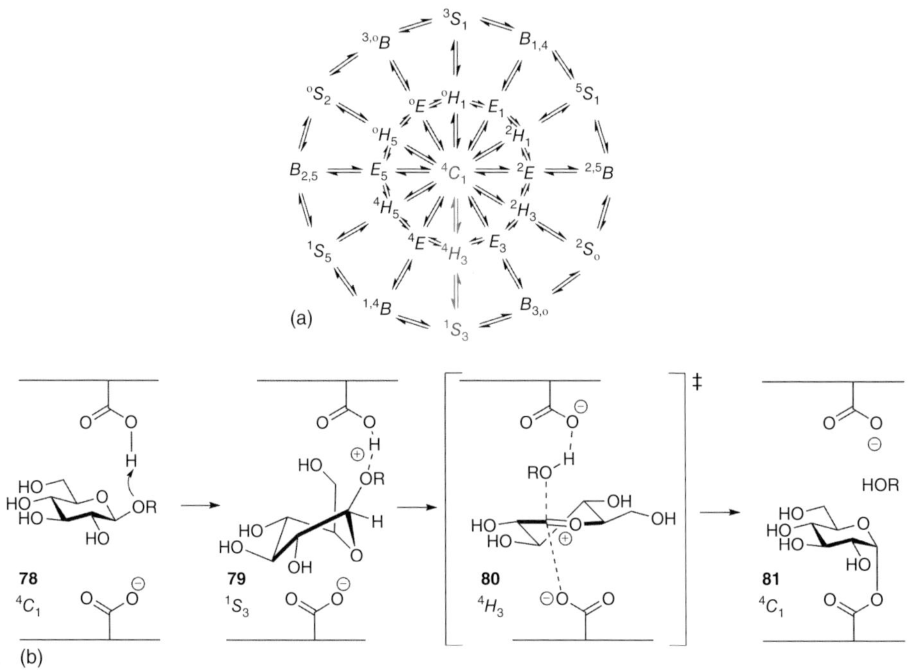

Figure 1.2 (a) Stoddart's hemisphere representation for conformational interconversions (only the Northern hemisphere is shown). (b) Conformational itinerary of the substrate as used by various β-glucosidases. The trajectory has been highlighted in the Stoddart diagram in (a) and was also calculated to be the lowest energy pathway of the ionization of a β-glycosyl triflate.

oxocarbenium ions preferentially take up an ^{4}E conformation **61**, where the corresponding benzylidene-mannose structure is most stable taking up a $B_{2,5}$-geometry (**62**) [32]. In the latter structure, both the through-space electron donation by the C3—O—Me ether and the hyperconjugative stabilization of the σ_{C2-H2} bond contribute to the stability of the ion. Recently, Hosoya *et al.* described a method that takes into account explicit solvent molecules in the determination of the stability of CIPs and solvent-separated oxocarbenium ions [10]. They first optimized the amount of solvent molecules required to obtain a reliable outcome while maintaining acceptable calculation costs, eventually using four dichloromethane molecules as an optimum. Using these solvent molecules, they were able to find the lowest energy CIPs and SSIPs. Stabilization of the developing charge was affected by the proper positioning of the solvent molecules: the hydrogen atoms of the dichloromethane molecules were capable of stabilizing the negative charge at the triflate leaving group, while the electron density around the chloride atoms of the dichloromethane molecules could be used to support the positive charge at the oxocarbenium ion. It was described that the stability of tetra-*O*-methyl glucopyranosyl oxocarbenium ions having a triflate associated at either the α- or β-face was quite similar in energy. The lowest energy oxocarbenium ion having the counterion associated on its α-face was found in a $^{2}H_3/E_3$ conformation (**63**/**64**) at +9.5 kcal mol^{-1} with respect to the lowest energy α-triflate. The β-CIP **65** took up a $^{4}H_3$ structure at +10.6 kcal mol^{-1}.[1] Interestingly, SSIPs that are lower in energy compared to the CIPs were found. A $^{2}H_3/^{2}S_O$ SSIP **71** was found to be the most stable, at +8.5 kcal mol^{-1}, where $^{4}H_3$ ion **72** was found at +10.6 kcal mol^{-1}. For tetra-*O*-methyl mannose, an α-CIP, with an $^{O}S_2/^{3}H_2$ structure (**66**/**67**), was determined to be the most stable ion pair. Although this oxocarbenium does not benefit from an optimal geometry of the C=O^{+} moiety, it can be stabilized by the σ_{C2-H2} bond and by electron donation from the *pseudo*-axial substituents at C3 and C4. These authors studied the 4,6-*O*-formylidene gluco- and mannopyranosyl oxocarbenium ion pairs to account for stereoselectivities obtained with benzylidene glucose and mannose donors (*vide infra*) [31]. For the formylidene glucose system, the lowest energy CIP turned out to be the cation **68** with a $^{4}E/^{4}H_3$ structure with the anion associated on its β-face (+13.2 kcal mol^{-1} with respect to the lowest energy covalent α-triflate). The lowest energy SSIP was found in an ^{4}E conformation (+10.9 kcal mol^{-1}), in line with the results of the Whitfield group. Also, for the formylidene mannose ion pairs, the SSIPs were found to be more stable compared to the lowest energy CIPs. A $B_{2,5}$ α-CIP (**69**) was found at +11.7 kcal mol^{-1} (with respect to the lowest energy α-triflate), where the lowest β-CIP **70** had an $^{1}S_5/B_{2,5}$ structure (+14.3 kcal mol^{-1}). The lowest energy SSIP also took up an $^{1}S_5/B_{2,5}$ conformation (**76**/**77**) and was significantly more stable (+9.8 kcal mol^{-1}). The latter geometry corresponds to the structure found by Whitfield [32].

Hünenberger and coworkers have computationally studied ion pairs derived from tetra-*O*-methylglucosyl triflate in different solvents [39]. Through a series of molecular dynamics and quantum mechanical calculations, they came to the hypothesis that different solvents affect the stability of pairs in a different manner. Their calculations suggest that in acetonitrile, the glucopyranosyl oxocarbenium ion preferentially takes up a $B_{2,5}$ structure, with the counterion associated on the α-face. In 1,4-dioxane, a $^{4}H_3$

1 The β-triflate was established to be only 0.8 kcal mol^{-1} higher in energy compared to its alpha-counterpart, translating to a 3 : 1 mixture of triflate anomers at equilibrium. Under experimental conditions, only the alpha-anomer has been observed for the tetra-*O*-benzylglucopyranosyl triflate.

structure proved to be most stable with the triflate ion coordinating on the β-face. The authors suggested that their calculations could provide an adequate explanation of the generally observed β- and α-directing effect of the solvents (acetonitrile and 1,4-dioxane, respectively) studied [39].

Because of the different computational approaches, care should be taken to compare the calculation methods described earlier. It is apparent, however, that similarities arise and that the minimum energy conformations determined with the different methods are close in conformational space. For tetra-*O*-benzyl (or methyl) glucose, a structure close to the $^4H_3$ half chair appears from all calculations. Here, all ring substituents take up a sterically favorable equatorial position. Stabilization of the cation is only provided by hyperconjugation of the C2—H2 bond. For the corresponding mannosyl cation, a similar structure arises, although the method described by Hosoya *et al.* indicates that cations having a rather different structure are also possible [10]. The introduction of a cyclic ketal restricts the conformational freedom of the cations, and the different methods collectively point to an E_4 envelope and $B_{2,5}$ boat structure as most stable oxocarbenium ions for the benzylidene (or formylidene) glucosyl and mannosyl cations, respectively.

1.4 Observation of Glycosyl Oxocarbenium Ions by NMR Spectroscopy

Where many anomeric triflates have been spectroscopically characterized, glycosyl oxocarbenium ions are too reactive to detect by straightforward NMR techniques. Very recently, Blériot and coworkers reported the use of a superacidic medium (HF/SbF_5) to generate glycosyl oxocarbenium ions and allow their spectroscopic investigation [40]. As depicted in Scheme 1.7, 2-deoxyglucosyl donor **82** was transformed into the ^4E oxocarbenium ion **83**, which proved to be stable in the super acid medium for several hours at −40 °C. The conformation of the ion was deduced from the coupling constants of the ring protons and corroborated by Density Functional Theory calculations and simulated spectra. The found structure is very close in conformational space to the fully substituted glucosyl oxocarbenium ions found in the described DFT calculation (Table 1.2). Quenching of the oxocarbenium ion by cyclohexane-d_{12} led to the selective formation of the α-deuterium 2-deoxy glucoside. The formation of this product can be accounted for by using the ^4E oxocarbenium ion as the product-forming

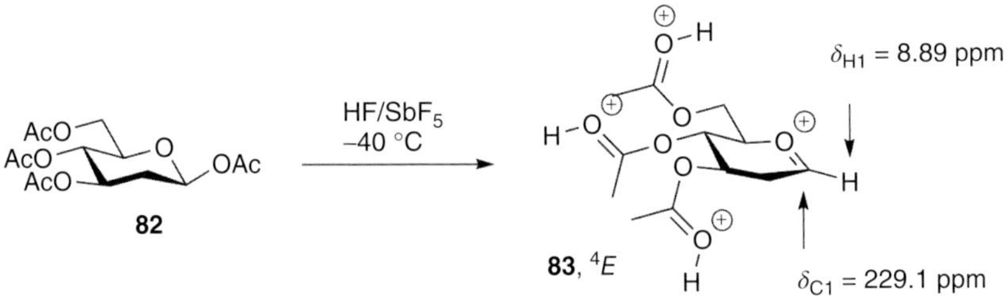

Scheme 1.7 Generation of a 2-deoxy glucosyl oxocarbenium ion in HF/SbF_5 allowing for its characterization by NMR.

intermediate and a favorable $^4E \rightarrow {^4C_1}$ reaction trajectory. Obviously, the solvent system used in this NMR experiment differs significantly from the solvents normally used in glycosylation reactions, and, therefore, care should be taken in the translation of the results obtained in the super acid medium to a "normal" glycosylation reaction. It does, however, provide valuable information on the conformation of glycosyl oxocarbenium ions. Expansion of these NMR studies to a broader pallet of carbohydrates with different functionalities will generate insight and spectroscopic proof for stereoelectronic substituent effects that determine the overall shape of the oxocarbenium ions.

1.5 Oxocarbenium Ion(-like) Intermediates as Product-Forming Intermediates in Glycosylation Reactions

The best-studied glycosylation system to date is the Crich β-mannosylation reaction [6, 41–43]. In this reaction (Scheme 1.8), a benzylidene (or related acetal)-protected mannosyl donor (such as **84**) is preactivated to provide an α-anomeric triflate **85**. The corresponding β-triflate is not observed because this species lacks the stabilizing anomeric effect, present in the α-anomer, and it places the anomeric substituent in an unfavorable Δ2-position. Addition of an acceptor to the activated donor then provides the β-linked product **89β**, generally with very high stereoselectivity [44]. While the construction of 1,2-*cis*-mannosides used to be one of the biggest challenges in synthetic carbohydrate chemistry, this linkage can now be installed with great fidelity using this methodology. In addition, the system has been a great inspiration to unravel the underlying mechanistic details to explain the observed stereoselectivity. Crich and coworkers have carried out a suite of studies to understand and learn from the mechanistic pathways operational in the system [45–48]. Secondary kinetic isotope effects, established in a glycosylation reaction with a 2,3,6-tri-*O*-benzyl glucosyl acceptor, revealed significant oxocarbenium ion character in the transition state leading to the β-linked product [45]. This led the authors to presume that the CIP is the actual reactive species in this glycosylation reaction. Using natural abundance ^{13}C primary kinetic isotope effects, in combination with computation methods, Crich, Pratt, and coworkers established the amount of carbocation character that develops in the transition state of glycosylations of *iso*-propanol with a benzylidene-protected glucosyl or mannosyl donor [46]. From the established values, corroborated by computational validation, it was established that the β-mannosyl products were formed through an associative pathway, in which the mannose ring adopts a $B_{2,5}$-structure in the transition state **87**. In contrast, the α-products originated from a more dissociative mechanism, involving a distinct oxocarbenium cation and triflate anion. In this case, computational studies suggested an $^4E/^4H_3$ structure for the intermediate oxocarbenium ion **86**. For the benzylidene glucose system, which is generally α-selective with carbohydrate acceptors, it was established that both the α- and β-isopropanol products **95α** and **95β** were formed through an S_N2-like mechanism [46]. Notably, upon preactivation of the benzylidene glucose donor, a single triflate is observed: the α-anomer **91** [49]. To account for the formation of the α-product, Crich and coworkers have proposed a Curtin–Hammett kinetic scenario in which the α- and β-triflates **91** and **92** are in rapid equilibrium [46]. Substitution of the most reactive of the two, that is, the β-triflate, then leads to the stereoselectivity observed in reactions of this

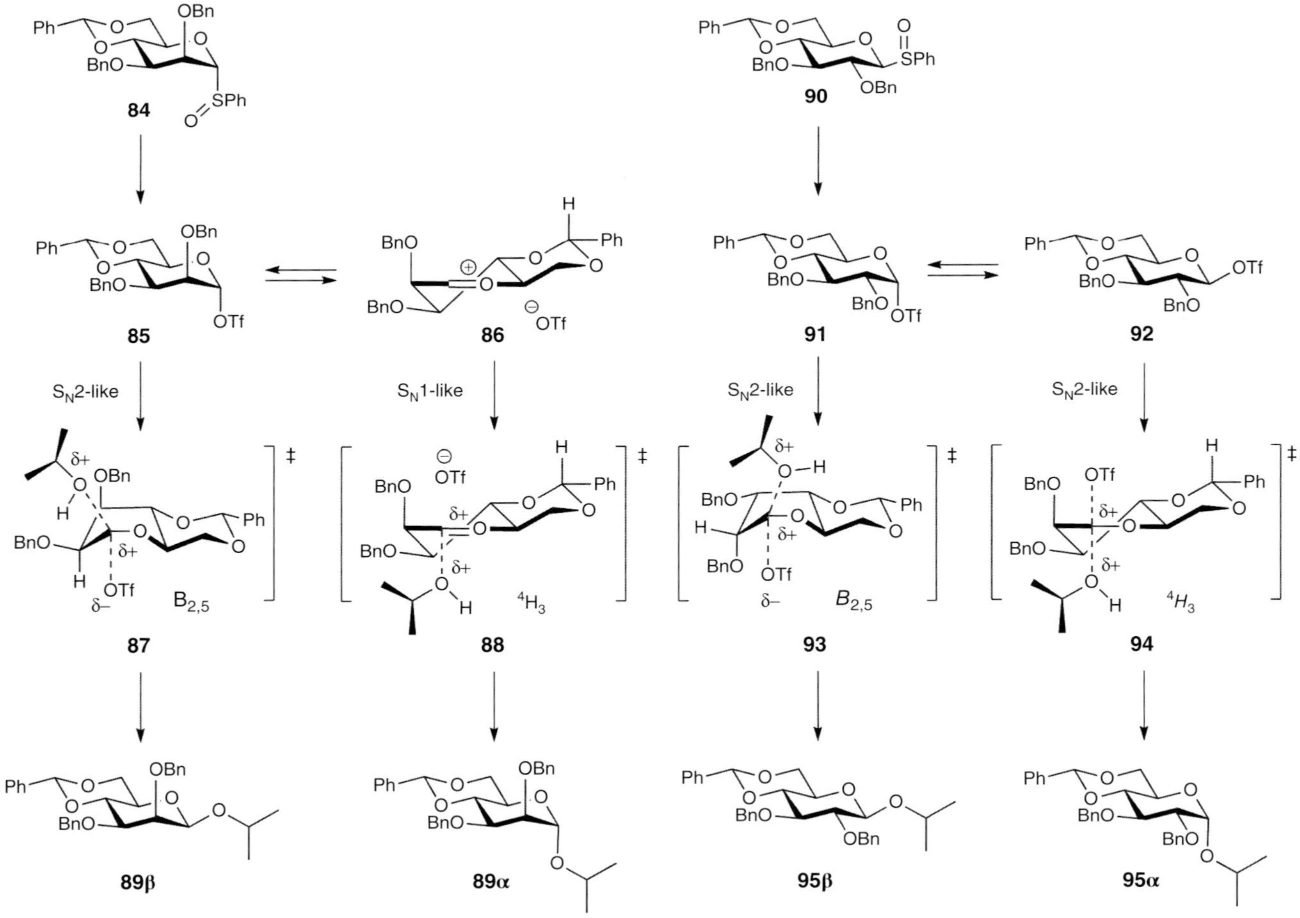

Scheme 1.8 Product-forming pathways for benzylidene mannosylations and glucosylations.

donor. As described next, equatorial glycosyl triflates have been observed, lending support to this scenario [50].

Using the cation clock methodology shown in Scheme 1.9, in which external nucleophiles (isopropanol or allyl trimethylsilane) are made to compete with an intramolecular nucleophile on mannosyl and glucosyl donors **96** and **97** (i.e., an allylsilane ether appended at C2), Crich and coworkers showed the O-glycosylation reactions to be more concentration-dependent compared to the corresponding C-glycosylation reactions [47, 48]. With this methodology, the finding that the formation of the *O*-glycosyl α- and β-benzylidene products results from different mechanistic pathways (an S_N1-like pathway for the former and an S_N2-like pathway for the latter) was corroborated. When trimethyl (methallyl)silane was employed as a nucleophile, only the β-*C*-allyl mannosyl and α-*C*-allyl glucosyl products (**101** and **104**, respectively) were obtained in a reaction that was relatively independent of the concentration of the nucleophile, indicating S_N1-characteristics in these reactions. Of note, formation of the trans-fused benzylidene mannosyl product **99** through intramolecular attack at the α-face indicates the intermediacy of an oxocarbenium ion adopting a $B_{2,5}$-conformation. The alternative $^4H_3$-half chair would not allow the *pseudo*-axially appended nucleophile to reach the α-face of the oxocarbenium ion.

Bols, Pedersen and coworkers have studied glycosylation reactions of the closely related 4,6-*O*-silylidene donor **105** as depicted in Scheme 1.10 [51]. They described that this donor provided the β-linked products with reasonable-to-good selectivity, regardless of the activation protocol (preactivation with BSP/Tf_2O or *in situ* activation with NIS/TfOH). Even when the *in situ* activation method (NIS/TfOH) was employed at room temperature, the β-product formation prevailed. This led the authors to propose the $B_{2,5}$ oxocarbenium ion as the actual reactive species.

The described studies on the benzylidene mannose and glucose systems provide an excellent example of the continuum of mechanisms that operate during a glycosylation reaction. Clearly, the different reaction paths are energetically very close to each other, making a clear-cut distinction between an S_N2- and an S_N1-type mechanism impossible. In fact, the structures of the glycosyl donor in the transition state are probably very similar to each other. The analysis presented by Crich and coworkers, as illustrated in Scheme 1.8, shows that the benzylidene mannose ring takes up $B_{2,5}$-like structure **87** in the S_N2-displacement of the covalent triflate, where the benzylidene glucose takes up a $^4H_3$ structure **94** [46]. These structures also represent the conformation of the most stable oxocarbenium ions of these donors. The exact amount of carbocation character in the transition states of these glycosylations will be determined by the difference in timing of the bond-breaking and bond-forming processes. This will critically depend on the nature of the nucleophile. When reactive nucleophiles (such as *iso*-propanol described above) are used, the formation of the new glycosidic linkage will be rather synchronous to the departure of the triflate. Unreactive nucleophiles (such as allyl-TMS and unreactive secondary carbohydrate alcohols) will react through a transition state further along the reaction coordinate, with departure of the triflate preceding the formation of the glycosidic linkage. In this scenario, more positive charge develops at the anomeric center, and the reaction becomes more S_N1-like.

The C6 oxidized analogs of mannosyl donors, mannuronic acid donors, have been found to be highly β-selective, when both preactivation and direct activation protocols are employed [52–54]. Although it was expected that this type of glycosyl donors would

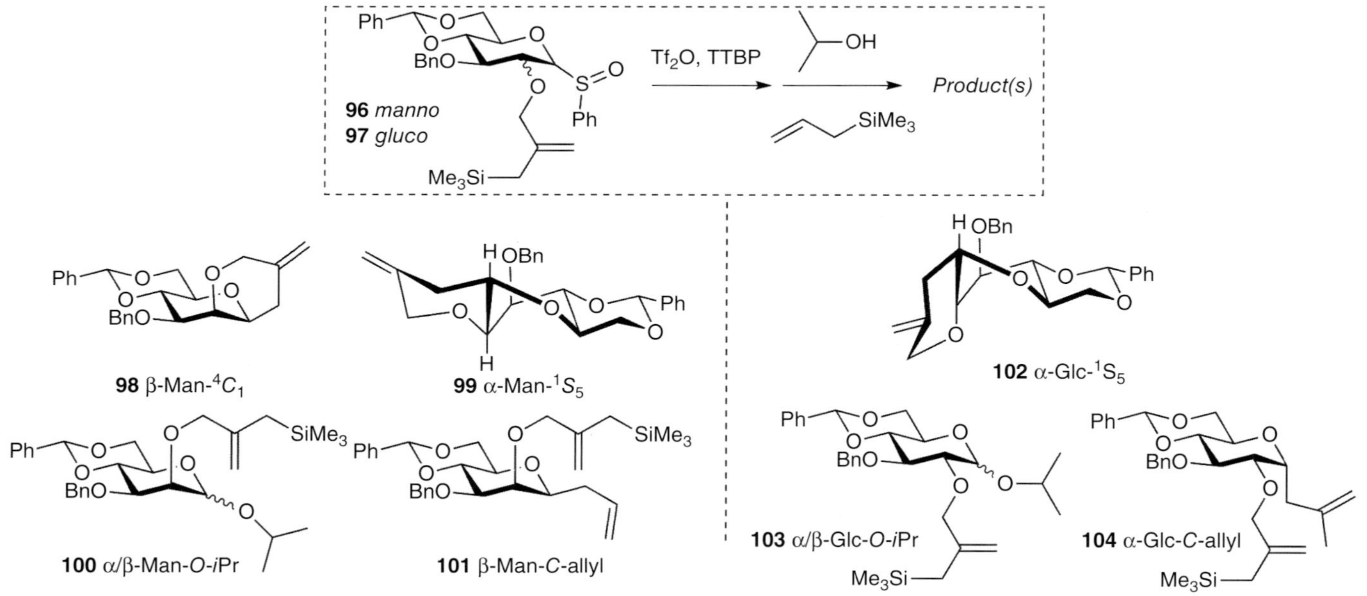

Scheme 1.9 Inter-/intramolecular competition reactions for O- and C-mannosylations and glucosylations.

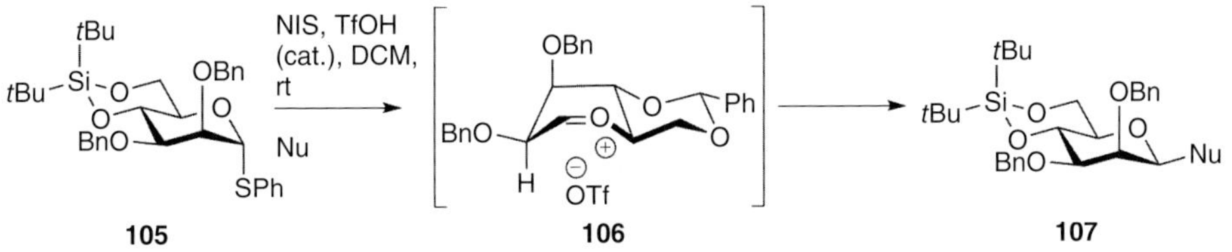

Scheme 1.10 β-Mannosylation methodology using a 4,6-*O*-silylidene donor.

be relatively inreactive, by virtue of the electron-withdrawing effect of the C5-carboxylate, these donors turn out to be rather reactive. This is manifested in competition reactions in which the reactivity of mannuronic acid donor **108** was shown to be in the same order as the reactivity of tetra-*O*-benzyl mannose donor **109** (Scheme 1.11a) [55]. Another illustration of the relatively high reactivity of these species is found in the stability of the anomeric triflates that are formed upon activation of the donors. These are stable up to

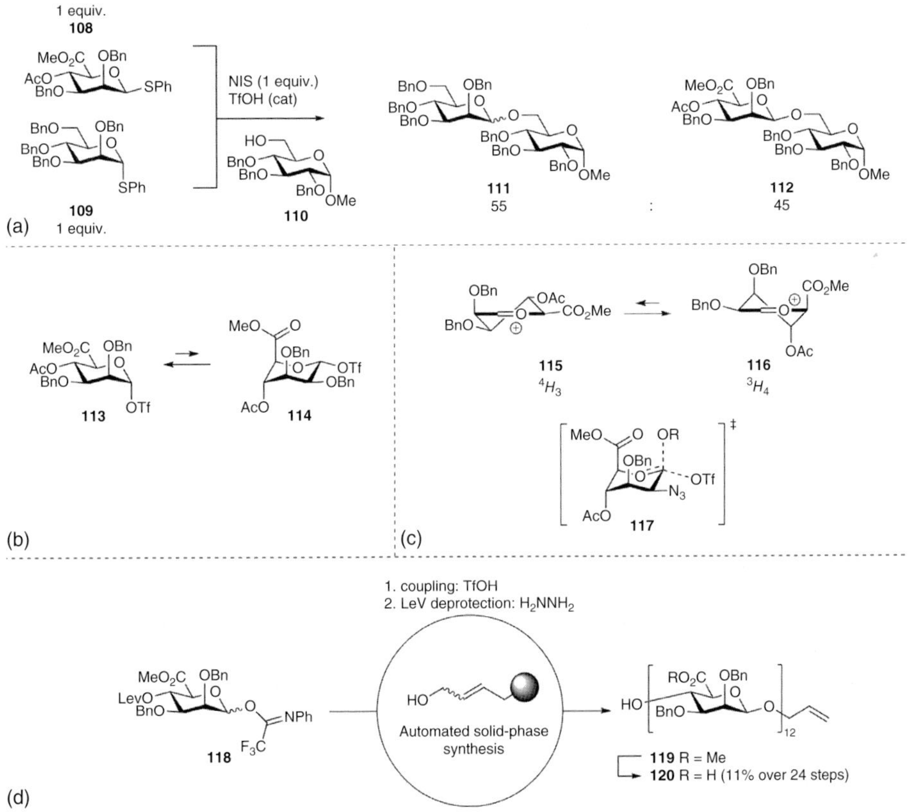

Scheme 1.11 Reactivity and selectivity in glycosylation of mannuronic acids. (a) Mannuronic acid donors are more reactive than expected. (b) Mannuronic acid triflates take up different ring conformations. (c) Mannuronic acid oxocarbenium ions and an exploded transition state leading to the β-mannuronic acid linkage. (d) Automated solid-phase assembly of mannuronic acid alginate fragments.

−40 °C, a decomposition temperature, which is below that of the 4,6-*O*-benzylidene mannosyl triflate [50]. Strikingly, low-temperature NMR spectroscopy showed the anomeric triflate derived from mannuronic acid donor **108** to take up two conformations at a low temperature, as both the $^4C_1$ and $^1C_4$ chairs products were present at −80 °C (**113** and **114**, respectively, see Scheme 1.11b). The "inverted" chair triflate places three of the ring substituents in a sterically unfavorable axial orientation. In addition, this structure places the anomeric triflate in an equatorial position, where it does not benefit from a stabilizing anomeric effect. This striking conformational behavior and the relatively high reactivity of these donors were rationalized by the hand of the structure of the oxocarbenium ion that can form from these donors (Scheme 1.11c). In $^3H_4$ half chair **116**, the C2, C3, and C4 substituents all take up optimal orientations to stabilize the electron-depleted anomeric center (*vide supra*). Model studies on pyranosides, having a single C5 carboxylate group, revealed that this substituent can provide stabilization of the oxocarbenium ion half chair when placed in a *pseudo*-axial position [53]. Thus, in the mannuronic acid $^3H_4$ half-chair oxocarbenium ion **116**, all substituents collaborate to stabilize the carbocation. This carbocation can also provide an explanation for the β-selectivity observed in glycosylations of these donors. Attack on the diastereotopic face of the half chair that leads to a transition state with a chair-like structure, that is, the β-face, accounts for this selectivity. The smaller size of the carboxylate in comparison to a methyloxybenzyl appendage, present in mannopyranosides, leads to diminished steric interactions of this group with the C3-substituent and the incoming nucleophile (as in Scheme 1.5) [53]. While the stereoselectivity of 4,6-*O*-benzylidene mannose donors decreased with small substituents (e.g., azides) at C2 and C3, the stereoselectivity of C2-azido and C2, C3-diazido mannuronic acid donors remained intact [56, 57]. We have accounted for this "robust" stereoselectivity by taking into account that both an S_N2-like substitution of the anomeric α-triflates **113** and **114** and an S_N1-pathway involving the $^3H_4$ half-chair oxocarbenium ion **116** lead to the β-product (Scheme 1.11c). To minimize steric interactions in this transition state, the mannuronic acid ring may adopt a closely related ^3E-structure. The mannuronic acid donors have been successfully employed in the construction of several bacterial oligosaccharides, as well as in the automated solid-phase synthesis of β-mannuronic acid alginate fragments (Scheme 1.11d) [58]. In the latter synthetic endeavor, mannuronic acid *N*-phenyl trifluoroacetimidate donor **118** was used to construct tetra-, octa-, and dodecasaccharide on a resin that was equipped with a butenediol linker system [59]. All glycosylation reactions were executed at a temperature just below the decomposition temperature of the intermediate triflates to allow for effective glycosylation reactions. The dodecasaccharide **120** was eventually obtained, after cleavage from the resin and saponification of the methyl esters, to allow for a straightforward purification, in 11% yield (±91% per step).

Studies on the C5 epimer of D-mannuronic acid, that is, L-guluronic acid, revealed a marked decrease in 1,2-*cis*-selectivity of these donors [60]. This was explained by taking into account that in the possible L-guluronic acid oxocarbenium ions, there would be conflicting "substituent interests": in the possible $^3H_4$ half chair, the C2, C3, and C4 substituents take up a stabilizing orientation, but the C5 carboxylate is positioned in an unfavorable *pseudo*-equatorial orientation. This situation is reversed in the opposite $^4H_3$ half chair. L-Gulose donors, such as 2,3,4,6-tetra-*O*-benzyl-L-gulosyl donor **121**, on the other hand, provide very selective glycosylation reactions also in the absence of any special stereodirecting functionalities (Scheme 1.12) [60]. In this case, the $^3H_4$ half-chair

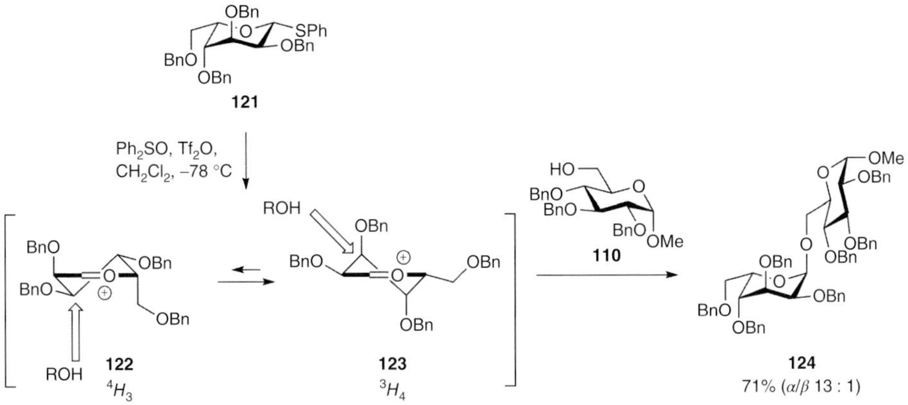

Scheme 1.12 Stereoselective glycosylation involving L-gulosyl donors.

oxocarbenium ion **123** represents a structure that benefits from the stabilization by the functionality at C2, C3, and C4, while minimizing steric interactions between the substituents (especially, the bulky C5 substituent). "Nonoxidized" gulose synthons have been successfully employed in the synthesis of L-guluronic acid alginates [60, 61], as well as "mixed" alginates, containing both β-D-mannuronic acid and α-L-guluronic acid residues [62].

Another L-sugar, renowned for its high 1,2-*cis*-selectivity in glycosylation reactions, is L-fucose, an important constituent of, among others, the blood group determinants [63]. Different methods are available for the introduction of the α-fucosyl linkage. One of the most common methods relies on the use of acyl groups at C3 and/or C4. Although there is ongoing debate on the role of the acyl functions, an often forwarded cited explanation is that they can provide "remote participation," generating species such as dioxolenium ion **126**, thereby shielding the bottom face of oxocarbenium ion and allowing the selective formation of the α-fucosidic linkage (see Scheme 1.13a) [64–66]. This approach has been successfully employed in the synthesis of fucoidan oligosaccharides [66, 67]. On the other hand, there are also numerous examples of fucosylation that do not rely on the presence of acyl groups at C3 and/or C4. For example, the highly reactive per-*O*-benzylated fucosyl donor **128** has been used by Wong and coworkers to synthesize Lewis^y hapten **131**, using a sequential reactivity-based one-pot strategy, in 44% yield (Scheme 1.13b) [14]. Although the use of the ethereal solvent mixture in this case can be important to install the α-fucosyl linkage, this is most likely not the only reason underlying the excellent selectivity observed in this glycosylation. Another example is shown in Scheme 1.13c where two fucose moieties have been introduced to synthesize pentasaccharide **134**, using perbenzylated thioethyl fucosyl donor **133**, in combination with methyl triflate (MeOTf), in dichloromethane at room temperature [68]. A possible explanation for the generally good α-selectivity observed in fucosylation reactions can be found in oxocarbenium ion half chair **136** (Scheme 1.13d). In this $^3H_4$ half-chair oxocarbenium ion, the axial C4 alkoxy substituent and the *pseudo*-equatorial H-2 can stabilize the positive charge of the oxocarbenium ion. The C5 methyl substituent is placed in a sterically favorable equatorial position. The high reactivity of perbenzylated fucosyl donors also supports the intermediacy of an oxocarbenium ion intermediate.

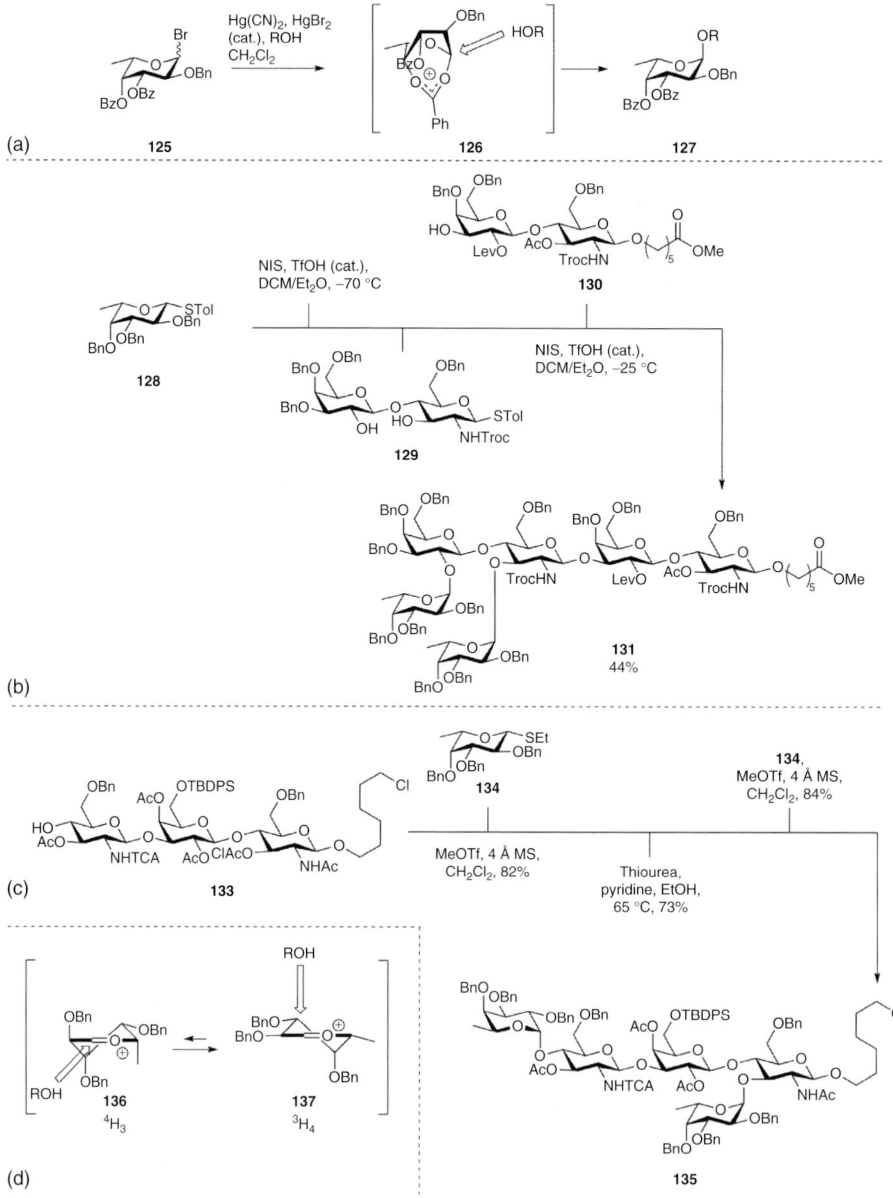

Scheme 1.13 (a) Remote participation to explain the α-selectivity of C4-acyl fucosyl donors. (b) Reactivity-based one-pot synthesis of a Lewis^y hexasaccharide using perbenzylated fucosyl thioglycosides. (c) MeOTf-promoted α-fucosylation in the synthesis of a Lewis^a–Lewis^x tumor-associated antigen. (d) Possible half-chair oxocarbenium ions involved in fucosylation reactions.

The use of cyclic protecting groups to conformationally restrict glycosyl donors has also been employed in furanosylation reactions. Almost simultaneously, the Ito, Boons, and Crich groups reported that locked arabinofuranoses can be used for the stereoselective construction of the β-arabinose bond [69–71]. Scheme 1.14a depicts the

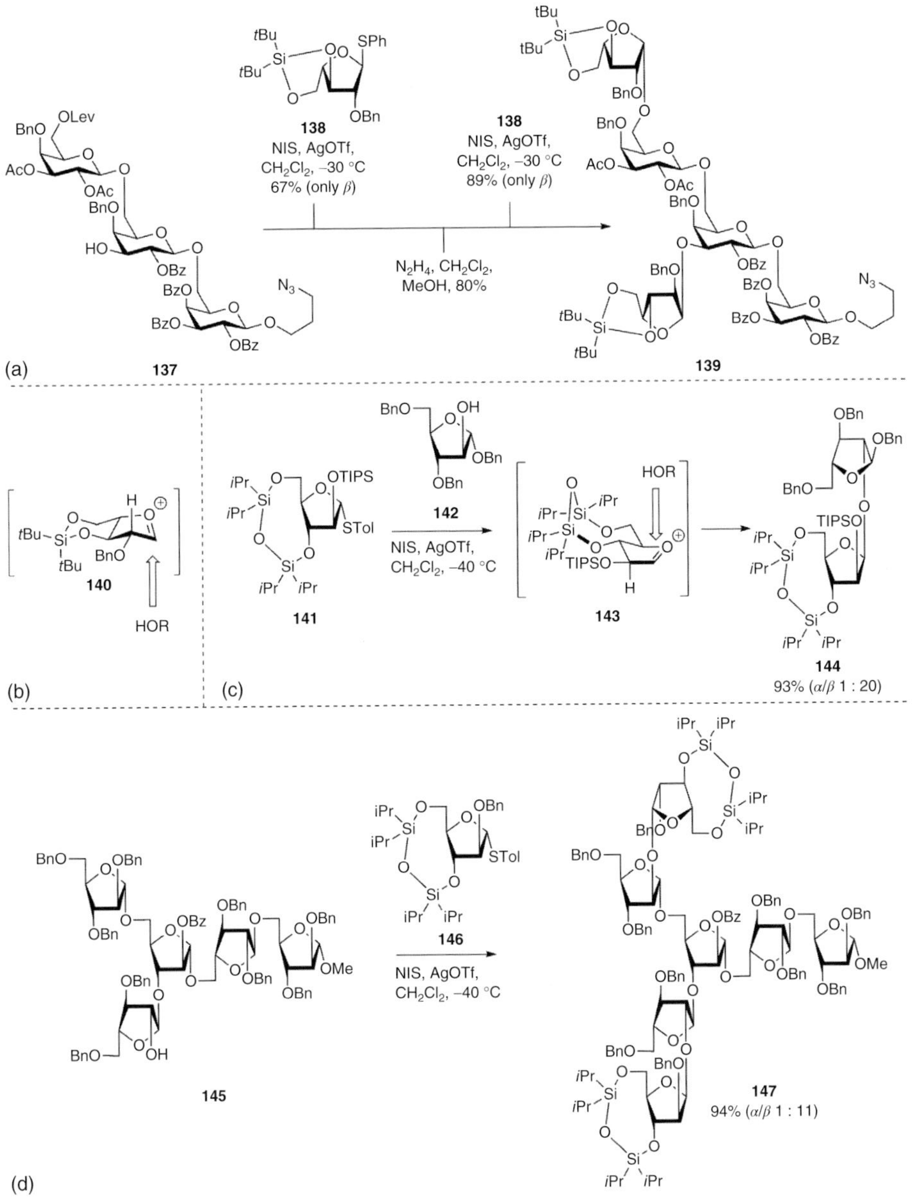

Scheme 1.14 Stereoselective arabinofuranosylations through the use of locked arabinofuranosyl donors. (a) Boons' synthesis of an arabinogalactan heptasaccharide. (b) Possible envelope oxocarbenium ions involved in arabinosylation reactions. (c) Ito's conformationally locked arabinoses to achieve high β-selectivities. (d) Ito's synthesis of oligo-arabinofuranosides.

synthesis of an arabinogalactan using 3,5-*O*-silylidene-protected arabinose donor **137** as reported by Boons and coworkers [70]. The high selectivity of donor **137** was rationalized by the intermediacy of oxocarbenium ion **139** (Scheme 1.14b), which is locked in the E_3 conformation by the cyclic protecting group. The "inside attack" model, described earlier, explains the facial selectivity for the attack of the incoming nucleophile. Ito and coworkers used a similar 3,5-*O*-di-(di-*iso*-propyl)siloxane-protected arabinosyl donor **140** (Scheme 1.14c), which also exhibits a very high β-selectivity [71]. The authors performed molecular modeling studies, which suggested that the total energy of β-glycosylation product **143** was about 3.7 kcal mol^{-1} lower than that of the alternative α-isomer. The β-glycosidic linkages take up a *pseudo*-axial orientation, thereby benefitting from a stabilizing anomeric effect and providing an explanation for the energy difference. Although the conditions used by the authors do not suggest thermodynamic control in these glycosylations, the energy difference of the products can already become somewhat apparent in the transition states leading to the products. A kinetic explanation for the observed selectivity based on the intermediate oxocarbenium ions is more likely. Analogous donor **145** was used for the introduction of terminal β-arabinofuranosidic residues in branched arabinan oligosaccharides (Scheme 1.14d), up to 22 monosaccharide units in length [71].

As described earlier, C-glycosylation reactions and reductions (addition of an *H*-nucleophile) on ribofuranosides generally proceed with excellent stereoselectivity to provide the product, resulting from nucleophilic attack at the α-face of the E_3 oxocarbenium ion [21]. Also, with other nucleophiles, this stereoselectivity is observed as described in Scheme 1.15a [72]. In their efforts to synthesize poly-(adenosine diphosphate-ribose) trisaccharide core **153** (a so-called supernucleoside), Kistemaker *et al.* used tribenzylated ribosyl donor **147** to form diriboside **149** in high yield and with complete α-selectivity [72]. Further elaboration on the trisaccharide was carried out with 5-*O*-TIPS-protected ribosyl donor **151**, leading to trisaccharide **152**, again with complete α-selectivity, in 57% yield.

The stereoselectivity of ribosyl donors can be reversed through the use of cyclic protecting groups to mask the bottom face of the ribofuranosyl oxocarbenium ion. Ichikawa *et al.* found that 2,3-*O*-(3-pentylidene)-protected ribosyl fluoride **157** reacted in a very β-selective manner with *C*-nucleophiles (Scheme 1.15b) [73, 74]. DFT computations showed that the intermediate oxocarbenium ion preferably adopts an E_3 conformation. Inside attack on this species is prohibited by the blocking 3-pentylidene-protecting group, leading to attack on the other side of the furanosyl ring. This approach was used in the total synthesis of the antibiotic (+)-caprazol **161** (Scheme 1.15c), where 2,3-*O*-(3-pentylidene)-5-azidoribosyl donor **157** gave near-complete β-selectivity in the glycosylation with acceptor **158**, while the less sterically demanding 2,3-*O*-propylidene analog **156** only gave modest β-selectivity [73].

1.6 Conclusion

The vast majority of glycosylation reactions take place somewhere along the continuum of mechanisms hemmed in between S_N1- and S_N2-type reactions, with product-forming transition states that have characteristics of both reaction types. Insight into and control over the place of a given glycosylation reaction in the continuum open up ways to control the stereochemical outcome of the glycosylation reaction at hand. Over the recent years – and spurred by the initial discovery of a covalent mannosyl triflate – much insight into reactive intermediates has been gathered. NMR spectroscopy has been used to characterize a multitude of covalent reactive intermediates such as anomeric

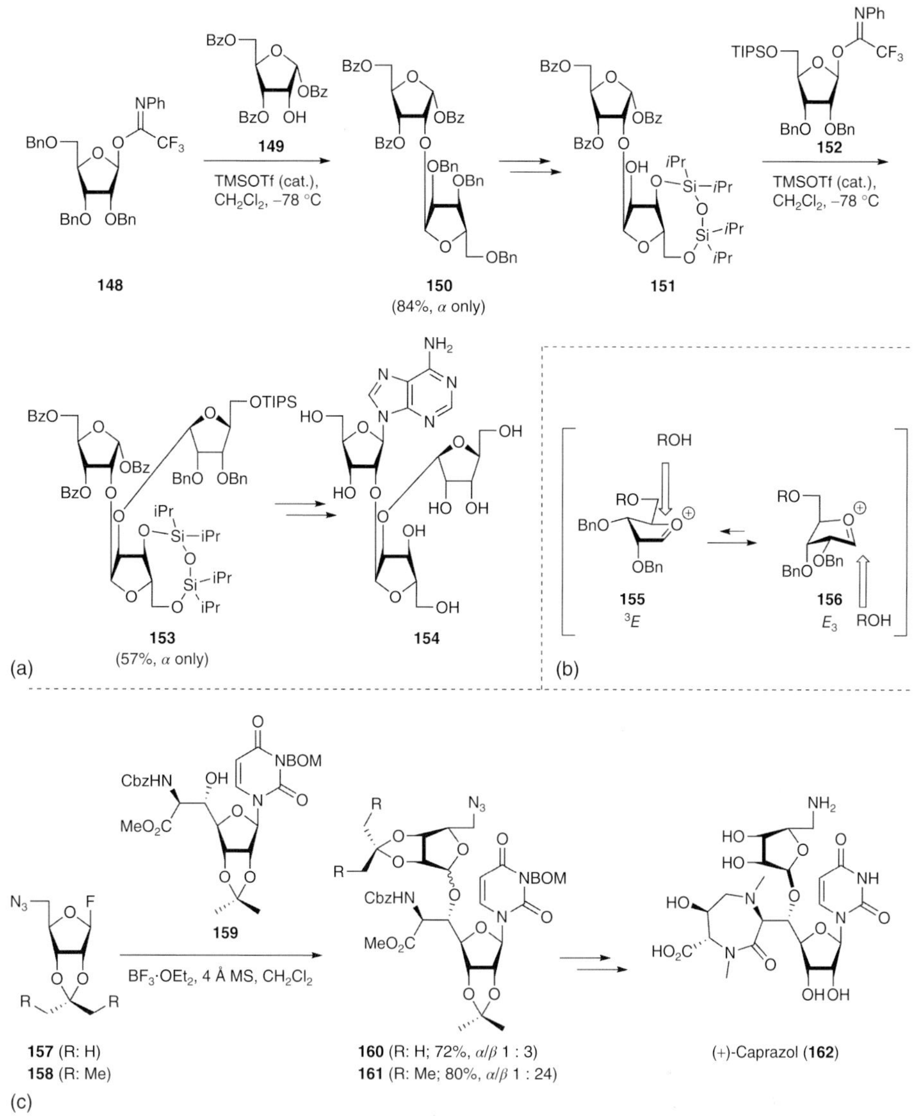

Scheme 1.15 Stereoselective ribosylations. (a) Synthesis of the "supernucleoside" **153**. (b) Possible envelope oxocarbenium ions involved in ribosylations. (c) Matsuda's total synthesis of (+)-caprazol **161**.

triflates. Often, only a single anomeric triflate can be observed spectroscopically, because the other anomer is highly unstable to allow its detection. To study glycosyl oxocarbenium ions, sophisticated and detailed DFT computational approaches have been presented to validate the experimental results. Very recently, NMR was added to the toolbox available to study glycosyl oxocarbenium ions. To pinpoint, the location of the mechanism of a given glycosylation reaction on the continuum of mechanisms, kinetic isotope effects, and cation clock methodology have been used to determine how much oxocarbenium character develops in the transition state of the glycosylation reaction. While the structure of noncharged covalent intermediates is primarily dictated by the steric requirements of the ring substituents, the shape of positively charged oxocarbenium-ion-like intermediates is governed by electronic (stabilizing or destabilizing) substituent effects. These effects will also be apparent in the transition state of a glycosylation reaction in which partial oxocarbenium ion character develops. The amount of positive charge at the anomeric center that can or has to develop for a given glycosylation reaction to occur depends not only on the nature of the donor but also on the nucleophilicity of the acceptor. Acceptors of high nucleophilicity will be able to displace covalent reactive intermediates, where poor nucleophiles require more oxocarbenium ion character. With the ever-growing insight into the reactivity of different reactive intermediates at play during a glycosylation reaction, more control over the stereoselective construction of glycosidic bonds will be gained, reducing the time- and labor-intensive trial-and-error component that has thwarted synthetic carbohydrate chemistry for so long.

References

1 (a) Zhu, X. and Schmidt, R.R. (2009) *Angew. Chem. Int. Ed.*, 48, 1900–1934; (b) Werz, D.B. and Seeberger, P.H. (2007) *Nature*, 446, 1046–1051; (c) Boltje, T.J., Buskas, T., and Boons, G.-J. (2009) *Nat. Chem.*, 1, 611–622.

2 Demchenko, A.V. (2008) Handbook of Chemical Glycosylation: Advances in Stereoselectivity and Therapeutic Relevance, Wiley-VCH Verlag GmbH.

3 Nigudkar, S.S. and Demchenko, A.V. (2015) *Chem. Sci.*, 6, 2687–2704.

4 Bohé, L. and Crich, D. (2011) *C.R. Chim.*, 14, 3–16.

5 Bohé, L. and Crich, D. (2015) *Carbohydr. Res.*, 403, 48–59.

6 Crich, D. (2010) *Acc. Chem. Res.*, 43, 1144–1153.

7 Amyes, T.L. and Jencks, W.P. (1989) *J. Am. Chem. Soc.*, 111, 7888–7900.

8 Sinnott, M.L. and Jencks, W.P. (1980) *J. Am. Chem. Soc.*, 102, 2026–2032.

9 Hosoya, T., Takano, T., Kosma, P., and Rosenau, T. (2014) *J. Org. Chem.*, 79, 7889–7894.

10 Hosoya, T., Kosma, P., and Rosenau, T. (2015) *Carbohydr. Res.*, 401, 127–131.

11 Saito, K., Ueoka, K., Matsumoto, K., Suga, S., Nokami, T., and Yoshida, J. (2011) *Angew. Chem. Int. Ed.*, 50, 5153–5156.

12 Douglas, N. L., Ley, S. V., Lücking, U., Warriner, S. L. (1998) *J. Chem. Soc.*, 51–66.

13 Zhang, Z., Ollmann, I.R., Ye, X.-S., Wischnat, R., Baasov, T., and Wong, C.-H. (1999) *J. Am. Chem. Soc.*, 121, 734–753.

14 Mong, K.-K.T. and Wong, C.-H. (2002) *Angew. Chem.*, 114, 4261–4264.

15 Larsen, C.H., Ridgway, B.H., Shaw, J.T., and Woerpel, K.A. (1999) *J. Am. Chem. Soc.*, 121, 12208–12209.

16 Shaw, J.T. and Woerpel, K.A. (1999) *Tetrahedron*, 55, 8747–8756.

17 Larsen, C.H., Ridgway, B.H., Shaw, J.T., Smith, D.M., and Woerpel, K.A. (2005) *J. Am. Chem. Soc.*, 127, 10879–10884.

18 Lucero, C.G. and Woerpel, K.A. (2006) *J. Org. Chem.*, 71, 2641–2647.

19 Romero, J.A.C., Tabacco, S.A., and Woerpel, K.A. (2000) *J. Am. Chem. Soc.*, 122, 168–169.

20 Ayala, L., Lucero, C.G., Romero, J.A.C., Tabacco, S.A., and Woerpel, K.A. (2003) *J. Am. Chem. Soc.*, 125, 15521–15528.

21 van Rijssel, E.R., van Delft, P., Lodder, G., Overkleeft, H.S., van der Marel, G.A., Filippov, D.V., and Codée, J.D.C. (2014) *Angew. Chem. Int. Ed.*, 53, 10381–10385.
22 van Rijssel, E.R., van Delft, P., van Marle, D.V., Bijvoets, S.M., Lodder, G., Overkleeft, H.S., van der Marel, G.A., Filippov, D.V., and Codée, J.D.C. (2015) *J. Org. Chem.*, 80, 4553–4565.
23 Rhoad, J.S., Cagg, B.A., and Carver, P.W. (2010) *J. Phys. Chem. A*, 114, 5180–5186.
24 Jensen, H.H., Nordstrøm, L.U., and Bols, M. (2004) *J. Am. Chem. Soc.*, 126, 9205–9213.
25 Frihed, T.G., Walvoort, M.T.C., Codée, J.D.C., van der Marel, G.A., Bols, M., and Pedersen, C.M. (2013) *J. Org. Chem.*, 78, 2191–2205.
26 Moumé-Pymbock, M., Furukawa, T., Mondal, S., and Crich, D. (2013) *J. Am. Chem. Soc.*, 135, 14249–14255.
27 Krumper, J.R., Salamant, W.A., and Woerpel, K.A. (2008) *Org. Lett.*, 10, 4907–4910.
28 Krumper, J.R., Salamant, W.A., and Woerpel, K.A. (2009) *J. Org. Chem.*, 74, 8039–8050.
29 Mayr, H., Bug, T., Gotta, M.F., Hering, N., Irrgang, B., Janker, B., Kempf, B., Loos, R., Ofial, A.R., Remennikov, G. *et al.* (2001) *J. Am. Chem. Soc.*, 123, 9500–9512.
30 Frihed, T.G., Bols, M., and Pedersen, C.M. (2015) *Chem. Rev.*, 115, 4963–5013.
31 Hosoya, T., Kosma, P., and Rosenau, T. (2015) *Carbohydr. Res.*, 411, 64–69.
32 Whitfield, D.M. (2007) *Carbohydr. Res.*, 342, 1726–1740.
33 Whitfield, D.M. (2012) *Carbohydr. Res.*, 356, 180–190.
34 Whitfield, D.M. (2015) *Carbohydr. Res.*, 403, 69–89.
35 Satoh, H. and Nukada, T. (2014) *Trends Glycosci. Glycotechnol.*, 26, 11–27.
36 Stoddard, J.F. (1971) Stereochemistry of Carbohydrates, Wiley-Interscience, Toronto.
37 Ardèvol, A., Biarnés, X., Planas, A., and Rovira, C. (2010) *J. Am. Chem. Soc.*, 132, 16058–16065.
38 Davies, G.J., Planas, A., and Rovira, C. (2012) *Acc. Chem. Res.*, 45, 308–316.
39 Satoh, H., Hansen, H.S., Manabe, S., van Gunsteren, W.F., and Hünenberger, P.H. (2010) *J. Chem. Theory Comput.*, 6, 1783–1797.
40 Martin, A., Arda, A., Désiré, J., Martin-Mingot, A., Probst, N., Sinaÿ, P., Jiménez-Barbero, J., Thibaudeau, S., and Blériot, Y. (2015) *Nat. Chem.* 8, 186–191.
41 Crich, D. and Sun, S. (1996) *J. Org. Chem.*, 61, 4506–4507.
42 El Ashry, E.S.H., Rashed, N., and Ibrahim, E.S.I. (2005) *Curr. Org. Synth.*, 2, 175–213.
43 Yang, L., Qin, Q., and Ye, X.-S. (2013) *Asian J. Org. Chem.*, 2, 30–49.
44 Crich, D. and Sun, S. (1998) *Tetrahedron*, 54, 8321–8348.
45 Crich, D. and Chandrasekera, N.S. (2004) *Angew. Chem.*, 116, 5500–5503.
46 Huang, M., Garrett, G.E., Birlirakis, N., Bohé, L., Pratt, D.A., and Crich, D. (2012) *Nat. Chem.*, 4, 663–667.
47 Huang, M., Retailleau, P., Bohé, L., and Crich, D. (2012) *J. Am. Chem. Soc.*, 134, 14746–14749.
48 Adero, P.O., Furukawa, T., Huang, M., Mukherjee, D., Retailleau, P., Bohé, L., and Crich, D. (2015) *J. Am. Chem. Soc.* 137, 10336–10345.
49 Crich, D. and Sun, S. (1997) *J. Am. Chem. Soc.*, 119, 11217–11223.
50 Walvoort, M.T.C., Lodder, G., Mazurek, J., Overkleeft, H.S., Codée, J.D.C., and van der Marel, G.A. (2009) *J. Am. Chem. Soc.*, 131, 12080–12081.
51 Heuckendorff, M., Bendix, J., Pedersen, C.M., and Bols, M. (2014) *Org. Lett.*, 16, 1116–1119.
52 van den Bos, L.J., Dinkelaar, J., Overkleeft, H.S., and van der Marel, G.A. (2006) *J. Am. Chem. Soc.*, 128, 13066–13067.

53 Codée, J.D.C., van den Bos, L.J., de Jong, A.-R., Dinkelaar, J., Lodder, G., Overkleeft, H.S., and van der Marel, G.A. (2009) *J. Org. Chem.*, 74, 38–47.
54 van den Bos, L.J., Codée, J.D.C., Litjens, R.E.J.N., Dinkelaar, J., Overkleeft, H.S., and van der Marel, G.A. (2007) *Eur. J. Org. Chem.*, 2007, 3963–3976.
55 Walvoort, M.T.C., de Witte, W., van Dijk, J., Dinkelaar, J., Lodder, G., Overkleeft, H.S., Codée, J.D.C., and van der Marel, G.A. (2011) *Org. Lett.*, 13, 4360–4363.
56 Walvoort, M.T.C., Lodder, G., Overkleeft, H.S., Codée, J.D.C., and van der Marel, G.A. (2010) *J. Org. Chem.*, 75, 7990–8002.
57 Walvoort, M.T.C., Moggré, G.-J., Lodder, G., Overkleeft, H.S., Codée, J.D.C., and van der Marel, G.A. (2011) *J. Org. Chem.*, 76, 7301–7315.
58 Walvoort, M.T.C., van den Elst, H., Plante, O.J., Kröck, L., Seeberger, P.H., Overkleeft, H.S., van der Marel, G.A., and Codée, J.D.C. (2012) *Angew. Chem. Int. Ed.*, 51, 4393–4396.
59 Andrade, R.B., Plante, O.J., Melean, L.G., and Seeberger, P.H. (1999) *Org. Lett.*, 1, 1811–1814.
60 Dinkelaar, J., van den Bos, L.J., Hogendorf, W.F.J., Lodder, G., Overkleeft, H.S., Codée, J.D.C., and van der Marel, G.A. (2008) *Chem. – Eur. J.*, 14, 9400–9411.
61 Chi, F.-C., Kulkarni, S.S., Zulueta, M.M.L., and Hung, S.-C. (2009) *Chem. – Asian J.*, 4, 386–390.
62 Zhang, Q., van Rijssel, E.R., Walvoort, M.T.C., Overkleeft, H.S., van der Marel, G.A., and Codée, J.D.C. (2015) *Angew. Chem. Int. Ed.*, 54, 7670–7673.
63 Storry, J.R. and Olsson, M.L. (2009) *Immunohematology*, 25, 48–59.
64 Park, J., Boltje, T.J., and Boons, G.-J. (2008) *Org. Lett.*, 10, 4367–4370.
65 Werz, D.B. and Vidal, S. (2013) Modern Synthetic Methods in Carbohydrate Chemistry: From Monosaccharides to Complex Glycoconjugates, Wiley VCH, pp. 135–143.
66 Gerbst, A.G., Ustuzhanina, N.E., Grachev, A.A., Khatuntseva, E.A., Tsvetkov, D.E., Whitfield, D.M., Berces, A., and Nifantiev, N.E. (2001) *J. Carbohydr. Chem.*, 20, 821–831.
67 Gerbst, A.G., Ustuzhanina, N.E., Grachev, A.A., Zlotina, N.S., Khatuntseva, E.A., Tsvetkov, D.E., Shashkov, A.S., Usov, A.I., and Nifantiev, N.E. (2002) *J. Carbohydr. Chem.*, 21, 313–324.
68 Guillemineau, M. and Auzanneau, F.-I. (2012) *J. Org. Chem.*, 77, 8864–8878.
69 Crich, D., Pedersen, C.M., Bowers, A.A., and Wink, D.J. (2007) *J. Org. Chem.*, 72, 1553–1565.
70 Zhu, X., Kawatkar, S., Rao, Y., and Boons, G.-J. (2006) *J. Am. Chem. Soc.*, 128, 11948–11957.
71 Ishiwata, A., Akao, H., and Ito, Y. (2006) *Org. Lett.*, 8, 5525–5528.
72 Kistemaker, H.A.V., Overkleeft, H.S., van der Marel, G.A., and Filippov, D.V. (2015) *Org. Lett.*, 17, 4328–4331.
73 Hirano, S., Ichikawa, S., and Matsuda, A. (2005) *Angew. Chem. Int. Ed.*, 44, 1854–1856.
74 Ichikawa, S., Hayashi, R., Hirano, S., and Matsuda, A. (2008) *Org. Lett.*, 10, 5107–5110.

2

Application of Armed, Disarmed, Superarmed, and Superdisarmed Building Blocks in Stereocontrolled Glycosylation and Expeditious Oligosaccharide Synthesis

Mithila D. Bandara, Jagodige P. Yasomanee, and Alexei V. Demchenko

2.1 Introduction: Chemical Synthesis of Glycosides and Oligosaccharides

From the building blocks of nature to disease-battling therapeutics and vaccines, carbohydrates have had a broad impact on evolution, society, economy, and human health. Numerous applications of these essential molecules in many areas of science and technology exist, foremost of which can be found in the areas of therapeutic-agent and diagnostic-platform development. Although carbohydrate oligomers, oligosaccharides, or glycans are desirable for biological and medical communities, these molecules remain very challenging targets for chemists. Among a number of hurdles including functionalization, elaborate protecting- and leaving-group manipulations, tedious purification, and sophisticated structure analysis, it is glycosylation, a coupling reaction performed between two monosaccharide units, that has proven particularly challenging to chemists. Nature flawlessly performs this reaction to obtain complex glycans and glycoconjugates [1]. Chemical glycosylation, however, remains challenging even with the aid of recent methodological breakthroughs [2] and modern technologies [3].

Many methods for chemical glycosylation have been developed, but it is the inability to control the stereoselectivity that has proven to be the major hurdle. The glycosylation typically follows a unimolecular S_N1 mechanism [4] via four distinct steps: activation, dissociation, nucleophilic attack, and proton transfer (Scheme 2.1) [4a]. In the case of a nonparticipating ether-type substituent at C-2, glycosylation proceeds via an oxocarbenium ion [2c]. The intermediacy of the flattened oxocarbenium ion typically results in the formation of anomeric mixtures in which 1,2-*cis*-glycosides [5] (for D-gluco/galacto series) are slightly favored due to the anomeric effect [6]. The goal of stereocontrolling glycosylation has been approached in many ways, and much effort has been dedicated to developing new leaving groups (LGs in schemes) and refining the reaction conditions. We know that leaving groups, temperature, promoter/additives, and the reaction solvent may have a significant effect on the reactivity of the reactants and the stereoselectivity of glycosylation [7]. However, since these factors still often fail to adequately control the outcome of many glycosylations that tend to proceed via the oxocarbenium ion, studies are refocusing on gaining a better understanding of the reaction mechanism.

Selective Glycosylations: Synthetic Methods and Catalysts, First Edition. Edited by Clay S. Bennett.

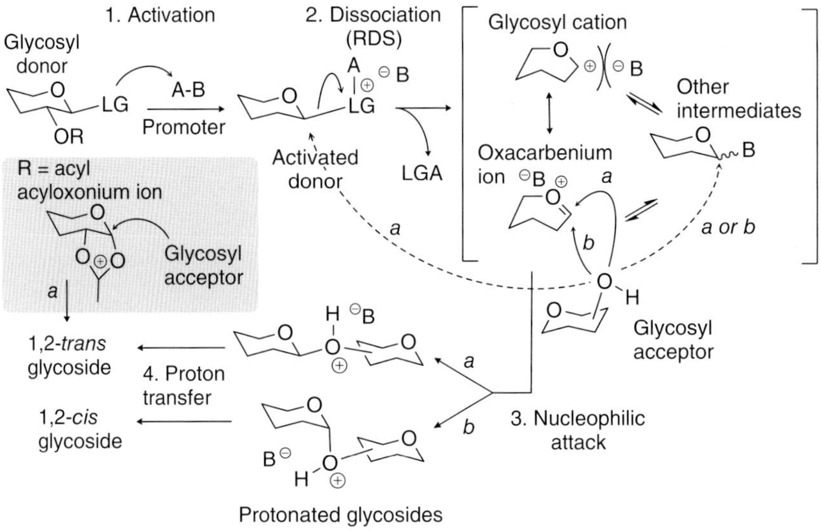

Scheme 2.1 Outline of the chemical glycosylation.

Protecting groups are extensively used in carbohydrate chemistry to mask additional sites of reactivity in polyfunctional compounds, including both donors and acceptors. Protecting groups, however, can affect the glycosylation in a variety of other ways. As stated by Fraser-Reid, "protecting groups do more than protect" [8]. Among the best-known effects of protecting groups is the neighboring acyl group participation [9], which remains one of the most powerful modes for controlling the stereoselectivity of glycosylation reaction (protecting groups do more than protect). A vast majority of 1,2-*trans*-glycosides is obtained from glycosyl donors equipped with a 2-acyl protecting group. These reactions proceed via the intermediacy of a cyclic acyloxonium ion, which is then opened with the glycosyl acceptor from the opposite (trans) face (Scheme 2.1).

In addition, protecting groups may have a profound effect on the conformation and stereoelectronics of the starting material, key reaction intermediates, and the products [10]. In the recent years, dedicated studies of these intermediates have led to the development of many stereocontrolled reactions, and the synthesis of β-mannosides via anomeric triflates by Crich and coworkers is arguably the best example of such a study [11]. Nevertheless, some linkages and targets remain challenging due to the

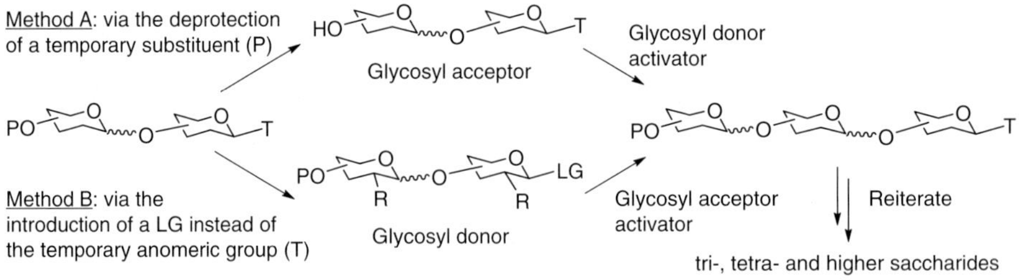

Scheme 2.2 Outline of conventional approaches to oligosaccharide synthesis.

requirement to achieve complete stereocontrol in each and every step [12] and suppress side reactions [7, 13].

Beyond that, the synthesis of oligosaccharides may require additional synthetic steps between glycosylations. In accordance with the traditional oligosaccharide synthesis, the disaccharide intermediate should be converted into a glycosyl acceptor or donor of the second generation. As depicted in Scheme 2.2, this can be accomplished via deprotection (Method A) or introduction of a leaving group instead of a temporary anomeric substituent (Method B). The modified disaccharide building blocks can then be reacted with other glycosyl donors or acceptors, resulting in the formation of a trisaccharide or larger saccharide if the convergent approach is incorporated. These synthetic steps can be then reiterated to obtain larger oligosaccharides.

A large number of additional synthetic steps between each glycosylation step typically lead to reduced yields and overall efficiency of the oligosaccharide assembly. Consequently, the past quarter of the century had witnessed the development of new strategies for oligosaccharide synthesis, among which selective and chemoselective concepts prevail [14]. Synthetic strategies based on selective activations make use of different leaving groups that are sequentially activated, and traditional selective activation [15], two-step activation [15a,16], active-latent concept [17], and orthogonal strategy [18] are just a few examples of such approaches [19]. Another general direction in expeditious oligosaccharide synthesis involves chemoselective activations. This strategy is based on the so-called armed–disarmed strategy introduced by Fraser-Reid *et al.* [20]. Building blocks used in chemoselective activations utilize only one type of a leaving group, and the building block reactivity is adjusted by the choice of protecting groups (protecting groups do more than protect). The next subchapter introduces this general strategy, and subsequent sections elaborate on the recent progress that has been made in the area of tuning the reactivity of building blocks and their application in stereoselective glycosylation and chemoselective oligosaccharide synthesis.

2.2 Fraser-Reid's Armed–Disarmed Strategy for Oligosaccharide Synthesis

Although the effect of protecting groups on reactivity has been known for many decades [21], Fraser-Reid was the first to describe a new mode by which the differential properties of protecting groups could be exploited [22]. It was noticed that ester protecting groups reduce the reactivity (disarm) of the *n*-pentenyl leaving group, in comparison to that of its alkylated (armed) counterpart (Scheme 2.3). One explanation for this phenomenon is that the increased electron withdrawal of ester groups decreases the electron density (nucleophilicity) of the anomeric heteroatom. This translates into a reduced leaving-group ability and works with the leaving groups capable of either direct (thioglycosides) or remote activation (*n*-pentenyl) [23]. To differentiate the reactivity, mild reaction conditions are required, and iodonium(di-γ-collidine)perchlorate (IDCP) was found to be a suitable mild activator for the armed *O*-pentenyl glycosyl donors allowing for direct chemoselective coupling between an activated (armed) glycosyl donor and a deactivated (disarmed) glycosyl acceptor. The disaccharide is then used directly in subsequent glycosidation, but the activation of its disarmed leaving group may require a stronger activator (N-iodosuccinimide/trifluoromethanesulfonic acid (NIS/TfOH) in case of

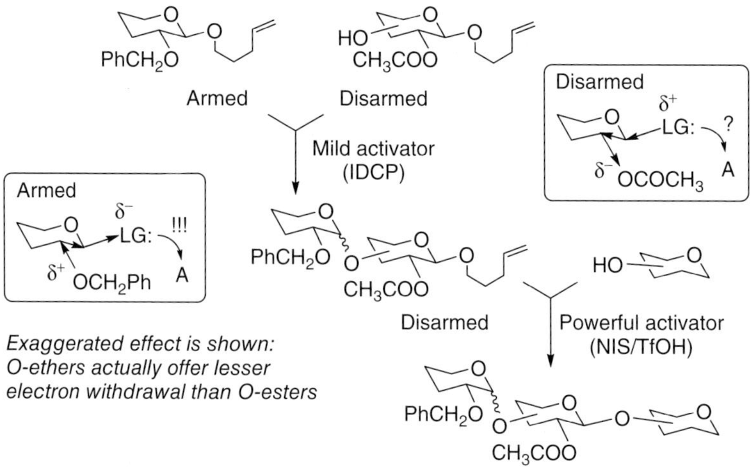

Scheme 2.3 Outline of the armed–disarmed strategy discovered by Fraser-Reid.

pentenyl leaving group). In a more general sense, the differentiation can be achieved by modulating the reaction conditions that in addition to the choice of promoter, reaction temperature or solvent could be exploited [24]. Discovered with *n*-pentenyl glycosides, this armed–disarmed strategy ultimately proved to be of a general nature and has been applied to many other classes of glycosyl donor (see subsequent sections) [25].

A different rationalization of the arming and disarming effects has emerged with the discovery of the "O-2/O-5 Cooperative Effect" [26]. In case of the armed donors, it is believed that the oxocarbenium ion intermediate is stabilized by integrated assistance from the lone pair of electrons on the adjacent ring oxygen (O-5, Figure 2.1). In the case of the disarmed donors, it is believed that the oxocarbenium ion intermediate is stabilized by charge distribution via acyloxonium ion intermediate formed through the acyl-type protecting group at O-2. The realization of the O-2/O-5 Cooperative Effect in glycosylation led to the discovery of electronically superarmed and superdisarmed glycosyl donors and acceptors (*vide infra*). Reinforcing early work by Isbell and Frush [27], Crich and Li emphasized that the 1,2-*trans* orientation of the 2-*O*-acyl and *S*-benzoxazolyl (SBox) leaving group is required for the anchimeric assistance to occur [28]. A similar conclusion was reached by Bols and Demchenko for *S*-phenyl glycosides (*vide infra*) [29]. Presumably, the stabilization takes place via the concerted displacement of the leaving group.

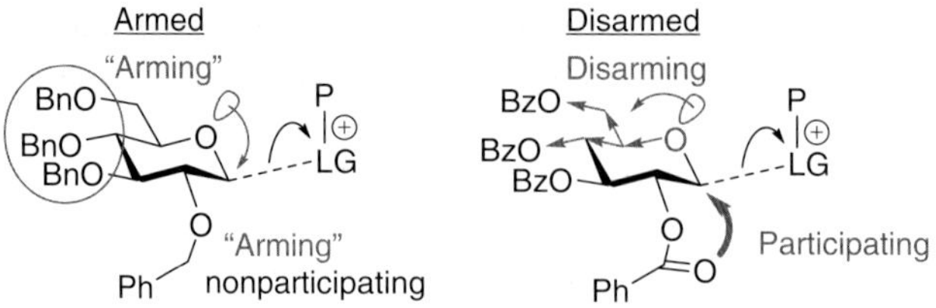

Figure 2.1 O-2/O-5 cooperative effect in glycosylation.

2.3 Many Reactivity Levels Exist between the Armed and Disarmed Building Blocks

From the early days, the researchers were devising different approaches to quantifying the relative reactivity of different building blocks. Following the pioneering study by Fraser-Reid dedicated to determining relative reactivities of variously protected pairs of the *n*-pentenyl glycosides [8], Ley and coworkers introduced a technique wherein building block reactivity could be "tuned" [30]. In a series of competitive experiments, wherein two glycosyl donors were set to compete for one glycosyl acceptor, a series of relative reactivity ratios were established (Scheme 2.4). For instance, the greatest disarming effect was seen from the 2-benzoyl substituent in compound **2** in comparison to that of 3-benzoyl and 4-benzoyl substituents (**3** and **4**). Dibenzoylated glycosyl donors **5** and **6** were less reactive than their monobenzoylated counterparts, with the disarmed perbenzoylated donor **7** being the least reactive in this series.

Wong and coworkers devised a comprehensive approach wherein a broad library of glycosyl donors and acceptors was assigned relative reactivity values (RRVs) [31]. The determination of RRVs was made using tolyl thioglycoside donors in the presence of an NIS/TfOH promoter system. More recently, Hung and Wong created a comprehensive database of RRVs for the series of d-glucose building blocks (Figure 2.2) [32]. According to their database, some tribenzylated acceptors **9–12** showed similar or even higher RRVs in comparison to the armed donor **8**. Not surprisingly, RRVs of monobenzoylated donors **13–16** were lower but still much higher than that of the disarmed acceptor **17** or disarmed donor **18**. The application of this approach to chemoselective oligosaccharide synthesis and determination of the RRVs of silylated donors will be discussed as follows.

Toshima and coworkers studied the effect of remote protecting groups of glycosyl donors of the 2,3-dideoxy series on the reactivity in glycosylations [33]. For this purpose, glycosylation reactions of glycosyl acetate donors **19** and **20** with acceptor **21** were performed in the presence of several Lewis and protic acid activators including trimethylsilyl trifluoromethanesulfonate (TMSOTf), tert-budtyldimethylsilyl trifluoromethanesulfonate (TBSOTf), $BF_3{\bullet}OEt_2$, TfOH, and montmorillonite K-10 (MK-10) (Scheme 2.5). It was found that glycosidation of donor **20** yielded disaccharide **23** with

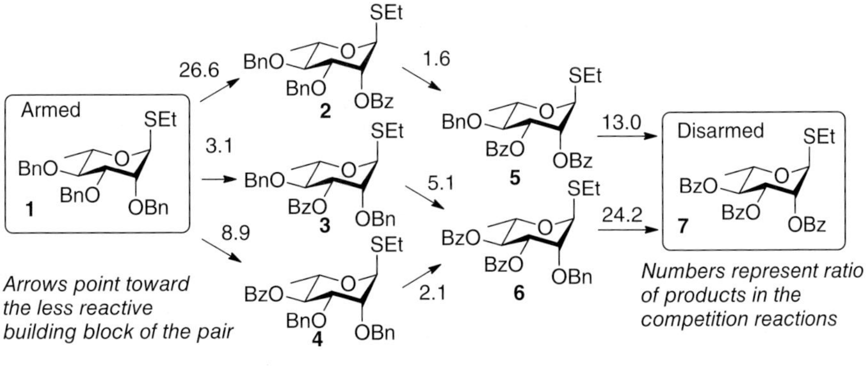

Scheme 2.4 Intermediate reactivity of a series of partially benzoylated rhamnosides.

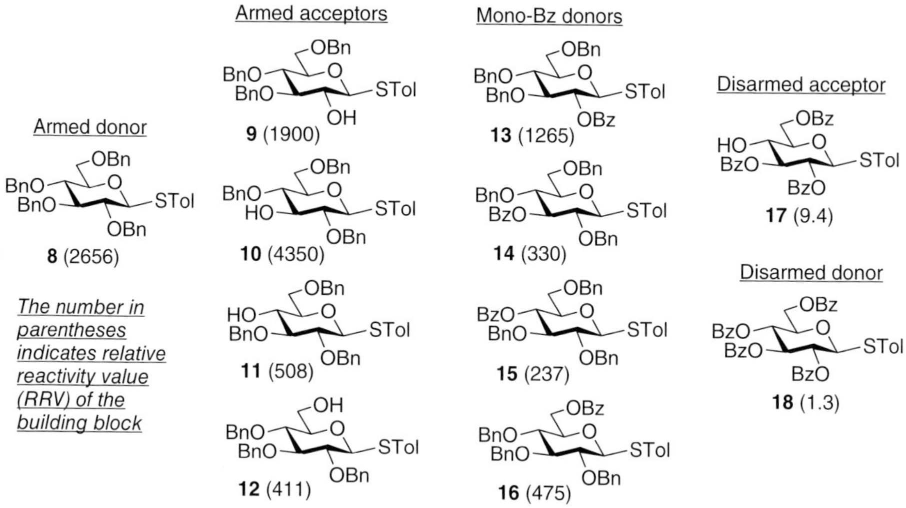

Figure 2.2 Relative reactivity values (RRVs) of differently protected STol glycosyl donors and acceptors.

excellent yield while donor **19** with 4,6-dibenzoyl protection gave disaccharide **22** in low yield under the same reaction conditions. A similar reactivity profile was determined for 4-benzoyl-6-benzyl donor [33].

Although most reactivity levels in the studies surveyed in this section fall between the traditional armed and disarmed building blocks, Wong's and Hung's studies revealed a number of building blocks that extend beyond this boundary. For example, 2-hydroxyl galactoside [31] and 3-hydroxyl glucoside [32] were found to be 3 and 1.5 times more reactive than their respective perbenzylated counterparts. This important discovery led to a new direction in studying building blocks, and a variety of new reactivity levels ranging from more reactive than the armed ones (superarmed) to even less reactive than the disarmed ones (superdisarmed) have been discovered. The studies arising from these two new directions will be discussed in the subsequent two subsections, respectively.

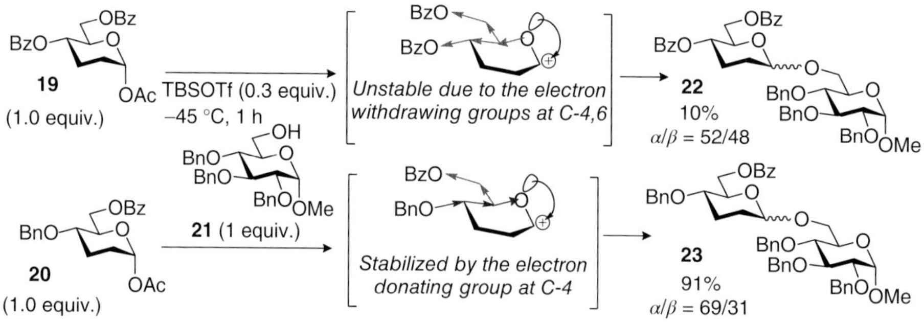

Scheme 2.5 Disarming effects on 2,3-dideoxy donors.

2.4 Modes for Enhancing the Reactivity: Superarmed Building Blocks

Uniformly protected perbenzylated glycosyl donors have become the benchmark for describing the armed glycosyl donors or reactivity levels associated with them. Over the years, benzyl groups had been almost exclusively used as arming ether protecting groups until Demchenko and coworkers showed that 2-*O*-picolinyl group has similar electronic properties and can also be used in chemoselective armed–disarmed activations [34]. Uniquely, the picolinyl group is capable of affecting the stereoselectivity of the reaction, which makes it suitable as an "arming participating group." More recently, *o*-cyano and *o*-nitrobenzyl have been introduced as arming participating groups [35].

Other recent improvements have revealed donors with reactivity levels that far exceed the reactivity of traditional uniformly benzylated armed building blocks. These "superarmed" glycosyl donors have further expanded the versatility of the armed–disarmed concept. Introduced by Bols for describing the reactivity of conformationally armed building blocks, the term superarmed is now used to describe all building blocks that are more reactive than perbenzylated armed building blocks. Building upon early studies of the effect of conformational changes on reactivity [36], Bols and coworkers hypothesized that the conformational change required to obtain flattened oxocarbenium intermediate that exists in a half-chair conformation will be facilitated in the axial-rich donor. If this conformational change could be facilitated, the activation energy of the rate-determining step (RDS) will decrease, and the donor reactivity will enhance. The conformational change of SPh glucoside **24** was induced via creating steric congestion with *tert*-butyldimethylsilyl (TBS) groups at the C-2, 3, and 4 positions, resulting in a skew-boat conformation of **25** (Scheme 2.6) [36, 37]. As a result, glycosyl donor **25** showed an estimated 20-fold increase in reactivity in comparison to the perbenzylated donor. This was ultimately translated into the direct chemoselective coupling of donor **25** with the armed acceptor **26** in the presence of NIS/TfOH at −78 °C to give disaccharide **27** in a high yield [37b,c].

The RRVs of partially silylated STol glycosyl donors reported by Hung and Wong clearly reinforce Bols' findings that the reactivity may vary drastically depending on the number of silyl groups present and their location in the sugar ring [32]. For example, monosilylated donors **28–31** express much higher reactivity than the standard armed donor **8** (Figure 2.3). Both TBS and their tri-isopropylsilyl (TIPS) counterparts have been studied, and the latter showed a marginally higher reactivity across the range of all

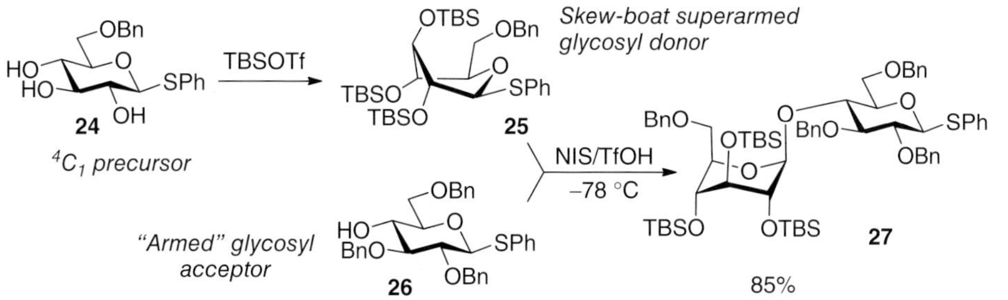

Scheme 2.6 Conformational superarming: conformational change to increase reactivity.

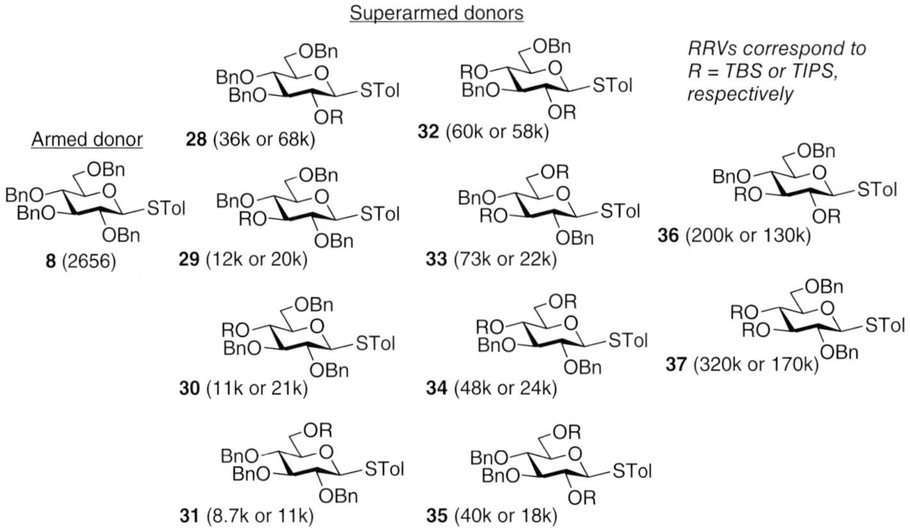

Figure 2.3 RRVs of partially silylated superarmed glycosyl donors.

monosilylated derivatives studied. Di-silylated thioglycosides **32–35**, in which silyl groups were remotely positioned to each other showed two to six times higher RRVs compared to their monosilylated counterparts. The greater reactivity enhancement was detected for disilylated derivatives **36** and **37** in which the two silyl substituents were placed at the neighboring *trans*-vicinal positions of the ring, 2,3 and 3,4 respectively. Interestingly, TBS substituents were more arming than TIPS in all disilylated donors **32–37**.

The scope of conformational superarming was broadened by the investigation of a series of glycosyl donors in which the axial-rich conformation was achieved via strategic tethering. Building upon their earlier studies of 2,4-diol tethering with di-*tert*-butyl silylene [37c] and Yamada's 3,6-*O*-(*o*-xylylene)-bridging [38], Bols and coworkers devised a series of novel glycosyl donors. Under this, 3,6-di-*tert*-butyl silylene tethering in gluco-, manno-, galacto-, and 2-azido-gluco pyranosides has been investigated [39]. All of these donors were found to adopt axial-rich $B_{1,4}$ boat or $^3S_1$ skew boat conformations. To determine the relative reactivity, direct activations of the new donors over the armed tribenzylated acceptor **26** have been conducted. As depicted in Scheme 2.7, chemoselective glycosylations with donors **38–40** afforded disaccharides **41–43** in 51–70% yields with preferential α-selectivity.

Demchenko and coworkers took a different approach for superarming glycosyl donors. The superarming of SBox and SEt glycosyl donors was based on the O-2/O-5 cooperative effect in glycosylation (*vide infra*) [26]. In this scenario, it is believed that the oxocarbenium ion intermediate is stabilized by the cooperative assistance from both the lone pair of electrons on the adjacent ring oxygen (O-5) and charge distribution via acyloxonium ion intermediate formed through the acyl-type protecting group at O-2 (Scheme 2.8). The 2-*O*-benzoyl-3,4,6-tri-*O*-benzyl derivatives gain extra stabilization upon activation through this O-2/O-5 cooperative mechanism and become a superarmed donor. As a result of the competitive glycosylation upon activation with dimethyl(thiomethyl)

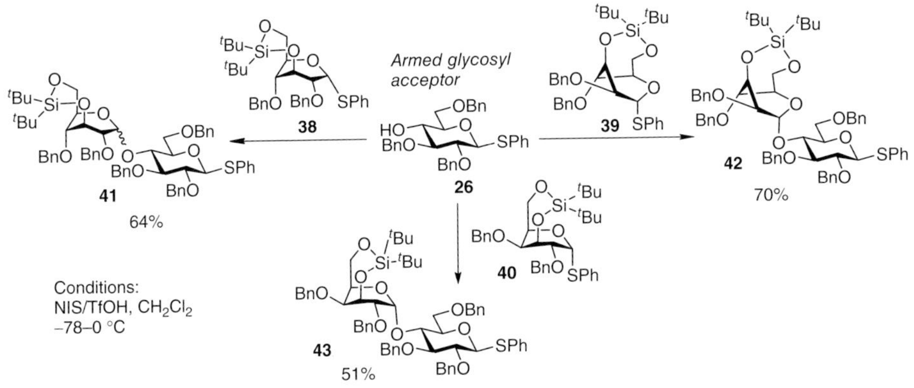

Scheme 2.7 Conformational superarming via 3,6-silylene tethering.

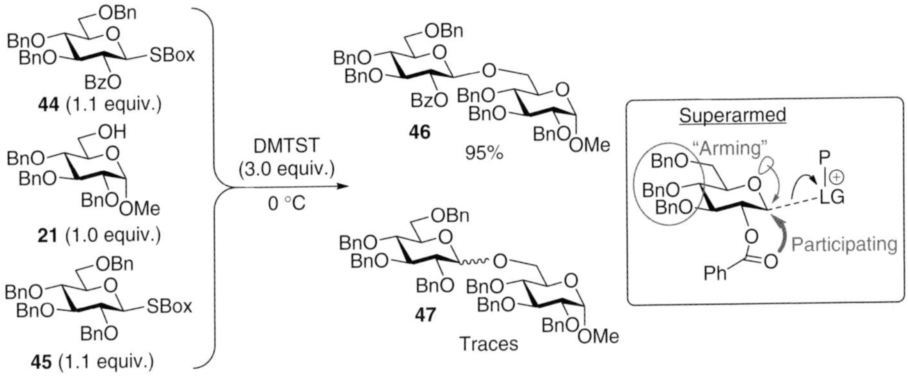

Scheme 2.8 Superarmed (**44**) and armed (**45**) glycosyl donors in competitive glycosylation.

sulfonium trifluoromethanesulfonate (DMTST), it was observed that donors equipped with 2-OBz-3,4,6-tri-OBn pattern are 10–20 times more reactive than their armed counterparts [40]. Thus, a competitive reaction of donors **44** and **45** with acceptor **21** gave disaccharide **46** derived from the superarmed donor **44** in 95% yield. Meanwhile, the disaccharide **47** derived from the armed donor **45** was found only in trace amount.

These results clearly showed that the superarmed donor is much more reactive than the armed counterpart. Further studies by the Demchenko group showed that the same trend of reactivity appears upon changing the leaving group. While efficient differentiation of glycosyl donors of the *S*-ethyl series could be efficiently achieved in the presence of iodine as promoter, the reactivity difference was notably lower for glycosyl donors of the SPh, *S*-tolyl, STaz (*S*-thiazolinyl), and *O*-pentenyl series [40c].

Through a collaboration between the Bols and Demchenko groups, the two different approaches to superarmed glycosyl donors were combined in one universal platform. Glycosylations with 2-OBz-3,4-di-OTBS donor **49** were swift, high yielding, and β-stereoselective (Scheme 2.9) [29]. In order to determine the relative reactivity of the new hybrid donor **49** in comparison to the previously investigated superarmed donors, a

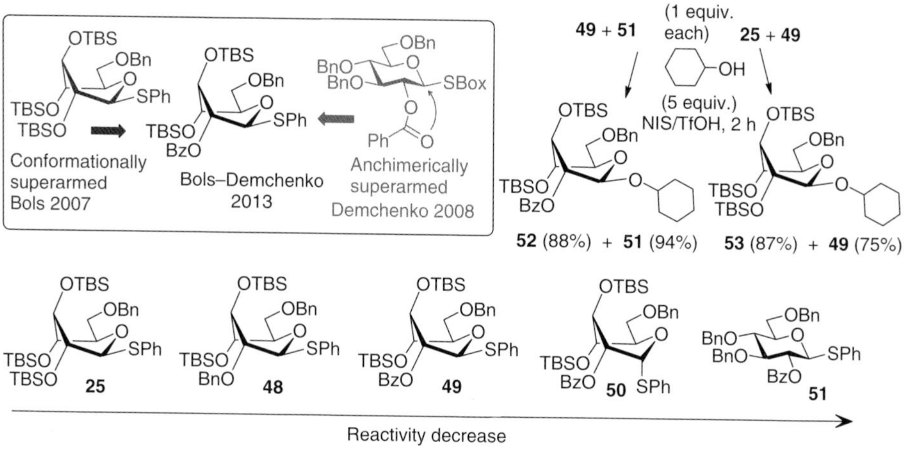

Scheme 2.9 Superarming by combined neighboring and conformational effects.

series of competitive experiments have been performed. Thus, a competition experiment between the hybrid donor **49** and the electronically superarmed donor **51** showed a 88% conversion of donor **49** to glycoside **52**, whereas unreacted donor **51** was recovered in 94%. When a similar competition experiment was performed between donors **49** and **25**, a higher conversion of donor **25** has been observed. A significance of the anomeric configuration brought up by Crich and Li [28] was further reinforced by this comparison study that showed higher reactivity of donor **49** than its α-linked counterpart **50**. It is also believed that the flipped skew–boat conformation of the donor may also diminish the anchimeric assistance due to the non-antiperiplanar nature of the 2-OBz group to the anomeric leaving group.

On the other hand, α-configured glycosyl donors equipped with a nonparticipating group at C-2 typically by far exceed the reactivity of their β-counterparts [29, 41]. This controversy reinforces the power of the anchimeric assistance that is able to invert the reactivity of α- and β-thioglycosides. Over the course of this study, the order of relative reactivity of various glycosyl donors was determined. It was also learned that the conformational arming is a very powerful tool to increase the reactivity and achieve excellent yields. The anchimeric superarming effects are weaker, but the participation ensures high 1,2-*trans* selectivity, which was unavailable with other conformationally superarmed donors. A recent relevant study revealed that SEt glycosyl donors follow a very similar relative reactivity trend, but all SEt glycosides are marginally more reactive than their SPh counterparts [42].

2.5 Modes for Decreasing the Reactivity: Superdisarmed Building Blocks

Standard disarmed building blocks are uniformly protected with benzoyls (perbenzoates). Madsen *et al.* have revealed that building blocks can be deactivated by placing a single electron-withdrawing group at a remote position [43]. Thus, they have shown

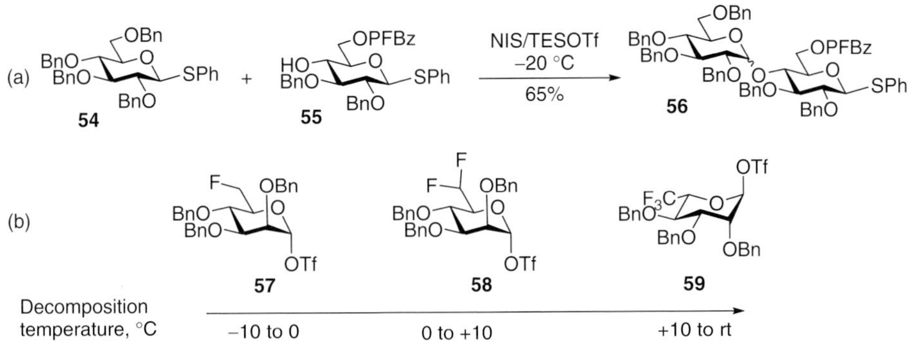

Scheme 2.10 Deactivation by strong electron withdrawal from C-6.

that 6-*O*-pentafluorobenzoyl (PFBz) group disarms the leaving group of **55**. The disarming effect is sufficient to selectively activate perbenzylated armed donor **54** over acceptor **55** to obtain disaccharide **56** (Scheme 2.10a). This concept was further extended to disarming building blocks of the galacto, manno, and mannosamine series. Crich and coworker have investigated the disarming effect that 6-fluoro-6-deoxy glycosyl donors in terms of reactivity, selectivity, and stability of the intermediate glycosyl triflates. Thus, it was observed that di- and trifluorinated D- or L-mannosyl triflates were more stable than their monofluorinated counterpart. This was accessed by comparing the decomposition temperatures specific to each glycosyl triflate as depicted in Scheme 2.10b [44].

The effect of cyclic acetals and ketals has also been studied toward expanding the scopes of armed–disarmed strategy. Fraser-Reid *et al.* [45] and Ley and coworkers [46] have shown that the presence of such cyclic protecting groups can deactivate the sugar derivative by increasing the rigidity of the sugar ring and thereby locking the $^4C_1$ chair conformation that may interfere with the formation of the flattened oxocarbenium ion [45, 47]. Extensive studies have led to the realization that reactivity of building blocks protected with cyclic groups can even be lower than that of the corresponding perbenzoylated disarmed building blocks. Thus, Boons and coworkers have shown that the presence of the cyclic 2,3-carbonate group disarms the thioglycosidic donor more than peracylation, making the former superdisarmed [48]. As a result, disarmed perbenzoylated glycosyl donor **60** could be selectively activated over the superdisarmed acceptor **61** to yield disaccharide **62** in a good yield (Scheme 2.11). Along similar lines, Demchenko and Kamat showed that even conventional benzylidene acetal groups can superdisarm building blocks of the SBox series [26].

The Bols group brought up the fact that the disarming effect of cyclic acetals can be not only due to the ring rigidity but also due to the electron-withdrawing effect of C-6 group, which is further enriched by its orientation [49]. From model studies, they have concluded that torsional effect, which greatly depends on the substituent orientation, plays a role in disarming the sugar moieties. More recently, Crich and coworkers investigated the effect of a 4,6-*O*-alkylidene acetal or its 7-carba analog on the rates of hydrolysis of methyl and 2,4-dinitrophenyl galactopyranosides in which the methoxy group adopts either an equatorial or an axial position according to the configuration

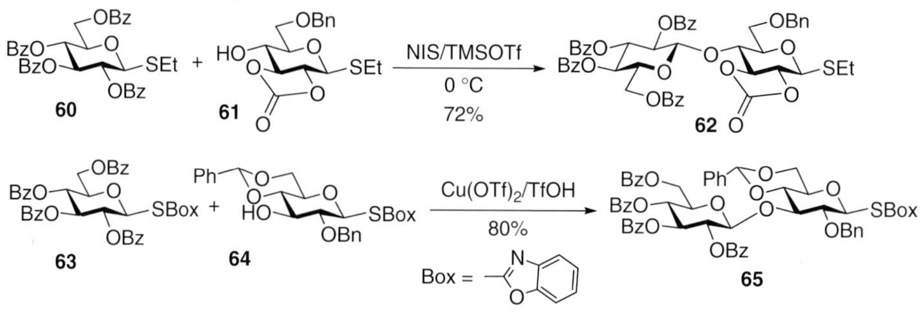

Scheme 2.11 Chemoselective activation of disarmed donors over acceptors bearing fused rings.

[50]. This study reinforced previous findings, and it was determined that the alkylidene acetal leads to decreased rates of hydrolysis with respect to comparable systems lacking the cyclic protecting group. A combination of the two effects, torsional and electronic, may be one of the reasons the donors containing the fused ring systems tend to be less reactive than the pure-electronically disarmed, acylated building blocks. In contrast, 7-carba analogs had practically no effect on the rates of hydrolysis.

Demchenko *et al.* discovered that SBox glycosyl donor protected with arming benzyl at C-2 and disarming benzoyl groups at C-3, 4, and 6 shows greatly diminished reactivity compared to both the armed perbenzylated and disarmed perbenzoylated glycosyl donors [26]. While glycosylation of acceptor **66** with SBox donors **45** and **63** in the presence of copper(II) triflate gave the disaccharides **67** and **68** in good yields, donor **69** remained totally unreactive (Scheme 2.12). Interestingly, this observation was contradictory to the previously predicted higher reactivity for such 2-O-benzylated glycosides compared to disarmed perbenzoylated donors [30, 31]. This discrepancy was rationalized by the "O-2/O-5 Cooperative Effect" [26]. In this application, the carbocation stabilization can be achieved neither from the endocyclic ring oxygen (O-5) as in the armed glycosyl donors nor from O-2 as in disarmed donors. As a result, this combination gave rise to the "superdisarming" protecting-group pattern overall.

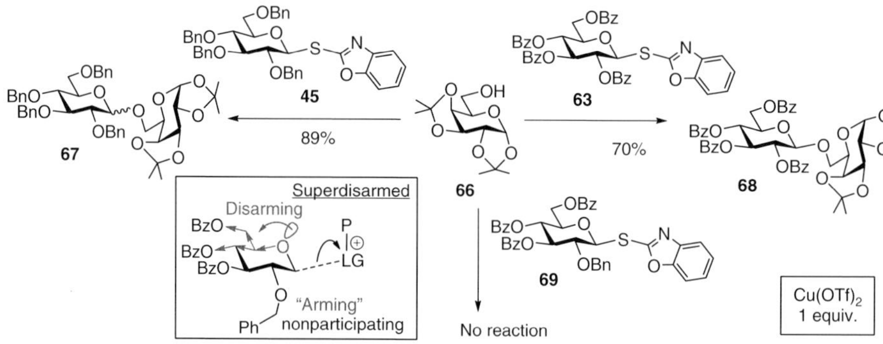

Scheme 2.12 Electronically superdisarmed building blocks.

Demchenko and coworkers discovered a conceptually new way of disarming the leaving groups. The conceptual difference of this approach from Fraser-Reid's armed–disarmed approach is that, herein the disarming is achieved by acylation of the leaving group itself, not by introducing the neighboring acyl substituents in the sugar moiety. This was investigated in application to *S*-benzimidazolyl (SBiz) leaving group versus N-anisoylated SBiz [51]. First, SBiz donor **70** was activated with MeI over the disarmed acceptor **71** to afford disaccharide **72** (Scheme 2.13). The disarmed N-anisoylated SBiz disaccharide was then glycosylated with the glycosyl acceptor **21** in the presence of AgOTf to give the trisaccharide **73** in 84% yield. It was noted that benzylated and benzoylated SBiz imidates can also be activated in the conventional armed–disarmed manner.

2.6 Application of Armed and Disarmed Building Blocks in Stereocontrolled Glycosylation

Armed and disarmed building blocks follow general stereoselectivity trends in glycosylations: armed per-benzylated derivatives provide some α-stereoselectivity, whereas disarmed benzoylated ones give complete β-selectivity due to the participation of 2-benzoyl. As mentioned, picolinyl can be used as an arming group at C-2, but the chemoselective activation of the 2-O-picolinylated donors leads to 1,2-*trans* glycosides, inverse stereoselectivity in comparison to that achieved with traditional benzylated armed glycosyl donors. Thus, upon activation with $Cu(OTf)_2$, the armed STaz glycosyl donor **74** gives the stable cyclic intermediate. The latter is subsequently glycosylated with the disarmed STaz acceptor **75** to give 1,2-*trans*-linked disaccharide **76** in 74% yield (Scheme 2.14). The resulting disarmed disaccharide donor **76** has been further

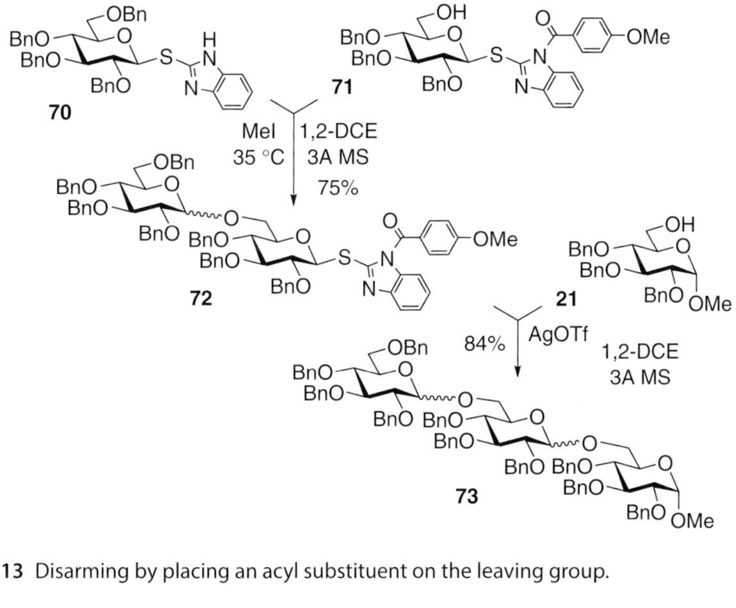

Scheme 2.13 Disarming by placing an acyl substituent on the leaving group.

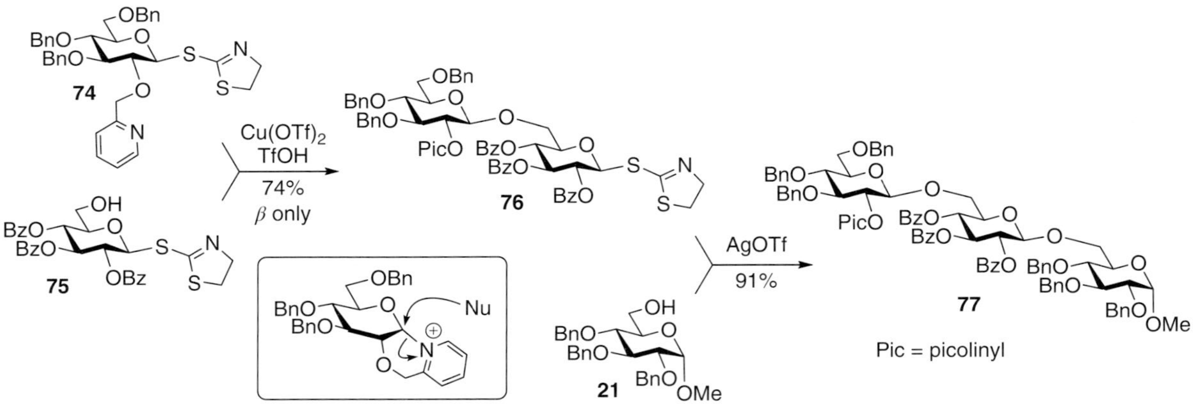

Scheme 2.14 Arming participating picolinyl-group-mediated 1,2-*trans*-glycosylation.

glycosylated with the acceptor **21** in the presence of stronger promoter AgOTf to obtain the *trans–trans*-linked trisaccharide **77** in 91% yield [34].

Mlynarski and coworkers investigated *ortho*-nitrobenzyl (NBn) as an arming participating group [35a]. They theorized that 2-*O*-*ortho*-nitrobenzyl will participate in glycosylation by stabilizing the oxocarbenium ion intermediate and hence blocking the bottom face from the nucleophilic attack (Scheme 2.15a). The activation of glucosyl donor **78** with Ph_2SO/Tf_2O/TTBP (2,4,6-tri-*tert*-butylpyrimidine) afforded disaccharide **79** in a high yield and preferential 1,2-*trans* selectivity. Liu and coworkers investigated another arming participating group, *o*-cyanobenzyl (CBn), at C-2 position of a glycosyl donor [35b]. The interesting feature of this glycosylation method is that a single glycosyl donor can yield either α- or β-linked products depending on the nature of the glycosyl acceptor. It is believed that the dual-directing effect of *o*-cyanobenzyl group is due to the equilibrium in reaction intermediates **A–C** (Scheme 2.15b). Thus, the activation of donor **80** will result in the formation of oxocarbenium ion **B** that is stabilized via *cis*-nitrilium ion **A.** The latter will direct the nucleophilic attack from the top face, hence offering β-directing effect that is preferred with reactive electron-rich glycosyl acceptors. Another mode by which CBn

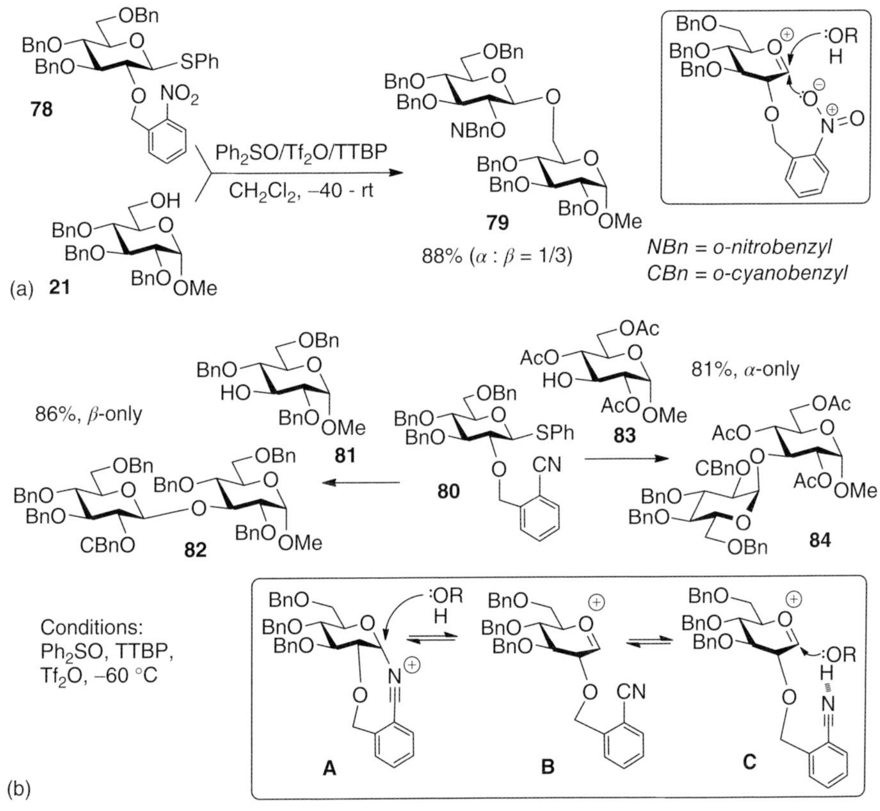

Scheme 2.15 Use of *o*-nitrobenzyl and *o*-cyanobenzyl arming participating groups.

group can react is via H-bond-mediated aglycone delivery (HAD). Discovered with remote picolinyl and picolyl groups, the HAD method has already yielded a number of highly selective syntheses and applications [52]. A similar HAD action can be envisaged for 2-*O*-ortho-cyanobenzyl group (see intermediate **C** in Scheme 2.15b). It was concluded that this mechanism of action leading to high α-selectivity is preferred with electron-deficient acceptors.

In the context of the stereoselective synthesis of α-glycosides, electronically superdisarmed glycosyl donors, which are also 2-O-benzylated, often provide higher stereoselectivity compared to their perbenzylated counterparts [5]. A valuable application of superdisarmed SEt donors has emerged with the synthesis and glycosidation of glycosyl sulfonium salts as activated intermediates of thioglycoside glycosidation. The mode of activation of thioglycosides has been proposed multiple times although direct evidence could only be acquired with superdisarmed thioglycosides due to high stability of the intermediate. Thus, direct alkylation of a superdisarmed ethylthioglycoside with MeOTf led to a sulfonium salt that could be isolated and characterized by NMR [41]. A number of other anomeric sulfonium salts, most notably those derived from the reaction with dimethyl(thiomethyl) sulfonium trifluoromethanesulfonate (DMTST), were obtained [41]. Various experiments with simple alcohol acceptors showed high 1,2-*cis* stereoselectivity of glycosidation of sulfonium salt intermediates.

Bromine-activated glycosylation of thioglycosides introduced by Demchenko *et al.* gave high stereoselectivity only with superdisarmed thioglycosides [53]. Thus, it was demonstrated that the 1,2-*trans* glycosyl bromide is the only intermediate leading to products while 1,2-*cis* bromide remains unreactive and the oxocarbenium intermediate does not form. Resultantly, the nucleophilic displacement of β-bromide takes place in the concerted bimolecular fashion leading to exclusive α-stereoselectivity. For instance, 3,4,6-tri-*O*-benzoyl-2-*O*-benzyl thioglycoside **86** was coupled with acceptor **21** in the presence of bromine to afford disaccharide **87** with exclusive 1,2-*cis* selectivity in 67% yield (Scheme 2.16). The use of α-thioglycoside starting material was found advantageous for generating β-bromide intermediate. The average yield, which is due to the competing isomerization of β- to α-bromide, could be further improved by using $HgBr_2$ as the copromoter, but this could also decrease the stereoselectivity. For comparison,

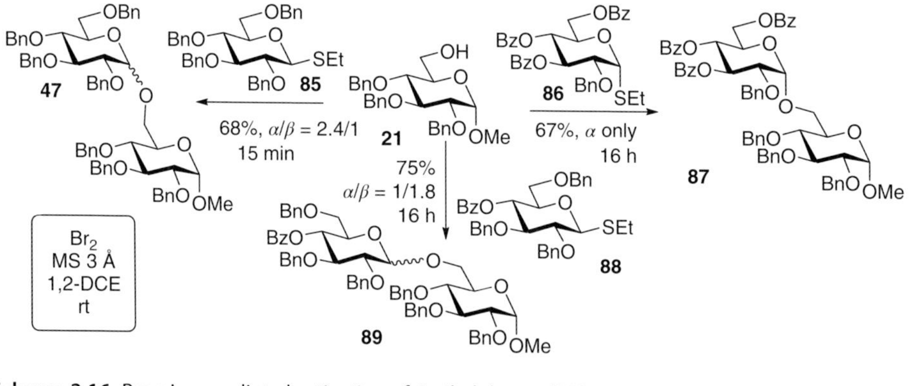

Scheme 2.16 Bromine-mediated activation of *S*-ethyl donors [53].

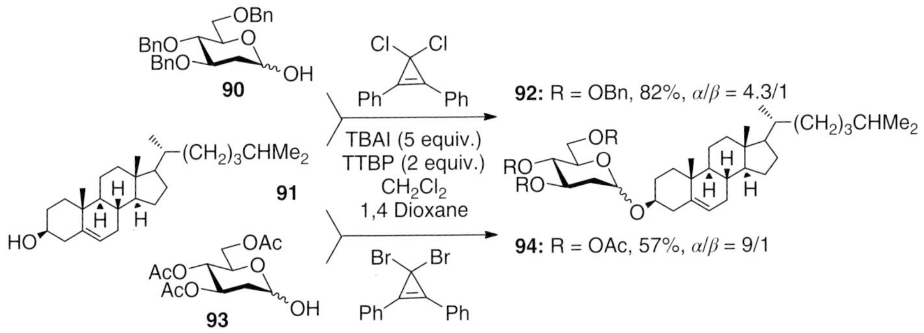

Scheme 2.17 Highly stereoselective glycosylation with armed and disarmed 2-deoxydonors.

armed donor **85** or 4-benzoyl donor **88**, gave no stereoselectivity for the formation of the disaccharides **47** and **89**, respectively.

Bennett and coworkers devised a new promoter system that provided high α-selectivity for both armed and disarmed glycosyl donors of the 2-deoxy series **90** and **93**, respectively (Scheme 2.17) [54]. NMR monitoring showed that the armed hemiacetal donor **90** is first converted into to the respective glycosyl chloride intermediate by the action of 3,3-dichloro-1,2-diphenylcyclopropene promoter. The latter is then converted into a mixture of α- and β-glycosyl iodides with tetrabutylammonium iodide, but predominantly, the more reactive β-iodide reacts with glycosyl acceptor **91**. Presumably, this displacement proceeds in the S_N2 fashion leading to glycosides in high α-selectivity. Interestingly, 3,3-dichloro-1,2-diphenylcyclopropene promoter failed to activate the disarmed donor **93** due to the high stability of the intermediate chloride. In this case, 3,3-dibromo-1,2-diphenylcyclopropene was found to be an effective promoter that provided glycoside **94** in excellent stereoselectivity.

Superarmed glycosyl donors allow to achieve various selectivities depending on their structure and the mode of superarming. Conformationally superarmed donor **25** introduced by Bols *et al.* is capable of providing high stereoselectivity that was attributed to the steric hindrance from the bulky 2-*O*-silyl substituents [36, 37]. Depicted in Scheme 2.6 is a β-stereoselective glycosylation of acceptor **26** that afforded disaccharide **27** in a high yield of 85% and complete selectivity. In application of the synthesis of rhamnosides, a drastic temperature effect on stereoselectivity was observed. The conformationally superarmed rhamnosyl donors produced modest β-selectivity at low temperatures, but increasing the temperature gave excellent α-selectivity [55].

A relaxed stereoselectivity achieved with conformationally superarmed donors was addressed by the introduction of a hybrid donor **49** in which the superarming was achieved by the combined anchimeric and conformational factors (2-*O*-Bz-3,4-di-*O*-TBS protection). All glycosylations with donor **49** were completely β-selective [29] as well as the electronically superarmed donors bearing 2-*O*-Bz-tri-*O*-Bn protection [40]. Very differently, glycosyl donors in which the superarming is achieved by 3,6-silicon tethering, such as donor **38**, provide predominantly α-selectivity, which was attributed to the steric hindrance of the top face of the oxocarbenium intermediate [39]. Table 2.1

Table 2.1 Survey of stereoselectivity in glycosylation with different donors.

Type of donors	Stereoselectivity	References
Armed perbenzyl	*Cis/trans*-mixtures, mainly 1,2-*cis*	[20, 22, 56]
Disarmed perbenzoyl	Complete 1,2-*trans*	[20, 22, 56]
Armed with 2-O-arming participating group	Complete 1,2-*trans*	[34, 35, 57]
Electronically superdisarmed	Moderate or high 1,2-*cis*	[26, 53]
Superdisarmed by fused ring systems	High or complete 1,2-*cis*[a]	[11d,g, 58]
Conformationally superarmed	*Cis/trans*-mixtures	[37a,b]
Conformationally and anchimerically superarmed	Complete 1,2-*trans*	[29]
Anchimerically superarmed	Complete 1,2-*trans*	[40]
Conformational superarming by tethering	High 1,2-*cis*	[39]

a) High stereoselectivity can be achieved in 4,6-benzylidene or 2,3-cyclocarbonyl/oxazolidinone systems. However, little has been investigated in the context of the chemoselective oligosaccharide synthesis.

surveys the currently known modes for arming/disarming and stereoselectivities that can be achieved by using these approaches.

2.7 Application of Armed/Superarmed and Disarmed Building Blocks in Chemoselective Oligosaccharide Synthesis

The expeditious preparation of complex oligosaccharides remains a significant challenge to synthetic organic chemistry. The combined demand of regio- and stereoselectivity in glycosidic bond formation has led to complex synthetic schemes and extensive protecting-group manipulations. As mentioned, the use of a chemoselective activation strategy avoids such extraneous manipulations, thus offering significant advantages for expeditious glycoside synthesis. Since the glycosidation of 2-O-acylated glycosyl donors typically proceeds via the formation of the bicyclic acyloxonium intermediate, the overall two-step armed–disarmed activation sequence leads to a *cis–trans*-patterned trisaccharide. Starting with Fraser-Reid's pentenyl-based synthesis [22], a number of relevant examples have emerged [25]. For instance, Hashimoto *et al.* activated the armed galactosyl donor **95** over the disarmed galactosyl acceptor **96** with TMSOTf at −46 °C to obtain disaccharide **97** in 85% yield (Scheme 2.18). The latter was then glycosidated with glycosyl acceptor **98** in the presence of TMSOTf at 0 °C to provide the requisite *cis–trans*-sequenced trisaccharide **99** in 85% yield [59]. While this synthesis is highly stereoselective, a majority of glycosylation with armed donors suffers from low stereoselectivity.

While the traditional armed–disarmed strategy provides a straightforward access to *cis–trans*-sequenced trisaccharides, other sequences cannot be directly accessed. It was quickly realized that for this excellent concept to become universally applicable, it

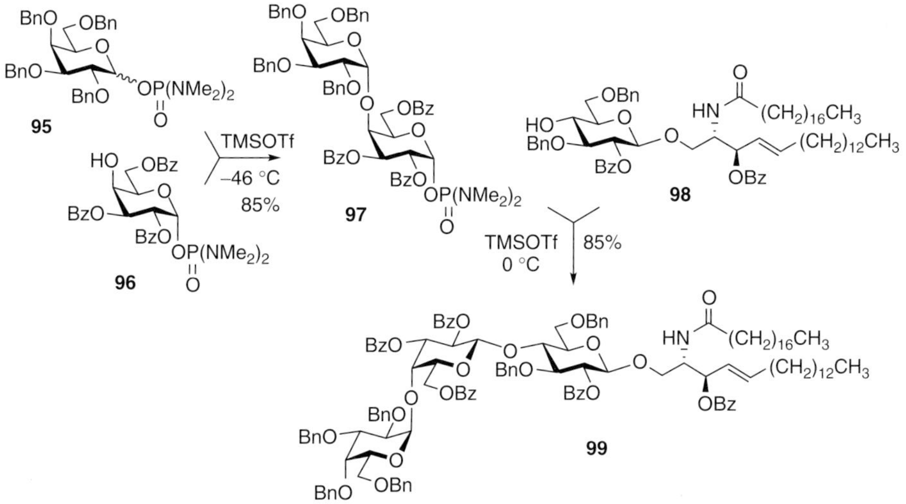

Scheme 2.18 Armed–disarmed synthesis [59] of glycosphingolipid **99**.

Table 2.2 A survey of oligosaccharide sequences that can be obtained by chemoselective activation.

Sequence[a]	Building blocks needed	References
Cis–trans	Traditional armed → disarmed	[22, 59]
Cis–trans–cis	Armed → disarmed → superdisarmed	[26]
Cis–trans–trans	Programmable strategy	[31]
Cis–cis	Armed → disarmed with the interim reprotection	[60]
	Armed → torsionally or anchimerically superdisarmed	[45, 57]
	Armed → 6-PFB disarmed	[43b]
Trans–trans	2-Pico armed → traditional disarmed (Scheme 2.14)	[34, 57]
	Anchimerically superarmed → armed	[40b]
Trans–cis	2-Pico armed → superdisarmed	[57]
	Anchimerically superarmed → disarmed	[40b]
Trans–cis–trans	Conformationally superarmed → armed → disarmed	[37a]
	Anchimerically superarmed → armed → disarmed	[40c]

a) Practically, any sequence can be achieved with the use of selective activations based on building blocks with different leaving groups. Preactivation concept pioneered by Huang and Ye is another way to achieve flexible sequencing.

should be expanded to a broader range of linkages, protecting-group patterns, and oligosaccharide sequences. Major improvements in this direction that have emerged in the past decade are summarized in Table 2.2. With the discovery of other levels of reactivity, a more flexible synthesis of a variety of oligosaccharide sequences using the chemoselective activation has become possible. For example, *cis–trans* oligosaccharide sequence obtained through the traditional armed and disarmed donors could be extended to *cis–trans–cis* sequence by adding a superdisarmed acceptor to this combination [26]. However, if the extension to another trans-linkage is desired, Wong's programmable strategy is the only way to achieve the chemoselective synthesis of *cis–trans–trans* oligosaccharides [31]. With many reactivity levels, the programmable strategy can be applied to many other targets and sequences, and some representative examples will be discussed as follows.

A number of approaches have been developed for the synthesis of *cis–cis*-sequenced oligosaccharides [43b, 45, 57, 60]. Arming participating 2-*O*-picolinyl and other similar groups can simplify the syntheses wherein the *trans*-linkage needs to be introduced first [34, 35, 57]. For instance, activation of 2-picolinyl donor over disarmed or superdisarmed acceptors can be used to obtain *trans–trans* or *trans–cis* oligosaccharide sequences, respectively [57]. The *trans–cis–trans* sequence has been obtained by the combination of either conformationally superarmed or anchimerically superarmed donor with armed and disarmed acceptors [37a, 40c]. Preactivation-based strategies, which tend to be classified as selective rather than chemoselective, also allow for obtaining many of these sequences. A few examples of such sequences are discussed as follows.

With the discovery of the anchimerically superdisarmed building blocks, it is now possible to produce the *cis–trans–cis* oligosaccharide sequence [26]. Thus, it was demonstrated that disarmed disaccharide **101**, obtained by classic armed–disarmed approach from building blocks **45** and **100**, could be chemoselectively activated over superdisarmed building block **102** (Scheme 2.19). This was affected in the presence of $Cu(OTf)_2$/TfOH to produce trisaccharide **103** (70% yield) [26],which can be used for further glycosylations directly.

The programmable strategy revealed many reactivity levels, which allow for modulating building blocks to obtain various sequences. An example of a sequence wherein no other chemoselective approaches could be used is depicted in Scheme 2.20. This

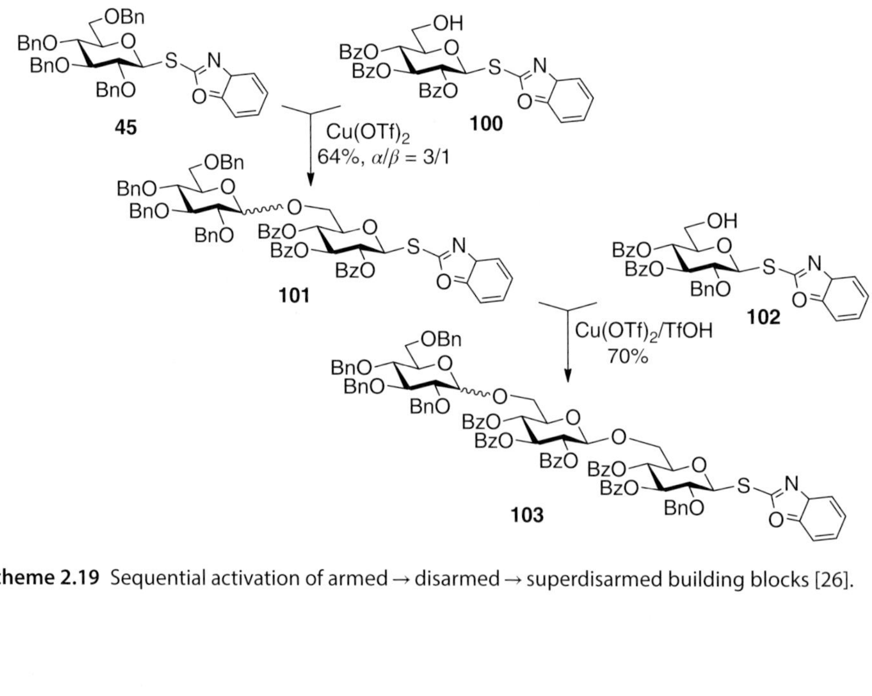

Scheme 2.19 Sequential activation of armed → disarmed → superdisarmed building blocks [26].

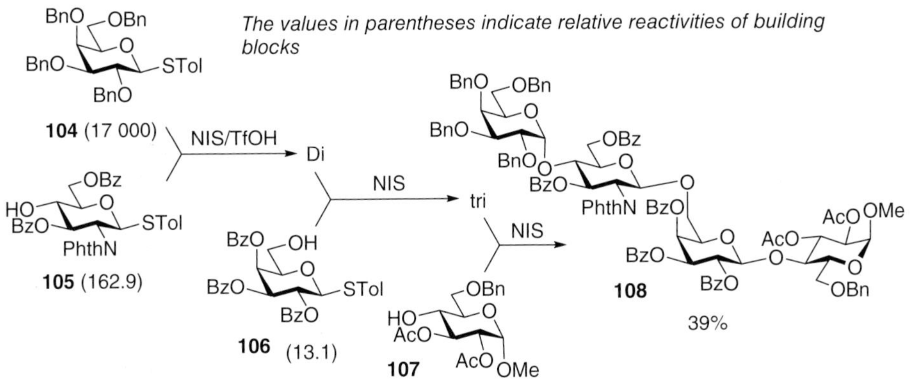

Scheme 2.20 Sequential activation in one-pot for the synthesis of *cis–trans–trans*-linked tetrasaccharide.

approach was conducted in one pot with no isolation and characterization of intermediate oligosaccharides. Armed glycosyl donor **104** was chemoselectively activated over glycosyl acceptor **105** in the presence of NIS/TfOH. The resulting disaccharide intermediate was then reacted with added disarmed glycosyl acceptor **106** to form the trisaccharide intermediate that was then glycosidated with added glycosyl acceptor **107** to provide *cis–trans–trans*-linked tetrasaccharide **108** in 39% overall yield [31].

Van Boom and coworkers invented a two-step method, glycosylation and protecting-group manipulation, to obtain *cis–cis*-linked oligosaccharides [60a]. Here, after the first armed–disarmed activation step, the resulting disaccharide was reprotected (OBz → OBn)

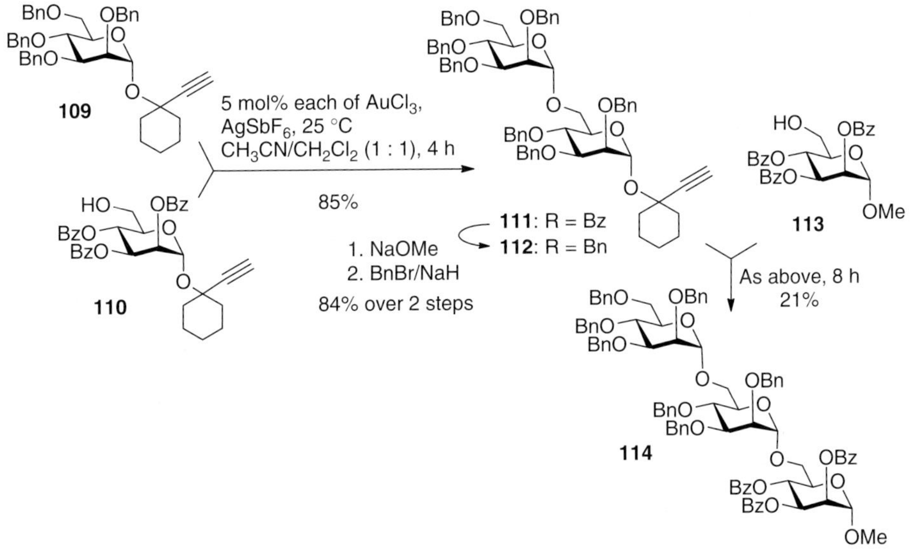

Scheme 2.21 Synthesis of *cis–cis*-patterned trisaccharide via the interim reprotection.

prior to the subsequent glycosidation. A more recent example of this strategy wherein propargyl mannosyl donors have been used is shown in Scheme 2.21 [60b]. Glycosylation of the armed mannosyl donor **109** with disarmed acceptor **110** was performed in the presence of 5 mol% of $AuCl_3$ and $AgSbF_6$ in CH_3CN/CH_2Cl_2 (1 : 1) at 25 °C. As a result, the disarmed disaccharide **111** obtained in 85% yield was then reprotected with benzyls to obtain the armed disaccharide **112** in 84% yield. The glycosylation between disaccharide **112** and disarmed acceptor **113** was performed under the same conditions to obtain the desired trisaccharide **114** in 21% yield. The utilization of the cooperative effect allows for the direct synthesis of *cis–cis*-linked oligosaccharides, similar to that discussed previously. In this application, the sequential activation of armed perbenzylated glycosyl donor over superdisarmed 3,4-di-*O*-benzoyl-2-*O*-benzyl protected STaz glycosyl acceptor led to the *cis–cis*-linked oligosaccharide [57]. The presence of the remote 6-*O*-pentafluorobenzoyl group disarms the glycosyl acceptor and also facilitates the synthesis of *cis–cis*-linked sequences [43b].

As mentioned, the application of the 1,2-*trans*-directing picolinyl functionality of armed glycosyl donor in activation over the standard disarmed glycosyl acceptor will allow for the synthesis of *trans–trans*-patterned oligosaccharides (see Scheme 2.14). The programmable strategy was also applied to the synthesis of *trans–trans*-linked oligosaccharide as shown in Scheme 2.22 [32]. Coupling of thioglycoside **115** and acceptor **116** in the presence of NIS/TMSOTf followed by quenching the activator with tripropargylamine and glycosylation with the lactosyl diol **117** in the presence of NIS/AgOTf afforded tetrasaccharide **118** in 40% yield in one-pot manner.

The application of the *trans*-directing 2-O-picolinylated armed glycosyl donors in activation over the superdisarmed acceptors allows one to obtain a *trans–cis* glycosylation pattern, which is opposite to the traditional armed–disarmed methodology [57]. Thus, glycosylation between STaz donor **74** and the disarmed acceptor **119** in the

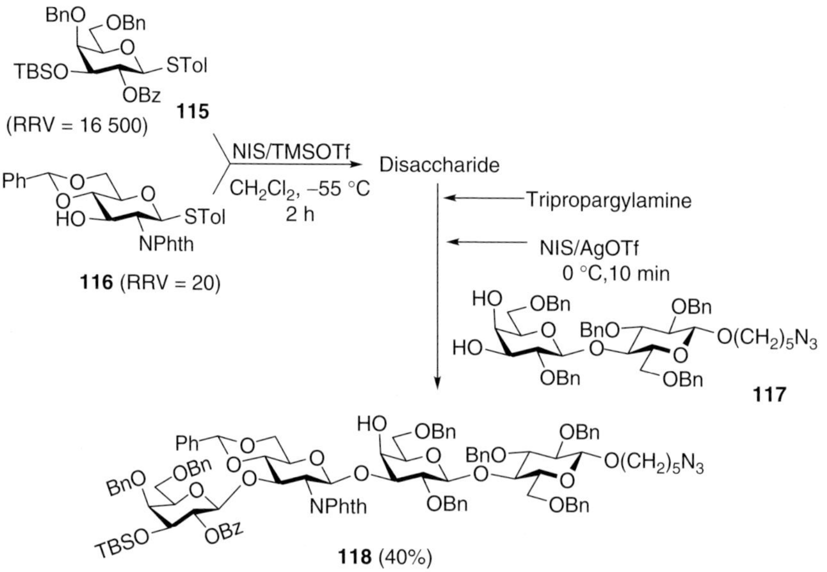

Scheme 2.22 Programmable strategy for the synthesis of trans–trans-linked oligosaccharide.

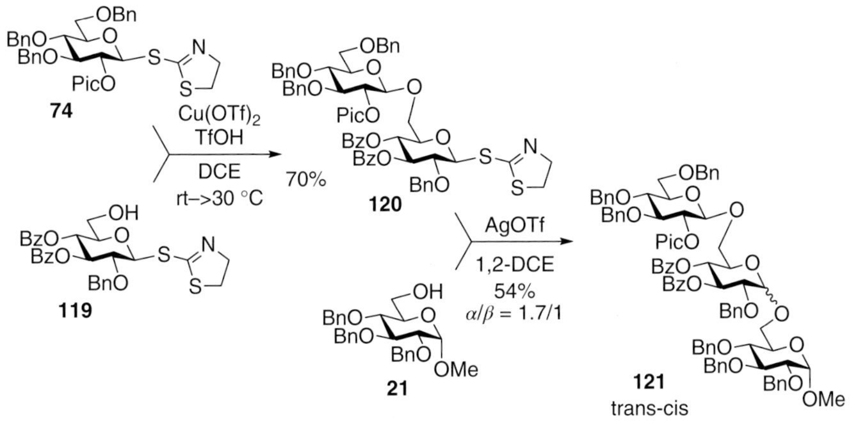

Scheme 2.23 Sequential activation of picolinylated armed → superdisarmed building blocks: synthesis of *trans–cis*-linked trisaccharide **121**.

presence of $Cu(OTf)_2$/TfOH afforded *trans*-linked disaccharide **120** in 70% yield (Scheme 2.23). Disaccharide **120** was then glycosylated with acceptor **21** in the presence of AgOTf to give the anticipated inverse-patterned *trans–cis*-linked trisaccharide **121** in 54% yield.

Conformational superarming concept has been successively proven by a one-pot glycosylation reaction performed between three building blocks **25, 122,** and **123**, which were placed in the same reaction vessel from the beginning (Scheme 2.24) [37a]. In this method, it is essential that all reaction components, not only glycosyl donors (**25** and **122**) but also glycosyl acceptors (**122** and **123**), have differential reactivity. The superarmed glycosyl donor **25** will be glycosylated with the more reactive primary glycosyl

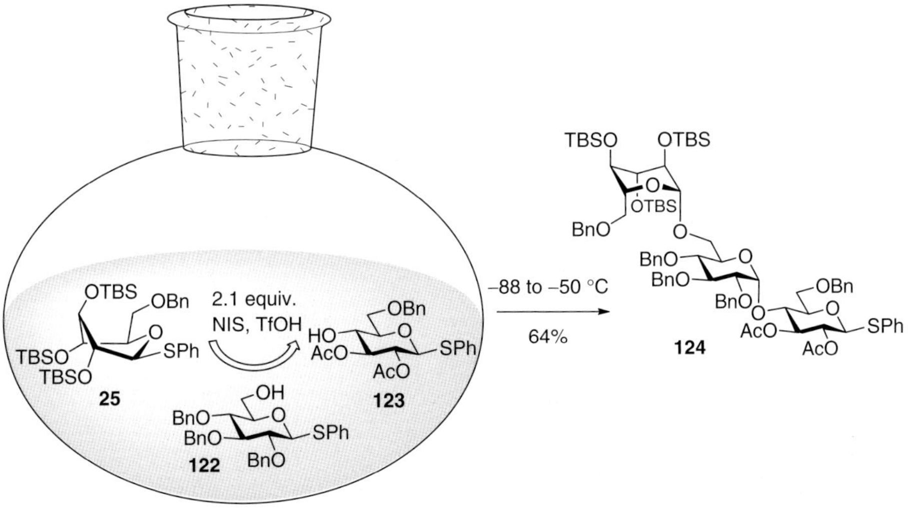

Scheme 2.24 Superarmed → armed → disarmed activation for the one-pot synthesis.

acceptor **122**. The resulting disaccharide intermediate will then react with the remaining glycosyl acceptor **123** to yield the desired trisaccharide **124** in 64% in the one-pot manner.

A similar sequence was obtained with the electronically superarmed glycosyl donors, but in this approach, a more conventional stepwise synthesis was performed. Thus, disaccharide **127** was obtained in 80% yield from the glycosylation between superarmed glycosyl donor **125** and armed acceptor **126** upon activation with iodine at −25 °C (Scheme 2.25). The resulting disaccharide **127** was glycosylated with the disarmed acceptor **128** in the presence of iodine at room temperature to afford trisaccharide **129** in 55% yield. Finally, trisaccharide **129** was glycosylated with glycosyl acceptor **21** in the presence of NIS/TfOH to obtain the desired tetrasaccharide **130** composed of the *trans–cis–trans* sequence.

Preactivation concept is independent of the building block reactivity since the leaving group of the glycosyl donor is first converted into a highly reactive species (preactivation) and then the acceptor is added [61]. Although this approach involves additional steps because the glycosylation herein is a two-step reaction, it offers more flexibility with the leaving and/or protecting groups. A relevant example illustrating this excellent concept in application to the synthesis of the tumor-associated carbohydrate antigen Globo H hexasaccharide is shown in Scheme 2.26 [62]. The fucosyl donor **131** was preactivated at −78 °C with *p*-TolSCl/AgOTf, and then the first acceptor **132** was added along with a sterically hindered base TTBP. The temperature was then raised to −20 °C to obtain the trisaccharide intermediate. Upon complete disappearance of acceptor **132**, the reaction mixture was cooled again to −78 °C followed by the sequential

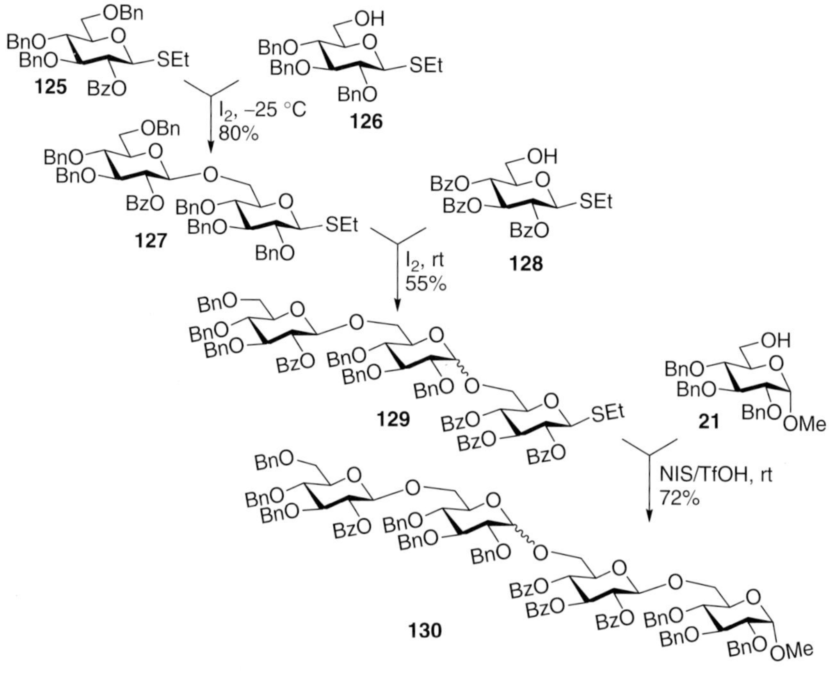

Scheme 2.25 Synthesis of tetrasaccharide with alternating *trans–cis* linkages.

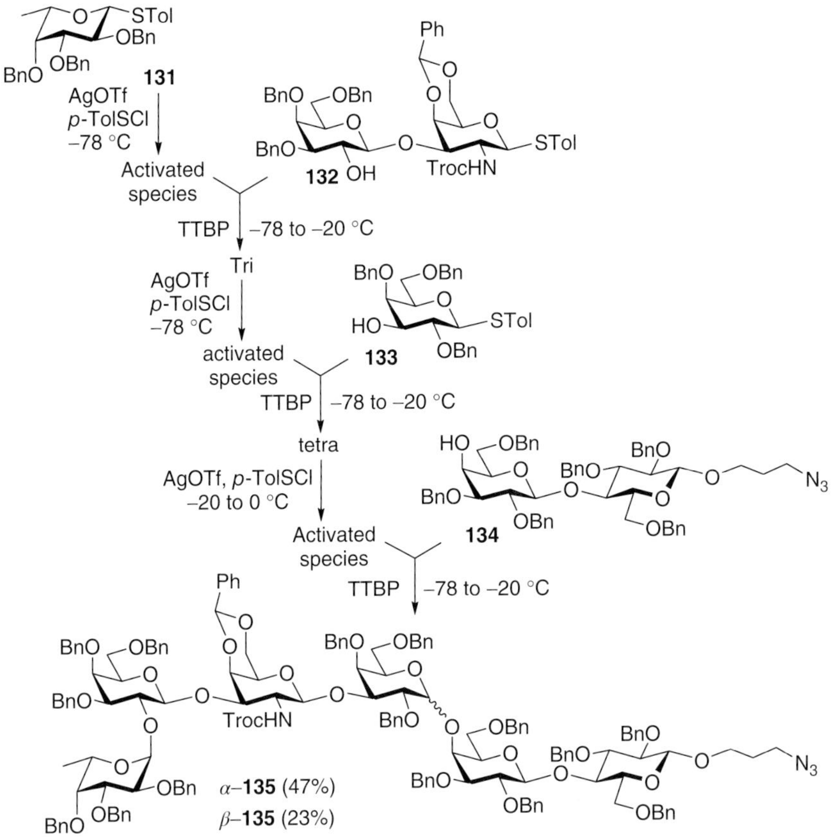

Scheme 2.26 One-pot synthesis of Globo H hexasaccharide based on preactivation.

addition of AgOTf, *p*-TolSCl, galactose acceptor **133**, and TTBP and then warming up the reaction to −20 °C. After complete disappearance of the acceptor **133**, the temperature was lowered to −78 °C, and the sequence was reiterated for glycosylation of lactoside **134**. The resulting Globo H hexasaccharide α-**135** was isolated in 47% overall yield based on the four-component one-pot reaction within 7 h.

Mong and coworkers investigated the possibility of using DMF-modulated glycosylation concept [63] in the so-called disarmed–armed glycosylation, which is also based on preactivation [64]. In accordance with this approach, the glycosyl donor produces a β-glycosyl imidinium triflate in the presence of DMF. It has been shown that dioxalenium ion and β-imidinium triflate are in equilibrium, and the acceptor can react with both of these species resulting in an *α*/*β*-mixture of glycosides. The β-selectivity is inversely correlated to the amount of DMF, and pure β-selectivity was achieved with 1.2 equiv. of DMF [63]. Since DMF-modulated glycosylations induce a preactivation step, this method opens the glycosyl donor to iterative glycosylations. To demonstrate the applicability of this method in oligosaccharide synthesis, several different trisaccharides, including the one-pot synthesis of the α-(1,2)-linked trisaccharide **139** shown

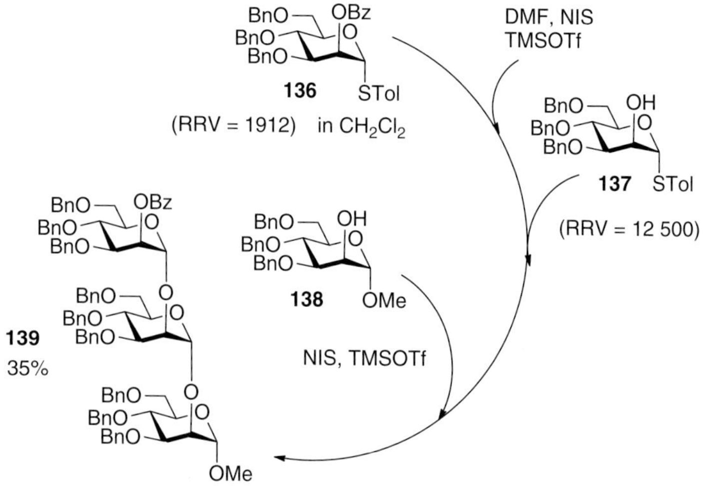

Scheme 2.27 DMF modulated disarmed–armed iterative glycosylation.

in Scheme 2.27, were described. Gildersleeve *et al.* have shown that the aglycone transfer side reaction may occur when the preactivation method is applied for the glycosylations of armed acceptors with the disarmed donors [65]. A number of approaches including the use of sterically hindered leaving groups (aglycones) [65, 66] have been invented to overcome the aglycone transfer in such preactivation glycosylations.

2.8 Conclusions and Outlook

Since the first glycosylation reactions were performed in the late 1800s, carbohydrate chemistry has evolved into a broad area of research that has persistently captured the interest of the scientific community. With recent advances in the rapidly expanding fields of glycosciences, the demand for reliable and stereocontrolled glycosylation methods has increased. Nevertheless, the installation of the glycosidic linkages and the assembly of oligosaccharide sequences remain cumbersome due to the lack of understanding of the mechanistic detail of glycosylation or the inability to translate such knowledge into practical execution. A number of excellent strategies that offer a reasonably efficient route to oligosaccharide assembly have already emerged, and the armed–disarmed approach for chemoselective oligosaccharide synthesis is undoubtedly among them. The search of new concepts continues, and the field of armed–disarmed glycosylations enjoyed an explosive expansion. New reactivity levels have been revealed, and a few new concepts for glycosyl donor activation have been introduced and tested in armed–disarmed strategies [67]. Many new sequences can now be achieved directly, but in a majority of applications, one needs to take care of protecting groups: protecting groups do more than protect. Although recent advancements discussed in this chapter have already significantly expanded the scope of the armed–disarmed methodology, it is clear that further development of efficient and general methods for the expeditious synthesis of complex carbohydrates will remain an important and active arena for scientific endeavors of the twenty-first century.

References

1 (a) Hehre, E.J. (2001) *Carbohydr. Res.*, 331, 347–368; (b) Muthana, S., Cao, H., and Chen, X. (2009) *Curr. Opin. Chem. Biol.*, 13, 573–581.

2 (a) Toshima, K. and Tatsuta, K. (1993) *Chem. Rev.*, 93, 1503–1531; (b) Zhu, X. and Schmidt, R.R. (2009) *Angew. Chem. Int. Ed.*, 48, 1900–1934; (c) Whitfield, D.M. (2009) *Adv. Carbohydr. Chem. Biochem.*, 62, 83–159; (d) Boltje, T.J., Kim, J.H., Park, J., and Boons, G.J. (2010) *Nat. Chem.*, 2, 552–557; (e) Huang, M., Garrett, G.E., Birlirakis, N., Bohe, L., Pratt, D.A., and Crich, D. (2012) *Nat. Chem.*, 4, 663–667; (f) Frihed, T.G., Bols, M., and Pedersen, C.M. (2015) *Chem. Rev.*, 115, 4963–5013.

3 (a) Seeberger, P.H. and Werz, D.B. (2005) *Nat. Rev.*, 4, 751–763; (b) Krock, L., Esposito, D., Castagner, B., Wang, C.-C., Bindschadler, P., and Seeberger, P.H. (2012) *Chem. Sci.*, 3, 1617–1622; (c) Vijaya Ganesh, N., Fujikawa, K., Tan, Y.H., Stine, K.J., and Demchenko, A.V. (2012) *Org. Lett.*, 14, 3036–3039; (d) Bennett, C.S. (2014) *Org. Biomol. Chem.*, 12, 1686–1698; (e) Nokami, T., Isoda, Y., Sasaki, N., Takaiso, A., Hayase, S., Itoh, T., Hayashi, R., Shimizu, A., and Yoshida, J. (2015) *Org. Lett.*, 17, 1525–1528; (f) Pistorio, S.G., Stine, K.J., and Demchenko, A.V. (2015) in Carbohydrate Chemistry: State-of-the-Art and Challenges for Drug Development (ed. L. Cipolla), Imperial College Press, London, pp. 247–276; (g) Seeberger, P.H. (2015) *Acc. Chem. Res.*, 48, 1450–1463; (h) Tang, S.L. and Pohl, N.L. (2015) *Org. Lett.*, 17, 2642–2645; (i) Hsu, C.H., Hung, S.C., Wu, C.Y., and Wong, C.H. (2011) *Angew. Chem. Int. Ed.*, 50, 11872–11923.

4 (a) Nukada, T., Berces, A., and Whitfield, D.M. (2002) *Carbohydr. Res.*, 337, 765–774; (b) Bohe, L. and Crich, D. (2015) *Carbohydr. Res.*, 403, 48–59.

5 Nigudkar, S.S. and Demchenko, A.V. (2015) *Chem. Sci.*, 6, 2687–2704.

6 (a) Juaristi, E. and Cuevas, G. (1995), The Anomeric Effect, CRC Press, Boca Raton, FL, pp. 1–48, 95–112, 173–194 and references therein; (b) Tvaroska, I. and Bleha, T. (1989) *Adv. Carbohydr. Chem. Biochem.*, 47, 45–123.

7 Demchenko, A.V. (2008), Handbook of Chemical Glycosylation: Advances in Stereoselectivity and Therapeutic Relevance, Wiley-VCH Verlag GmbH, Weinheim.

8 Fraser-Reid, B., Jayaprakash, K.N., López, J.C., Gómez, A.M., and Uriel, C. (2007) in Frontiers in Modern Carbohydrate Chemistry, ACS Symposium Series, vol. 960 (ed. A.V. Demchenko), Oxford University Press, pp. 91–117.

9 Goodman, L. (1967) *Adv. Carbohydr. Chem. Biochem.*, 22, 109–175.

10 (a) Manabe, S. and Ito, Y. (2009) *Curr. Bioact. Compt.*, 4, 258–281; (b) Guo, J. and Ye, X.S. (2010) *Molecules*, 15, 7235–7265.

11 (a) Crich, D. and Sun, S. (1996) *J. Org. Chem.*, 61, 4506–4507; (b) Crich, D. and Sun, S. (1997) *J. Am. Chem. Soc.*, 119, 11217–11223; (c) Crich, D. and Sun, S. (1998) *J. Am. Chem. Soc.*, 120, 435–436; (d) Crich, D. and Cai, W. (1999) *J. Org. Chem.*, 64, 4926–4930; (e) Crich, D. (2002) *J. Carbohydr. Chem.*, 21, 667–690; (f) Crich, D., Banerjee, A., and Yao, Q. (2004) *J. Am. Chem. Soc.*, 126, 14930–14934; (g) Crich, D. and Vinogradova, O. (2006) *J. Org. Chem.*, 71, 8473–8480; (h) Nokami, T., Shibuya, A., Tsuyama, H., Suga, S., Bowers, A.A., Crich, D., and Yoshida, J. (2007) *J. Am. Chem. Soc.*, 129, 10922–10928; (i) Crich, D. (2011) *J. Org. Chem.*, 76, 9193–9209.

12 (a) Mydock, L.K. and Demchenko, A.V. (2010) *Org. Biomol. Chem.*, 8, 497–510; (b) Crich, D. (2010) *Acc. Chem. Res.*, 43, 1144–1153.

13 Christensen, H.M., Oscarson, S., and Jensen, H.H. (2015) *Carbohydr. Res.*, 408, 51–95.

14 Smoot, J.T. and Demchenko, A.V. (2009) *Adv. Carbohydr. Chem. Biochem.*, 62, 161–250.

15 (a) Koto, S., Uchida, T., and Zen, S. (1973) *Bull. Chem. Soc. Jpn.*, 46, 2520–2523; (b) Nicolaou, K.C. and Ueno, H. (1997) in Preparative Carbohydrate Chemistry (ed. S. Hanessian), Marcel Dekker, Inc., New York, pp. 313–338.

16 (a) Nicolaou, K.C., Dolle, R.E., Papahatjis, D.P., and Randall, J.L. (1984) *J. Am. Chem. Soc.*, 106, 4189–4192; (b) Nicolaou, K.C., Caulfield, T., Kataoka, H., and Kumazawa, T. (1988) *J. Am. Chem. Soc.*, 110, 7910–7912; (c) Williams, L.J., Garbaccio, R.M., and Danishefsky, S.J. (2000) in Carbohydrates in Chemistry and Biology, vol. 1 (eds B. Ernst, G.W. Hart, and P. Sinay), Wiley-VCH Verlag GmbH, Weinheim, New York, pp. 61–92.

17 (a) Roy, R., Andersson, F.O., and Letellier, M. (1992) *Tetrahedron Lett.*, 33, 6053–6056; (b) Boons, G.J. and Isles, S. (1994) *Tetrahedron Lett.*, 35, 3593–3596; (c) Allen, J.G. and Fraser-Reid, B. (1999) *J. Am. Chem. Soc.*, 121, 468–469; (d) Kim, K.S., Kim, J.H., Lee, Y.J., Lee, Y.J., and Park, J. (2001) *J. Am. Chem. Soc.*, 123, 8477–8481.

18 Kanie, O., Ito, Y., and Ogawa, T. (1994) *J. Am. Chem. Soc.*, 116, 12073–12074.

19 Kaeothip, S. and Demchenko, A.V. (2011) *J. Org. Chem.*, 76, 7388–7398.

20 Fraser-Reid, B., Wu, Z., Udodong, U.E., and Ottosson, H. (1990) *J. Org. Chem.*, 55, 6068–6070.

21 Paulsen, H. (1982) *Angew. Chem., Int. Edit. Engl.*, 21, 155–173.

22 Mootoo, D.R., Konradsson, P., Udodong, U., and Fraser-Reid, B. (1988) *J. Am. Chem. Soc.*, 110, 5583–5584.

23 Ranade, S.C. and Demchenko, A.V. (2013) *J. Carbohydr. Chem.*, 32, 1–43.

24 Lahmann, M. and Oscarson, S. (2000) *Org. Lett.*, 2, 3881–3882.

25 Premathilake, H.D. and Demchenko, A.V. (2011) in Topics in Current Chemistry: Reactivity Tuning in Oligosaccharide Assembly, vol. 301 (eds B. Fraser-Reid and J.C. Lopez), Springer-Verlag, Berlin, Heidelberg, pp. 189–221.

26 Kamat, M.N. and Demchenko, A.V. (2005) *Org. Lett.*, 7, 3215–3218.

27 Isbell, H.S. and Frush, H.L. (1949) *J. Res. Nat. Bur. Stand.*, 43, 161–171.

28 Crich, D. and Li, M. (2007) *Org. Lett.*, 9, 4115–4118.

29 Heuckendorff, M., Premathilake, H.D., Pornsuriyasak, P., Madsen, A.Ø., Pedersen, C.M., Bols, M., and Demchenko, A.V. (2013) *Org. Lett.*, 15, 4904–4907.

30 Douglas, N.L., Ley, S.V., Lucking, U., and Warriner, S.L. (1998) *J. Chem. Soc., Perkin Trans. 1*, 51–65.

31 Zhang, Z., Ollmann, I.R., Ye, X.S., Wischnat, R., Baasov, T., and Wong, C.H. (1999) *J. Am. Chem. Soc.*, 121, 734–753.

32 Hsu, Y., Lu, X.A., Zulueta, M.M., Tsai, C.M., Lin, K.I., Hung, S.C., and Wong, C.H. (2012) *J. Am. Chem. Soc.*, 134, 4549–4552.

33 Tomono, S., Kusumi, S., Takahashi, D., and Toshima, K. (2011) *Tetrahedron Lett.*, 52, 2399–2403.

34 Smoot, J.T., Pornsuriyasak, P., and Demchenko, A.V. (2005) *Angew. Chem. Int. Ed.*, 44, 7123–7126.

35 (a) Buda, S., Gołębiowska, P., and Mlynarski, J. (2013) *Eur. J. Org. Chem.*, 3988–3991; (b) Hoang, K.L.M. and Liu, X.-W. (2014) *Nat. Commun.*, 5, 5051. doi: 10.1038/ncomms6051

36 Pedersen, C.M., Marinescu, L.G., and Bols, M. (2010) *C.R. Chim.*, 14, 17–43.

37 (a) Jensen, H.H., Pedersen, C.M., and Bols, M. (2007) *Chem. Eur. J.*, 13, 7576–7582; (b) Pedersen, C.M., Nordstrom, L.U., and Bols, M. (2007) *J. Am. Chem. Soc.*, 129, 9222–9235; (c) Pedersen, C.M., Marinescu, L.G., and Bols, M. (2008) *Chem. Commun.*, 2465–2467; (d) Heuckendorff, M., Pedersen, C.M., and Bols, M. (2010) *Chem. Eur. J.*, 16, 13982–13994.

38 Okada, Y., Asakura, N., Bando, M., Ashikaga, Y., and Yamada, H. (2012) *J. Am. Chem. Soc.*, 134, 6940–6943.

39 Heuckendorff, M., Pedersen, C.M., and Bols, M. (2013) *J. Org. Chem.*, 78, 7234–7248.

40 (a) Mydock, L.K. and Demchenko, A.V. (2008) *Org. Lett.*, 10, 2103–2106; (b) Mydock, L.K. and Demchenko, A.V. (2008) *Org. Lett.*, 10, 2107–2110; (c) Premathilake, H.D., Mydock, L.K., and Demchenko, A.V. (2010) *J. Org. Chem.*, 75, 1095–1100.

41 Mydock, L.K., Kamat, M.N., and Demchenko, A.V. (2011) *Org. Lett.*, 13, 2928–2931.

42 Bandara, M.D., Yasomanee, J.P., and Demchenko, A.V. (2016) (submitted).

43 (a) Clausen, M.H. and Madsen, R. (2003) *Chem. Eur. J.*, 9, 3821–3832; (b) Schmidt, T. and Madsen, R. (2007) *Eur. J. Org. Chem.*, 3935–3941.

44 Crich, D. and Vinogradova, O. (2007) *J. Am. Chem. Soc.*, 129, 11756–11765.

45 Fraser-Reid, B., Wu, Z., Andrews, C.W., and Skowronski, E. (1991) *J. Am. Chem. Soc.*, 113, 1434–1435.

46 Boons, G.J., Grice, P., Leslie, R., Ley, S.V., and Yeung, L.L. (1993) *Tetrahedron Lett.*, 34, 8523–8526.

47 Ley, S.V., Baeschlin, D.K., Dixon, D.J., Foster, A.C., Ince, S.J., Priepke, H.W.M., and Reynolds, D.J. (2001) *Chem. Rev.*, 101, 53–80.

48 Zhu, T. and Boons, G.J. (2001) *Org. Lett.*, 3, 4201–4203.

49 Jensen, H.H., Nordstrom, L.U., and Bols, M. (2004) *J. Am. Chem. Soc.*, 126, 9205–9213.

50 Moumé-Pymbock, M., Furukawa, T., Mondal, S., and Crich, D. (2013) *J. Am. Chem. Soc.*, 135, 14249–14255.

51 Hasty, S.J., Kleine, M.A., and Demchenko, A.V. (2011) *Angew. Chem. Int. Ed.*, 50, 4197–4201.

52 (a) Yasomanee, J.P. and Demchenko, A.V. (2012) *J. Am. Chem. Soc.*, 134, 20097–20102; (b) Pistorio, S.G., Yasomanee, J.P., and Demchenko, A.V. (2014) *Org. Lett.*, 16, 716–719; (c) Yasomanee, J.P. and Demchenko, A.V. (2014) *Angew. Chem. Int. Ed.*, 53, 10453–10456; (d) Kayastha, A.K., Jia, X.G., Yasomanee, J.P., and Demchenko, A.V. (2015) *Org. Lett.*, 17, 4448–4451; (e) Visansirikul, S., Yasomanee, J.P., and Demchenko, A.V. (2015) *Russ. Chem. Bull.*, 1107–1118; (f) Yasomanee, J.P. and Demchenko, A.V. (2015) *Chem. Eur. J.*, 21, 6572–6581; (g) Yasomanee, J.P., Parameswar, A.R., Pornsuriyasak, P., Rath, N.P., and Demchenko, A.V. (2016) *Org. Biomol. Chem.*, 14, 3159–3169; (h) Liu, Q.-W., Bin, H.-C., and Yang, J.-S. (2013) *Org. Lett.*, 15, 3974–3977; (i) Ruei, J.-H., Venukumar, P., Ingle, A.B., and Mong, K.-K.T. (2015) *Chem. Commun.*, 51, 5394–5397; (j) Xiang, S., Hoang Kle, M., He, J., Tan, Y.J., and Liu, X.W. (2015) *Angew. Chem. Int. Ed.*, 54, 604–607.

53 Kaeothip, S., Yasomanee, J.P., and Demchenko, A.V. (2012) *J. Org. Chem.*, 77, 291–299.

54 Nogueira, J.M., Issa, J.P., Chu, A.-H.A., Sisel, J.A., Schum, R.S., and Bennett, C.S. (2012) *Eur. J. Org. Chem.*, 2012, 4927–4930.

55 Heuckendorff, M., Pedersen, C.M., and Bols, M. (2012) *J. Org. Chem.*, 77, 5559–5568.

56 (a) Mootoo, D.R. and Fraser-Reid, B. (1989) *Tetrahedron Lett.*, 30, 2363–2366; (b) Fraser-Reid, B., Udodong, U.E., Wu, Z.F., Ottosson, H., Merritt, J.R., Rao, C.S., Roberts, C., and Madsen, R. (1992) *Synlett*, 927–942 and references therein.

57 Smoot, J.T. and Demchenko, A.V. (2008) *J. Org. Chem.*, 73, 8838–8850.

58 (a) Crich, D. and Chandrasekera, N.S. (2004) *Angew. Chem. Int. Ed.*, 43, 5386–5389 and references therein; (b) Crich, D. and Sharma, I. (2008) *Org. Lett.*, 10, 4731–4734; (c) Manabe, S., Ishii, K., and Ito, Y. (2006) *J. Am. Chem. Soc.*, 128, 10666–10667; (d) Manabe, S., Ishii, K., and Ito, Y. (2011) *Eur. J. Org. Chem.*, 497–516; (e) Satoh, H., Manabe, S., Ito, Y., Lüthi, H.P., Laino, T., and Hutter, J. (2011) *J. Am. Chem. Soc.*, 133, 5610–5619.

59 Hashimoto, S.I., Sakamoto, H., Honda, T., Abe, H., Nakamura, S.I., and Ikegami, S. (1997) *Tetrahedron Lett.*, 38, 8969–8972.

60 (a) Veeneman, G.H. and van Boom, J.H. (1990) *Tetrahedron Lett.*, 31, 275–278; (b) Kayastha, A.K. and Hotha, S. (2013) *Beilstein J. Org. Chem.*, 9, 2147–2155.

61 Huang, X., Huang, L., Wang, H., and Ye, X.S. (2004) *Angew. Chem. Int. Ed.*, 43, 5221–5224.

62 Wang, Z., Zhou, L., Ei-Boubbou, K., Ye, X.S., and Huang, X. (2007) *J. Org. Chem.*, 72, 6409–6420.

63 Lu, S.R., Lai, Y.H., Chen, J.H., Liu, C.Y., and Mong, K.K. (2011) *Angew. Chem. Int. Ed.*, 50, 7315–7320.

64 Lin, Y.H., Ghosh, B., and Mong, K.-K.T. (2012) *Chem. Commun.*, 48, 10910–10912.

65 Li, Z. and Gildersleeve, J. (2006) *J. Am. Chem. Soc.*, 128, 11612–11619.

66 (a) Crich, D. and Li, W. (2007) *J. Org. Chem.*, 72, 7794–7797; (b) Peng, P., Xiong, D.C., and Ye, X.S. (2014) *Carbohydr. Res.*, 384, 1–8.

67 (a) Wever, W.J., Cinelli, M.A., and Bowers, A.A. (2013) *Org. Lett.*, 15, 30–33; (b) He, H. and Zhu, X. (2014) *Org. Lett.*, 16, 3102–3105; (c) Nokami, T., Saito, K., and Yoshida, J. (2012) *Carbohydr. Res.*, 363, 1–6.

3

Solvent Effect on Glycosylation

KwoK-Kong Tony Mong[a], Toshiki Nokami[b], Nhut Thi Thanh Tran[a], and Pham Be Nhi[a]

3.1 Introduction

Glycosylation is the key reaction in the synthesis of oligosaccharides and glycosylated natural products [1], which invokes the coupling of a glycosyl donor and an acceptor in the presence of a promoter [2]. The acceptor can be a carbohydrate or noncarbohydrate component. The regio- and stereochemistry of glycosylation are strictly controlled to avoid the formation of undesired isomers. The control of regiochemistry can be taken care of by orthogonal protection of hydroxyl groups in carbohydrate substrates, and such processes can be expedited via a one-pot protection strategy [3]. In contrast, the control of stereochemistry is by no means trivial due to the diverse structure of carbohydrate substrates [4].

Until now, glycosylation chemistry has still been dominated by the concept of Koenigs–Knorr glycosylation [5]. In Koenigs–Knorr-type glycosylation, a glycosyl donor bearing an anomeric leaving group is activated by an electrophilic promoter to form a glycosyl oxocarbenium ion. The glycosyl oxocarbenium ion is accompanied with a counterion as an ion pair, which may exist as a close-contact ion pair (CCIP), a solvent-separated ion pair (SSIP), and/or a covalent glycoside adduct (Scheme 3.1) [2]. In the absence of a participatory protecting group, each of these ion pairs or covalent adduct may react with an acceptor to form the glycosylation product. Consequently, the stereochemical outcome of a new glycosylation reaction is difficult to predict.

It is generally agreed that the stereochemical outcome of glycosylation can be affected by multiple factors, which include (i) structure and conformation of glycosyl substrates [6], (ii) glycosylation reagents [7] or promoters [8], (iii) nature of solvent [9–11], (iv) presence of a participatory [12–14] or a chiral auxiliary protecting group [15, 16], (v) presence of conformationally locked protecting group [17, 18], (vi) presence of a glycosyl acceptor tethering group [19, 20], and/or (vii) presence of an exogenous nucleophilic additive [21–26].

Among the aforementioned factors, the influence of solvent on glycosylation has received considerable attention. As the solvent can be removed by simple workup, the use of solvent to control the stereochemistry of glycosylation is an attractive yet

Selective Glycosylations: Synthetic Methods and Catalysts, First Edition. Edited by Clay S. Bennett.

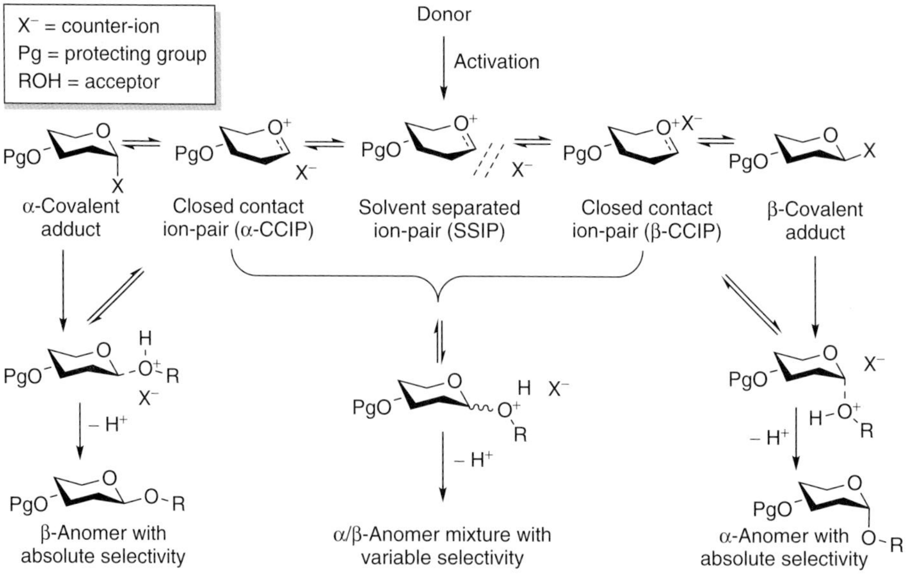

Scheme 3.1 General mechanism of glycosylation.

practical strategy. This strategy eliminates the use of specialized protecting groups and simplifies the preparation of carbohydrate building blocks for assemblage of oligosaccharide structure. The objective of this chapter is to discuss the effect of solvent on glycosylation with respect to the physical properties of the solvent.

3.2 General Properties of Solvents Used in Glycosylation

The century-old doctrine in chemistry is "like dissolves like" (similia similibus solvuntur). This guideline is highlighted by the dissolution function of solvent in glycosylation. In addition to dissolution of glycosyl substrates and reagents, particular physical properties of solvent, including dielectric constant (ε_r), molecular dipole (μ), donicity number (DN), and Lewis basicity ($-\Delta H^0_{\text{D-BF3}}$), are crucial because they are closely related to the mechanism of glycosylation that determines the stereochemistry and yield of the reaction (Table 3.1) [30].

Most of glycosylation methods work effectively in a particular range of reaction temperatures. In the iterative one-pot glycosylation, a low temperature such as −60 °C was needed in the activation step of the glycosyl donor, but the subsequent coupling with a glycosyl acceptor was conducted at a higher temperature. Diethyl ether (Et_2O) (mp −117 °C) and dichloromethane (CH_2Cl_2) (mp −95 °C) with a subzero melting point are suitable for such a one-pot glycosylation method. In some glycosylations, promoted by a halide ion, high reaction temperatures are required to effect an S_N2-like coupling reaction; under such circumstances, benzene, 1,2-dichloroethane ($C_2H_4Cl_2$), and toluene are the solvents of choice because of their relatively higher bp.

A binary or quaternary solvent system contains two or more solvent components and is common in glycosylation. For examples, Mong *et al.* used a 1:2:1 CH_2Cl_2–CH_3CN–EtCN solvent mixture in the low-concentration glycosylation method (see Section 3.3). This

Table 3.1 mp, bp, dielectric constant, molecular dipole, donicity number, and Lewis basicity of several organic solvents that are used in various glycosylation contexts.

Solvent	mp, bp (°C)	Dielectric constant (ε_r) [27]	Molecular dipole (μ) [27]	Donicity number (DN) [28]	Lewis basicity index ($-\Delta H^0_{D\text{-}BF3}$) [29]
Dioxane	12, 101	2.25	0.45	0.38	74.1
Benzene	5, 80	2.27	0	—	—
Toluene	−95, 111	2.38	0.43	—	—
Et_2O	−117, 35	4.33	1.3	0.49	78.8
THF	−109, 66	7.58	1.75	0.52	90.4
CH_2Cl_2	−95, 40	8.93	1.55	—	—
$C_2H_4Cl_2$	−36, 84	10.36	6.1	0.0	10.0
CH_3NO_2	−28, 101	35.9	3.54	0.07	37.6
CH_3CN	−44, 82	37.5	3.45	0.36	60.4
EtCN	−93, 98	28.6	—	0.41	—

solvent system does not "freeze" at a temperature as low as −78 °C, and at such a temperature, the nitrile solvent effect works effectively [11]. In another example, the synergistic α-directing effect of a binary $CHCl_3$ and ether solvent mixture has been demonstrated by Ito and coworkers (see Section 3.4) [31].

Besides mp and bp, the polarity of a solvent plays an important role in glycosylation. Solvent polarity is a bulk property, which is characterized by the dielectric constant (ε_r) and molecular dipole (μ). Generally, the solvation of ion pairs in weakly polar solvents is poor; thus, they are regarded as nondissociating solvents. In a nondissociating solvent, CCIP intermediates are stabilized. Dioxane (ε_r 2.25), Et_2O (ε_r 4.33), and benzene (ε_r 2.27) are nonpolar or weakly polar and nondissociating solvents, while nitromethane (CH_3NO_2) (ε_r 35.9) and CH_3CN (ε_r 37.5) are highly polar and dissociating solvents. The polarity of CH_2Cl_2 (ε_r 8.93) lies between those of dioxane and CH_3CN; therefore, a particular ion pair of glycosyl oxocarbenium ion and its counterion can exist in CH_2Cl_2 at a suitable temperature range but may be decomposed at a higher temperature.

Although different forms of glycosyl oxocarbenium counterion ion pairs may exist in a weakly polar solvent, the closeness of these ion pairs depends on the exact polarity of a solvent as well as the stereoelectronic features of the sugar substrates. Thus far, most glycosyl covalent adducts generating from the reaction of (i) close-contact oxocarbenium ion and counterion or (ii) oxocarbenium ion and nucleophile were identified in moderately polar CD_2Cl_2 or $CDCl_3$ at subzero temperatures. Examples include 4,6-*O*-benzylidene α-mannosyl triflate [32], 2-azido-2-deoxygalactosyl imidinium ion [26], and 2-azido-2-deoxy-glucosyl sulfonium ions [33] (Figure 3.1).

As the dielectric constant (ε_r) of a solvent is a bulk property, the electron-donating capacity refers to a specific interaction between solvent and solute molecules. Participation of an electron-pair-donating solvent toward a glycosyl oxocarbenium ion has long been exploited in glycosylation. DN and Lewis basicity index have been used to characterize the electron-donating property of a solvent [28, 29]. $C_2H_4Cl_2$ has a DN of

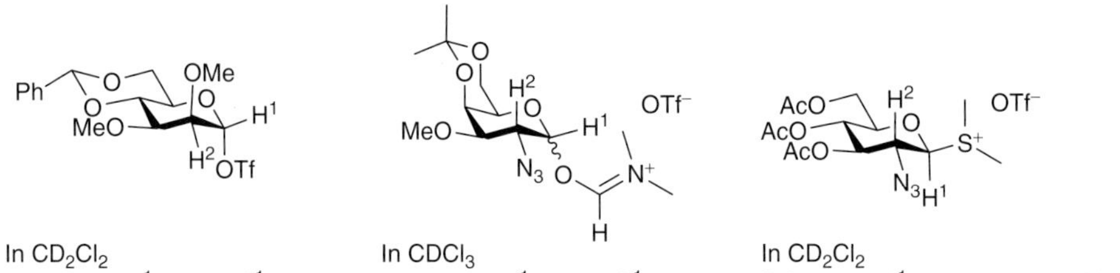

Figure 3.1 (a) α-Mannosyl triflate in β-mannosylation. (b) α- and β-Imidinium ion adduct in DMF-modulated glycosylation. (c) β-Glucosyl sulfonium triflate covalent adduct.

"zero" and a Lewis basicity index of 10 and is regarded as the solvent having the least electron-pair-donating capacity. In principle, the lone electron pair in Et_2O (DN 0.49; $-\Delta H^0_{D\text{-}BF3}$ 78.8), dioxane (DN 0.38; $-\Delta H^0_{D\text{-}BF3}$ 74.1), CH_3CN (DN 0.36; $-\Delta H^0_{D\text{-}BF3}$ 60.4), EtCN (DN 0.41), and THF (DN 0.52; $-\Delta H^0_{D\text{-}BF3}$ 90.4) can coordinate with the anomeric center of a glycosyl oxocarbenium cation.

Based on the polarity and electron-pair-donating capacity, this chapter classifies the solvents used in glycosylation to (i) weakly polar and noncoordinating solvents such as benzene and toluene; (ii) moderately polar and noncoordinating solvents such as CH_2Cl_2, CH_3NO_2, and $C_2H_4Cl_2$; (iii) highly polar and coordinating solvents such as CH_3CN and EtCN; and (iv) weakly polar and coordinating solvents such as diethyl ether (Et_2O) and dioxane. The effects of these solvents in glycosylation are described in the following sections.

3.3 Polar and Noncoordinating Solvents in Glycosylation

Typical Koenigs–Knorr glycosylation employed a glycosyl bromide or chloride as a glycosyl donor, which was usually activated by a heavy metal salt (Lewis acid) and then coupled with a glycosyl acceptor to yield a glycosylation product. Metal salts such as Ag_2CO_3 [34], Ag_2O [35], $Hg(CN)_2$ [36], $HgBr_2$ [36], AgOTf [37], $AgClO_4$ [37] were exploited as promoters. To provide adequate solubility of the metal salt, highly polar CH_3NO_2 was employed in these reactions. Furthermore, CH_3NO_2 has a bp of 101 °C that allows the glycosylation at a higher reaction temperature. The following are two selected examples for the use of CH_3NO_2 in glycosylation.

In the synthesis of Lewis A blood antigen determinant, Lemieux prepared Gal-β-(1,3)-GlcNAc disaccharide **3** from the glycosylation of 2-acetamido-2-deoxy-β-D-glucoside **2** with α-galactosyl bromide **1** using mercuric cyanide [$Hg(CN)_2$] as a promoter, and the reaction was conducted in CH_3NO_2 at 60 °C (Scheme 3.2a) [23]. Kadokawa *et al.* studied the glycosylation of cyclohexanol **5** with 2-acetamido-2-deoxy-glucosyl diphenyl-phosphine **4** using trimethylsilyl trifluoromethanesulfonate (TMSOTf) as a promoter [38]. The reaction occurred at 100 °C in CH_3NO_2, and 2-acetamido-2-deoxy-α-glucoside **6** was obtained as a single anomer (Scheme 3.2b).

Before the 1970s, oligosaccharide synthesis was almost dominated by the use of glycosyl halide donors. Since that time, a number of new glycosyl donors with different

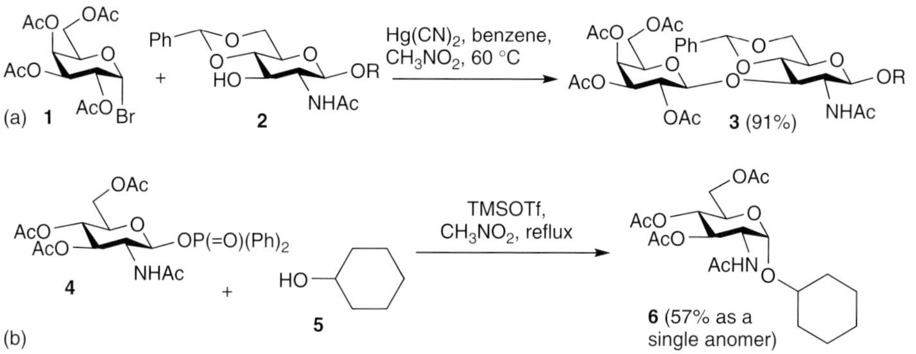

Scheme 3.2 (a) Use of $Hg(CN)_2$ promoter in CH_3NO_2 and (b) use of CH_3NO_2 for glycosylation at high temperature.

anomeric leaving groups have emerged. These "new" glycosyl donors could be activated with nonmetallic Lewis acid or Brønsted acid. Such a change to nonmetallic promoter enables the use of less polar solvents in glycosylation.

Being moderately polar, CH_2Cl_2 is not a good solvent for metallic salts, but it provides good solubility for ordinary glycosyl substrates. This appealing feature allows for conducting glycosylation at temperatures ranging from −80 to 40 °C. Furthermore, its low Lewis basicity index (presumably approximate to $C_2H_4Cl_2$, see Table 3.1) renders CH_2Cl_2 a noncoordinating solvent that would not interfere with the participatory effect of a specialized protecting group or an exogenous additive. For example, in the formamide-modulated glycosylation method, α- and β-glycosyl oxocarbenium imidinium triflate ion pairs **7a** and **7b** were generated from the reaction of thioglycoside donor 7 and formamide nucleophile **8** in CH_2Cl_2 (Scheme 3.3). In CH_3CN, such an imidinium triflate ion pair was not observed. This should be attributed to the higher polarity of CH_3CN that dissociates the imidinium triflate ion pair.

3.4 Weakly Polar and Noncoordinating Solvents in Glycosylation

The use of a nonpolar and noncoordinating solvent such as benzene and toluene in glycosylation is not common due to the modest solubility of glycosyl substrates in these solvents. For a particular glycosylation method that required a higher reaction

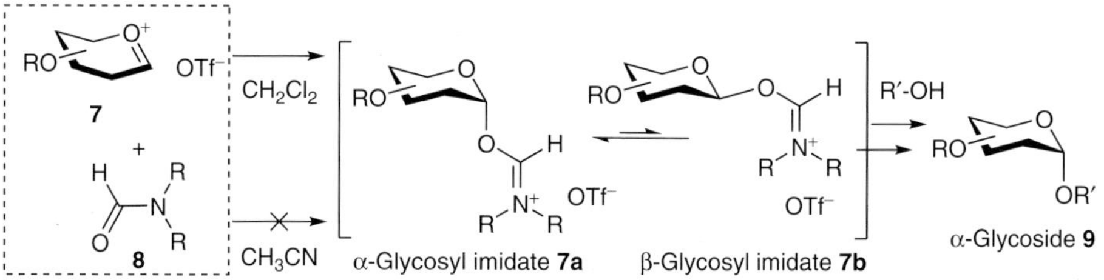

Scheme 3.3 Formamide-modulated glycosylation in noncoordinating solvent CH_2Cl_2 and coordinating solvent CH_3CN.

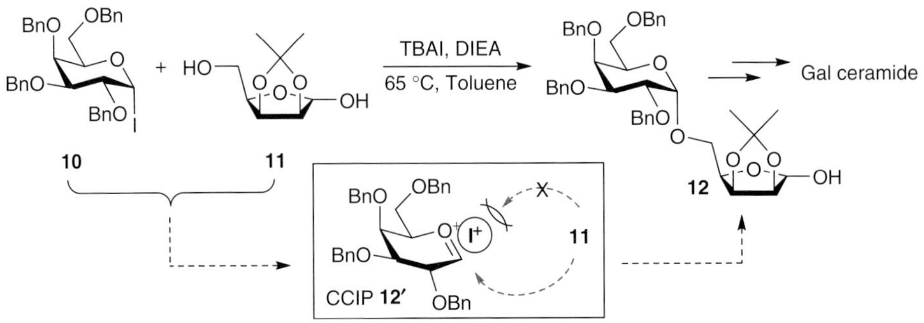

Scheme 3.4 Use of weakly polar benzene as solvent in iodide-promoted α-glycosylation.

temperature, however, benzene or toluene is the solvent of choice. For example, in the synthesis of α-galactosyl ceramide by Luo *et al.*, 2,3-*O*-isopropylidene-protected D-lyxose **11** was reacted with α-galactosyl iodide **10** in toluene at 65 °C to afford α-linked disaccharide **12** in excellent selectivity (Scheme 3.4). A CCIP **12**′, that is, galactosyl oxocarbenium iodide, is presumably the key intermediate in the reaction. The weakly polar nature of toluene stabilizes the CCIP **12**′; thus, it can react with glycosyl acceptor **11** in an S_N2-like mechanism at refluxing temperature [39].

3.5 Polar and Coordinating Solvents in Glycosylation

The participating effect of polar and coordinating solvents has been documented since 1970s. Sinaÿ *et al.* reported the participation of a nitrile solvent toward a glycosyl oxocarbenium ion through the entrapment of a glucosyl nitrilium ion intermediate by 2-chlorobenzoic acid in a Ritter-like mechanism [11, 40]. However, the correct assignment of the α-configuration of the glucosyl nitrilium ion was achieved in 1990 by Fraser-Reid *et al.* (Scheme 3.5) [41]. In their studies, 4-pentenyl glucoside **13** was

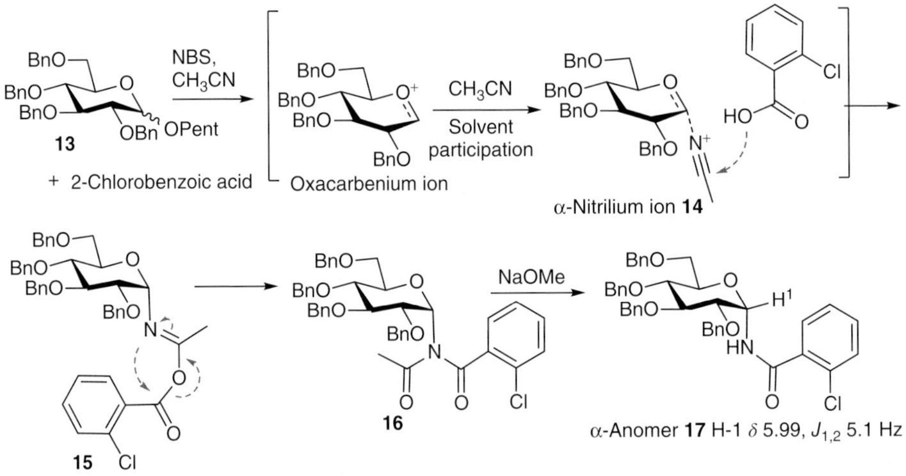

Scheme 3.5 Entrapment of the α-glucosyl nitrilium ion intermediate **14**.

activated in CH_3CN to form a α-nitrilium ion intermediate **14,** which reacted with 2-chlorobenzoic acid giving an α-imidate **15**. The α-imidate **15** underwent rearrangement to furnish *N,N*-diacetamido derivative **16,** which was hydrolyzed in basic conditions to furnish the α-*N*-glucosylamide **17** for NMR characterization. To confirm the configuration of **17**, its corresponding β-anomer was prepared by a known procedure.

Over the years, the nitrile solvent effect for the construction of a 1,2-*trans* β-glycosidic bond formation has been applied to nonparticipating glycosyl donors with different leaving groups. These donors include glycosyl fluorides (Scheme 3.6a and b) [10], glycosyl trichloroacetimidate (Scheme 3.6c) [42], glycosyl phosphates (Scheme 3.6d)

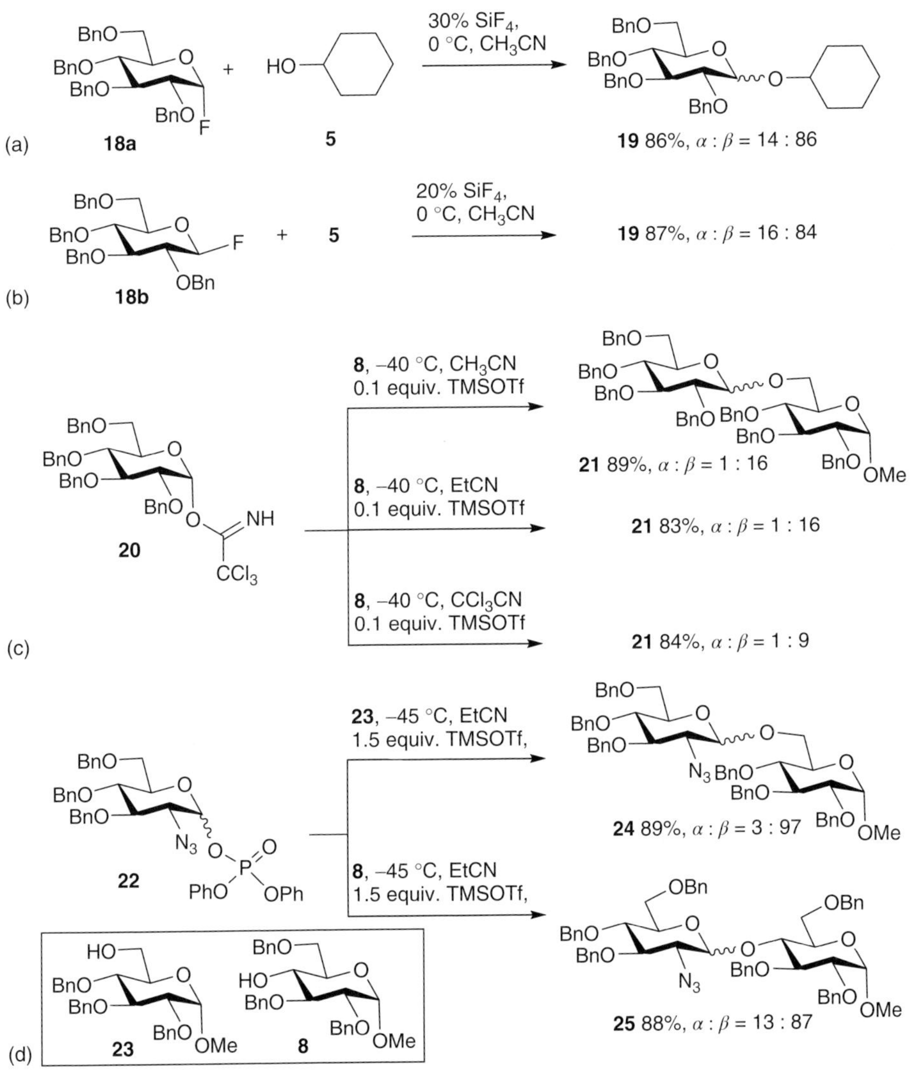

Scheme 3.6 (a and b) Glycosylation with α-/β-glucosyl fluorides (**18a** and **18b**) in nitrile solvent. (c) Glycosylation with glycosyl trichloroacetimidate **20** in the nitrile solvent. (d) Glycosylation with glycosyl phosphate **22** in the nitrile solvent.

Table 3.2 Selectivity of glycosylations at different concentrations of **26** and **27**.

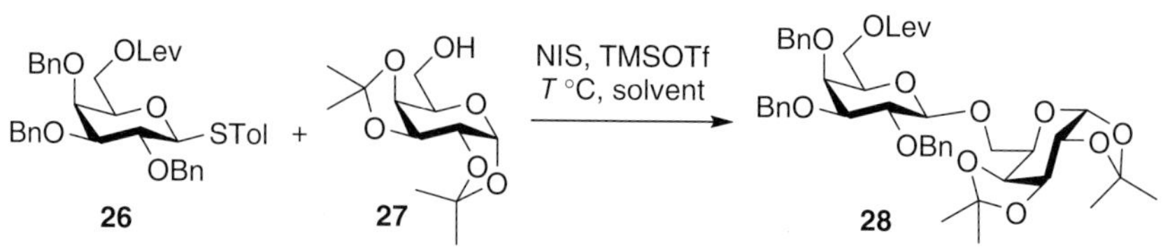

Entry	[**26**] mM	[**27**] mM	Solvent[a]	*T* (°C)	**28**, yield (%) (α:β)[b]
1	240	120	A	−55	83 (1:6)
2	120	100	A	−55	80 (1:9)
3	60	50	A	−55	84 (1:10)
4	12	10	A	−55	83 (1:13)
5	12	10	A	−70	84 (1:19)
6	12	10	B	−70	81 (1:19)

a Solvent system: A = 1 : 3 v/v CH_2Cl_2—CH_3CN; B = 1 : 2 : 1 v/v CH_2Cl_2—CH_3CN—EtCN.
b α:β-Anomeric ratio was determined by HPLC analysis of crude product mixture.

[43], and thioglycosides [11, 43, 44] (Table 3.2). In general, the observed β-selectivity does not depend on the configuration of the anomeric leaving group, implicating that the reaction follows a SN1-like mechanism (Scheme 3.6a and b). The β-selectivity and the yield of glycosylations are higher for reactive primary glycosyl acceptors than for less reactive secondary glycosyl acceptors (Scheme 3.6d). The electron-pair-donating property of nitrile solvent was inferred by the glycosylation in trichloroacetonitrile (CCl_3CN), which is electron-deficient and therefore poor electron-pair donor. As a result, the β-selectivity of the glycosylation in CCl_3CN was diminished; however, the study does not take into account the steric bulk of the CCl_3 group (Scheme 3.6c).

Further investigation on the nitrile solvent effect in 2009 revealed the relationship between the concentration of sugar substrates and β-selectivity [11]. Thiogalactoside donor **26** at different concentrations (240 to 12 mM) was reacted with galactose acceptor **27** in 1:3 CH_2Cl_2:CH_3CN mixture at −55 °C. The α:β ratio of glycosylation product **28** was found to increase from 1:6 to 1:13 as the concentration of **26** decreased (Table 3.2, entries 1–4). Further improvement of α:β ratio to 1:19 was achieved at −70 °C in either 1:3 CH_2Cl_2:CH_3CN or 1:2:1 CH_2Cl_2:CH_3CN:EtCN mixtures (Table 3.2, entries 5 and 6). It should be noted that such a concentration and selectivity relationship is also found in Yu's studies on the Au(I)-mediated glycosylation of 2-deoxyribofuranosyl *o*-hexynylbenzoates with purine and pyrimidine acceptors [45].

Although the β-selectivity observed from D-gluco and D-galacto donors in the aforementioned studies can be explained by the formation of an α-nitrilium ion intermediate, this does not adequately account for the lack of β-selectivity for a D-manno donor [45]. In the continuous mechanistic study of the low-concentration glycosylation method, the participation role of the C2 benzyl ether group in CH_3CN was examined (Scheme 3.7) [46].

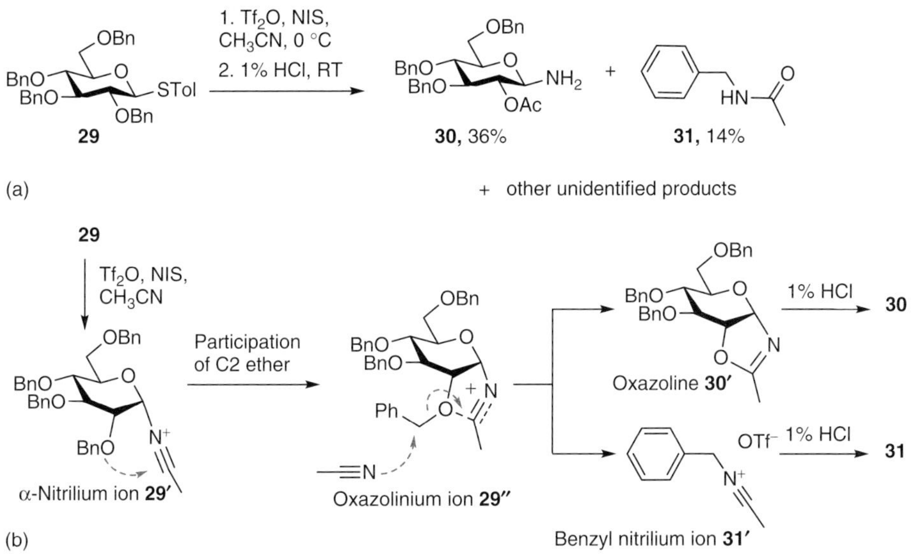

Scheme 3.7 (a) Formation of debenzylation products glycosylamine **30** and benzylacetamide **31** in nitrile solvent. (b) Proposed mechanism of the participation of C2 benzyl ether function of **29**.

To this end, thioglucoside **29** was activated with triflic anhydride (Tf_2O) and *N*-iodosuccinimide (NIS) in CH_3CN in the absence of a glycosyl acceptor. From the reaction mixture, the debenzylated products, namely β-2-*O*-acetyl-3,4,6-tri-*O*-benzyl-glucosylamine **30** and *N*-benzylacetamide **31,** were isolated together with some unidentified compounds (Scheme 3.7a). It was reasoned that the formation of **30** and **31** was attributed to the acid hydrolysis of glucosyl oxazoline **30**′ and benzyl nitrilium ion **31**′ (Scheme 3.7b). The oxazoline **30**′ and benzyl nitrilium ion **31**′ were in turn derived from the nucleophilic attack of a nitrile solvent molecule on oxazolinium ion **29**″. Oxazolinium ion **29**″ was generated from the participation of the C2 ether function in α-nitrilium ion **29**′. It should be mentioned that Fraser-Reid also noted the formation of debenzylation products in their entrapment experiments, but no further studies were followed [41]. Finally, in addition to simple glycosides, this low-concentration nitrile solvent effect has been applied to one-pot glycosylation [47], utilized for the total synthesis of a tetrasaccharide glycoglycerolipid [48].

Other than six carbon sugar substrates, the nitrile solvent effect has also been applied to glycosylation with higher carbon glycosyl donors including thiosialosides and Kdo glycal donors. For example, the glycosylation of 1-octanol with *N*-acetyl-5-*N*,4-*O*-oxazolidinone-protected adamantanyl thiosialoside donor **32** in 1:1 CH_2Cl_2:CH_3CN solvent mixture at −78 °C gave the α-linked sialoside product **33** in 95% yield as the sole anomer (Scheme 3.8). It has been pointed out that the presence of the oxazolidinone function is also essential for the α-selectivity of the reaction [49].

Interestingly, the stereodirecting effect of a nitrile solvent for 4,5:7,8-di-*O*-isopropylidene-protected Kdo glycal donor **34** is opposite to that of sialic acid donor **32**, and in the former case, a β-selective glycosylation was achieved in the nitrile solvent mixture. The glycosylation of glycosyl acceptor **23** with **34** in 1:1 CH_2Cl_2—CH_3CN gave disaccharide **35** with an excellent $\alpha:\beta$ ratio of 1:>20, but a moderate 1:5 $\alpha:\beta$ ratio was

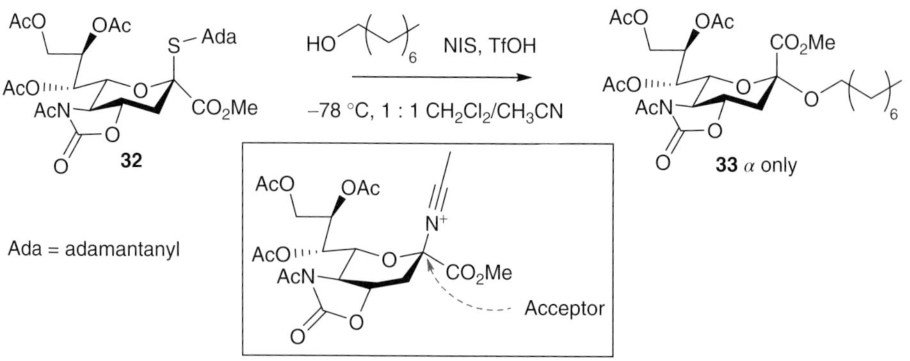

Scheme 3.8 Nitrile solvent effect in glycosylation of octanol with thiosialoside donor **32**.

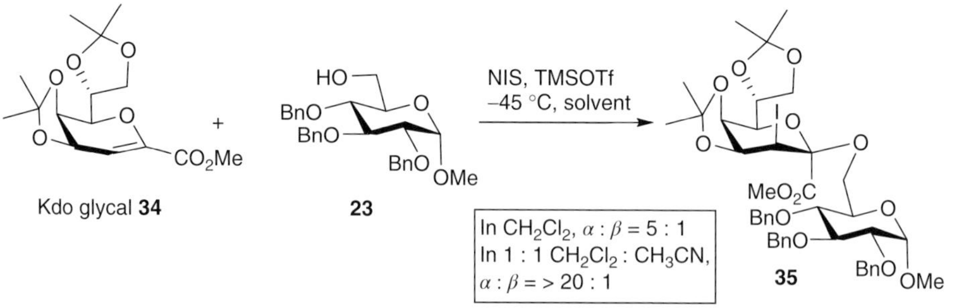

Scheme 3.9 Nitrile solvent effect in glycosylation of acceptor **23** with Kdo donor **34**.

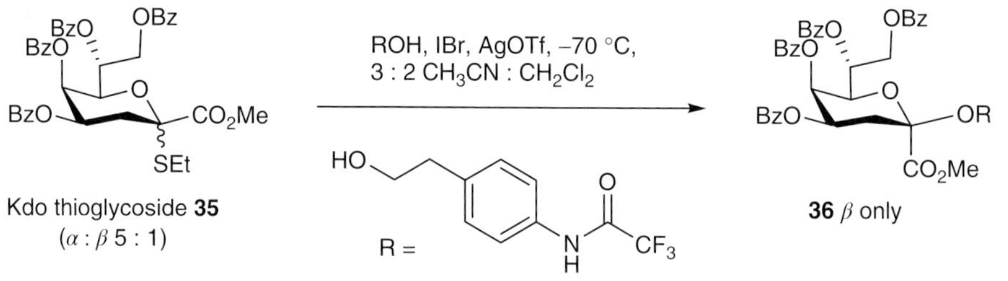

Scheme 3.10 Nitrile solvent effect in glycosylation with perbenzoyl Kdo thioglycoside donor **35**.

achieved in CH_2Cl_2 as a sole solvent (Scheme 3.9) [50]. A similar β-directing property of nitrile solvent was reported by Oscarson when per-*O*-benzoyl-protected Kdo thioglycoside donor **35** was used in glycosylation (Scheme 3.10) [51].

3.6 Weakly Polar and Coordinating Solvents in Glycosylation

Although ethers and alkyl nitriles are classified as coordinating (or participating) solvents due to their high Lewis basicity index, the molecular dipole (μ) and dielectric constant (ε_r) of ether are far smaller than those of alkyl nitriles (see Table 3.1). The solvent polarity is correlated to the capacity of ion-pair dissociation in solution. Et_2O and 1,4-dioxane can be regarded as nondissociating (or weakly dissociating) due to their small dielectric constant. In other words, an ion pair would be more stable in ether solvents than in nitrile solvents. This argument agrees well with some experimental findings. However, there remains no physical evidence for coordination of ether solvent in glycosylations.

In 1968 and 1972, Hasegawa *et al.* reported the α-directing effect of dioxane in glycosylation [52]. Such a property was attributed to the coordination of dioxane to the glucosyl oxocarbenium ion. They proposed an incipient onium adduct as the intermediate responsible for the α-selectivity obtained. Even though no evidence for the mechanism was given, the pioneering works sparked the exploitation on the α-directing effect of ether for the synthesis of α-glycosides. Following are some selected examples.

Schuerch *et al.* described the α-stereodirecting effect of Et_2O in glycosylation of MeOH with α-glucopyranosyl tosylates **37** and **38** (Table 3.3) [53]. In this study, the observed α-selectivity of glycosylation was significantly higher in Et_2O than in CH_3CN (Table 3.3, entries 1 vs 2). Ether with a bulkier substituent such as diisopropyl ether (iPr_2O) or dioxane was less effective in α-glycosylation (Table 3.3, entries 2 vs 3 and 4). The presence of an *N*-phenylcarbamate group at C6 position of glycosyl donor **37** increased the α-selectivity of glycosylation (Table 3.3, entries 2 and 6). The authors suggested that this was due to the electron-withdrawing group at C6 of glycosyl donor **38** stabilizing the β-configured ion-pair intermediate **40**″ (Scheme 3.11).

Schuerch also outlined a possible mechanism of the glycosylation as shown in Scheme 3.11. α-Glucosyl tosylate donor **38** underwent the exocyclic C—O cleavage to generate at first close-contact α-glycosyl oxocarbenium tosylate ion pair (α-CCIP) **41**′, which was in equilibrium with β-glycosyl oxocarbenium tosylate ion pair (β-CCIP) **42**″. It was reasoned that **42**″ was more stable in Et_2O than in CH_3CN, and therefore, subsequent reaction of **42**″ with MeOH afforded the α-anomer of glucoside **40**. Examination

Table 3.3 Ether solvent effect for glucosyl tosylates **37** and **38**.

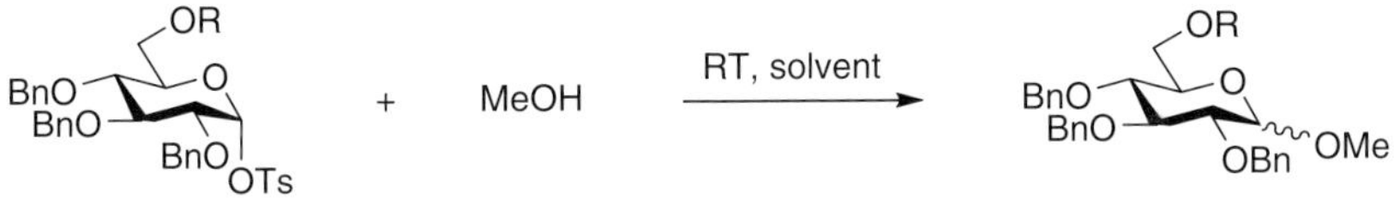

37 R = Bn
38 R = C(=O)NHPh

39 R = Bn
40 R = C(=O)NHPh

Entry	Donor	Solvent	Product	α-Anomer (%)	β-Anomer (%)
1	**37**	CH_3CN	**39**	60	40
2	**37**	Et_2O	**39**	81	19
3	**37**	iPr_2O	**39**	57	43
4	**37**	Dioxane	**39**	69	31
5	**38**	CH_3CN	**40**	80	20
6	**38**	Et_2O	**40**	100	0

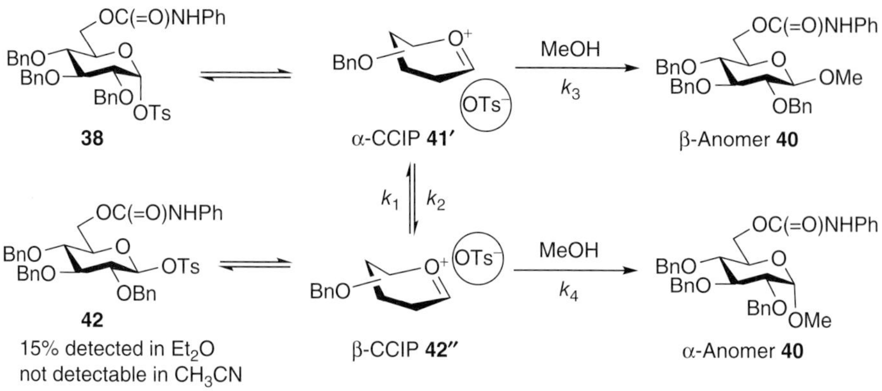

Scheme 3.11 Mechanism of ether solvent effect proposed by Schuerch *et al.*

with NMR spectroscopy revealed the presence of 15% of β-glucosyl tosylate **42** in the reaction mixture of Et_2O solvent, inferring the presence of β-glycosyl oxocarbenium tosylate ion pair (β-CCIP) **42**″, which is presumably the precursor of **42**.

Demchenko and Boons investigated the effect of ether-containing binary solvent systems in glycosylation between thioglucosyl donor **43** and galactose acceptor **29** (Table 3.4) [54]. In the study, iodonium dicollidine perchlorate (IDCP) and

Table 3.4 Ether-containing binary solvent system in glycosylation.

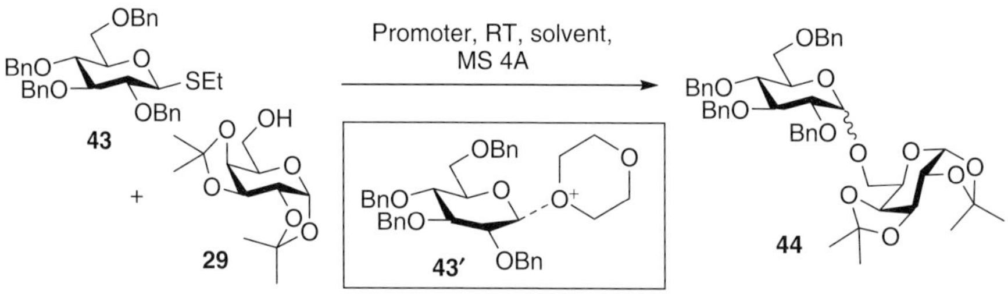

Entry	Solvent	Promoters	$\alpha:\beta$ of **44**	Remarks
1	CH_2Cl_2	IDCP	0.7 : 1	—
2	1 : 1 CH_2Cl_2 : Et_2O	IDCP	2.0 : 1	—
3	1 : 4 CH_2Cl_2 : Et_2O	IDCP	3.5 : 1	—
4	1 : 2.4 toluene : Et_2O	IDCP	12.0 : 1	Sluggish reaction
5	1 : 2 toluene : dioxane	IDCP	15.3 : 1	—
6	1 : 1 toluene : dioxane	NIS, TMSOTf	1.8 : 1	—
7	1 : 2 toluene : dioxane	NIS, TMSOTf	7.3 : 1	—
8	2 : 1 toluene : dioxane	NIS, TMSOTf	5.5 : 1	Slower reaction
9	1 : 3 toluene : dioxane	NIS, TMSOTf	8.6 : 1	—

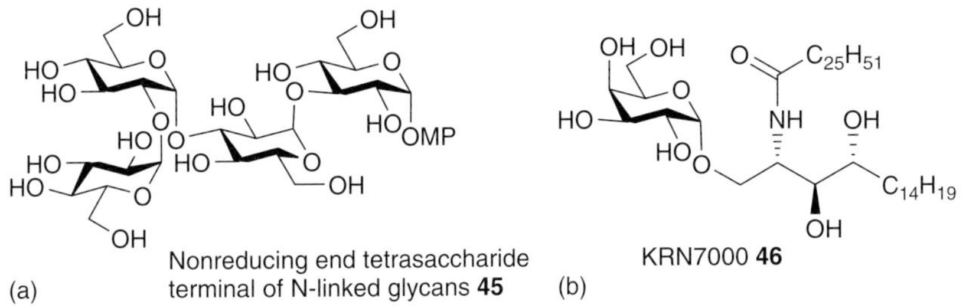

Figure 3.2 (a) Biologically relevant tetrasaccharide terminal of N-linked glycan **45**. (b) Glycolipid KRN7000 **46**.

NIS/TMSOTf promoter systems were employed, and all reactions took place at room temperature. The α-stereodirecting effect of Et_2O and dioxane was clearly observed when IDCP was used as the promoter (Table 3.4, entries 1–5). The glycosylation was slower in the binary solvent mixture than in CH_2Cl_2 alone. Moreover, the α-selectivity of glycosylation was higher with the IDCP promoter than with the NIS and TMSOTf promoter system (Table 3.4, entries 5 and 7). A higher α-selectivity of glycosylation was afforded in toluene–dioxane solvent mixture (weakly polar) than $CH_2Cl_2:Et_2O$ (moderately polar) (Table 3.4, entries 1–4). Similar to the studies by Hasegawa [52], the authors also suggested that β-onium adduct **43′** (see inset on Table 3.4) was formed in ether solvent. Such an ether-containing binary solvent system was employed in the construction of tetrasaccharide **45** [31] and glycolipid KRN7000 **46** (Figure 3.2) [55].

3.7 Solvent Effect of Ionic Liquid on Glycosylation

The toxicity of CH_2Cl_2, dioxane, and CH_3CN to living organisms and contamination of environment renders their use as solvents in laboratory increasingly prohibitive and unaffordable. Therefore, the development of a green chemistry approach for oligosaccharide synthesis is needed. Ionic liquids that can reasonably dissolve glycosyl substrates and promoters can be used as green glycosylation solvents [56]. Toshima and coworkers applied the ionic liquid for the first time as a solvent for glycosylation. In their study, perbenzyl-protected glucosyl phosphite **47** in 1-*n*-hexyl-3-methylimidiazolium salt (C_6mim[X]) was reacted with cyclohexylmethanol **48** in a glycosylation promoted by the conjugate acid of the ionic liquid (Table 3.5) [57].

C_6mim[counterion] salts with different counterions were examined (Table 3.5, entries 1–3). In the absence of a participating group, the reaction produced the desired glucoside **49** in good yields, and moderate β selectivity was observed (Table 3.5, entries 1–3). The combination of C_6mim[NTf_2] and $HNTf_2$ was further investigated by decreasing the amount of the acid (Table 3.5, entries 4 and 5). Eventually, 1.0 M of C_6mim[NTf_2] salt (for 0.1 mmol donor) and 1.0 mol% of $HNTf_2$ (with respect to C_6mim[NTf_2] salt) were applied to the glycosylation of carbohydrate acceptors **8** and **23** (Table 3.5, entries 6 and 7). In general, no significant stereoselectivity was observed, and the yield of glycosylation was lower for secondary carbohydrate acceptor (Table 3.5, entry 7). In

Table 3.5 Use of ionic liquid as solvent in glycosylation.

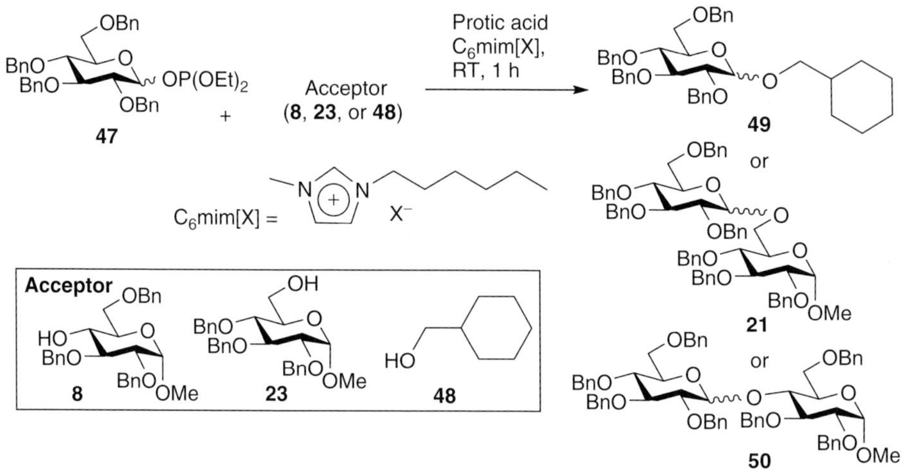

Entry	Acceptor	Ionic liquid	Protic acid (equiv.)	Yield (%)	$\alpha:\beta$
1	**47**	C_6mim[BF_4]	HBF_4 (1.0)	**49**, 72	22:78
2	**47**	C_6mim[OTf]	HOTf (1.0)	**49**, 89	26:74
3	**47**	C_6mim[NTf_2]	$HNTf_2$ (1.0)	**49**, 81	18:82
4	**47**	C_6mim[NTf_2]	$HNTf_2$ (0.1)	**49**, 80	18:82
5	**47**	C_6mim[NTf_2]	$HNTf_2$ (0.01)	**49**, 86	16:84
6	**8**	C_6mim[NTf_2]	HNTf[a]	**21**, 84	23:77
7	**23**	C_6mim[NTf_2]	HNTf[a]	**50**, 63	57:43

a Exact amounts of acid and ionic salt refer to Ref.[57].

addition to the phosphite donor **47**, ionic liquid solvents have also been used in glycosylations with glycosyl trichloroacetimidate [58] and thioglycosyl donors [59].

The high dielectric constant of ionic liquid renders it a strong ion-pair dissociating solvent. In such an ionic solvent, formation of close-contact oxocarbenium counterion ion pair or covalent adduct is not possible. The aforementioned stereodirecting mechanisms that benefit from solvent participation or CCIP would not occur in ionic solvent. The lack of stereochemical control in ionic solvents is therefore not unexpected.

In 2006, Poletti and coworkers reported a study on the glycosylation of isopropanol with α- and β-2,3,4,6-tetra-*O*-benzyl glucopyranose trichloroacetimidate in the presence of 0.01 equiv. of Me_3SiOTf in different solvents including CH_2Cl_2, 1-ethyl-3-methylimidazolium triflate ([emim][OTf]), and 1-butyl-3-methylimidazolium triflate ([bmim][PF6]) [60]. It is intriguing that the reaction proceeds even in the absence of Me_3SiOTf in [emim][OTf]. Generally, the stereochemical outcome of glycosylation (inversion configuration of the anomeric leaving group) stems from the stereochemistry of the starting trichloroacetimidate donor; however, in [emim][OTf], the β-isomer is the major product from both α- and β-isomer of the trichloroacetoimidate donor.

Low-temperature NMR experiments revealed that the corresponding α-glycosyl triflate is generated in the presence of a catalytic amount of boron trifluoride diethyl etherate as the Lewis acid (0.1–0.45 equiv.) in CD_2Cl_2 with ionic liquid (0.6–1.8 equiv.). These results suggested that the observed β-selectivity in glycosylation may attribute to the α-configuration of the glycosyl triflate intermediate.

3.8 Solvent Effect on Electrochemical Glycosylation

Acetonitrile is an ideal solvent of electrolysis because of its high stability under electrochemical oxidative conditions and its solubility of supporting electrolytes including lithium and sodium salts. In the pioneering work of electrochemical glycosylation of arylglycoside donors by Noyori and Kurimoto, the desired glycosylation products with methyl or benzyl protecting groups were obtained β-selectively (up to 78%) in CH_3CN in the presence of $LiClO_4$ as a supporting electrolyte (Scheme 3.12) [61]. Amatore, Sinaÿ, and coworkers reported the same level of β-selectivities (up to 80%) in the glycosylation of thioglycosides with benzyl protecting groups in CH_3CN in the presence of Bu_4NBF_4 as a supporting electrolyte [62]. A higher level of β-selectivity (up to 96%) was also observed by performing reaction with the same CH_3CN/Bu_4NBF_4 system at low temperatures (−20 °C) [63]. Although the formation of a reactive α-nitrilium intermediate is suggested to account for the selectivity, none of these studies provided evidence of such an intermediate. Conversely, electrochemically generated α-glycosyl triflates were observed by low-temperature NMR measurements of the sample prepared by anodic oxidation in the CH_2Cl_2/Bu_4NOTf system (Scheme 3.13) [64]. In this case, Bu_4NOTf works as both a supporting electrolyte and a reservoir of triflate anion. The observed β-selectivity may stem from α-configuration of glycosyl triflates.

3.9 Molecular Dynamics Simulations Studies on Solvent Effect

Both the entrapment of α-glycosyl nitrilium or oxazolinium ion intermediates and the isolation of the debenzylation product of the oxazolinium ion provide experimental evidence for nitrile solvent participation. The mechanism of the ether solvent effect

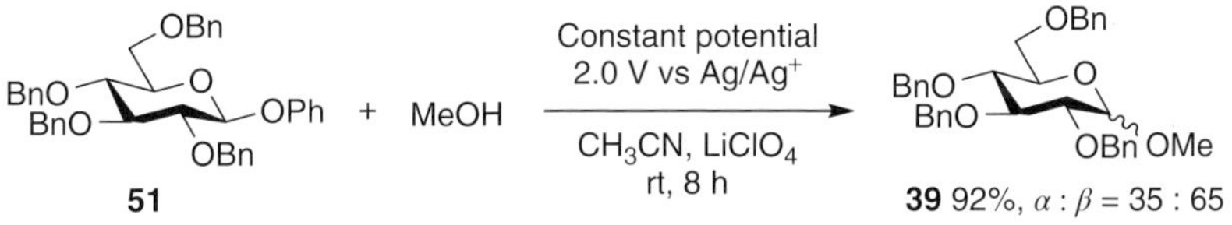

Scheme 3.12 β-Selective electrochemical glycosylation in CH_3CN.

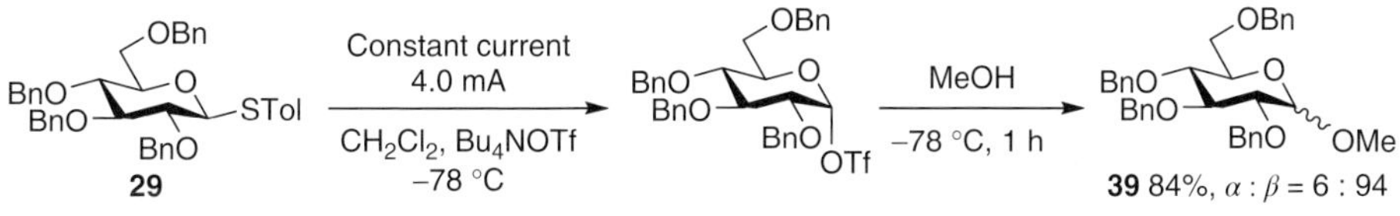

Scheme 3.13 β-Selective glycosylation via electrochemically generated α-glycosyl triflate in CH_2Cl_2.

Table 3.6 Population of glucosyl conformers in different solvent systems.

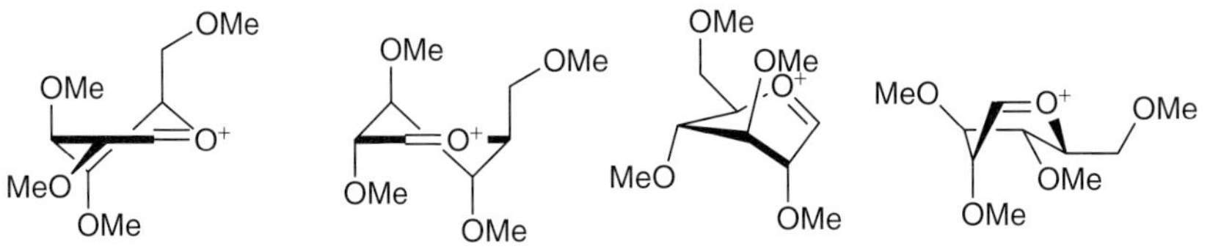

52 $^5H_4$ **53** $^3H_4$ **54** $^0S_2$ **55** $B_{2,5}$

	Ratios of conformers (%)			
Solvent	**52**	**53**	**54**	**55**
CH_3CN	0.0	36.5	60.7	2.7
Et_2O	0.0	50.1	47.6	2.3
Toluene	0.0	70.9	27.7	1.3
Dioxane	0.0	61.4	37.0	1.6
Vacuum	0.1	68.8	29.6	1.5

remains elusive because similar entrapment experiments are not possible in ether solvents. In 2010, Satoh and Hunenberger used quantum mechanical and molecular dynamic simulations to probe the mechanism of solvent effect [65]. In their study, the conformer population of a model per-*O*-methyl glucosyl oxocarbenium ion was elucidated via molecular dynamics simulation study (Table 3.6).

The population of various conformers is found closely related to the solvent polarity. In polar CH_3CN, skew–boat (OS_2) conformer **54** is the major component, and in weakly polar Et_2O, nonpolar dioxane, or toluene, the glucosyl oxocarbenium ion adopts a half-chair ($^3H_4$) conformation, that is, **53**. Furthermore, the distribution of triflate counterion relative to the oxocarbenium ion was also calculated. In less polar solvents (Et_2O, dioxane, and toluene), more (>50%) of the triflate counterion was clustered at the β-side of the oxocarbenium ion, whereas in polar CH_3CN, more of the triflate counterion was found at the α-side of the oxocarbenium ion. On the basis of the aforementioned data, a conformer and counterion hypothesis was established to account for the nitrile and ether solvent effect.

3.10 Conclusions

From a practical point of view, ether and nitrile solvent effects have been utilized to control selectivity in oligosaccharide synthesis; however, these solvent effects have limitations. For example, selectivity of the nitrile solvent effect is higher for reactive acceptors than for less reactive acceptors, while such a trend is reversed for the ether solvent effect.

To study the mechanism of the solvent effect is difficult because the interaction between solvent and solute is a complicate physical process. Several physical properties of a solvent act simultaneously together to influence the stereochemical outcome of a glycosylation. Focusing on a particular physical property may overlook the importance of the others. An example of this is the different mechanistic models of the nitrile solvent effect suggested by experimental and computational chemists. In carrying out such studies, it is important to keep in mind that the scope of study is often limited to particular substrates and experimental settings. Accordingly, care should be taken when examining any hypothesis derived from a particular model system, as it is highly unlikely that it will prove to be general.

References

1 (a) Kharel, M.K., Pahari, P., Shepherd, M.D., Tibrewal, N., Nybo, S.E., Shaaban, K.A., and Rohr, J. (2012) *Nat. Prod. Rep.*, 29, 264–325; (b) Newton, C.G. and Sherburn, M.S. (2015) *Nat. Prod. Rep.*, 32, 865–876; (c) Yang, Y., Zhang, X., and Yu, B. (2015) *Nat. Prod. Rep.* doi: 10.1039/c5np00033e

2 (a) Bohe, L. and Crich, D. (2011) *C.R. Chim.*, 14, 3–16; (b) Mydock, L.K. and Demchenko, A.V. (2010) *Org. Biomol. Chem.*, 8, 497–510.

3 Wang, C.-C., Lee, J.-C., Luo, S.-Y., Kulkarni, S.S., Huang, Y.-W., Lee, C.-C., Chang, K.-L., and Hung, S.-C. (2007) *Nature*, 446, 896–899.

4 (a) Mulani, S.K., Hung, W.-C., Ingle, A.B., Shiau, K.-S., and Mong, K.-K.T. (2014) *Org. Biomol. Chem.*, 12, 1184–1197; (b) Nigudkar, S.S. and Demchenko, A.V. (2015) *Chem. Sci.*, 6, 2687–2704.

5 Koenigs, W. and Knorr, E. (1901) *Ber. Dtsch. Chem. Ges.*, 34, 957–981.

6 Yang, M.T. and Woerpel, K.A. (2009) *J. Org. Chem.*, 74, 545–553.

7 Cai, F. and Yang, F. (2014) *J. Carbohydr. Chem.*, 33, 1–19.

8 Issa, J.P. and Bennett, C.S. (2014) *J. Am. Chem. Soc.*, 136, 5740–5744.

9 Pougny, J.R. and Sinay, P. (1976) *Tetrahedron Lett.*, 17, 4073–4076.

10 Hashimoto, S., Hayashi, M., and Noyori, R. (1984) *Tetrahedron Lett.*, 25, 1379–1382.

11 Chao, C.-S., Li, C.-W., Chen, M.-C., Chang, S.-S., and Mong, K.-K.T. (2009) *Chem. Eur. J.*, 15, 10972–10982.

12 Frush, H.L. and Isbell, H.S. (1941) *J. Res. Nat. Bur. Stand.*, 27, 413–428.

13 Paulsen, H. and Herold, C.-P. (1970) *Chem. Ber.*, 103, 2450–2462.

14 Hoang, K.L.M., Zeng, J., and Liu, X.-W. (2014) *Nat. Commun.* doi: 10.1038/ncomms6051

15 Kim, J.H., Yang, H., and Boons, G.J. (2005) *Angew. Chem. Int. Ed.*, 44, 947–949.

16 Kim, J.H., Yang, H., Khot, H., Whitfield, D., and Boons, G.J. (2006) *Eur. J. Org. Chem.*, 12, 5007–5028.

17 Huang, M., Garrett, G.E., Birlirakis, N., Boh, L., Pratt, D.A., and Crich, D. (2012) *Nat. Chem.*, 4, 663–667.

18 Jensen, H.H., Nordstro, L.U., and Bols, M. (2004) *J. Am. Chem. Soc.*, 126, 9205–9213.

19 Yasomanee, J.P. and Demchenko, A.V. (2013) *J. Am. Chem. Soc.*, 134, 20097–20102.

20 Ruei, J.-H., Patteti, V., Ingle, A.B., Mong, K.-K.T. (2015) Chem. Commun., 51, 5394–5397.

21 Walk, J.T., Buchan, Z.A., and Montgomary, J. (2015) *Chem. Sci.*, 6, 3448–3453.
22 Lemieux, R.U., Hendriks, K.B., Stick, R.V., and James, K. (1975) *J. Am. Chem. Soc.*, 97, 4056–4062.
23 Lemieux, R.U. and Driguez, H. (1975) *J. Am. Chem. Soc.*, 97, 4063–4068.
24 Hadd, M.J. and Gervay, J. (1999) *Carbohydr. Res.*, 320, 61–69.
25 Lu, S.-R., Lai, Y.-H., Chen, J.-H., Liu, C.-Y., and Mong, K.-K.T. (2011) *Angew. Chem. Int. Ed.*, 50, 7315–7320.
26 Ingle, A.B., Chao, C.-S., Hung, W.-C., and Mong, K.-K.T. (2013) *Org. Lett.*, 15, 5290–5293.
27 Wypych, G. (ed.) (2001) Handbook of Solvents (Solvent Database on CD-ROM), Chem Tec Publishing, Toronto, and William Andrew Publishing, New York.
28 (a) Marcus, Y., Maria, P.-C., and Gal, J.-F. (1985) *J. Phys. Chem.*, 89, 1296–1304; (b) Maria, P.-C., Gal, J.-F., de Franceschi, J., and Fargin, E. (1987) *J. Am. Chem. Soc.*, 109, 483–492.
29 Persson, I. (1986) *Pure Appl. Chem.*, 58, 1153.
30 Reichardt, C. (ed.) (2009) *Solvents and Solvent Effects in Organic Chemistry*, 3rd edn, Wiley-VCH Verlag GmbH, Weinheim, pp. 5–56.
31 Ishwata, A., Munemura, Y., and Ito, Y. (2008) *Tetrahedron*, 64, 92–102.
32 (a) Crich, D. and Sun, S. (1996) *J. Am. Chem. Soc.*, 118, 9239–9248; (b) Crich, D. (2002) *J. Carbohydr. Chem.*, 21, 663–686.
33 Nokami, T., Shibuya, A., Manabe, S., Ito, Y., and Yoshida, J. (2009) *Chem. Eur. J.*, 15, 2252–2255.
34 Wulff, G. and Rohle, G. (1974) *Angew. Chem. Int. Ed.*, 13, 157–170.
35 Igarashi, K. (1977) *Adv. Carbohydr. Chem. Biochem.*, 34, 243–283.
36 Helferich, B. and Wedemeyer, K.-F. (1949) *Liebigs Ann. Chem.*, 563, 139–145.
37 (a) Lemieux, R.U., Takeda, T., and Chung, B.Y. (1976) *ACS Symp. Ser.*, 39, 90–115; (b) Bredereck, H., Wagner, A., Kuhr, H., and Ott, H. (1960) *Chem. Ber.*, 93, 1201–1206.
38 Kadokawa, J., Nagaoka, T., Ebaba, J., Tagaya, H., and Chiba, K. (2000) *Carbohydr. Res.*, 327, 341–344.
39 Yen, Y.F., Kulkarni, S.S., Chang, C.W., and Luo, S.Y. (2013) *Carbohydr. Res.*, 368, 35–39.
40 Braccini, I., Derouet, C., Esnault, J., Herve du Penhoat, C., Mallet, J.-M., Michon, V., and Sinaÿ, P. (1993) *Carbohydr. Res.*, 246, 23–41.
41 Ratcliffe, A.J. and Fraser-Reid, B. (1990) *J. Chem. Soc. Perkin Trans 1*, 747–749.
42 Schmidt, R.R., Behrendt, M., and Toepfer, A. (1990) *Synlett*, 11, 694–696.
43 Tsuda, T., Nakamura, S., and Hashimoto, S. (2004) *Tetrahedron*, 60, 10711–10737.
44 Mong, K.-K.T., Yen, Y.-F., Hung, W.-C., Lai, Y.-H., and Chen, J.-H. (2012) *Eur. J. Org. Chem.*, 15, 3009–3017.
45 Yang, F., Zhu, Y., and Yu, B. (2012) *Chem. Commun.*, 48, 7097–7099.
46 Chiao, C.-S., Lin, C.Y., Mulani, S., Hung, W.C., and Mong, K.K.T. (2011) *Chem. Eur. J.*, 17, 12193–12202.
47 Chao, C.-S., Yen, Y.-F., Hung, W.-C., and Mong, K.-K.T. (2011) *Adv. Synth. Catal.*, 353, 879–884.
48 Ghosh, B., Lai, Y.-H., Shih, Y.-Y., Pradhan, T.K., Lin, C.-H., and Mong, K.-K.T. (2013) *Chem. Asian J.*, 8, 3191–3199.
49 Crich, D. and Li, W.-J. (2007) *J. Org. Chem.*, 72, 7794–7797.
50 Pradhan, T.K., Lin, C.C., and Mong, K.-K.T. (2014) *Org. Lett.*, 16, 1474–1477.
51 Mannerstedt, K., Ekelof, K., and Oscarson, S. (2007) *Carbohydr. Res.*, 342, 631–637.

52 (a) Hasegawa, A., Kurihara, N., Nishimura, D., and Nakajima, M. (1968) *Agric. Biol. Chem.*, 32, 1130–1134; (b) Hasegawa, A., Kurihara, N., Nishimura, D., and Nakajima, M. (1972) *Agric. Biol. Chem.*, 36, 1767–1772.
53 Eby, E. and Schuerch, C. (1974) *Carbohydr. Res.*, 34, 79–90.
54 Demchenko, A.V., Stauch, T., and Boons, G.J. (1997) *Synlett*, 7, 818–820.
55 Koshiba, M., Suzuki, N., Arihara, R., Tsuda, T., Nambu, H., Nakamura, S., and Hashimoto, S. (2008) *Chem. Asian J.*, 3, 1664–1677.
56 Review of ionic liquid: Ghandi, K. (2014) *Green Sustainable Chem.*, 4, 44–53.
57 Sasaki, K., Nagai, H., Matsumura, S., and Toshima, K. (2003) *Tetrahedron Lett.*, 44, 5605–5608.
58 Rencurosi, A., Lay, L., Russo, G., Caneva, E., and Poletti, L. (2005) *J. Org. Chem.*, 70, 7765–7768.
59 Zhang, J. and Ragauskas, A. (2006) *Beilstein J. Org. Chem.*, 2 (12). doi: 10.1186/1860-5397-2-12
60 Rencurosi, A., Lay, L., Russo, G., Caneva, E., and Poletti, L. (2006) *Carbohydr. Res.*, 341, 903–908.
61 Noyori, R. and Kurimoto, I. (1986) *J. Org. Chem.*, 51, 4320–4322.
62 Amatore, C., Jutand, A., Mallet, J.-M., Meyer, G., and Sinaÿ, P. (1990) *J. Chem. Soc., Chem. Commun.*, 718–719.
63 Mallet, J.-M., Meyer, G., Yvelin, F., Jutand, A., Amatore, C., and Sinaÿ, P. (1993) *Carbohydr. Res.*, 244, 237–246.
64 Nokami, T., Shibuya, A., Tsuyama, H., Suga, S., Bowers, A.A., Crich, D., and Yoshida, J. (2007) *J. Am. Chem. Soc.*, 129, 10922–10928.
65 Satoh, H., Hansen, H.S., Manabe, S., van Gunsteren, W.F., and Hunenberger, P.H. (2010) *J. Chem. Theory Comput.*, 6, 1783–1797.

Part II

Stereocontrolled Approaches to Glycan Synthesis

4

Intramolecular Aglycon Delivery toward 1,2-*cis* Selective Glycosylation

Akihiro Ishiwata and Yukishige Ito

4.1 Introduction

1,2-*cis* Glycosidic linkages, such as β-mannopyranoside and α-glucopyranoside, are often found in natural glycans such as glycoproteins, glycolipids, proteoglycans, microbial polysaccharides, and bioactive natural products. Unlike 1,2-*trans* isomers, which can be synthesized through neighboring group participation of an acyl group at the C-2 position, the formation of 1,2-*cis* glycosides is far less straightforward [1]. The key factors that control the stereoselectivity of glycosylation are largely understood; however, completely selective formation of these 1,2-*cis* glycosides is generally difficult [2] especially in the case of 1,2-*cis*-β (equatorial) linkage formation such as β-mannopyranosylation [3].

Among a number of strategies [4] explored for the synthesis of 1,2-*cis* glycosides, approaches based on intramolecular aglycon delivery (IAD) (Figure 4.1) are of special promise. Stereoselectivity of glycosylation through IAD may well be independent of the structure of the acceptor allowing the exclusive formation of 1,2-*cis* glycosides under kinetic control [5]. In the synthesis of complex molecules, tethering is an essential step to perform intramolecular reactions in a regio- and stereoselective manner [6]. In this context, various mixed acetal linkages have been developed for IAD, and a number of leaving groups on glycosyl donors have been examined.

In the case of IAD with 2-axial-oriented substrates, for example, β-mannopyranosylation, the stereochemical outcome of IAD is clear since the pathway producing the corresponding α-glycosides is essentially prohibited. Namely, the axial orientation of the tether at the C-2 position guarantees the formation of the β-glycoside. The pathway toward 1,2-*trans* furanosides through IAD is also disfavored, because it requires the intermediacy of a transfused [3,3,0] bicycle. On the other hand, the stereochemical outcome of two-equatorial-oriented 1,2-*cis* pyranosides, for example, α-gluco- or α-galactopyranoside formation as shown in Figure 4.1b for the naphthylmethyl (NAP) ether-mediated IAD, is less obvious [7].

Selective Glycosylations: Synthetic Methods and Catalysts, First Edition. Edited by Clay S. Bennett.

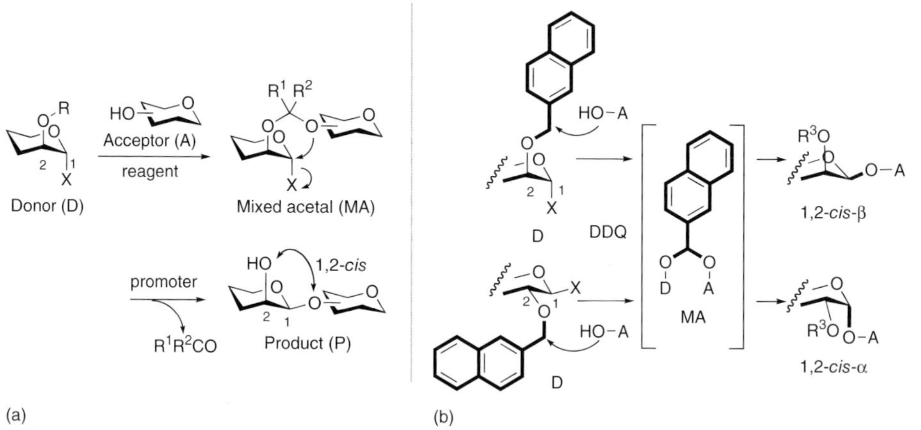

Figure 4.1 (a) Intramolecular aglycon delivery (IAD) and (b) NAP ether-mediated IAD (R^1 = Naph, R^2 = H).

4.2 Ketal Type Tethers

In 1991, an IAD approach to the β-mannoside was reported by Barresi and Hindsgaul [8]. Their strategy featured the initial formation of mixed dimethylketal **4** (O–CR_2–O, R = CH_3) under acidic conditions between aglycon **3** and isopropenyl ether **2**, which was prepared from acetate **1** with Tebbe's reagent (Figure 4.2a and Scheme 4.1). Subsequent thioglycoside activation by *N*-iodosuccinimide (NIS) in the presence of di-*t*-butylmethyl-pyridine (DTBMP) gave the corresponding disaccharide **5** in a highly stereocontrolled manner. The stereospecificity of the intramolecular transfer was strictly controlled by a kinetically favored approach fixed by the acetal tether. The mixed acetal (ketal) intermediates have been obtained using NIS as an alternative electrophile through iodoetherification of exomethylene compounds as reported by Fairbanks *et al.* [9]. The use of a 4-methoxybenzoate instead of an acetate was also examined as an alternative precursor of exomethylene [9b].

4.3 Silicon Tethers

Stork *et al.* [10] developed a strategy that employed silyl acetal (O–SiR_2–O, R = CH_3) as a tether for β-mannosylation (Figure 4.2b), which was first reported in 1992. The requisite silyl acetal was obtained in nearly quantitative yield from phenyl 1-thio-3,4,6-tri-*O*-benzyl-mannoside (**6**) and chlorodimethylsiloxy derivative **7**. After oxidation of the phenylthio group of **8**, the resulting sulfoxide **9** was subjected to activation with triflic anhydride and DTBMP [11] (0.05 M in CH_2Cl_2) to give β-mannopyranoside **10** (Scheme 4.2).

The dimethylsilyl-acetal-based IAD approach was applied to the practical synthesis of 1,2-*cis*-α-glucoside and α-galactoside structures [12]. Recently, Montgomery *et al.* reported an approach to synthesize dimethylsilyl acetal using chlorodimethylsilane [13], which was applied to the stereoselective glycosylations through IAD. As another silyl acetal linkage, diisopropylsilaketal (Figure 4.2b, R = $CH(CH_3)_2$) has been developed

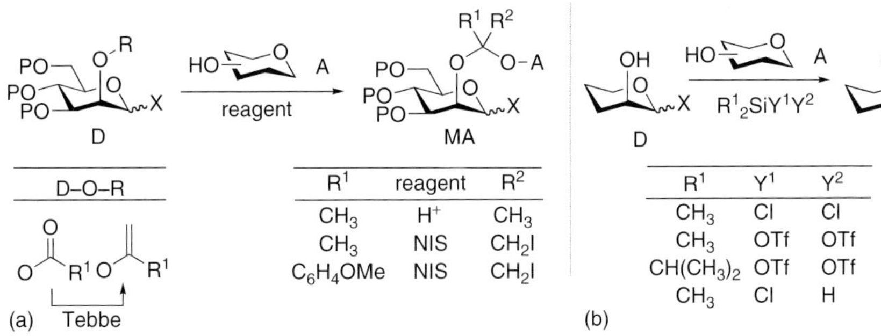

Figure 4.2 (a) Ketal-tethered MA for IAD and (b) silicon-tethered MA for IAD.

and applied to the introduction of the β-glycoside of D-mycosamine [14]. In addition, Bols *et al.* have examined the scope of silyl-acetal-based strategy using long-range intramolecular glycosylation (Scheme 4.3). IAD of 3-*O*-tethered glucosyl donor (MA **11**) with a simple octanol as an acceptor was not completely selective (**12**α:**12**β = 1 : 4). On the other hand, IAD of 5-*O*-tethered ribofuranosyl donor (MA **13**) afforded the *cis*-transferred adduct **14** with complete stereoselectivity [12e].

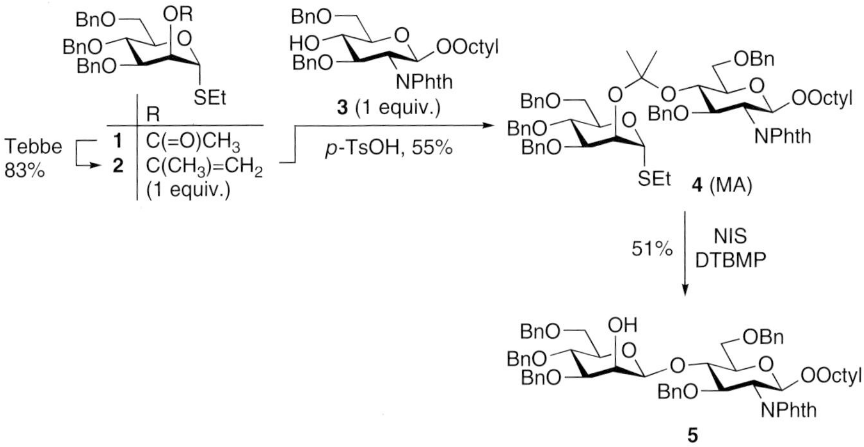

Scheme 4.1 Ketal-tethered MA for IAD.

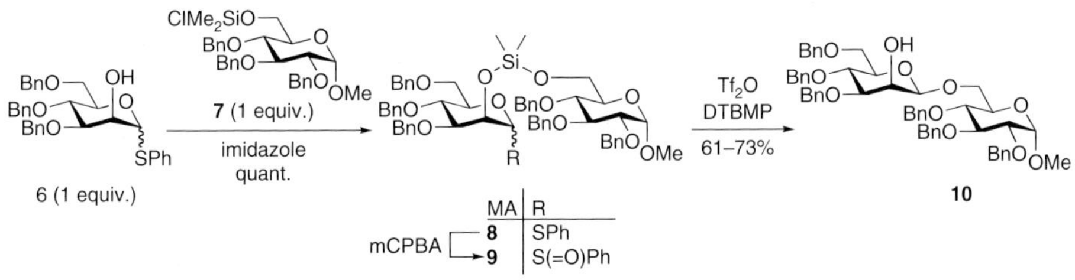

Scheme 4.2 Silicon-tethered MA for IAD.

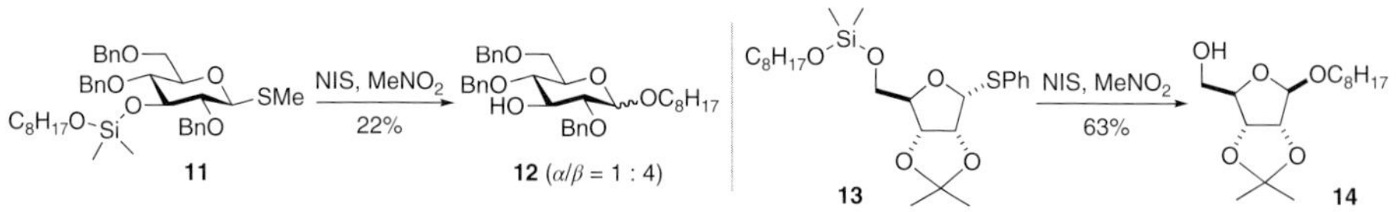

Scheme 4.3 IAD through dimethylsilaketal-tethered intermediates.

4.4 2-Iodoalkylidene Acetals as Tether

In order to maximize the efficiency of the bridged intermediate formation, the intermediacy of a less hindered alkylidene acetal (O–CHR–O) was extensively tested by Fairbanks *et al.* (Figure 4.3). They first investigated the use of an allyl ether [15] as the precursor of the enol ether, which was converted into the acetal with NIS or iodonium dicollidine triflate. To achieve this, the allyl ether **15** was converted to 1-propenyl ether **16** by treatment with $(Ph_3P)_3RhCl$-*n*BuLi (Scheme 4.4a). The iodoethylidene mixed acetal **18** was prepared from **16** with **17** in 80% yield, which in turn was subjected to IAD with I_2-AgOTf in the presence of DTBMP. The corresponding disaccharide β-Man-(1→4)-GlcNAc **19** was obtained in 66% yield [15c]. A combination of I_2-AgOTf in the presence of 2,6-di-*tert*-butyl-4-methylpyridine (DTBMP) was found to be optimal for mixed acetal formation [15d], while the subsequent glycosylation was best achieved by Me_2S_2, Tf_2O, and DTBMP [16]. The allyl-ether-mediated IAD was applied to the synthesis of β-rhamnopyranoside [17a], α-glucofuranoside [17a], and α-Glc-(1→2)-α-Glc-(1→3)-α-Glc-(1→3)-Man [17b,c]. This latter compound corresponds to the non-reducing terminal structure of triglucosylated high-mannose-type oligosaccharide $Glc_3Man_9GlcNAc_2$, a biosynthetic precursor of *N*-glycans.

Next, vinyl ethers as a precursor to a tether with minimum steric hindrance were examined (Figure 4.3) [18]. As shown in Scheme 4.4b, the vinyl ether **21** could be prepared from alcohol **20** by transvinylation [19]. This compound was then linked with an acceptor **22** to afford the 2-iodoethylidene acetal intermediate **23**. The resultant mixed acetal (MA) **23** was activated with I_2-AgOTf in the presence of base to give the corresponding disaccharide **24**.

Finally, the use of a propargyl ether as a precursor for MA formation was also examined (Scheme 4.4c) [20]. Propargyl ether of thiomannoside **25** was isomerized to allenyl

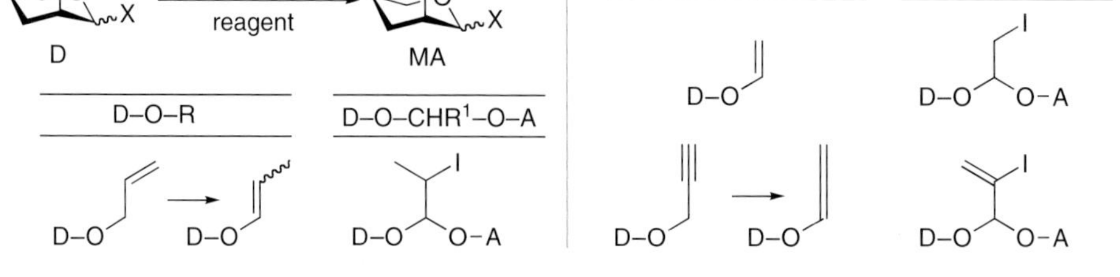

Figure 4.3 2-Iodoalkylidene-tethered intermediate for allyl-, vinyl-, and propargyl-ether-mediated IAD.

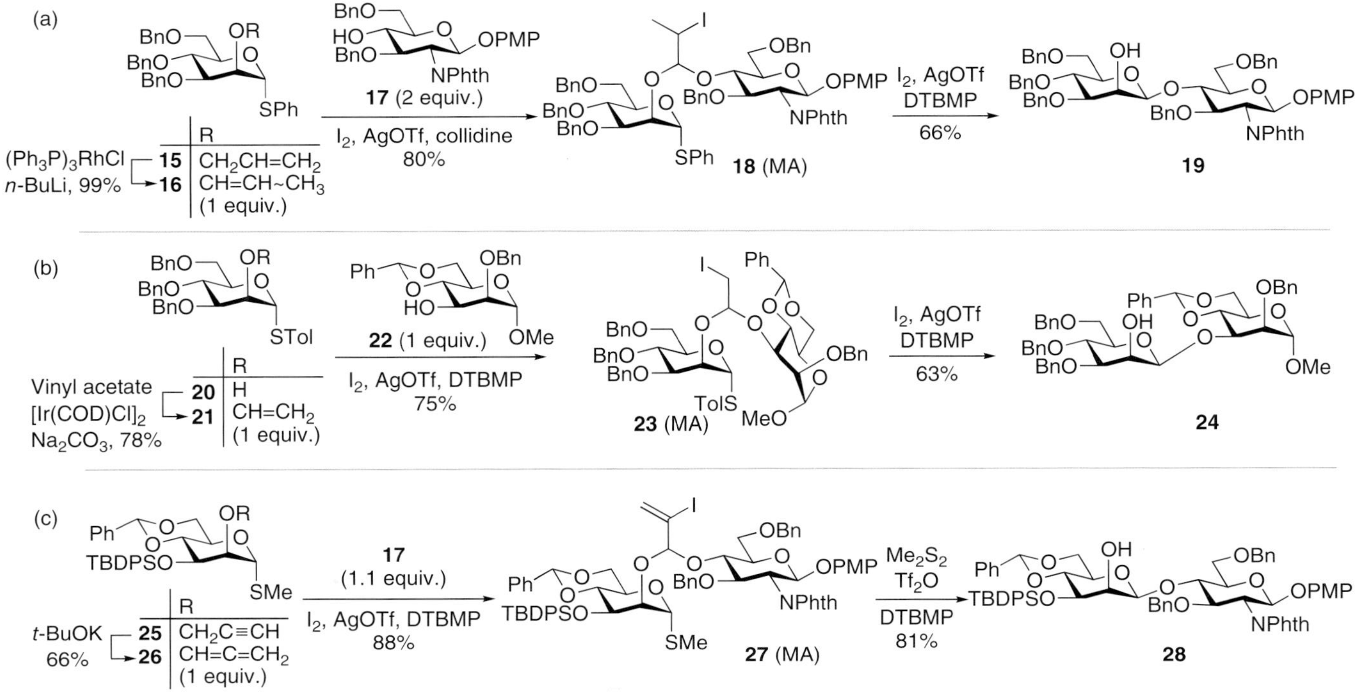

Scheme 4.4 β-Mannosylation through (a) allyl-, (b) vinyl-, and (c) propargyl-ether-mediated IADs.

ether **26**, which was then converted to the 2-iodo-2-propenylidene mixed acetal **23** by reaction with acceptor **17**. Subsequent glycosylation gave β-Man-(1 → 2)-GlcNAc disaccharide **28**, which was successfully converted to *N*-linked glycan core pentasaccharide $Man_3GlcNAc_2$.

4.5 Benzylidene Acetals as Tether

In 1994, Ito and Ogawa developed an alternative approach using *p*-methoxybenzylidene acetal (**O–CHR–O, R** = aryl) as the MA intermediate. This compound was obtained from the corresponding *p*-methoxybenzyl (PMB) ether (Figure 4.4) [21]. Treatment of the donor–acceptor mixture with DDQ cleanly afforded the mixed acetal. Subsequent activation of the anomeric position of donor moiety (e.g., thioglycoside [22] or fluoride [23]) gave the desired β-mannopyranoside. In addition to detailed mechanic studies [23], the efficiency of the PMB-assisted β-mannosylation was reported to be optimum when a 4,6-*O*-cyclic protective group was present on the donor (**30–33**). By comparison, the use of donor **29**, which did not possess the constraining protecting group, led to a diminished yield of the desired anomer (Scheme 4.5) [23c]. Notably, the practicality of the PMB-IAD approach was demonstrated in the context of synthetic studies toward biologically important glycoprotein glycans such as high mannose [24] and complex [25] type *N*-glycans. Oligomannosyl donors could be used for fragment coupling to afford core structures of *N*-glycans [22c] whose conversion to bisecting and fucose-containing $Fuc_1Man_3GlcNAc_2$ [22b] and a glycosphingolipid from *Protostomia phyla* [26] were also reported.

The approach has also been extended to polymer-supported synthesis. When thiomannoside was connected to polyethylene glycol (PEG, MW ~ 5000) [27] via a *p*-alkoxybenzyl linker (**34**), the mixed acetal **35** was isolated easily by precipitation and unreacted acceptor was removed by filtration at this stage (Scheme 4.6). After IAD, the resultant β-mannoside **36** was specifically released into the nonpolymeric phase.

The PMB-mediated IAD has been widely adopted by the synthetic community for the construction of difficult glycoside linkages. Examples include its application to the stereoselective constructions of various 1,2-*cis* linkages of β-Man [28], Gal*p* [29], fructofuranoside (Fru*f*) [30], arabinofuranoside (Ara*f*) [31], and fucofuranoside (Fuc*f*) [32]. Furthermore, a variant of this approach, the dimethoxybenzyl (DMB)-ether-mediated IAD, was effectively applied to the stereospecific synthesis of (1 → 1)-disaccharides. Examples include the construction of targets such as glucosyl- (**40**), galactosyl-, and

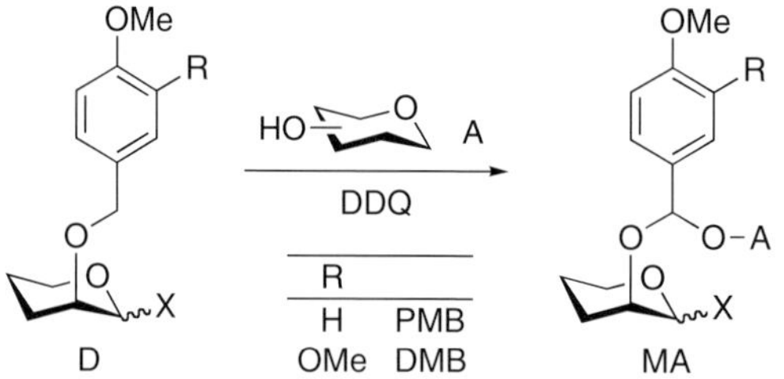

Figure 4.4 PMB and DMB ether-mediated IAD.

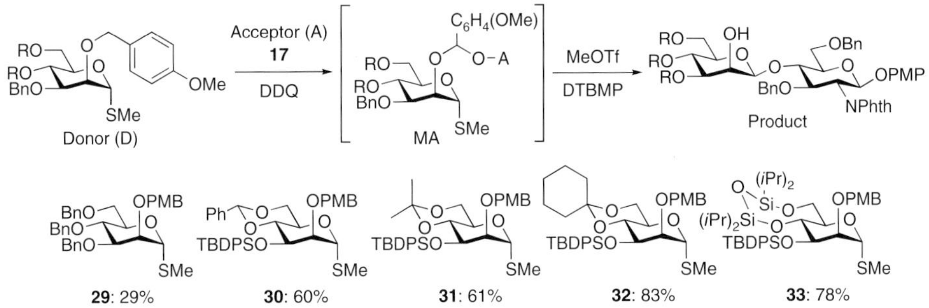

Scheme 4.5 Effect of protective groups on PMB-mediated IAD with **46**.

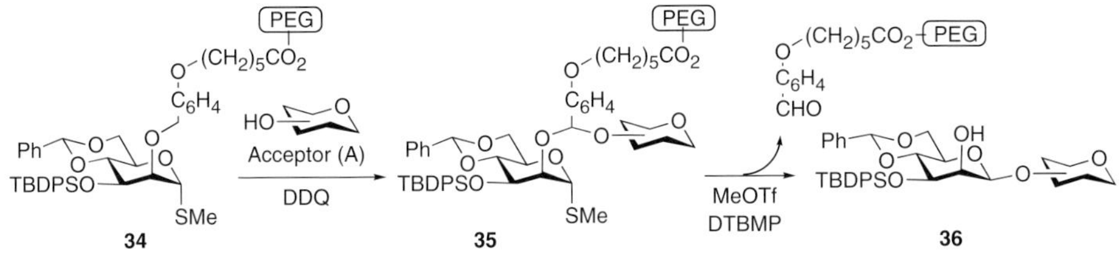

Scheme 4.6 Polymer-supported version of PMB-mediated IAD.

xylosyl-trehaloses through MAs obtained from the reaction of **37** with **38**, **41**, and **42**, respectively (Scheme 4.7) [33a]. This approach permitted the rapid construction of the *Mycobacterium tuberculosis* sulfolipid-1 containing glucosyl-trehalose [33b].

As another variant of the PMB-based strategy, recent studies have focused on the use of the 2-NAP-ether-mediated IAD (Figure 4.1b) [9]. The formation of naphthylidene acetal **44** with 2-*O*-NAP-protected thiomannoside **43** and the acceptor **17** proceeded quantitatively. Subsequent IAD mediated by MeOTf–DTBMP cleanly gave β-Man*p*, which was isolated as **45** after acetylation in 90% yield over three steps (Scheme 4.8). 4,6-*O*-Cyclic protection, shown to be required for achieving the highest efficiency in the PMB-mediated IAD, was not required for NAP-IAD. This permitted the synthesis of β-Man*p*-Xyl minimum structural repeat motif **48** through the reaction between **46** and acceptor [**47**] (Scheme 4.8) [34].

The power of the NAP MA approach is illustrated by the reaction of **46** with a chitobiose acceptor **49**, which has been a challenging acceptor for IAD without using a cyclic-acetal-protected donor. Furthermore, even with the acetal, not all classes of IAD are able to deliver this target in good yield. This problem was partially solved using PMB acetal-protected donor **32**, possessing a cyclic-protecting group. When the reaction was carried out using NAP-protected donor **46**, however, it proceeded without difficulty, giving the desired core trisaccharide derivative ($Man_1GlcNAc_2$, **50**) in a highly satisfactory yield of 83% (Table 4.1).

The β-Ara*f* linkages have also been stereoselectively constructed using 1,2-*cis* induction from IAD to afford both D- and L-Ara*f* derivatives from mycobacterial arabinan [35] and plant glycoproteins [36], respectively. For example, two β-Ara*f* moieties have been effectively introduced at the nonreducing terminal end of a mycobacterial arabinan diol using NAP-IAD. Larger furanoside donors can also be used in the reaction.

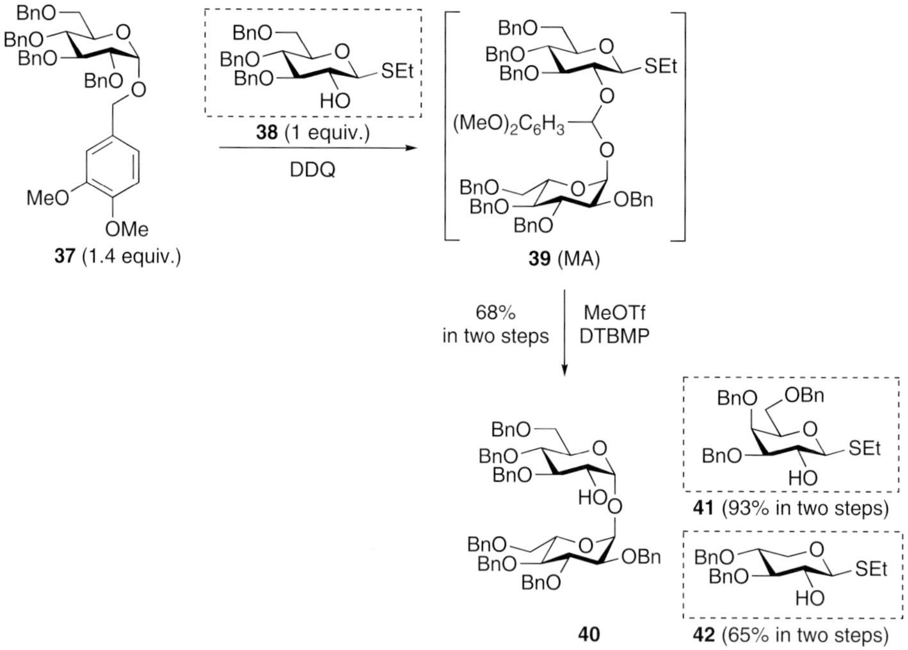

Scheme 4.7 DMB-mediated IAD for β-Man*p*.

An example is the synthesis of the plant extensin hydrophilic, repeating motif α-L-Ara*f*-(1→3)-β-L-Ara*f*-(1→2)-β-L-Ara*f*-(1→2)-β-L-Ara*f*-Hyp [36c]. Fragment coupling using NAP-IAD between disaccharide donor **52** and NAP-protected acceptor **51** through MA **53** has been also achieved for the stereoselective synthesis of derivative **54** in good overall yield (Scheme 4.9).

For 1,2-*cis*-α-Glc formation, the mixed acetal formations from both 2-*O*-NAP-protected Glc donor **55a** and Glc acceptor **56a** (Approach A) and from 2-*O*-unprotected donor **55b** and *O*-NAP-protected acceptor **56b** (Approach B) (Scheme 4.10) were examined. Both approaches A and B are basically possible for all IAD reactions except in the special cases where the stereochemistry of mixed acetal might affect the intramolecular transfer. The subsequent intramolecular glycosylation of mixed acetal **57** afforded the desired 1,2-*cis* glycoside, which was isolated as pentaacetate **58a** after acidic treatment and acetylation. Addition of silyl hydrides, especially $(TMS)_3SiH$, was found to be effective to reduce the benzylic cation of the IAD product, giving the glycoside product having both 2-*O*-NAP and 4,6-*O*-cyclohexylidene protection (**58b**), which was then deprotected to afford **59**. These findings extended the scope of the NAP-IAD, as exemplified by the synthesis of α-Glc-(1→2)-α-Glc-(1→3)-α-Glc-(1→3)-Man **63** from **55c** and **60** through the intermediacy of **61** and **62** [37].

One of the most challenging glycosidic linkages to construct selectively is β-L-rhamnopyranoside (β-L-Rha). Most significantly, the direct glycosylation strategy developed for β-D-Man is not effective for β-L-Rha, because the 6-deoxy structure of Rha excludes the use of 4,6-*O*-cyclic (e.g., benzylidene) protection, which is essential for the stereoselective β-D-Man formation [38]. The application of NAP-mediated IAD to solve this problem was investigated. Formation of the mixed acetal from

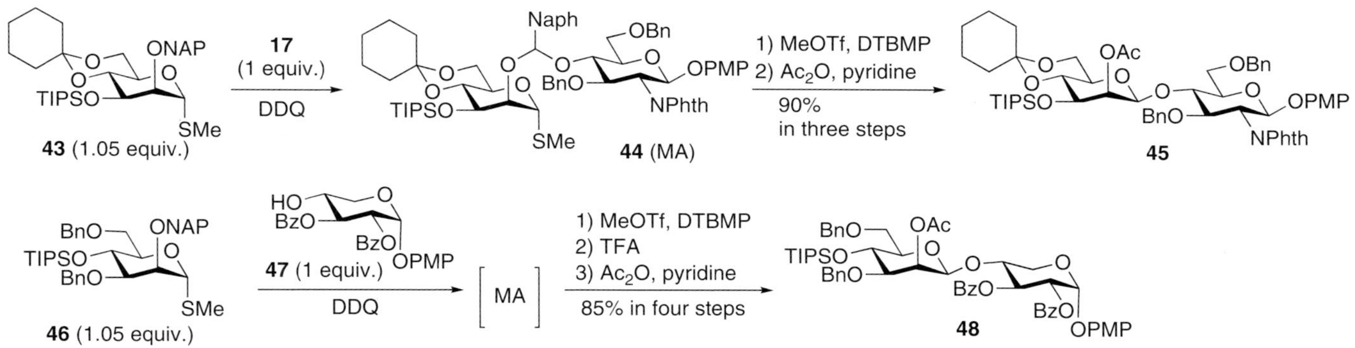

Scheme 4.8 NAP-mediated IAD for β-Man*p*.

Table 4.1 β-Mannosylation with chitobiose acceptors through IAD.

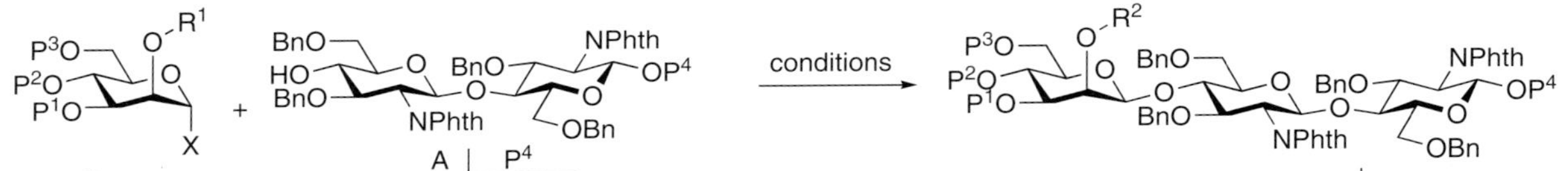

A	P^4
49a	octyl
49b	PMP

P	P^4	R^2
50a	octyl	H
50b	PMP	H
50c	PMP	Ac

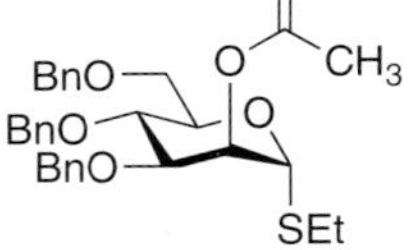

2

(1) TsOH
(2) NIS, DTBMP

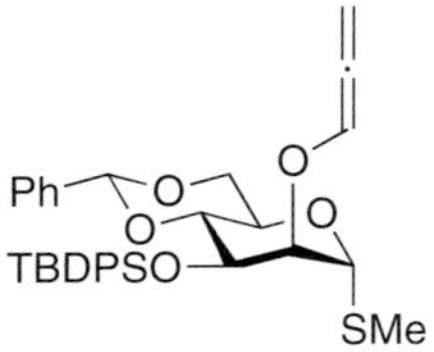

26

(1) I_2, AgOTf, DTBMP
(2) Me_2S_2, Tf_2O
DTBMP

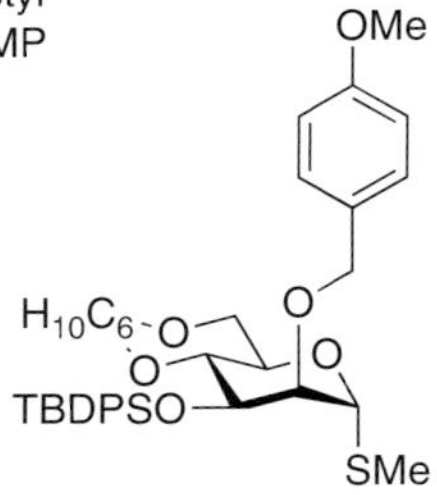

32

(1) DDQ
(2) MeOTf, DTBMP

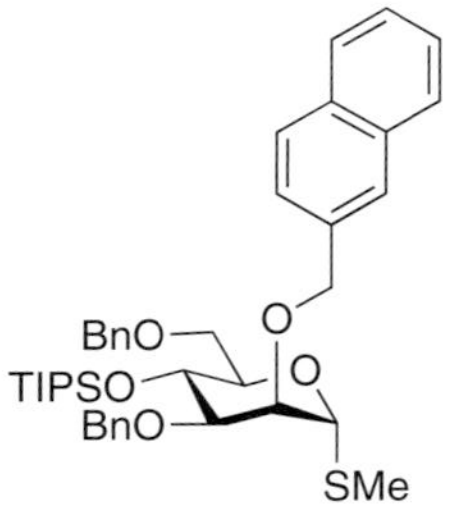

46

(1) DDQ
(2) (i) MeOTf, DTBMP
(ii) TFA
(iii) Ac_2O, pyridine

Entry	A	D	P	Yield (%)
1	**49a**	**2**	**50a**	10
2	**49b**	**26**	**50b**	22
3	**49b**	**32**	**50b**	85
4	**49b**	**46**	**50c**	83

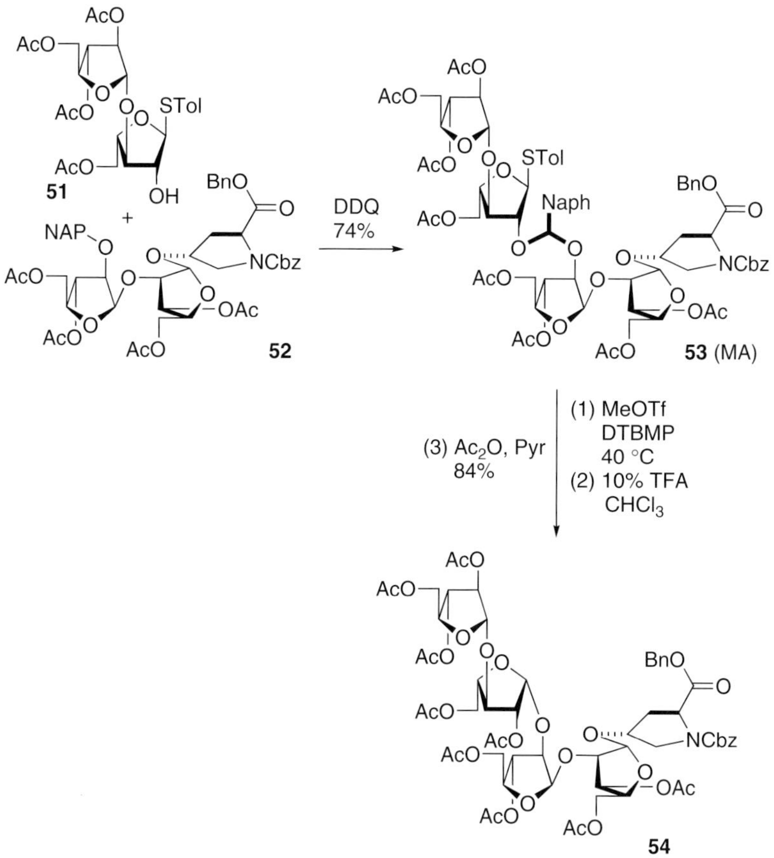

Scheme 4.9 NAP-mediated IAD for β-Ara*f*.

2-*O*-NAP-protected L-Rha donor **64a** and Glc acceptor **65**, followed by IAD under our standard conditions, afforded β-L-rhamnosides **66a** in 72% yield as a single isomer after acidic treatment and acetylation (Scheme 4.11) [39]. When 3-*O*-TMS-protected donor **64b** was used for NAP-IAD, the naphthylidene cyclic acetal **66b** with endo stereochemistry was obtained in 68% yield. In order to approach α-L-Rha*p*-(1→3)-β-L-Rha*p*-(1→4)-Glc, the substructure of *Sphaerotilus natans* polysaccharide, β-L-Rha*p* naphthylidene acetal of **66b,** was regioselectively opened with DIBAL-H to liberate the 3-OH **67** of β-L-Rha*p*.

The approach has been extended to direct construction of β-D-manno-heptosides. In many cases, the use of a 4,6-benzylidine acetal is not compatible with the target structures, which do not possess oxygenation at C6. This makes them another class of structures that are difficult to access directly; however, IAD offers a potential solution to this problem. For example, Kenfack *et al.* reported the direct stereoselective synthesis of the repeating unit of homopolymeric capsular polysaccharide consisting of β-D-manno-heptoside **71** from *Burkholderia pseudomallei* and *Burkholderia mallei* using NAP-mediated IAD approach from **71** and **69** through **70** (Scheme 4.12) [40].

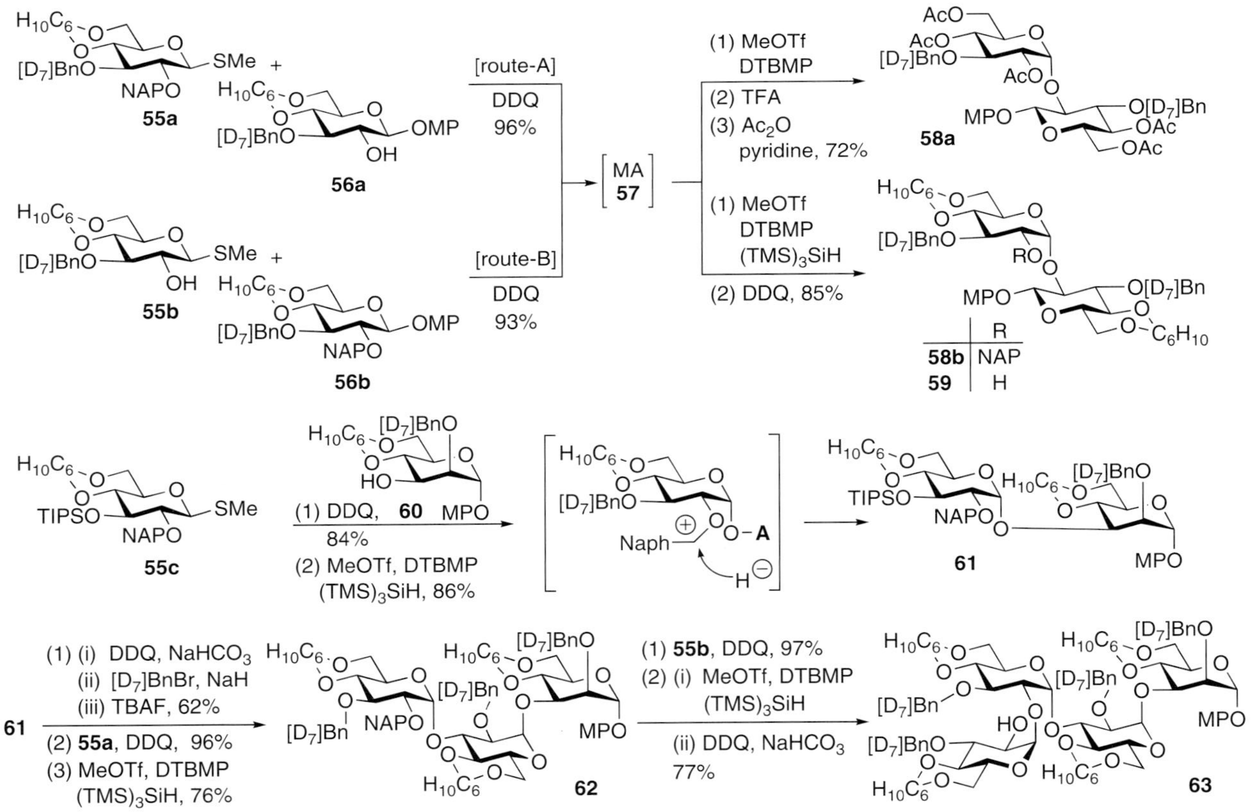

Scheme 4.10 NAP-mediated IAD for α-Glc*p*.

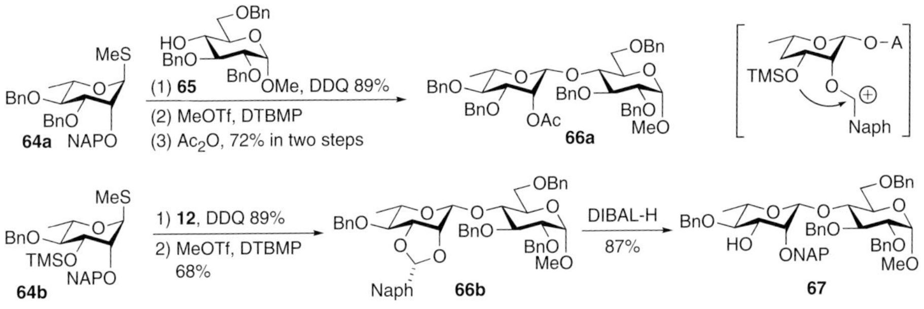

Scheme 4.11 NAP-mediated IAD for β-L-Rhap.

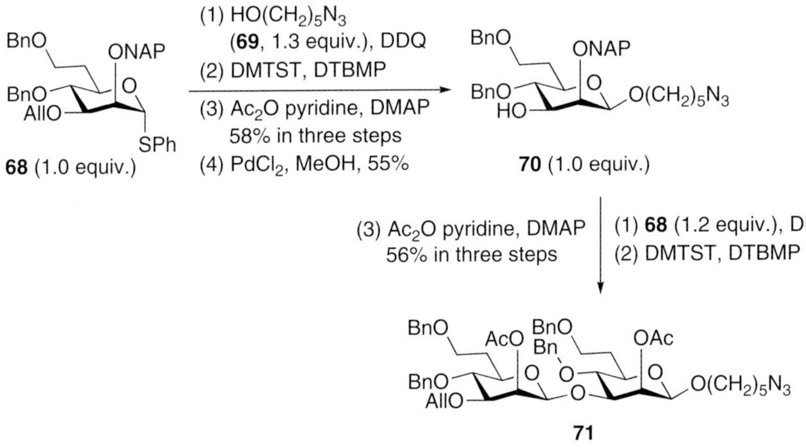

Scheme 4.12 NAP-mediated IAD for β-D-manno-heptoside.

4.6 IAD through Hemiaminal Ethers

Another class of glycosides that are difficult to synthetize are the 1,2-*cis*-α-2-deoxy-2-amino sugars commonly found in many glycoproteins and carbohydrate antigens. The use of IAD provides a unique opportunity to provide access to these structures. For example, the Knapp group demonstrated the first example of applying IAD as an alternative route for the stereoselective synthesis of α-linked D-glucosamine derivatives **75** from *M. tuberculosis*. In this approach, they used tethered donor **73** and acceptor **72** to afford MA aminal **74** (Scheme 4.13). Activation of the MA under standard conditions then afforded the target **75** in excellent yield as a single isomer [41].

4.7 Conclusions

The IAD methodology through various mixed acetal linkages has been extensively studied, due to its potential to provide a solution for stereoselective 1,2-*cis* glycoside formations. These approaches have enabled the stereospecific construction of various

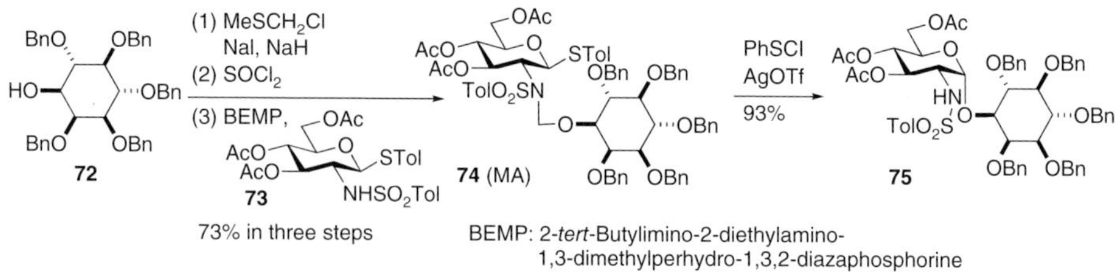

Scheme 4.13 Aminal-tethered MA for IAD to give α-linked D-glucosamine derivative.

1,2-*cis* glycosidic linkages such as β-mannopyranoside as well as other linkages including β-L-rhamnoside and β-D-manno-heptoside almost always with consistently high levels of selectivity. These methodologies have also been applied successfully to the construction of complex iterative 1,2-*cis* linkages such as the α-Glc-(1 → 2)-α-Glc-(1 → 3)-α-Glc-(1 → 3)-Man and α-L-Ara*f*-(1 → 3)-β-L-Ara*f*-(1 → 2)-β-L-Ara*f*-(1 → 2)-β-L-Ara*f*-Hyp derivatives. Finally, the approach is not limited to donors possessing oxygenation at C-2 and has been used for the stereoselective synthesis of 2-deoxy-2-amino sugars. Although tethering and intramolecular glycosylation procedure for IAD is indirect, it is one of the most reliable methods to obtain 1,2-*cis* glycoside linkages and has an excellent track record of being used to synthesize various biologically active and complex glycans.

References

1 (a) Toshima, K. and Tatsuta, K. (1993) *Chem. Rev.*, 93, 1503–1531; (b) Schmidt, R.R. (1986) *Angew. Chem. Int. Ed. Engl.*, 25, 212–235.

2 (a) Demchenko, A.V. (2003) *Synlett*, 1225–1240; (b) Mydock, L.K. and Demchenko, A.V. (2010) *Org. Biomol. Chem.*, 8, 497–510.

3 For example, see; Tsuda, T., Arihara, R., Sato, S., Koshiba, M., Nakamura, S., and Hashimoto, S. (2005) *Tetrahedron*, 61, 10719–10733.

4 (a) Ernst, B., Hart, G.W., and Sinaÿ, P. (eds) (1999) Carbohydrates in Chemistry and Biology, vol. 1 & 2, Wiley-VCH Verlag GmbH, Weinheim; (b) Fraser-Reid, B., Tatsuta, K., and Thiem, J. (eds) (2001) Glycoscience, I-III, Springer, Berlin.

5 For reviews, see; (a) Jung, K., Müller, M., and Schmidt, R.R. (2000) *Chem. Rev.*, 100, 4423–4442; (b) Gridley, J.J. and Osborn, M.I. (2000) *J. Chem. Soc., Perkin Trans. 1*, 1471–1491; (c) Davis, B.G. (2000) *J. Chem. Soc., Perkin Trans. 1*, 2137–2160; (d) Cumpstey, I. (2008) *Carbohydr. Res.*, 343, 1553–1573; (e) Carmona, A.T., Moreno-Vargas, A.J., and Robina, I. (2008) *Curr. Org. Synth.*, 5, 33–63; (f) Ishiwata, A. and Ito, Y. (2009) *Trends Glycosci. Glycotechnol.*, 21, 266–289; (g) Ishiwata, A., Lee, Y.J., and Ito, Y. (2010) *Org. Biomol. Chem.*, 8, 3596–3608; (h) Goto, K. (2010) *Trends Glycosci. Glycotechnol.*, 22, 207–208; (i) Ishiwata, A. and Ito, Y. (2012) *J. Synth. Org. Chem. Jpn.*, 70, 382–394.

6 (a) Cox, L.R. and Ley, S.V. (1999) in Templated Organic Synthesis (eds F. Diederich and P.J. Stang), Wiley-VCH Verlag GmbH, Weinheim, pp. 275–395; (b) Bols, M. and Skrydstrup, T. (1995) *Chem. Rev.*, 95, 1253–1277.

7 Ishiwata, A., Munemura, Y., and Ito, Y. (2008) *Eur. J. Org. Chem.*, 2008, 4250–4263.

8 (a) Barresi, F. and Hindsgaul, O. (1991) *J. Am. Chem. Soc.*, 113, 9376–9377; (b) Barresi, F. and Hindsgaul, O. (1992) *Synlett*, 1992, 759–761; (c) Barresi, F. and Hindsgaul, O. (1994) *Can. J. Chem.*, 72, 1447–1465.
9 (a) Ennis, S.C., Fairbanks, A.J., Tennant-Eyles, R.J., and Yeates, H.S. (1999) *Synlett*, 1999, 1387–1390; (b) Ennis, S.C., Fairbanks, A.J., Slinn, C.A., Tennant-Eyles, R.J., and Yeates, H.S. (2001) *Tetrahedron*, 57, 4221–4230.
10 (a) Stork, G. and Kim, G. (1992) *J. Am. Chem. Soc.*, 114, 1087–1088; (b) Stork, G. and La Clair, J.L. (1996) *J. Am. Chem. Soc.*, 118, 247–248.
11 Kahne, D., Walker, S., Cheng, Y., and Van Engen, D. (1989) *J. Am. Chem. Soc.*, 111, 6881–6882.
12 (a) Bols, M. (1992) *J. Chem. Soc., Chem. Commun.*, 913–914; (b) Bols, M. (1993) *J Chem. Soc., Chem. Commun.*, 791–792; (c) Bols, M. (1993) *Tetrahedron*, 49, 10049–10060; (d) Bols, M. (1996) *Acta Chem. Scand.*, 931–937; (e) Bols, M. and Hansen, H.C. (1994) *Chem. Lett.*, 1049–1052.
13 (a) Buchan, Z.A., Bader, S.J., and Montgomery, J. (2009) *Angew. Chem. Int. Ed.*, 48, 4840–4848; (b) Partridge, K.M., Bader, S.J., Buchan, Z.A., Taylor, C.E., and Montgomery, J. (2013) *Angew. Chem. Int. Ed.*, 52, 13647–13650.
14 Packerd, K. and Rychnovsky, S.D. (2001) *Org. Lett.*, 3, 3393–3396.
15 (a) Seward, C.M.P., Cumpstey, I., Aloui, M., Ennis, S.C., Redgrave, A.J., and Fairbanks, A.J. (2000) *Chem. Commun.*, 1409–1410; (b) Cumpstey, I., Fairbanks, A.J., and Redgrave, A.J. (2001) *Org. Lett.*, 3, 2371–2374; (c) Aloui, M., Chambers, D.J., Cumpstey, I., Fairbanks, A.J., Redgrave, A.J., and Seward, C.M.P. (2002) *Chem. Eur. J.*, 8, 2608–2621; (d) Cumpstey, I., Chayajarus, K., Fairbanks, A.J., Redgrave, A.J., and Seward, C.M.P. (2004) *Tetrahedron: Asymmetry*, 15, 3207–3221.
16 Tatai, J. and Fügedi, P. (2007) *Org. Lett.*, 9, 4647–4650.
17 (a) Cumpstey, I., Fairbanks, A.J., and Redgrave, A.J. (2004) *Tetrahedron*, 60, 9061–9074; (b) Fairbanks, A.J. (2003) *Synlett*, 1945–1958; (c) Attolino, E., Cumpstey, I., and Fairbanks, A.J. (2006) *Carbohydr. Res.*, 341, 1609–1618.
18 Chayajarus, K., Chambers, D.J., Chughtai, M.J., and Fairbanks, A.J. (2004) *Org. Lett.*, 6, 3797–3800.
19 Olimoto, Y., Sakaguchi, S., and Ishii, Y. (2002) *J. Am. Chem. Soc.*, 124, 1590–1591.
20 (a) Attolino, E. and Fairbanks, A.J. (2007) *Tetrahedron Lett.*, 48, 3061–3064; (b) Attolino, E., Ridsing, T.W.D.F., Heidecke, C.D., and Fairbanks, A.J. (2007) *Tetrahedron: Asymmetry*, 18, 1721–1734.
21 Ito, Y. and Ogawa, T. (1994) *Angew. Chem. Int. Ed. Engl.*, 33, 1765–1767.
22 (a) Dan, A., Ito, Y., and Ogawa, T. (1995) *J. Org. Chem.*, 60, 4680–4681; (b) Dan, A., Ito, Y., and Ogawa, T. (1996) *Carbohydr. Lett.*, 1, 469–474; (c) Dan, A., Lergenmüller, M., Amano, M., Nakahara, Y., Ogawa, T., and Ito, Y. (1998) *Chem. Eur. J.*, 4, 2182–2190; (d) Dan, A., Ito, Y., and Ogawa, T. (1995) *Tetrahedron Lett.*, 36, 7487–7490.
23 (a) Lergenmüller, M., Nukada, T., Kuramochi, K., Dan, A., Ogawa, T., and Ito, Y. (1999) *Eur. J. Org. Chem.*, 1999, 1367–1376; (b) Ito, Y., Ando, H., Wada, M., Kawai, T., Ohnishi, Y., and Nakahara, Y. (2001) *Tetrahedron*, 57, 4123–4132; (c) Ito, Y., Ohnishi, Y., Ogawa, T., and Nakahara, Y. (1998) *Synlett*, 1998, 1102–1104.
24 (a) Matsuo, I., Wada, M., Manabe, S., Yamaguchi, Y., Otake, K., Kato, K., and Ito, Y. (2003) *J. Am. Chem. Soc.*, 125, 3402–3403; (b) Matsuo, I. and Ito, Y. (2003) *Carbohydr. Res.*, 338, 2163–2168; (c) Matsuo, I., Kashiwagi, T., Totani, K., and Ito, Y. (2005) *Tetrahedron Lett.*, 46, 4197–4200; (d) Matsuo, I. and Ito, Y. (2005) *Trends Glycosci.*

Glycotechnol., 17, 85–95; (e) Matsuo, I., Totani, K., Tatami, A., and Ito, Y. (2006) *Tetrahedron*, 62, 8262–8277; (f) Koizumi, A., Matsuo, I., Takatani, M., Seko, A., Hachisu, M., Takeda, Y., and Ito, Y. (2013) *Angew. Chem. Int. Ed.*, 52, 7426–7431; (g) Fujikawa, K., Koizumi, A., Hachisu, M., Seko, A., Takeda, Y., and Ito, Y. (2015) *Chem. Eur. J.*, 21, 3224–3233.

25 (a) Seifert, J., Lergenmüller, M., and Ito, Y. (2000) *Angew. Chem. Int. Ed.*, 39, 531–534; (b) Ohnishi, Y., Ando, H., Kawai, T., Nakahara, Y., and Ito, Y. (2000) *Carbohydr. Res.*, 328, 263–276; (c) Nakano, J., Ohta, H., and Ito, Y. (2006) *Bioorg. Med. Chem. Lett.*, 16, 928–933; (d) Nakano, J., Ishiwata, A., Ohta, H., and Ito, Y. (2007) *Carbohydr. Res.*, 342, 675–695.

26 (a) Ohtsuka, I., Hada, N., Sugita, M., and Takeda, T. (2002) *Carbohydr. Res.*, 337, 2037–2047; (b) Ohtsuka, I., Hada, N., Ohtaka, H., Sugita, M., and Takeda, T. (2002) *Chem. Pharm. Bull.*, 50, 600–604.

27 Ito, Y. and Ogawa, T. (1997) *J. Am. Chem. Soc.*, 119, 5562–5566.

28 Gannedi, V., Ali, A., Singh, P.P., and Vishwakarma, R.A. (2014) *Tetrahedron Lett.*, 55, 2945–2947.

29 Bernlind, C., Homans, S.W., and Field, R.A. (2009) *Tetrahedron Lett.*, 50, 3397–3399.

30 (a) Krog-Jensen, C. and Oscarson, S. (1996) *J. Org. Chem.*, 61, 4512–4513; (b) Krog-Jensen, C. and Oscarson, S. (1998) *J. Org. Chem.*, 63, 1780–1784.

31 (a) Désiré, J. and Prandi, J. (1999) *Carbohydr. Res.*, 317, 110–118; (b) Bamhaoud, T., Sanchez, S., and Prandi, J. (2000) *Chem. Commun.*, 659–660; (c) Sanchez, S., Bamhaoud, T., and Prandi, J. (2000) *Tetrahedron Lett.*, 41, 7447–7452; (d) Marotte, K., Sanchez, S., Bamhauold, T., and Prandi, J. (2003) *Eur. J. Org. Chem.*, 2003, 3587–3598; (e) Shinohara, H. and Matsubayashi, Y. (2013) *Plant Cell Physiol.*, 54, 369–374; (f) Okamoto, S., Shinohara, H., Mori, T., Matsubayashi, Y., and Kawaguchi, M. (2013) *Nat. Commun.*, 4, 2191. doi: 10.1038/ncomms3191

32 Gelin, M., Ferriéres, V., Lefeuvre, M., and Plusquellec, D. (2003) *Eur. J. Org. Chem.*, 2003, 1285–1293.

33 (a) Pratt, M.R., Leigh, C.D., and Bertozzi, C.R. (2003) *Org. Lett.*, 5, 3185–3188; (b) Leigh, C.D. and Bertozzi, C.R. (2008) *J. Org. Chem.*, 73, 1008–1017.

34 Ishiwata, A., Sakurai, A., Nishimiya, Y., Tsuda, S., and Ito, Y. (2011) *J. Am. Chem. Soc.*, 113, 19524–19535.

35 Ishiwata, A. and Ito, Y. (2011) *J. Am. Chem. Soc.*, 133, 2275–2291.

36 (a) Kaeothip, S., Ishiwata, A., and Ito, Y. (2013) *Org. Biomol. Chem.*, 11, 5892–5907; (b) Kaeothip, S., Ishiwata, A., Ito, T., Fushinobu, S., Fujita, K., and Ito, Y. (2013) *Carbohydr. Res.*, 382, 95–100; (c) Ishiwata, A., Kaeothip, S., Takeda, Y., and Ito, Y. (2014) *Angew. Chem. Int. Ed.*, 53, 9812–9816.

37 (a) Ishiwata, A. and Ito, Y. (2005) *Tetrahedron Lett.*, 46, 3521–3524; (b) Ishiwata, A., Munemura, Y., and Ito, Y. (2008) *Tetrahedron*, 64, 92–102.

38 (a) Crich, D. and Sun, S. (1996) *J. Org. Chem.*, 61, 4506–4507; (b) Crich, D. and Sun, S. (1998) *J. Am. Chem. Soc.*, 120, 435–436; (c) Kim, K.S., Kim, J.H., Lee, Y.J., Lee, Y.J., and Park, J. (2001) *J. Am. Chem. Soc.*, 123, 8477–8481; (d) Baek, J.Y., Choi, T.J., Jeon, H.B., and Kim, K.S. (2006) *Angew. Chem. Int. Ed.*, 45, 7436–7440.

39 Lee, Y.J., Ishiwata, A., and Ito, Y. (2008) *J. Am. Chem. Soc.*, 130, 6330–6331.

40 Kenfack, M.T., Blériot, Y., and Gauthier, C. (2014) *J. Org. Chem.*, 79, 4615–4636.

41 Ajayi, K., Thakur, V.V., Lapo, R.C., and Knapp, S. (2010) *Org. Lett.*, 12, 2630–2633.

5

Chiral Auxiliaries in Stereoselective Glycosylation Reactions

Robin Brabham and Martin A. Fascione

5.1 Introduction

The synthesis of 1,2-*trans* glycosides can be achieved with complete stereoselectivity by virtue of neighboring group participation (NGP). In this strategy, following glycosyl donor activation, the resulting oxocarbenium ion can be intercepted by a nucleophilic moiety suitably close to the anomeric carbon, often at the O-2 position (Scheme 5.1) [1]. The intermediate dioxolanium species impedes 1,2-*cis* attack by the desired glycosyl acceptor, resulting in exclusive formation of the 1,2-*trans* glycoside.

Esters are most commonly used for this purpose, with the carbonyl oxygen atom being sufficiently nucleophilic to engage in NGP attack of the oxocarbenium ion, although amide carbonyls [2] and sulfides [3], among others, are also known to function effectively. The stereoselectivity of O-2 NGP makes 1,2-*trans* glycoside synthesis amenable to solid-phase oligosaccharide synthesis [4], as the lack of purification of intermediates in solid-phase synthesis means that all coupling steps must be highly stereoselective. Indeed, complex 1,2-*trans* oligosaccharides have been synthesized in this way [5]. However, the stereoselective synthesis of 1,2-*cis* glycosides is less straightforward with many synthetic strategies suffering from poor yields or anomeric contamination. Traditionally, the use of nonparticipating groups at O-2 is required to afford 1,2-*cis* glycosides, predominantly by virtue of the anomeric effect, but despite this established stereoelectronic preference [6], in practice, 1,2-*trans* glycosides are also formed.

5.2 Neighboring Group Participation of O-2 Chiral Auxiliaries

One of the most elegant and imaginative solutions to the problems posed by stereoselective glycoside synthesis was an O-2 chiral auxiliary-based approach first disclosed by Boons and coworkers in 2005 [7]. The authors hypothesized that a chiral participating/protecting group based on mandelic acid could be potentially used at the O-2 position of a glucose or galactose donor, to direct glycosyl acceptors to a particular face of the donor

Selective Glycosylations: Synthetic Methods and Catalysts, First Edition. Edited by Clay S. Bennett.

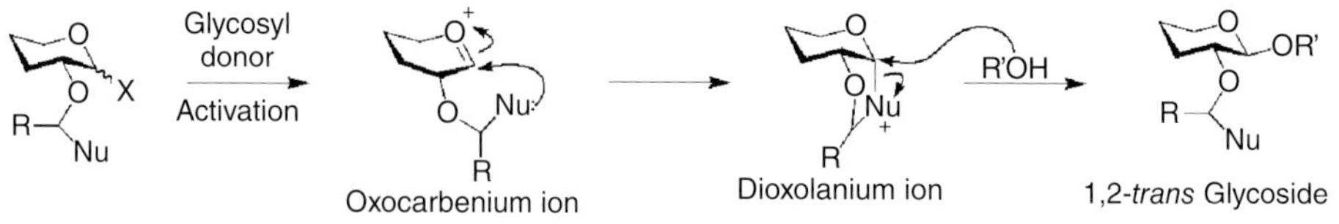

Scheme 5.1 Neighboring group participation (NGP) in the synthesis of 1,2-*trans* glycosides.

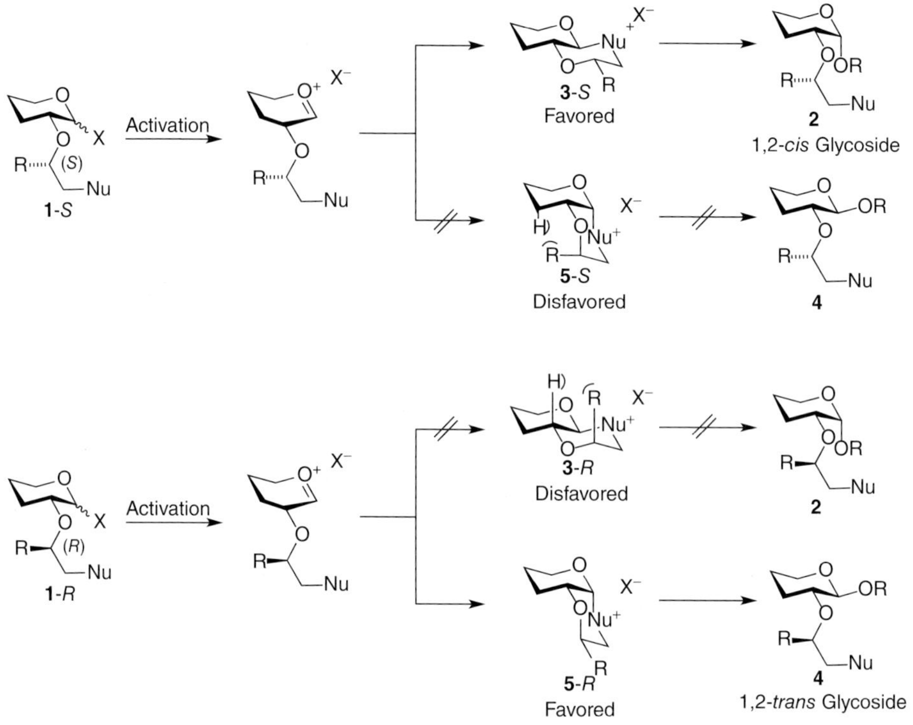

Scheme 5.2 Schematic summary of Boons' chiral auxiliary-based approach to stereoselective glycoside synthesis.

(1,2-*cis* or α-face (for D-sugars), or 1,2-*trans or* β-face (for D-sugars), Scheme 5.2). Depending on which diastereomeric donor **1-*S*** or **1-*R*** was used, the chiral auxiliary could either direct the formation of 1,2-*cis* glycosides **2,** via glycosylation of a "quasi-stable" *trans*-decalin intermediate **3-*S*** or **3-*R*,** or formation of 1,2-*trans* glycosides **4,** via glycosylation of a "quasi-stable" *cis*-decalin intermediate **5-*S*** or **5-*R*** (Scheme 5.2).

Both the *R*- or *S*-configured chiral auxiliaries were easily accessible through the use of readily available enantiopure ethyl mandelate **6*R*/*S*** and epoxide **7** (Scheme 5.3). Precursor **8*R*/*S*** was protected and activated in three routine steps, affording glycosyl donors **9*R*** and **9*S*** in about 40% cumulative yield.

Activation of *S*-configured trichloroacetimidate donor **9-*S*** with TMSOTf at −78 °C afforded excellent yields of disaccharides with a range of acceptors, including

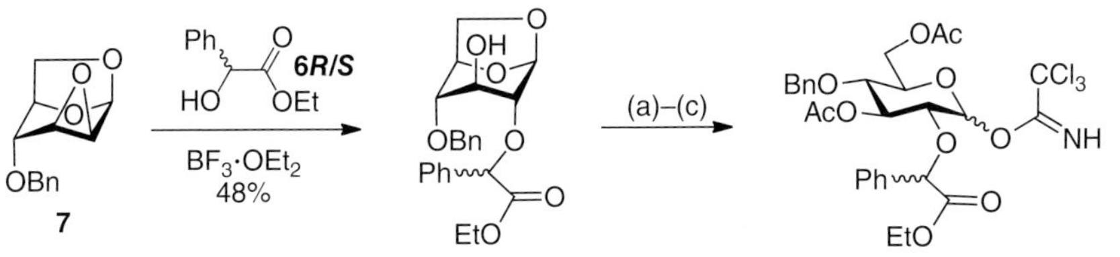

Scheme 5.3 Four-step synthesis of glycosyl donor **9-R/S**. (a) TMSOTf, Ac_2O (93%); (b) $H_2NNH_2{\cdot}HOAc$ (95%); and (c) Cl_3CCN/DBU (92%).

glycosylation of secondary alcohol **10** in 95% yield (Table 5.1). Impressively, stereoselective formation of 1,2-*cis*-α-glycosides was achieved across the range of acceptors presumably via the putative *trans*-decalin intermediate **3-*S***, although selectivity was reduced in the case of more reactive alcohols. Glycosylation of the analogous *R*-configured trichloroacetimidate donor afforded equally impressive yields of the disaccharides at low temperatures, with the caveat that the stereoselectivity achieved with the *R*-configured auxiliary was uniformly lower than that with the *S*-configured auxiliary. This less-pronounced selectivity is thought to result from decreased stability of the putative *cis*-decalin intermediate **5-*R*** due to inherent destabilizing 1,3-diaxial interactions. Further studies on the glycosylation reaction were carried out to optimize reaction conditions and protecting groups to maximize stereoselectivity, which revealed the importance of (i) diluting the reaction mixture and (ii) O-3 ester protecting groups in achieving high levels of stereoselectivity [8]. Dilution can be rationalized mechanistically as decreasing the likelihood of the oxocarbenium ion being intercepted by the glycosyl acceptor prior to dioxolanium ion formation, while the authors postulated that the electron-withdrawing ester-protecting group at O-3 may destabilize the oxocarbenium ion more than the O-3 ether-protecting groups and therefore favor glycosylation via decalin intermediates **3-*S*** and **5-*R***. Density functional theory (DFT) quantum mechanical calculations subsequently confirmed that the respective *trans*- and *cis*-decalin intermediates **3-*S*** and **5-*R*** are preferentially more stable compared to the corresponding oxocarbenium ion when O-3 is ester-protected [8]. This first-generation auxiliary was unfortunately hampered by both the limited nucleophilicity of the carbonyl oxygen atom in the ethyl mandelate auxiliary and the greater entropic penalty of associated six-membered dioxolanium ring formation compared to the five-membered counterpart [9]. Nevertheless, Boons' initial foray into this field revealed the potential of decalin-type structures in stereoselective glycosylations, particularly in the synthesis of 1,2-*cis* glycosides.

Boons' second-generation auxiliary attempted to improve on the ethyl mandelate ether by utilizing a (1*S*)-phenyl-2-(phenylsulfanyl) ethyl chiral auxiliary where the aryl sulfur moiety acted as the intercepting nucleophile *in lieu* of a carbonyl [10]. Synthetically, this strategy boasts a more facile installation via a convergent synthesis using the more readily accessible acetate-protected starting material **24Bn/Ac**, albeit requiring a four-step synthesis of the auxiliary precursor **23*S*** from ethyl mandelate enantiomer **20*S*** (Scheme 5.4). Lewis-acid-catalyzed episulfonium ion formation, followed by S_N2 attack by a free sugar hydroxyl group, permitted installation with retention of auxiliary

Table 5.1 Yields and anomeric stereoselectivity for glycosylations of donors **9-*R*/*S*** with acceptor alcohols **10–14** using TMSOTf in DCM at −78 °C to form glycosides **15–19**.

Acceptor	Product	Yield *R*	Yield *S*	α/β *R*	α/β *S*
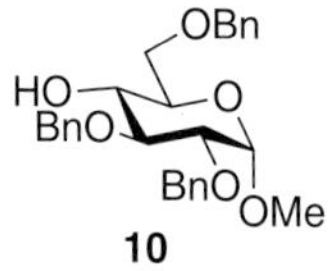 **10**	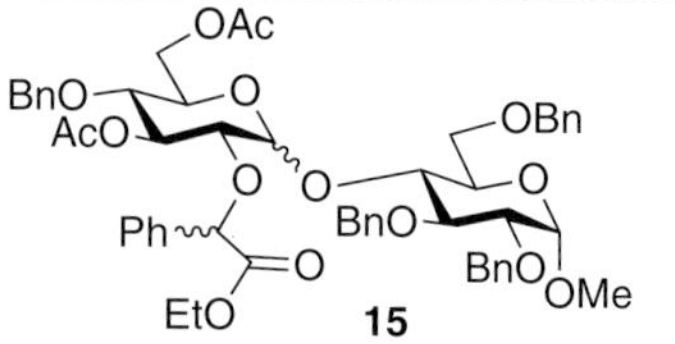 **15**	93	95	1:5	20:1
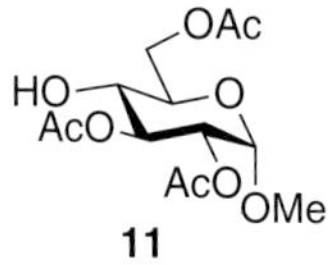 **11**	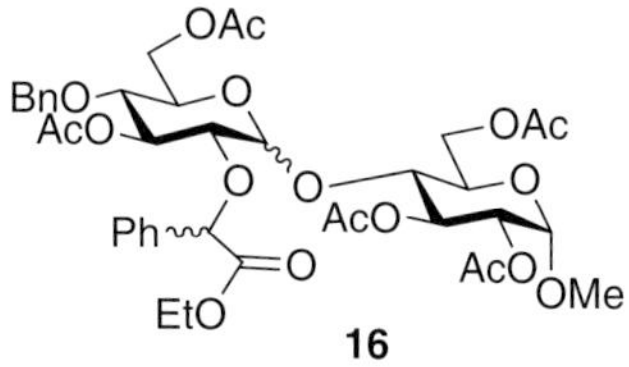 **16**	89	94	1:1	18:1
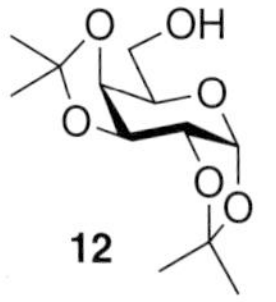 **12**	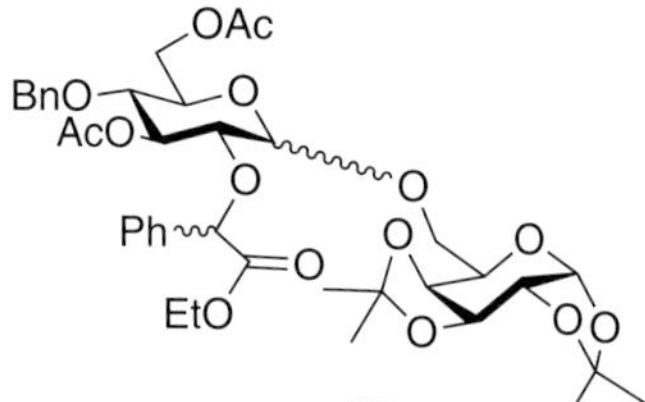 **17**	88	92	1:3	12:1
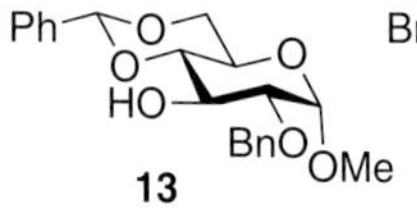 **13**	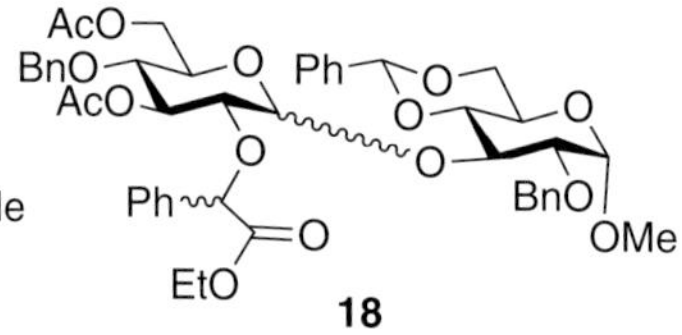 **18**	94	95	1:8	10:1
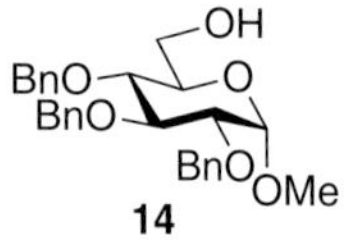 **14**	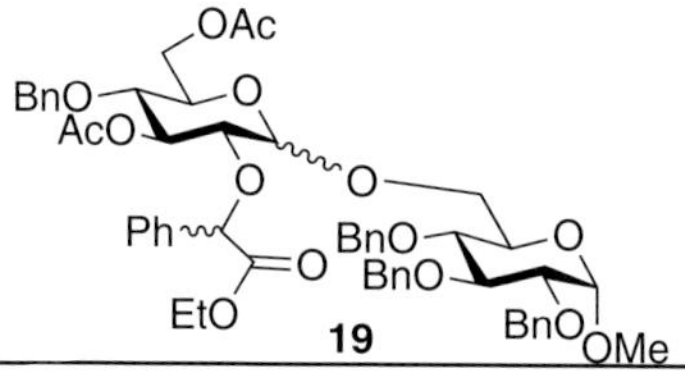 **19**	89	94	1:8	4:1

stereochemistry in good yield of **25Bn/Ac**, followed by standard protecting group chemistry to afford trichloroacetimidate donors **26Bn/Ac**.

Boons' and coworkers hypothesized that the chiral auxiliary could potentially intercept the oxocarbenium ion and form an arylsulfonium ion. Low-temperature ^{1}H-NMR studies of the glycosylation reaction using **26-Bn** subsequently confirmed the exclusive formation of this intermediary *trans*-decalin sulfonium ion **27,** driven by the *S*-configuration of the chiral auxiliary, which then reacts on the addition of methanol to afford the α-methyl glycoside **28** as the sole product of the reaction (Scheme 5.5). This glycosylation confirmed the validity of the original hypothesis for stereocontrol (Scheme 5.1), the observed

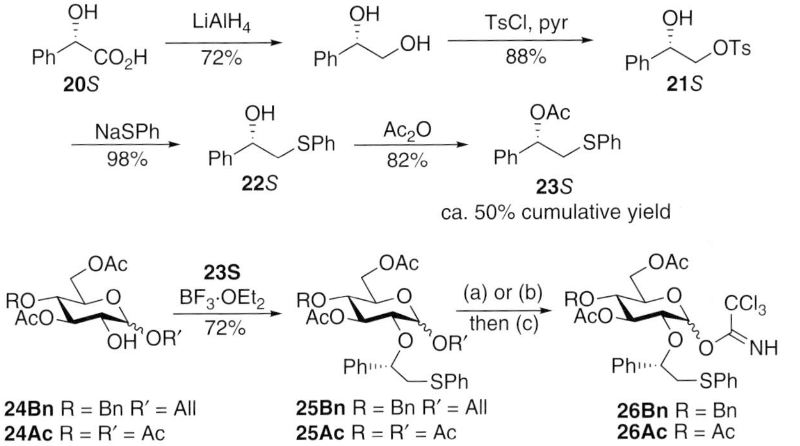

Scheme 5.4 Synthesis of chiral auxiliary precursor **23-*S***, followed by installation and activation. (a) $Pd(PPh_3)_4$ (**25Bn**, 90%); (b) $H_2NNH_2{\cdot}HOAc$ (**25-Ac**, 83%); and (c) Cl_3CCN/DBU (**26Bn**, **26Ac**, 94%).

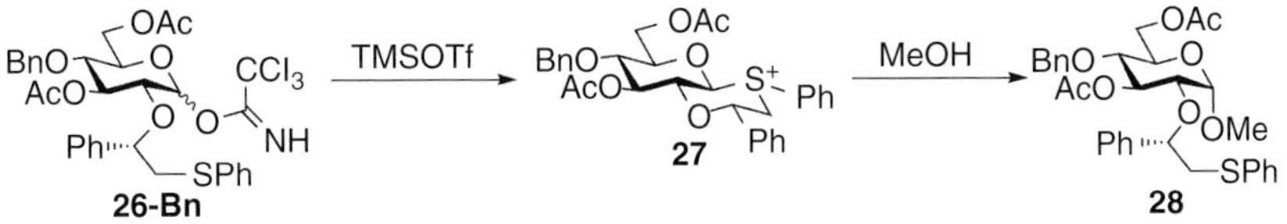

Scheme 5.5 α-Methyl glycoside **28** is formed via aryl sulfonium ion intermediate species **27**.

absolute selectivity likely a result of the increased propensity of SPh to participate in *trans*-decalin formation over the ethyl mandelate moiety in donor **9-*S***.

Upon activation with TMSOTf at −78 °C, glycosylations using trichloroacetimidate donors **26-Ac** and **26-Bn** afforded 1,2-*cis* glycosides in excellent yields, always in excess of 80%, with several tri-(benzyl/benzoyl) α-methyl glycosyl acceptors, including an O-4 acceptor. Notably, ester functionality at O-3 was conserved in both donors. Attempts to synthesize 1,2-*trans*-glycosides selectively using the analogous donor bearing an *R*-configured thiophenyl auxiliary, however, were again unsuccessful: using acceptor **15**, an 88% yield of the product glycoside was achieved but as a 1 : 1 mixture of anomers, reinforcing the reduced stability of putative *cis*-decalin intermediate **5-*R***. Following glycosylations, auxiliary cleavage could be carried out orthogonally using the same Lewis-acid-catalyzed strategy required for installation, with possible recovery of **23-*S***. Synthesis of a trisaccharide, containing a 1,2-*cis* and a 1,2-*trans* linkage, was also performed with the chiral auxiliary unaffected by NIS/TMSOTf activation of the thioglycoside utilized in the 1,2-*trans* glycosylation, prior to 1,2-*cis* glycosylation.

Following these initial investigations, ambiguity remained over the true nature of the stereodirecting intermediate in the presence of the thiophenyl auxiliary (oxocarbenium ion vs sulfonium ion), and this was investigated by Woerpel and coworkers using a model system [11]. Direct displacement of an observed *trans*-decalin β-sulfonium ion such as **27** with an acceptor theoretically proceeds with an inversion of configuration to afford α-glycosides. However, Woerpel presented evidence against this "S_N2-like" glycosylation of β-sulfonium ions by direct comparison with an analogous β-ethyl sulfonium ion **29** (Scheme 5.6) [11a].

Activation of a model donor **30** afforded an oxocarbenium ion **31**, which then cyclized to afford β-ethyl sulfonium ion **29** at −40 °C, observed as the only intermediate by ^{1}H-NMR spectroscopy (similar to the *trans*-decalin sulfonium ion **29** observed by Boons). However, following the addition of an allyl silane nucleophile, the reaction only proceeded when the temperature reached −10 °C and afforded a mixture of *C*-glycosides ($\alpha:\beta$ 11:89). Any reaction through direct displacement of the β-ethyl sulfonium ion **29** would afford *trans*-glycoside **32α** only; however, in this case, *cis*-glycoside **32β** is formed as the major product. This result implies, along with the discrepancy in reaction temperatures, that the products may arise from stereoelectronically controlled attack of oxocarbenium ion **31**, rather than from the attack of β-ethyl sulfonium ion **29**. Woerpel and coworkers observed a very similar result with the analogous thiophenyl-substituted donor, when reaction occurred at a lower temperature of −45 °C (signifying an increase in reactivity relative to the alkyl sulfonium ion). The results of this model study are interpreted as indicative of glycosylation via the higher energy oxocarbenium ion **31**, despite the major component of the equilibrium being the more stable β-ethyl sulfonium ion, a so-called Curtin–Hammett kinetic scenario [12]. The alternative reasoning for the complete stereocontrol of glycosylations with Boons' *trans*-decalin sulfonium intermediate **27** and/or its parent oxocarbenium ion was not discussed.

Subsequently, Boons and coworkers showcased the practical utility of the chiral auxiliary method by adaptation to solid-phase glycosylation [13], culminating in the synthesis of complex branched pentasaccharide **33**. The glycosylation–deprotection repeating cycle of such a strategy is highly amenable to solid support, but only if the glycosylation step is totally regio- and stereoselective. Given that acceptor-bound glycosylation offers the highest yields [14], the building blocks required have to be orthogonally protected glycosyl donor precursors (Scheme 5.7). The branching in **33** mandated the use of orthogonal protection-deprotection strategies, and hence, the protection pattern of monosaccharides **34–37** was crucial in achieving a 25% yield of pentasaccharide **33** over 13 steps, an average of 90% yield per step, with complete anomeric control. The same procedure was repeated using a D-galactose counterpart of **36** to produce the galactose–tetraglucose counterpart of **33** in 13% yield. The only hindrance reported was that the O-2 chiral auxiliary needed to be cleaved and O-2 acetylated prior

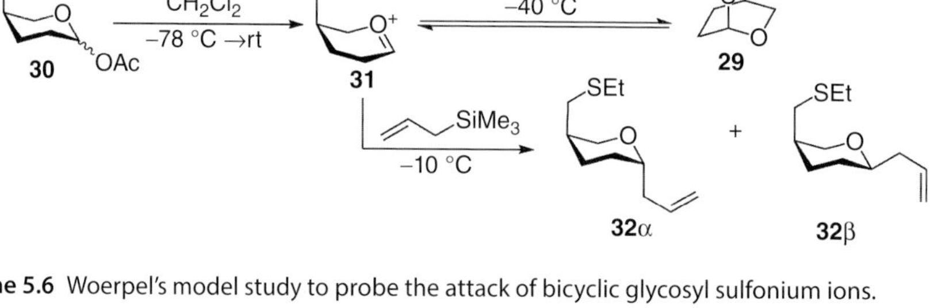

Scheme 5.6 Woerpel's model study to probe the attack of bicyclic glycosyl sulfonium ions.

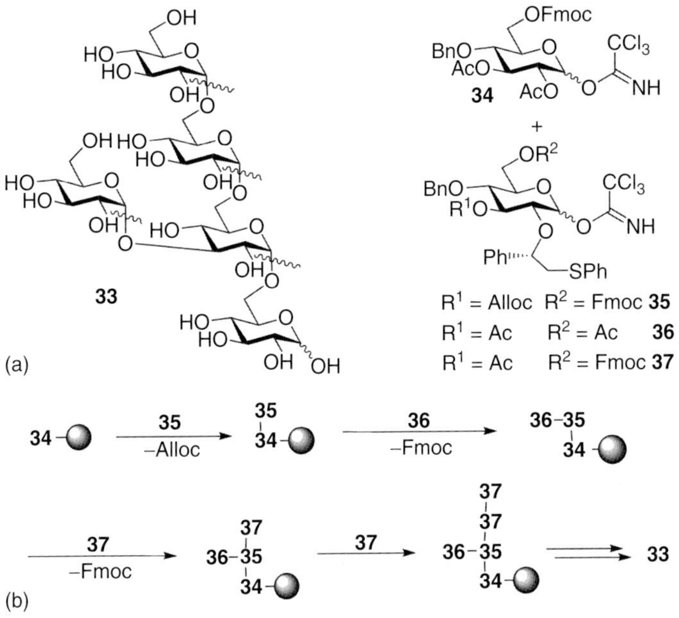

Scheme 5.7 Boons' second-generation chiral auxiliary is amenable to solid-phase synthetic strategies, such as that of pentasaccharide **33**.

to O-3 glycosylation, as the presence of the phenyl group reduced O-3 reactivity and decreased the yield of this glycosylation.

As is often the case with glycosylation reactions, minor protocol optimizations can have a significant impact on yields and stereoselectivity. One pitfall of this strategy is that stereocontrol was diminished when using armed protecting groups, with tribenzyl-protected donors exhibiting little selectivity [15]. The β-sulfonium ion is still detectable but presumably exists in equilibrium with the oxocarbenium ion, which is sufficiently stabilized by the three electron-rich benzyl ether groups.

5.3 Neighboring Group Participation of O-2 Achiral Auxiliaries

Following on from the success of the chiral auxiliary approach, the Fairbanks group attempted to achieve stereoselective glycosylation using NGP of achiral analogues of Boons' auxiliaries. The first-generation Fairbanks' auxiliary consisted of a (2-thiophenyl)methyl ether [16], which the authors hoped would form a similar *trans* decalin-type structure during glycosylation but without the need for a chiral center. Starting from a glycosyl orthoester **38,** installation of the chiral auxiliary was achieved via Williamson conditions using a simple precursor **39**, yielding glucosyl donors **40a–c** (Scheme 5.8).

However, the resultant glycosylation reactions using these achiral auxiliary bearing donors exhibited disappointingly low stereoselectivities in comparison with their chiral auxiliary counterparts (Table 5.2). Comparing a range of glycosylation conditions and

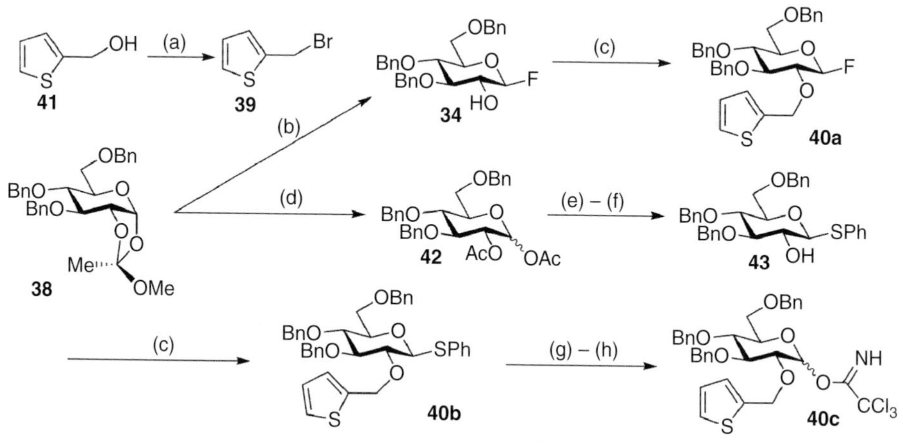

Scheme 5.8 Synthesis of glycosyl donors **40a–c**. (a) HBr/AcOH (85%); (b) unavailable; (c) NaH, **39** (**40a**, 74%; **40b**, 78%); (d) AcOH (aq.), then Ac_2O, DMAP, pyr (81%); (e) PhSH, BF_3OEt_2 (89%); (f) NaOMe/MeOH (85%); (g) TFA, NIS (84%); and (h) Cl_3CCN, DBU (80%).

Table 5.2 Reaction conditions, yields, and anomeric ratios of glycosylations using donors **40a–c** or **44** with acceptor **12.**

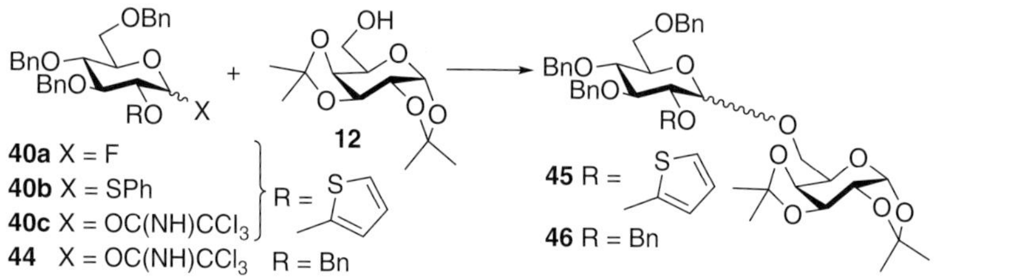

Donor	Product	Equiv. TMSOTf	Equiv. NIS	Equiv. $BF_3 \cdot OEt_2$	*t* (h)	*θ* (°C)	Yield (%)	*α*/*β*
40a	**45**	—	—	1.5	0.5	0	45	1:1.25
40a	**45**	—	—	1.5	1.0	-41	40	1:1
40a	**45**	—	—	1.5	1.5	-78	36	2:1
40b	**45**	0.5	1.5	—	1.0	0	36	1.5:1
40b	**45**	0.5	1.5	—	2.0	-41	51	2:1
40b	**45**	0.5	1.5	—	3.5	-78	48	6:1
40c	**45**	0.1	—	—	0.5	0	93	1:1.25
40c	**45**	0.1	—	—	1.0	-41	84	9:1
40c	**45**	0.1	—	—	1.5	-78	71	8:1
44	**46**	0.1	—	—	—	-41	80	1:4

activating groups, the highest $\alpha:\beta$ ratio obtained was 9:1, with O-2-benzyl control donor **44** affording 1:4 stereoselectivity for comparison (Table 5.2). As preliminary glycosylations proved to be more successful using trichloroacetimidate donors, these conditions were extended to O-2- and O-3-mannose acceptors, which exhibited much greater selectivity, about 30:1 ($\alpha:\beta$), but with lower yields. However, this study was the first demonstration that participatory achiral auxiliaries could be used for stereoselective glycosylation, with the donor series' advantage being that chiral reagents were not required in synthesis.

Fairbanks and coworkers subsequently investigated whether the use of alternative nucleophilic atoms in achiral auxiliaries could achieve more stereoselective glycosylations compared to those previously reported [17]. The authors highlighted that selenium and iodine were potentially nucleophilic enough to intercept the oxocarbenium during glycosylation and offer attractively facile installation. For example, the *O*-2-allyl-protecting ether of **47Ac/Bn** and **48Ac/Bn** was interconverted into the corresponding 2-iodo- or 2-phenylselenoethyl ethers **51–54Ac/Bn** (Scheme 5.9), followed by further activation to yield donors **55Ac/Bn** and **56Ac/Bn.** Once again, trichloroacetimidate donors afforded higher yields of glycoside disaccharide products compared to glycosyl fluorides, ranging from 60% to 90%. However even under optimized conditions, stereoselectivity for benzylated donors **55Bn** and **56Bn,** $\alpha:\beta$ ratio approximately 4:1, and acetylated donors **55Ac** and **56Ac** approximately 1:2, was low. ^{1}H-NMR studies were

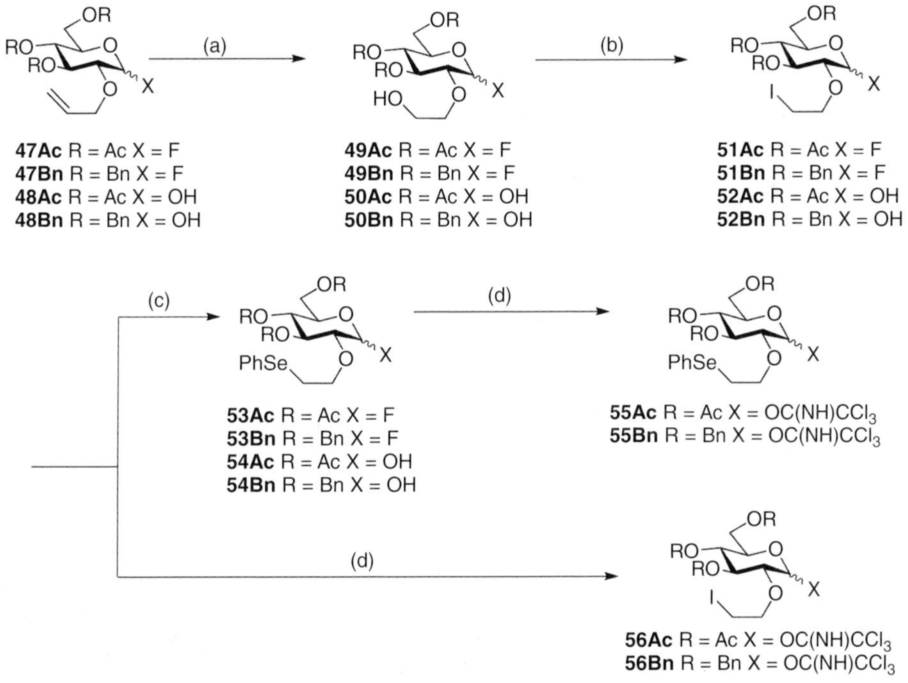

Scheme 5.9 Synthesis of 2-iodo- and 2-selenophenyl donors **51Ac/Bn**, **53Ac/Bn**, **55Ac/Bn,** and **56Ac/Bn**. (a) O_3, then $NaBH_4$ (83%); (b) I_2, PPh_3, imidazole (83%); (c) PhSeH, NaH (83%); and (d) Cl_3CCN, DBU (86%).

used to probe the mechanistic detail of glycosylation with evidence for NGP being the focus. Although signal shifts for protons adjacent to the nucleophile (i.e., H-1 and H-8, after activation) were noted and were indicative of interception of the oxocarbenium ion, this was only observed for the 2-selenophenyl series and not for the 2-iodo series (Scheme 5.9).

An analogous study by Liu and coworkers also investigated the utility of cyanobenzyl ethers as achiral auxiliaries in stereoselective glycosylations [18], with the obvious advantages of straightforward installation and cleavage. The 2-cyanobenzyl ether auxiliary was designed to facilitate 1,2-β-*trans* glycosylation via formation of an α-benzylnitrilium ion, analogous to acetonitrile participation in glycosylation [19]. As anticipated glycosylation of donor **57** with butan-1-ol **58** and armed benzylated acceptors **59a/b** afforded 1,2-β-*trans* glycosides **60** and **61a/b** in upward of 80% yield with complete stereoselectivity. However, disarmed acceptors such as 2,2,2-trifluoroethanol **62** and acetylated glycosides **63a/b** were conversely completely 1,2-*cis*-α-selective in the synthesis of glycosides **64** and **65a/b** (Table 5.3).

The dual-directing effect of the cyanobenzyl auxiliary is attributed to two scenarios (Figure 5.1): the aforementioned benzylnitrilium species facilitating 1,2-*trans*-β-attack [19] and hydrogen bonding to the disarmed acceptor by the nitrile group [20], placing the acceptor in line for 1,2-*cis*-α-attack. The latter scenario requires the acceptor to be a poor nucleophile, with nucleophilicity subsequently enhanced by hydrogen bonding and hence, the 1,2-*cis*-α-selectivity with electron-poor acceptors such as **63a/b**. Although stereoselective, this glycosylation strategy is dependent upon protecting groups and acceptor nucleophilicity.

5.4 Preconfigured Chiral Auxiliaries

As demonstrated in Boons' early work on chiral auxiliaries (*vide supra*), NGP strategies toward 1,2-*cis* glycosylation are limited by the requirement of forming a six-membered cyclic species *in situ*, which is more challenging than the formation of the corresponding five-membered cyclic species in 1,2-*trans* glycosylation. A preconfigured glycosyl donor, already locked in a six-membered ring, therefore circumvents this limitation, potentially reducing the entropic penalty of stereoselective glycosylation. Turnbull and coworkers described the first example of a preconfigured bicyclic oxathiane glycosyl donor, where the *trans*-decalin oxathiane ring is formed prior to glycosylation [21]. The authors hypothesized that activation of the latent oxathiane thioglycoside **66** could then afford a *quasi*-stable sulfonium ion **67** (Scheme 5.10), akin to the species **27** (Scheme 5.5) formed transiently during the Boons' chiral auxiliary glycosylation, which could be then displaced stereoselectively by the glycosyl acceptor to afford 1,2-*cis* glycoside **68.** In this strategy, therefore, the sulfur containing auxiliary acts as both the anomeric leaving group and the participating group.

Donor synthesis began from commercially available pentaacetate **69,** a one-pot transformation from which afforded thioglycoside **70,** via an intermediate glycosyl isothiouronium salt (Scheme 5.11) [22]. Concomitant Zemplén deprotection and tandem acetal formation/stereoselective cyclization then yielded *trans*-decalin oxathiane **71,** which was subsequently protected as the ketal **72Ac/Bn**.

Table 5.3 Yields and anomeric ratios of glycosides synthesized from *O*-2-(2-cyanobenzyl) glycosyl donor 57.

Acceptor	Product	Yield (%)	α/β
58	**60**	90	β only
59a	**61a**	86	β only
59b	**61b**	89	β only
62	**64**	71	α only
63a	**65a**	81	α only
63b	**65b**	87	α only

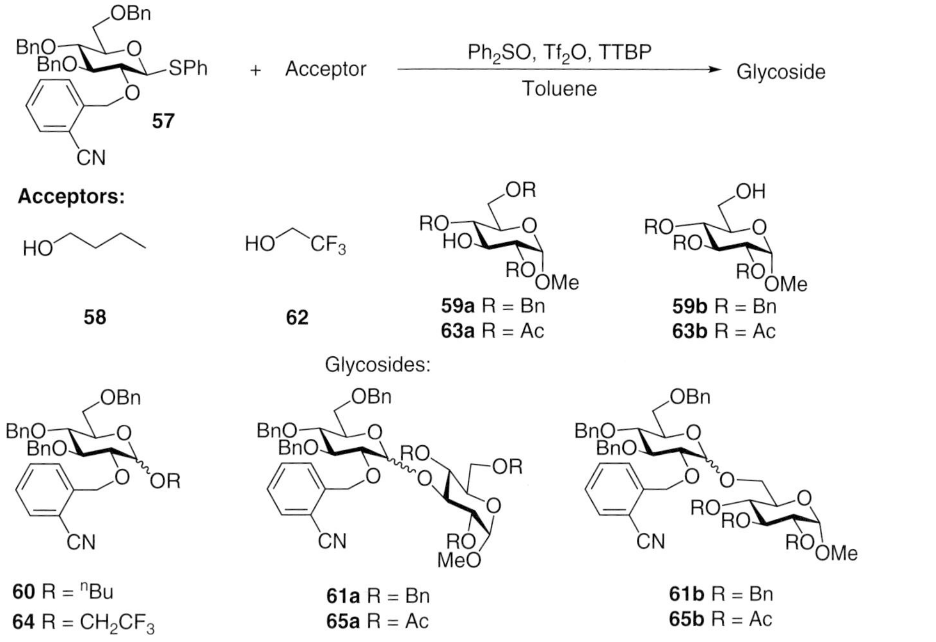

A number of thioglycoside activation strategies were screened in initial glycosylation reactions using the oxathiane ketals. Methylation afforded a *trans*-decalin methylsulfonium ion; however, the activated structure was so inert that its crystal structure could be obtained, the first ever of a glycosyl sulfonium salt [23]. Turnbull and coworkers also demonstrated that arylation via benzyne was achievable [24], but the arylsulfonium ion was intercepted by acetate from the co-reagent $Pb(OAc)_4$ before glycosyl acceptors could react. The most successful activation method described by the Turnbull group, however, was a Kahne-type [25] activation of oxathiane sulfoxides (synthesized by mCPBA oxidation of the oxathiane ketal) with triflic anhydride and trimethoxybenzene [21]. This combination of reagents facilitated an *in situ* S_EAr-like arylation of the

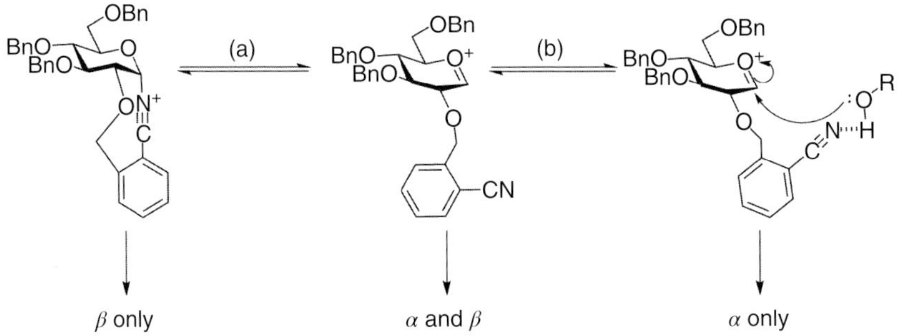

Figure 5.1 Interception of the oxocarbenium ion occurring (a) intramolecularly or (b) via intermolecular attack guided by hydrogen bonding.

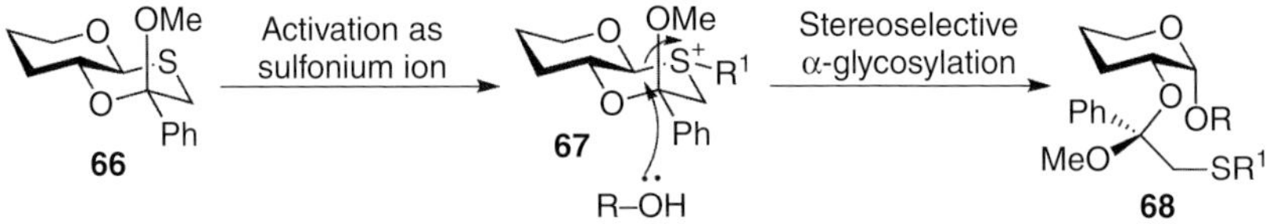

Scheme 5.10 Activation of the preformed six-membered oxathiane ring results in selective 1,2-*cis* glycosylation.

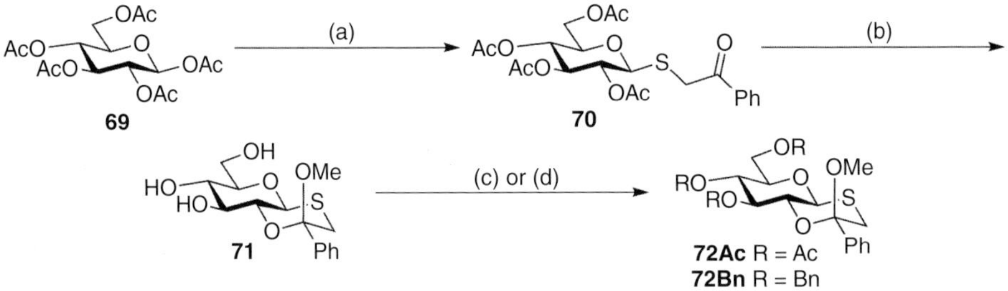

Scheme 5.11 Three-step protected oxathiane synthesis. (a) $BF_3{\cdot}OEt_2$, thiourea, then Et_3N, 2-bromoacetophenone (64%); (b) NaOMe/MeOH, then *p*-TsOH (61%); (c) Ac_2O, pyridine (**72Ac**, 76%); and (d) NaH, then BnBr (**61Bn**, 87%).

oxathiane sulfoxides **73Ac**/**Bn** to afford trimethoxyphenyl (TMP) sulfonium ions **74Ac**/**Bn** (Table 5.4), which then reacted with glycosyl acceptors **12** and **59b**.

Yields ranged from moderate to high, but importantly, only 1,2-*cis* glycoside formation was observed even at 50 °C. Unfortunately, after glycosylation, the resulting acyclic O-2 ketal linkage was labile, leading to *in situ* loss of methanol and interception of the glycosyl donor to afford methyl glycosides, reducing the yield of the desired product when using less reactive glycosyl acceptors. Turnbull and coworkers sought to circumvent this limitation by reducing the oxathiane ketal to an oxathiane ether [23, 24], in order to prevent methanol release during glycosylation. Glycosylations using these next-generation oxathiane ether donors, however, suffered from decreased

Table 5.4 Yields and anomeric ratios of products **75Ac/Bn and 76Ac/Bn** synthesized using donors **73Ac/Bn** and acceptors **12** and **59b**, proceeding via activated aryl sulfonium ion intermediate **74**.

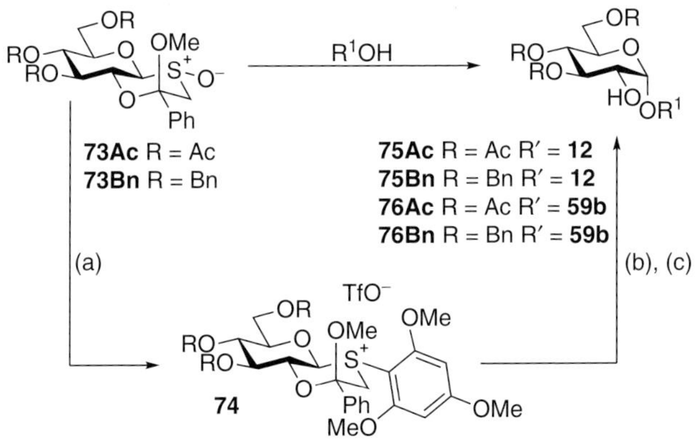

Donor	Acceptor	Product	Yield (%)	*α*/*β*
62Ac	**12**	**75Ac**	85	> 98:2
62Bn	**12**	**75Bn**	89	> 98:2
62Ac	**59b**	**76Ac**	44	> 98:2
62Bn	**59b**	**76Bn**	72	> 98:2

(a) Trimethoxybenzene, DIPEA, Tf_2O; (b) R^1OH, DIPEA; and (c) $BF_3{\cdot}OEt_2$.

stereoselectivity [23]. This reduction in stereoselectivity implied an unexpected level of complexity in the mechanism by which these TMP sulfonium donors react with glycosyl acceptors, even though, once again, ^{1}H-NMR studies confirmed that an arylated β-sulfonium ion is formed during glycosylation.

B3LYP functional DFT calculations were used to investigate whether the glycosylation proceeded through an S_N1 or S_N2 pathway [23]. Using simplified model *methyl-ether-protected* TMP sulfonium ions of the oxathiane ketal **77** and ether **78**, a comparison of the $\Delta G^‡$ of the S_N2 (**79** and **81,** Figure 5.2) and S_N1 (**80** and **82**) transition states revealed that the S_N2 transition state for the oxathiane ketal is 5.8 kcal mol^{-1} ($\Delta\Delta G^‡$) lower in energy compared to the S_N1 transition state while only 4.4 kcal mol^{-1} lower in energy for the oxathiane ether. It is assumed that the S_N2 pathway is stereospecific for inversion, that is, can only afford 1,2-*cis* products, as opposed to an S_N1 mechanism, which could theoretically result in 1,2-*trans* by-product formation; therefore, the calculated greater flux through the S_N2 pathway for the oxathiane ketal could account for the greater stereoselectivity over the oxathiane ether. Calculations for the *acetylated* model TMP sulfonium ions highlighted an even greater energetic preference for the S_N2 pathway compared to when ether-protecting groups were modeled, which was also consistent with the experimentally observed increase in stereoselectivity for the acetylated

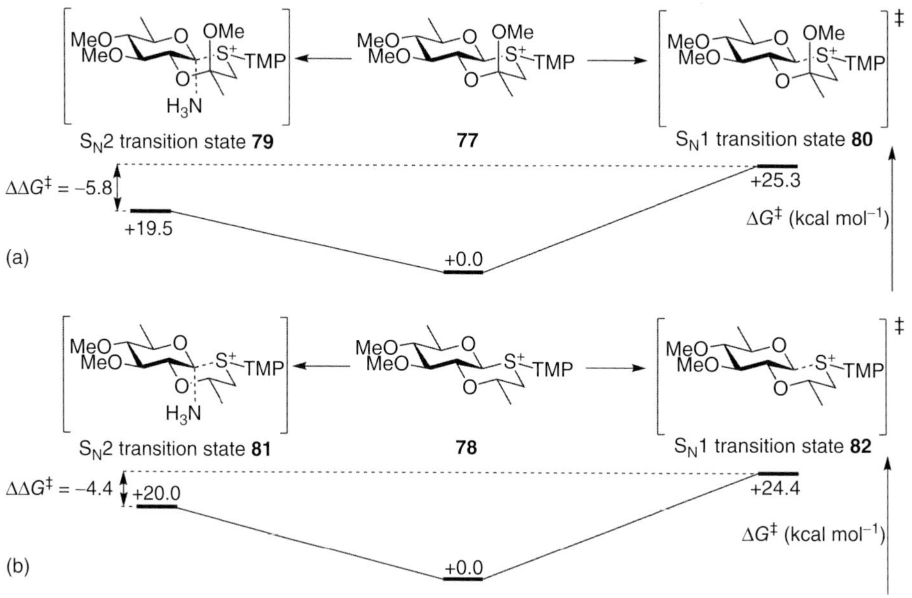

Figure 5.2 DFT calculations of energies of transition states **79–82** for S_N1 and S_N2 attack of model arylsulfonuim (a) ketal **77** and (b) ether **78** by nucleophile NH_3. TMP: trimethoxyphenyl.

donors over benzylated donors. However, further DFT calculations using both ether- and acetyl-protected phenyl sulfonium ions, akin to Boons' transiently observed intermediate, which affords 1,2-*cis* glycosides with complete stereoselectivity experimentally, indicated that this system would have the lowest calculated energetic preference for the S_N2 pathway, while *in vitro* it is more stereoselective compared to the TMP oxathiane ether donors. Therefore, the authors concluded that despite the calculated balance between S_N1 and S_N2 accounting for some trends in observed stereoselectivity, the calculations were not consistent with this hypothesis when taking a holistic view of all experimental data. Thus, an alternative two-conformer hypothesis was explored to explain the difference in stereoselectivity [11a, 26], where product formation was considered a result of acceptor attack upon oxocarbenium ion intermediates, in an S_N1-like mechanism. In this hypothesis, the position of equilibrium between the two oxocarbenium ion $^3H_4$ and $^4H_3$ half-chair conformations governs the level of stereocontrol, as the inherent facial selectivity of attack on each conformer is generally accepted to afford only one stereoisomer (1,2-*trans* products from attack on the $^3H_4$ conformer, and 1,2-*cis* products from attack on the $^4H_3$ conformer) [27]. The authors utilized DFT calculations (M06 functional) once more to probe the energetic landscape for $^3H_4$ and $^4H_3$ oxocarbenium ion conformers for the TMP and phenyl sulfonium ions of both oxathiane ether and ketal models; however, once again, the calculated populations were not consistent with a holistic view of all the experimentally observed stereoselectivities. The ambiguous nature of the results of these investigations therefore leaves the mechanism of the oxathiane glycosylation reaction still open for debate. However, it seems plausible that the real scenario is arguably a combination of both theories, with multiple reaction pathways likely accessible.

Despite this ongoing mechanistic ambiguity over their reactivity, variations of the oxathiane ether donors [28] have proven very useful in more challenging synthetic endeavors. Boons and coworkers built on the work of the Turnbull group and demonstrated the power of differentially protected oxathiane ether donor **83** in a range of glycosylations with acceptors **84–88** (Table 5.5) [29], where 1,2-*cis* glycosides **89–93** were produced in high yields and excellent stereoselectivity. The orthogonally protected oxathiane ether donors **94–95** were finally utilized in consecutive stereoselective glycosylations, *en route* to the synthesis of branched tetrasaccharide **105** in an impressive overall yield and, importantly, with absolute 1,2-*cis*-α stereoselectivity (Scheme 5.12).

5.5 Conclusion

While multiple profound and promising strategies for 1,2-*cis* stereoselective glycosylation have been explored within the literature [30], it is clear that at present there is no single general strategy capable of the total stereoselectivity, high yields, and large

Table 5.5 Yields and anomeric ratios of product glycosides **89–93** synthesized using oxathiane ether donor **83**.

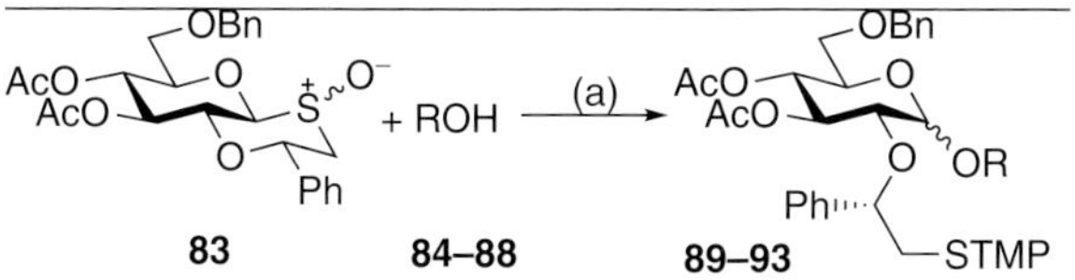

83 **84–88** **89–93**

Acceptor	Product	Yield / %	α:β
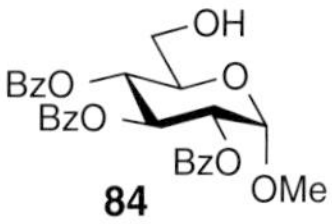 **84**	**89**	91	>15:1
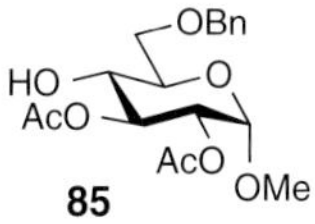 **85**	**90**	62	>25:1
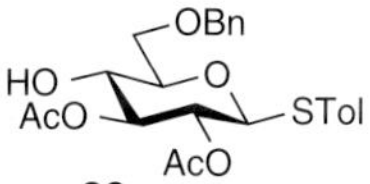 **86**	**91**	67	>15:1
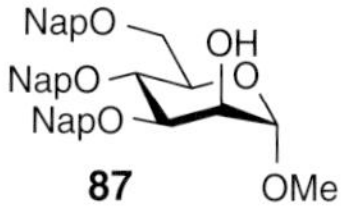 **87**	**92**	72	α only
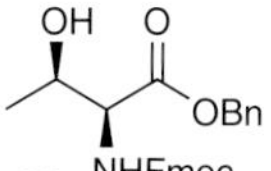 **88**	**93**	89	α only

TMB = 2,4,6-trimethoxyphenyl. (a) Trimethoxybenzene, Tf_2O, DIPEA.

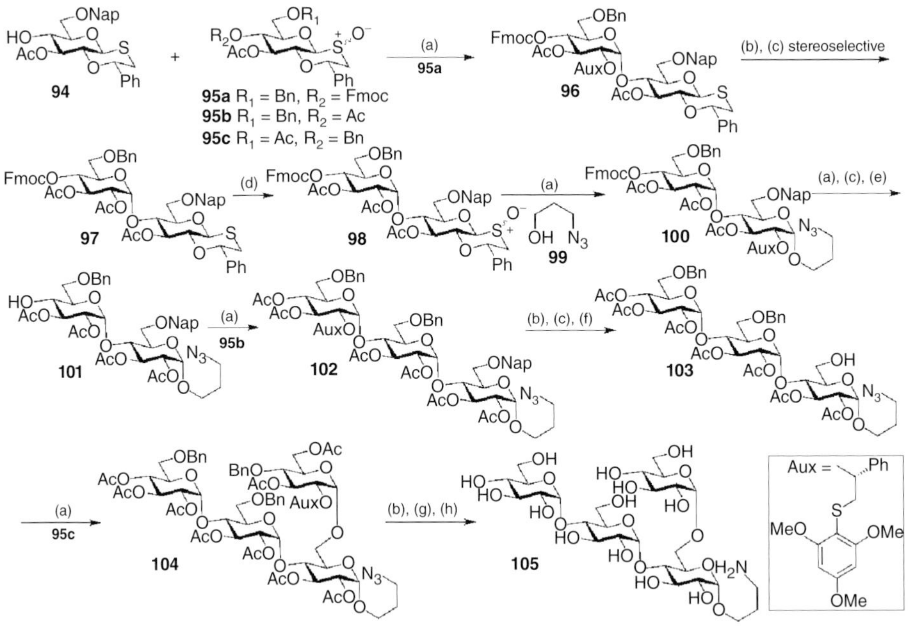

Scheme 5.12 Synthetic route to glucan **105** from building blocks **94–95**. (a) Trimethoxybenzene, Tf_2O, DTBMP (about 74% yield, α only); (b) TFA (about 87%); (c) Ac_2O, pyridine (about 88%); (d) mCPBA (98%); (e) NMP (91%); (f) DDQ (92%); (g) NaOMe/MeOH (83%); and (h) H_2, Pd/C (83%).

substrate scope, which can compare to the general methods for 1,2-*trans* glycosylation. However, in this chapter, we have highlighted a number of inventive and functional synthetic strategies, which have employed chiral auxiliaries to control stereoselectivity, with notably more success compared to their achiral counterparts in glycosylation reactions. The auxiliaries pioneered by the Boons and Turnbull groups have led the field in this area, often achieving the desired stereoselectivity and conversion. However, the continually demonstrated diversity of glycans in nature and, thus, the myriad of different glycosylations required to synthesize even a fraction of these structures place restrictions on the extent to which these strategies can be considered truly general.

References

1 Demchenko, A.V. (ed.) (2008) Handbook of Chemical Glycosylation: Advances in Stereoselectivity and Therapeutic Relevance, Wiley-VCH.

2 Debenham, J.S., Madsen, R., Roberts, C., and Fraser-Reid, B. (1995) *J. Am. Chem. Soc.*, 117, 3302–3303.

3 Stalford, S.A., Kilner, C.A., Leach, A.G., and Turnbull, W.B. (2009) *Org. Biomol. Chem.*, 7, 4842–4852.

4 Krock, L., Esposito, D., Castagner, B., Wang, C.-C., Bindschadler, P., and Seeberger, P.H. (2012) *Chem. Sci.*, 3, 1617–1622.

5 (a) Schmidt, D., Schuhmacher, F., Geissner, A., Seeberger, P.H., and Pfrengle, F. (2015) *Chem. Eur. J.*, 21, 5709–5713; (b) Weishaupt, M.W., Matthies, S., and Seeberger, P.H. (2013) *Chem. Eur. J.*, 19, 12497–12503; (c) Calin, O., Eller, S., and Seeberger, P.H. (2013) *Angew. Chem. Int. Ed.*, 52, 5862–5865.
6 Lemieux, R.U. (1971) *Pure Appl. Chem.*, 25, 527–548.
7 Kim, J.-H., Yang, H., and Boons, G.-J. (2005) *Angew. Chem. Int. Ed.*, 44, 947–949, S947/941–S947/916.
8 Kim, J.-H., Yang, H., Khot, V., Whitfield, D., and Boons, G.-J. (2006) *Eur. J. Org. Chem.*, 2006, 5007–5028.
9 Casadei, M.A., Galli, C., and Mandolini, L. (1984) *J. Am. Chem. Soc.*, 106, 1051–1056.
10 Kim, J.H., Yang, H., Park, J., and Boons, G.J. (2005) *J. Am. Chem. Soc.*, 127, 12090–12097.
11 (a) Beaver, M.G., Billings, S.B., and Woerpel, K.A. (2008) *J. Am. Chem. Soc.*, 130, 2082–2086; (b) Beaver, M.G., Billings, S.B., and Woerpel, K.A. (2008) *Eur. J. Org. Chem.*, 2008, 771–781.
12 Seeman, J.I. (1983) *Chem. Rev.*, 83, 83–134.
13 Boltje, T.J., Kim, J.-H., Park, J., and Boons, G.-J. (2010) *Nat. Chem.*, 2, 552–557.
14 Seeberger, P.H. (2008) *Chem. Soc. Rev.*, 37, 19–28.
15 Boltje, T.J., Kim, J.-H., Park, J., and Boons, G.-J. (2011) *Org. Lett.*, 13, 284–287.
16 Cox, D.J. and Fairbanks, A.J. (2009) *Tetrahedron: Asymmetry*, 20, 773–780.
17 Cox, D.J., Singh, G.P., Watson, A.J.A., and Fairbanks, A.J. (2014) *Eur. J. Org. Chem.*, 2014, 4624–4642.
18 Le Mai Hoang, K. and Liu, X.-W. (2014) *Nat. Commun.*, 5, 5014. doi: 10.1038/ncomms6051
19 (a) Schmidt, R.R., Behrendt, M., and Toepfer, A. (1990) *Synlett*, 1990, 694–696; (b) Ishiwata, A., Munemura, Y., and Ito, Y. (2008) *Tetrahedron*, 64, 92–102.
20 (a) Pistorio, S.G., Yasomanee, J.P., and Demchenko, A.V. (2014) *Org. Lett.*, 16, 716–719; (b) Yasomanee, J.P. and Demchenko, A.V. (2012) *J. Am. Chem. Soc.*, 134, 20097–20102.
21 Fascione, M.A., Adshead, S.J., Stalford, S.A., Kilner, C.A., Leach, A.G., and Turnbull, W.B. (2009) *Chem. Commun.*, 39, 5841–5843.
22 Ibatullin, F.M., Shabalin, K.A., Janis, J.V., and Shavva, A.G. (2003) *Tetrahedron Lett.*, 44, 7961–7964.
23 Fascione, M.A., Kilner, C.A., Leach, A.G., and Turnbull, W.B. (2012) *Chem. Eur. J.*, 18, 321–333.
24 Fascione, M.A. and Turnbull, W.B. (2010) *Beilstein J. Org. Chem.*, 6 (19).
25 Kahne, D., Walker, S., Cheng, Y., and Van Engen, D. (1989) *J. Am. Chem. Soc.*, 111, 6881–6882.
26 (a) Yang, M.T. and Woerpel, K.A. (2009) *J. Org. Chem.*, 74, 545–553; (b) Billings, S.B. and Woerpel, K.A. (2006) *J. Org. Chem.*, 71, 5171–5178.
27 Stevens, R.V. (1984) *Acc. Chem. Res.*, 17, 289–296.
28 Fascione, M.A., Webb, N.J., Kilner, C.A., Warriner, S.L., and Turnbull, W.B. (2012) *Carbohydr. Res.*, 348, 6–13.
29 Fang, T., Mo, K.-F., and Boons, G.-J. (2012) *J. Am. Chem. Soc.*, 134, 7545–7552.
30 Nigudkar, S.S. and Demchenko, A.V. (2015) *Chem. Sci.*, 6, 2687–2704.

6

Glycosylation with Glycosyl Sulfonates

Luis Bohé and David Crich

6.1 Introduction

Although first prepared by Helferich and Gootz in 1929 [1] and studied extensively by Schuerch and coworkers in the 1970s and 1980s [2–7], it is only within the last 20 years that the prowess of the glycosyl sulfonates as glycosyl donors has been recognized and exploited. This chapter focuses on the chemistry of the glycosyl sulfonates subsequent to a review that covered the area up to and including the year 2000 [8, 9], to which the reader is referred for earlier work. The chapter is not intended to be comprehensive but is written with the goal of placing the glycosyl sulfonates and their reactivity in a comprehensive mechanistic framework to aid in future exploitation in synthesis.

6.2 Formation of Glycosyl Sulfonates

Early work focused on the synthesis of glycosyl sulfonates by substitution of glycosyl halides with silver sulfonate salts and enabled the formation of a number of such substances and, in the hands of Scheurch and his coworkers, the study of their reactivity toward nucleophiles [1–9]. Per-*O*-benzyl-D-glucopyranosyl triflate (assumed to have the α-configuration) formed in this manner at −78 °C was found to be highly reactive and could not be characterized owing to its instability [2]. Consistent with the work of Helferich and Gootz, however, reaction of per-*O*-benzyl-α-D-glucopyranosyl bromide or chloride with silver toluenesulfonate in acetonitrile at room temperature, followed by filtration of the silver halide, concentration, and dissolution in deuterochloroform, gave a glycosyl toluenesulfonate that was sufficiently stable for NMR spectra to be recorded at room temperature; the anomeric chemical shift and coupling constant of δ 6.1 and $^3J_{H1,H2} = 3.5$ Hz were consistent with the α-configuration [3]. Application of the analogous protocol to the 2,3,4-tri-*O*-benzyl-6-*O*-(*N*-phenylcarbamoyl)-D-glucopyranosyl series gave a 85 : 15 mixture of the α- and β-toluenesulfonates, with the latter characterized by δ 5.5 and $^3J_{H1,H2} = 8.0$ Hz for the anomeric proton in $CDCl_3$ [3]. A range of galactopyranosyl, mannopyranosyl, and rhamnopyranosyl sulfonates protected by multiple benzyl ethers and either 6-*O*-carbamoyl groups (galactose series) or 2-*O*-sulfonyl

Selective Glycosylations: Synthetic Methods and Catalysts, First Edition. Edited by Clay S. Bennett.

groups (mannose and rhamnose series) were prepared by comparable methods, characterized by NMR spectroscopy, and their reactivity toward simple alcohols was studied [4–7]. It was also noted that anomeric toluenesulfonates formed from silver toluenesulfonate and peracetyl or perbenzoyl glucopyranosyl halides were considerably more stable [3]. The generation of a glycosyl triflate by reaction of a glycosyl fluoride with TMS triflate at low temperature has been demonstrated by NMR spectroscopy [10].

The reaction of a variety of anomeric hemiacetals with sulfonylating agents (sulfonic anhydrides and sulfonyl halides) was studied earlier by a number of groups but was generally found to be impractical and to yield either glycosyl halides, orthoesters, or 1,1′-disaccharides depending on the protecting systems and additives employed [8, 9, 11–20]. Indeed, a careful ^{19}F NMR study of the dehydrative self-coupling of 2,3,4,6-tetra-*O*-benzyl glucopyranose employing triflic anhydride as reagent in dichloromethane conducted by Pavia and Ung-Chhun suggested the mechanism to involve simple acid-catalyzed dehydration, rather than glycosyl triflate formation, with removal of water from the equilibrium in the form of the insoluble hydronium triflate salt [20]. Dehydrative activation of hemiacetals by means of triflic anhydride and diaryl sulfoxides results in the formation of glycosyloxy sulfonium salts rather than glycosyl triflates [10].

Subsequent work by Bennett and coworkers has established that the clean formation and characterization of glycosyl sulfonates from anomeric hemiacetals may be accomplished by the formation of the potassium salt of the hemiacetal followed by reaction with toluenesulfonic anhydride or better toluenesulfonyl 4-nitroimidazolide (Scheme 6.1) [21, 22]. 3,4,6-Tri-*O*-benzyl-2-deoxy-α-D-glucopyranosyl toluenesulfonate generated in this manner was shown to have anomeric chemical shifts of δ_H 6.11 and δ_C 102.3 in perdeuterotetrahydrofuran and to be stable below −10 °C, at which point it decomposed to the glucal [22]. Kim and coworkers have also established by NMR spectroscopy the formation of the α-glycosyl triflate upon the activation of 4,6-*O*-benzylidene-2,3-*O*-benzyl-D-mannopyranose by phthalic anhydride followed by triflic anhydride [23].

The activation of glycosyl sulfoxides with triflic anhydride at −78 °C in dichloromethane provides a rapid and clean entry into the glycosyl triflates (Scheme 6.2) [24, 25], which can then be characterized by low-temperature NMR spectroscopy and their decomposition

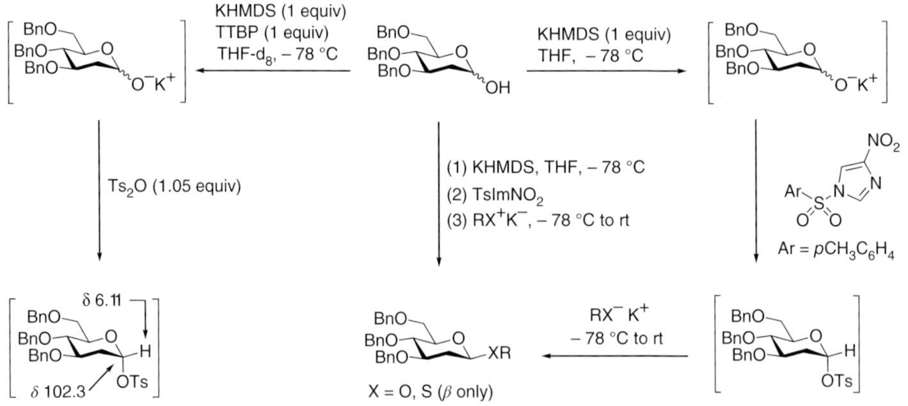

Scheme 6.1 Glycosyl sulfonate generation from hemiacetals.

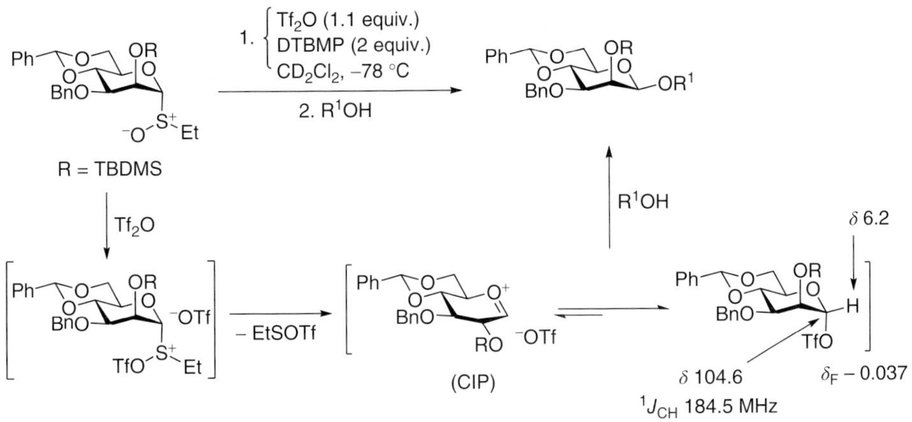

Scheme 6.2 Glycosyl triflate generation from glycosyl sulfoxides with triflic anhydride.

temperatures estimated by Variable Temperature NMR spectroscopy (VT-NMR). Note, however, that the use of catalytic triflic anhydride has been demonstrated to promote the rearrangement of the glycosyl sulfoxide to the corresponding glycosyl sulfenate, which is a much less potent glycosylating agent than the glycosyl triflate [26].

Arenesulfenyl triflates, generated *in situ* from the sulfenyl chloride and silver triflate, are powerful activating agents for thioglycosides and convert them to glycosyl triflates as demonstrated by Crich and Sun [27], and subsequently broadly exploited by the Huang and Ye groups [25, 28, 29]. In the 2-deoxypyranose series, however, it has been demonstrated that this protocol converts the thioglycoside into the glycosyl chloride, which is nevertheless a powerful glycosylating agent [30]. Subsequently, Crich and Smith developed the combinations of *S*-(methoxyphenyl)benzenethiosulfenate (MPBT) [31] and the more powerful *N*-benzenesulfinylpiperidine (BSP) [32, 33] with triflic anhydride for the conversion of thioglycosides to glycosyl triflates. The even more powerful combination of diphenyl sulfoxide (DPSO) and triflic anhydride was introduced by the van Boom group [34, 35]. Yet, further variations on the theme include the activation of thioglycosides by the combination of triflic anhydride with *N*-benzenesulfenyl caprolactam [36], *N*-benzenesulfinylmorpholine [37] and, most noteworthy, because of the commercial availability of the reagent, dimethyl disulfide [38]. The BSP and DPSO methods have been widely adopted, but it must be noted that the reagents themselves are nucleophilic and will displace the glycosyl triflate if used in excess as first demonstrated by Gin [10]. Electrochemical methods using a divided cell and tetrabutylammonium triflate as supporting electrolyte also provide a clean source of glycosyl triflates from thioglycosides [39].

The formation of 2,3,4,6-tetra-*O*-benzyl-α-D-glucopyranosyl triflate from the corresponding trichloroacetimidate precursors has been recorded by NMR spectroscopy on activation by TMSOTf and $BF_3{\cdot}OEt_2$ in mixtures of CD_2Cl_2 and the ionic liquid ethyl methylimidazolinium triflate at −78 °C [40]. The formation of glycosyl triflates on activation of the corresponding *N*-phenyl trifluoroacetimidates by triflic acid has also been confirmed by NMR spectroscopy [41]. The problem of competing aglycone transfer in the preactivation of thioglycosides can be minimized by the use of bulky thioglycosides such as the *o*-tolylthioglycosides [42].

6.3 Evidence for Glycosyl Sulfonates

As indicated in the aforementioned discussion, the main evidence for glycosyl sulfonates derives from NMR spectroscopy with many glycosyl sulfonates, predominantly triflates, so identified in recent years. Compilations, including the influence of protecting groups and configuration on their decomposition temperatures, have been published [43, 44]. Typically, the axial sulfonate very strongly predominates, no doubt because the strongly electron-withdrawing effect of the sulfonate group, and especially of the triflate group, accentuates the anomeric effect. Equatorial glycosyl triflates have been identified as the major species formed upon activation of thioglycosides of mannuronate esters by van der Marel, Codée, and their coworkers [45, 46]. It has been suggested on the basis of chemical shift arguments that these equatorial triflates are stabilized by an unusual hydrogen bond from the axial ester in the $^1C_4$ conformer of the sugar to the axial anomeric hydrogen (Scheme 6.3) [47].

With ester-protected glycosyl donors, the situation is more complex. Thus, Crich and Sun reported that upon activation of phenyl 2,3,4,6-tetra-*O*-acetyl-β-D-thioglucopyranoside sulfoxide with triflic anhydride at −78 °C in CD_2Cl_2, two products were observed by NMR spectroscopy in the ratio of 1.4 : 1, which were assigned as the α- and β-triflates, respectively, with the latter adopting a twist boat conformation [24]. In contrast, activation of phenyl 2,3,4-tri-*O*-benzoyl-β-D-thioxylopyranoside with benzenesulfenyl triflate in CD_2Cl_2 at −78 °C resulted in ^1H and ^{13}C NMR spectra that were difficult to interpret owing to dynamic broadening of the signals, suggestive of a rapid equilibrium between two or more species. With the aid of a sample enriched in ^{13}C in each of the carbonyl carbons, the major product in this equilibrium mixture was assigned as a bridging

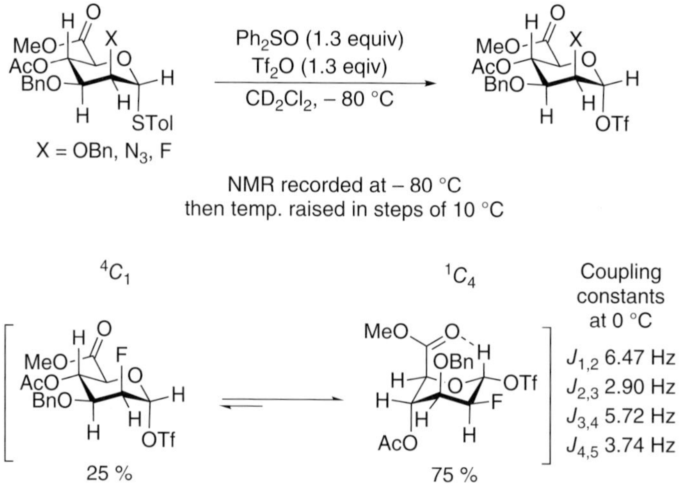

Scheme 6.3 Formation and characterization of equatorial glycosyl triflates in the mannuronate esters.

dioxolenium ion [48]. Subsequently, Huang and coworkers using NMR methods in $CDCl_3$ at −60 °C identified the α-galactosyl triflate as the major species formed upon activation of tolyl 2,3,4,6-tetra-*O*-benzoyl-β-D-thiogalactopyranoside with toluenesulfenyl triflate [49]. When the more armed 3,4,6-tri-*O*-benzyl-2-*O*-benzoyl thiogalactoside was activated under the same conditions, the bridged dioxalenium ion was identified as the major product. Finally, activation of tolyl 2,3,4,6-tetra-*O*-acetyl-β-D-thioglucopyranoside under the same conditions revealed the formation of an approximately equimolar ratio of the α-glycosyl triflate, consistent with the work of Crich and Sun [24], and the bridged dioxolenium ion [49]. Upon warming to −20 °C, the equilibrium between these two species shifted to the extent that only the glycosyl triflate was visible, but upon recooling to −60 °C, the original ratio was restored. Thus, on the basis of this limited data set, it appears that in more strongly disarmed systems, the covalent anomeric triflate is favored over bridging dioxolenium ions formed by participation of a 2-*O*-acyl group. With less disarmed systems, the dioxalenium ion is more favored consistent with the greater stabilization of the positive charge. Covalent triflates are in dynamic equilibrium with bridging dioxolenium ions, and the former are more strongly favored at higher temperatures (Scheme 6.4).

The ease of activation of a given glycosyl donor and its conversion to a glycosyl triflate is a function of the protecting group array, consistent with the general concept of armed and disarmed donors. Simple low and VT-NMR experiments readily reveal if a particular activation cocktail converts a given donor to the anomeric sulfonate and can be of great help in optimizing reaction temperature. This is well illustrated by the work of Lowary and coworkers, who showed that the action of triflic anhydride on a furanosyl sulfoxide gave a mixture containing a number of intermediates at −78 °C, which were only converted slowly to the triflate. However, upon warming to −40 °C, all species were shown to rapidly yield the triflate. Consistent with the presence of a single glycosylating agent at −40 but not at −78 °C, subsequent glycosylation reactions were more selective at the higher temperature (Scheme 6.5) [50].

All glycosyl sulfonates for which ^{13}C NMR spectra have been recorded to date [43, 44, 50] display chemical shifts close to δ 100 for the anomeric carbon, which is both typical for acetal carbons and far from the range $\delta \sim 220–250$ recorded for the sp^2-hybridized carbon in simple oxocarbenium ions, whether in superacid media or dichloromethane [51], and from the $\delta \sim 229$ reported for the 3,4,6-tri-*O*-acetyl-2-deoxyglucopyranosyl oxocarbenium ion in $SbF_5/HSbF_6$ [52], and the 2,4,6-tri-*O*-methyl-2-deoxyglucopyranosyl oxocarbenium ion in SO_3 and SbF_5 [53]. The chemical shift of $\delta \sim 100$ therefore confirms the glycosyl triflates to be covalent entities in which the triflate is firmly bound to the anomeric carbon. Furthermore, the anomeric chemical shift indicates that any dynamic equilibria between the covalent triflates and the oxocarbenium ion/triflate ion pairs very strongly favor the former [24].

6.4 Location of the Glycosyl Sulfonates in the General Glycosylation Mechanism

The mechanism of glycosylation reactions is best viewed as an example of the Winstein ion pair formalization [54] of nucleophilic substitution as originally discussed by Vernon [55], Schuerch [5], and Lemieux *et al.* [56]. Accordingly, for donors lacking the

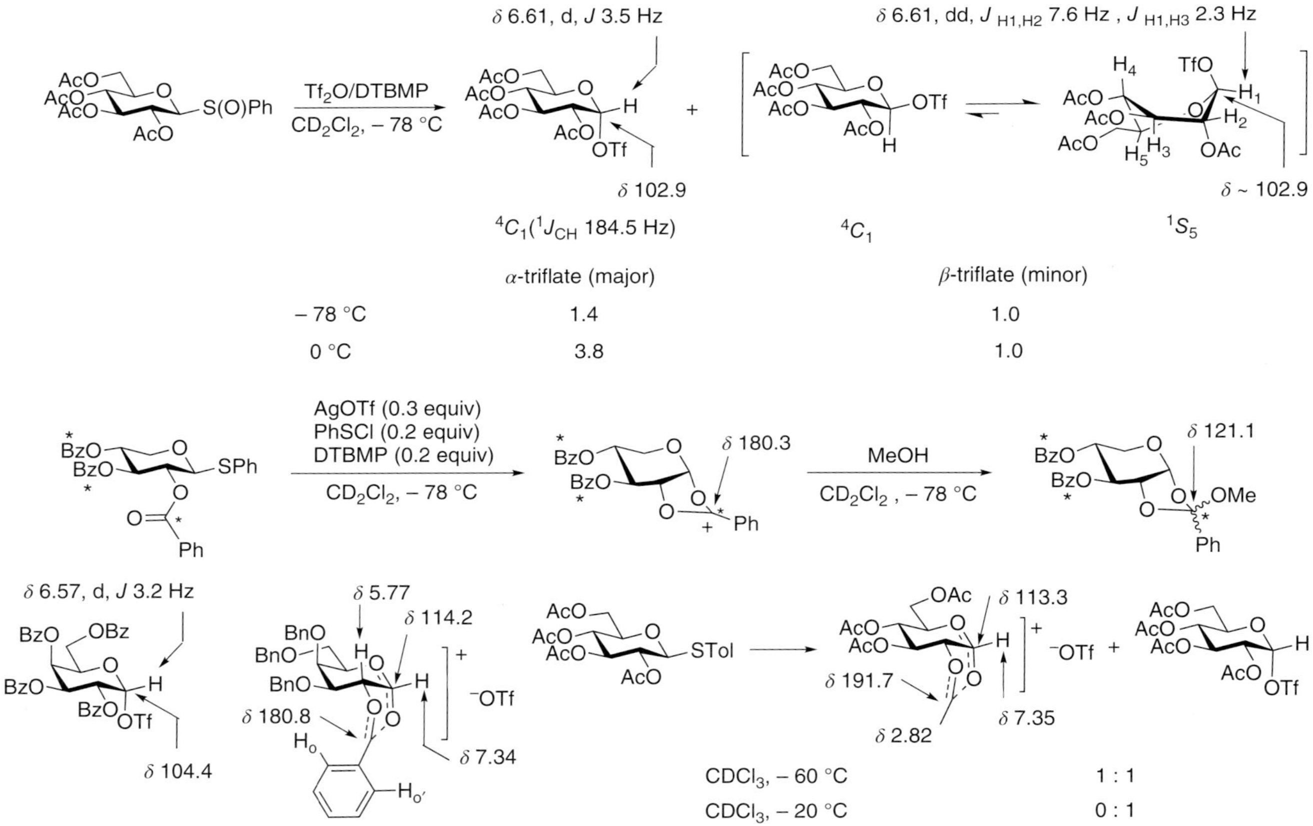

Scheme 6.4 Influence of acyl groups on the formation of anomeric triflates.

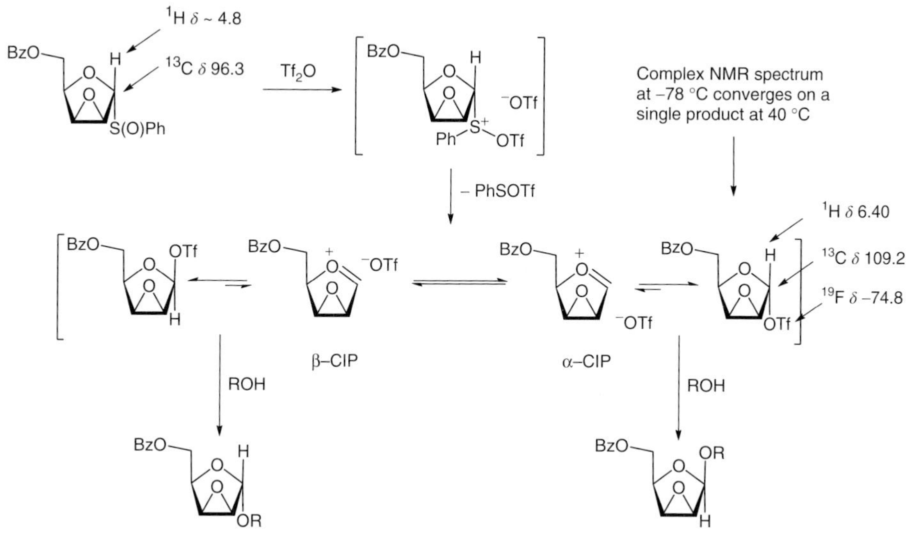

Scheme 6.5 Use of VT NMR to optimize glycosyl triflate formation.

possibility of participation by ester or other groups, glycosylation can be represented as involving a series of donors, activated donors, contact ion pairs (CIPs), solvent-separated ion pairs (SSIPs), and, in extreme circumstances, free oxocarbenium ions (Scheme 6.6) [51, 57, 58]. Reaction of glycosyl acceptors with these intermediates spans the entire range of possibilities from bimolecular S_N2 to unimolecular S_N1.

A central element of the general mechanism is the change in the equilibrium position between the various transient ion pairs and the covalent intermediates with protecting groups and conditions. This determines in large part the degree of association in the

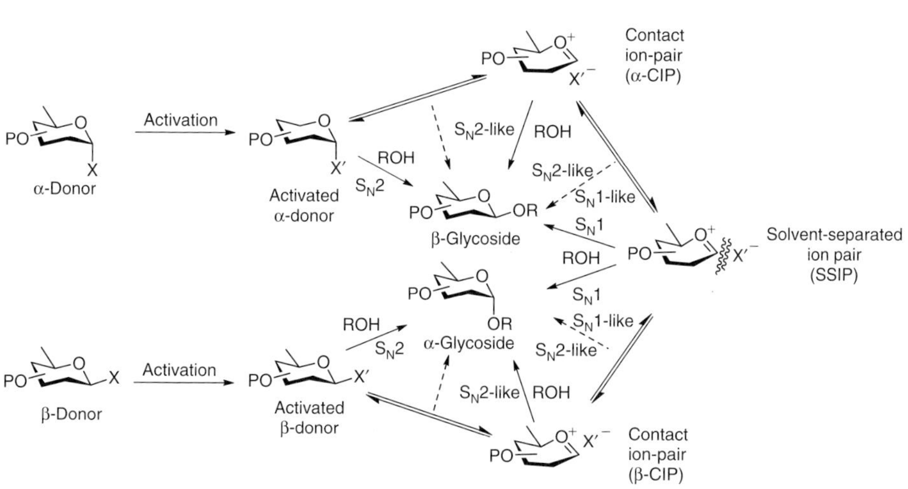

Scheme 6.6 General glycosylation mechanism.

substitution mechanism, that is, the extent to which a given glycosylation can be considered S_N2-like or S_N1-like, and, consequently, is a major factor in the stereoselectivity of the process. Essentially, the closer the equilibrium lies toward the covalent donor, the more the glycosylation reaction will resemble an S_N2 process. Conversely, the looser the ion pair, the more S_N1-like the reaction will be and the greater the relevance of steric and stereoelectronic considerations in the attack of the incoming nucleophile on the free oxocarbenium ion [59–61]. The concept of arming and disarming protecting groups in glycosylation is well established [43, 62–68] and has its roots in the influence of different protecting groups on the ability of the donor to support positive charge in the form of the glycosyl oxocarbenium ion. Solvent effects are similarly understood in terms of the ability of the system to support charge separation and, in certain cases, participation.

The role of the leaving group or counterion in the covalent intermediate/ion pair equilibrium is less appreciated but revolves around the strength of the C1-leaving group bond (bond dissociation energy) and the ability of the leaving group to support negative charge (electron affinity; pK_a of the conjugate acid). Sulfonates are one among many classes of leaving group whose reactivity is defined by the acidity of the sulfonic acids and by the relatively strong covalent C1—O bond. Indeed, it is the very strength of the C1—O bond, as compared to, say, the relatively weak C1—I bond in glycosyl iodides, that makes the covalent glycosyl sulfonates, and in particular the glycosyl triflates (and also glycosyl perchlorates [69]), so readily observable and relevant to the glycosylation mechanism. Similarly, it is the very strength of the C1—O bond in the glycosyl sulfonates that underlies their involvement in many glycosylation reactions even when only catalytic amounts of triflate are present. While the BDE of the C1—O bond will not vary much from one glycosyl sulfonate to another, the pK_a of the conjugate acid will change considerably from ~ −1.9 for methanesulfonic acid to ~ −12 for triflic acid. This gives the experimentalist considerable scope for rational improvement of conditions. Thus, for a given protecting group system, changing from triflate to mesylate will increase the S_N2 character of the glycosylation reaction and, in the absence of other factors such as Curtin–Hammett kinetic scenarios involving minor diastereomers of the covalent donor, increase the stereoselectivity of the process. Of course, reducing the leaving group ability also implies retarding the reaction rate, perhaps to an impractical extent, such that an appropriate balance has to be found. This is readily appreciated in the work of the Bennett laboratory on the synthesis of 2-deoxy-β-glycosides: the greater stability of the 2-deoxy-glycosyl oxocarbenium ion ruled out the use of triflates and led to the use of the toluenesulfonates; the relatively poor reactivity of the α-glycosyl toluene sulfonate required the activation of the acceptor alcohol in the form of its potassium salt for the reaction to proceed at a practical rate [21, 22].

Computational support for the intervention of sulfonate anions in ion pairs with glycosyl oxocarbenium ions has only recently become possible with the discovery that explicit solvation of the sulfonate with three molecules of dichloromethane prevents the otherwise spontaneous collapse to the covalent glycosyl sulfonate [70, 71]. When applied to the 4,6-*O*-benzylidene-directed mannosylation reaction, these calculations provide strong support for the formation of the β-anomers by an S_N2-like process involving an α-CIP [72]. Experimental evidence for S_N2-like pathways in the displacement of glycosyl triflates by alcohols has been provided, at least for the 4,6-*O*-benzylidene-directed β-mannosylations and related glucosylations, by ^{13}C primary kinetic isotope effect measurements [73], with further support derived from a series of cation clock reactions that clearly demonstrate the dependence of glycosylation rate on acceptor concentration [74, 75].

6.5 Applications in *O*-Glycoside Synthesis

Among the numerous applications of the use of glycosyl sulfonates in synthesis [25, 76], the most widespread has undoubtedly been the 4,6-*O*-benzylidene-directed β-mannosylation reaction discovered by Crich and Sun [77] and subsequently applied to the synthesis of numerous mannans and other mannose-containing glycans [78–80]. Application of the process by the Wong group in the synthesis of a β-(1 → 4)-mannan is illustrative (Scheme 6.7) [81].

In the L-rhamnopyranosides, when the use of a stereodirecting 4,6-*O*-benzyliene acetal is not possible, moderate-to-good β-selectivity may be achieved through the combined use of a nonparticipating but electron-withdrawing 2-*O*-sulfonate ester and an electron-withdrawing 4-*O*-ester. Other powerfully electron-withdrawing but non-participating protecting groups at O2 may also be employed to stabilize the anomeric triflate and induce β-rhamnoside formation (Scheme 6.8) [82, 83]. The function of the two electron-withdrawing groups in this protocol is the destabilization of the glycosyl oxocarbenium ion and the consequent favoring of the S_N2-like displacement of the intermediate α-rhamnosyl triflate.

The β-D-rhamnopyranosides are best approached through the use of a 4-*O*,6-*S*-(α-cyanobenzylidene)-protected 6-thiomannosyl donor, readily accessible from D-mannose, with desulfurization by Raney nickel following the glycosylation reaction (Scheme 6.9) [84].

As first demonstrated by the van Boom group [35], and recently exploited by the Boons laboratory in the context of a synthesis of the *Staphylococcus aureus* type 5 trisaccharide repeating unit (Scheme 6.10) [85], the 4,6-*O*-benzylidene effect can also be used in the β-selective formation of 2-deoxy-2-amino mannosides provided the amino group is protected in the form of an azide. In the 4,6-*O*-benzylidene-directed synthesis

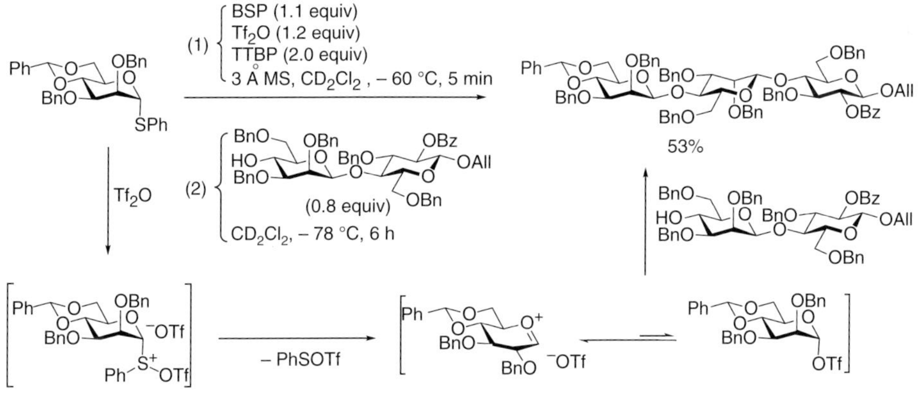

Scheme 6.7 Application of glycosyl sulfonates to the synthesis of a mannan.

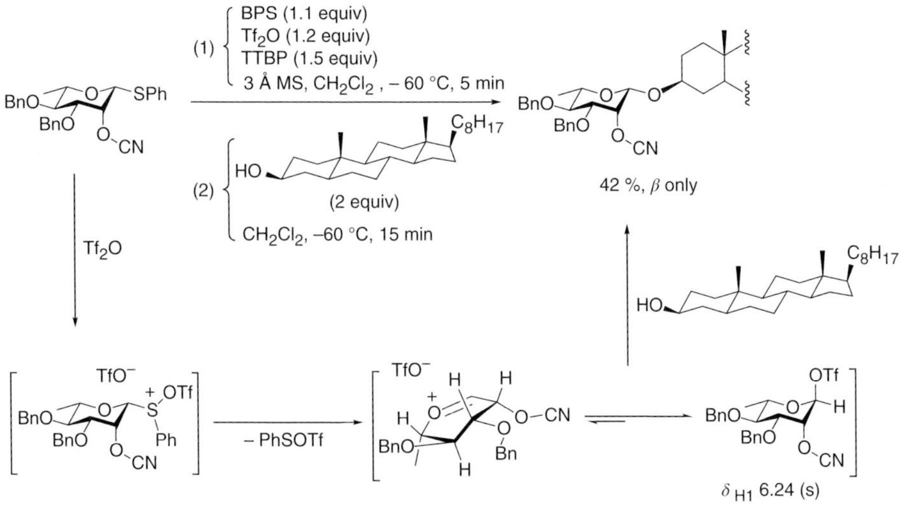

Scheme 6.8 Application of glycosyl sulfonates to the synthesis of a β-L-rhamnoside.

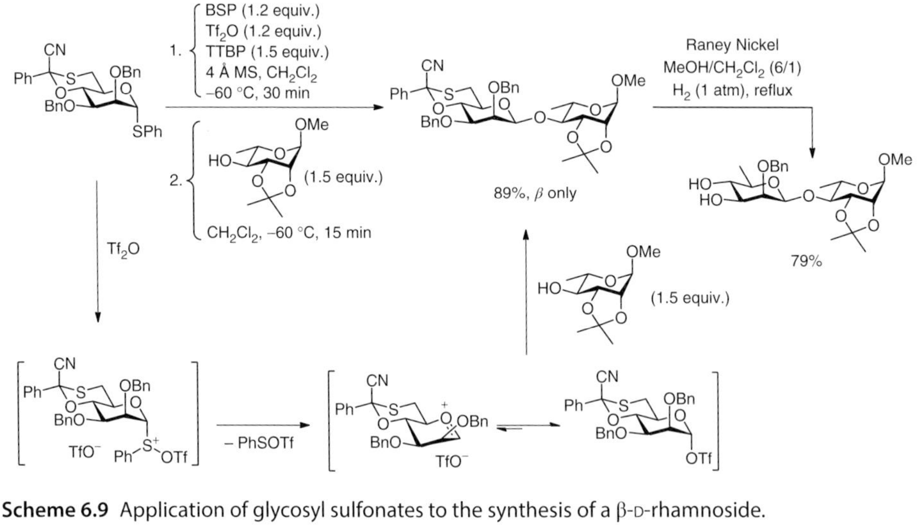

Scheme 6.9 Application of glycosyl sulfonates to the synthesis of a β-D-rhamnoside.

of 3-amino-3-deoxy-β-mannosides, however, the amine function is best protected in the form of a 4-trifluoromethylbenzylidene imine [86].

In the glucopyranose series, the 4,6-*O*-benzylidene acetal is α-directing [87–89]. As supported by kinetic isotope effect measurements [73], cation clock protocols [74, 75], and computations [72], the reaction apparently proceeds through a Curtin–Hammett kinetic scheme in which a reactive β-glucosyl triflate is displaced, more rapidly than the more stable α-triflate, in an S_N2-like manner by the incoming alcohol. An illustrative example is taken from the work of the Maycock group (Scheme 6.11) [90]. In another highly selective synthesis of α-glucopyranosides, this time employing a per-*O*-benzyl-protected

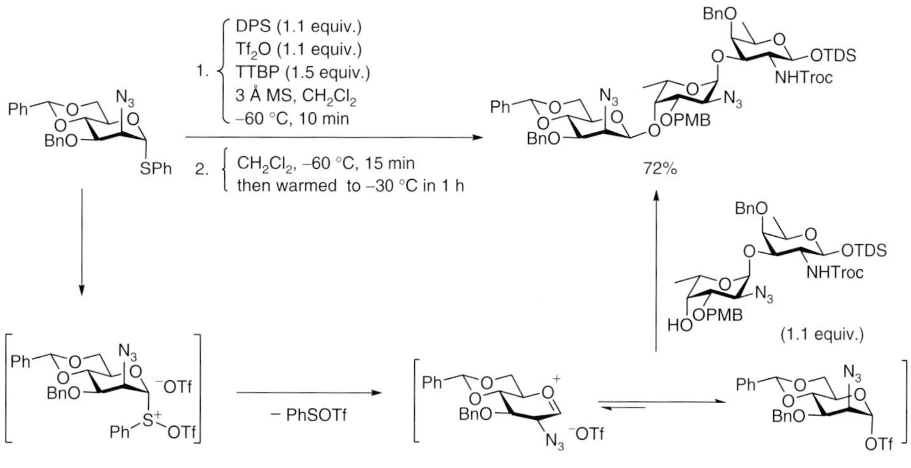

Scheme 6.10 Application of glycosyl sulfonates to the synthesis of a 2-amino-2-deoxy-β-D-mannoside.

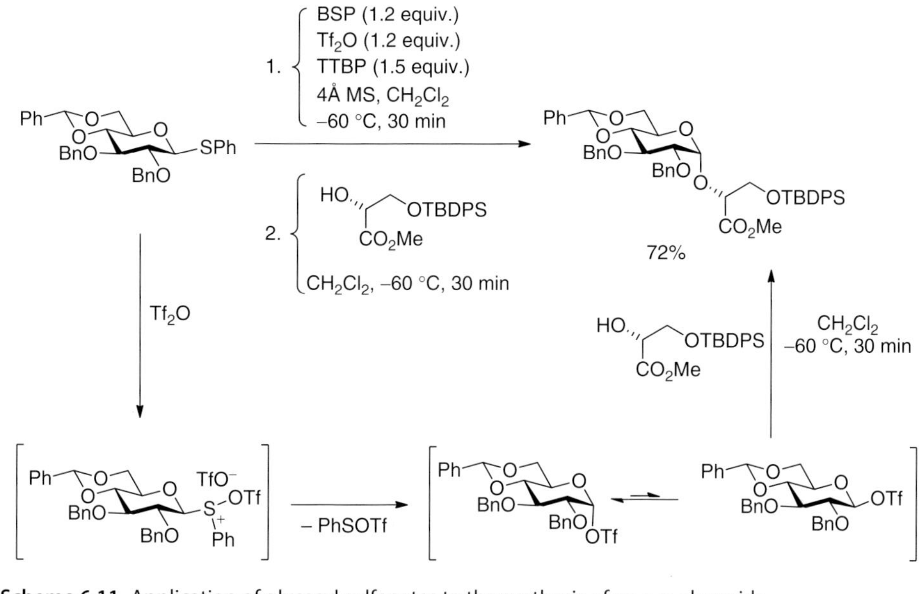

Scheme 6.11 Application of glycosyl sulfonates to the synthesis of an α-D-glucoside.

thioglycoside as donor, an initially generated glucosyl triflate serves as a precursor to the α-glucosyl iodide, which is the effective donor [91].

As discussed earlier (Scheme 6.1), the 2-deoxy-β-glucosides may be expeditiously accessed by the *in situ* generation of a 2-deoxy-α-glucosyl toluenesulfonate, with subsequent displacement by the acceptor alcohol activated as the alkoxide [21, 22]. Scheme 6.12 illustrates the application of this protocol in the context of antibiotic synthesis [92]. The use of alkoxides in this manner to augment the nucleophilicity of the acceptor and so promote S_N2-like reactivity is rare but finds precedent in the

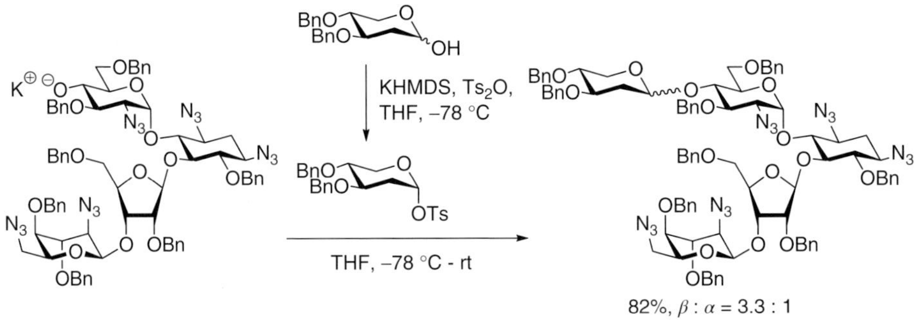

Scheme 6.12 Application of glycosyl sulfonates to the synthesis of a complex 2-deoxy β-D-glucoside.

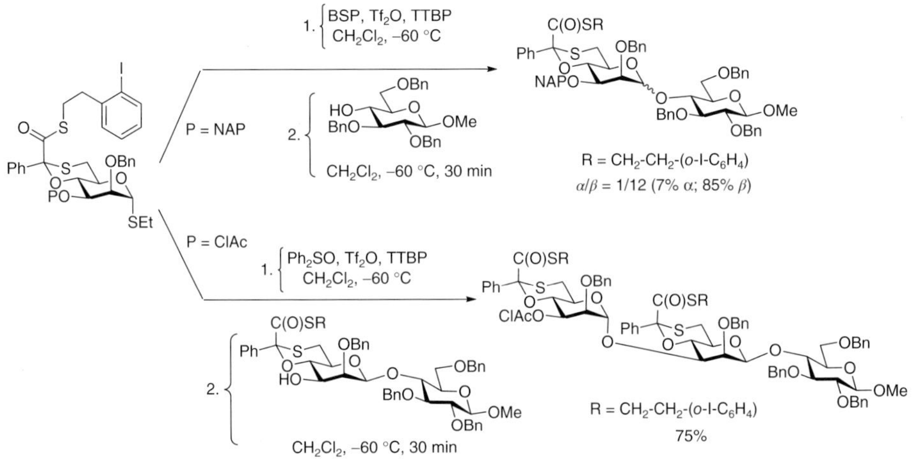

Scheme 6.13 Influence of the 3-O-protecting group on the stereoselectivity of acetal-protected mannosyl donors.

displacement of 2-deoxy-α-glucosyl dithiophosphates with the selective formation of 2-deoxy-β-glucosides [93].

For reasons not yet fully understood [94], but likely not because of remote participation [95], the presence of an ester at the 3-position of a 4,6-*O*-benzylidene-protected mannosyl donor results in an inversion of the usual high β-selectivity and affords the α-mannosides in high yield and selectivity [96]. The striking influence of the protecting group at the 3-position on the anomeric selectivity of 4,6-*O*-acetal-protected mannosyl donors is nicely illustrated by the example of Scheme 6.13 [97].

As discussed earlier, in the mannuronic acid series, the α-glycosyl triflate adopts two equilibrating conformations in which the inverted $^1C_4$ conformer with the equatorial triflate is favored. Such mannuronyl triflates undergo very highly β-selective glycosylations as reported by van der Marel and coworkers (Scheme 6.14) [45, 46, 98–100]. The selectivity of this process is such that, subsequent to glycosylation, reduction of the ester group provides a useful entry into the β-mannosides [101].

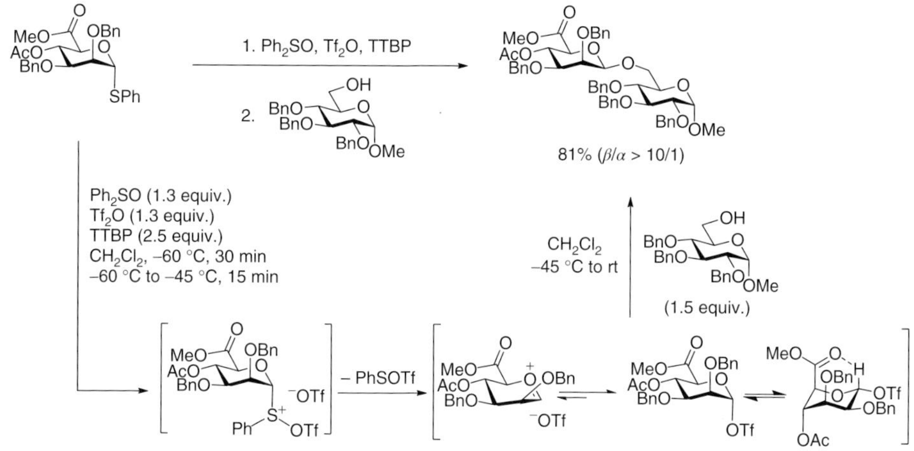

Scheme 6.14 Application of glycosyl sulfonates to the synthesis of an alginate.

Application of neighboring group participation for the control of anomeric stereochemistry in the sulfoxide glycosylation method is beautifully illustrated by Kahne's synthesis of the moenomycin pentasaccharide. In this synthesis, three of the four glycosidic bonds are assembled by the sulfoxide method with the aid of a stereodirecting auxiliary at the 2-position, while the α-selective final linkage makes use of a benzyl ether at that position (Scheme 6.15) [102].

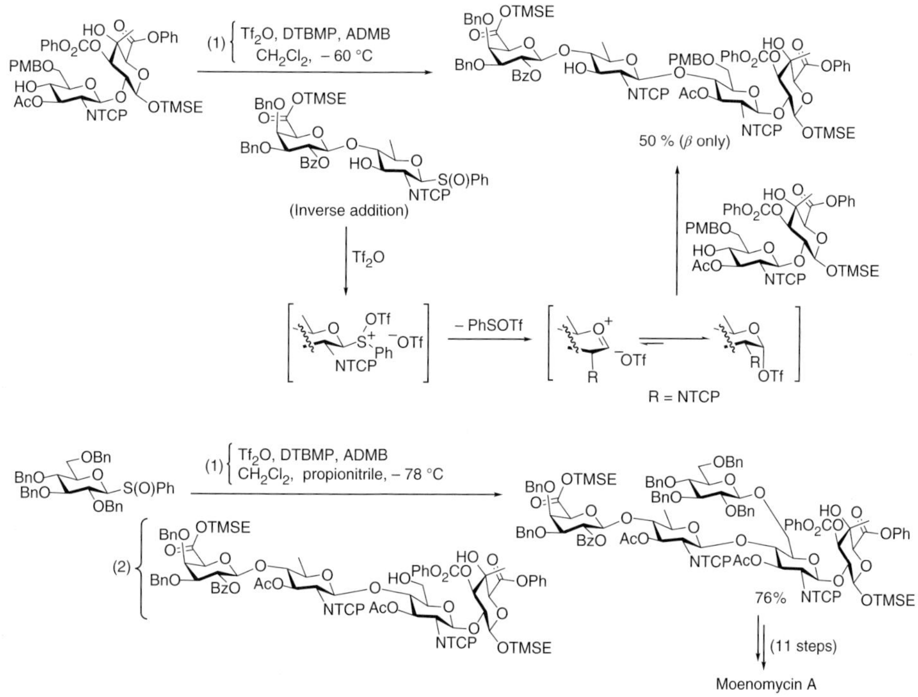

Scheme 6.15 Repeated application of glycosyl sulfonates in the synthesis of the moenomycin pentasaccharide.

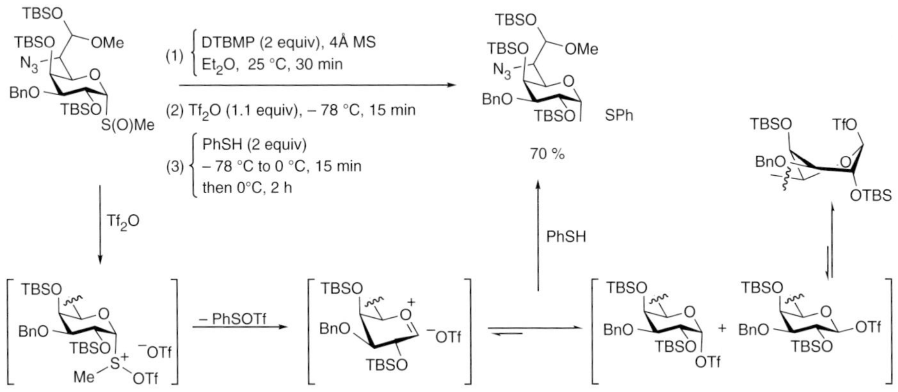

Scheme 6.16 Application of glycosyl sulfonates to the synthesis of S-glycosides.

Although numerous studies have been published on the preactivation of sialyl donors with triflic anhydride, they have typically been conducted in the presence of an excess of a diaryl sulfoxide and/or in the presence of acetonitrile. As such, the activated intermediates are likely to be the sialyl oxysulfonium salts or sialyl nitrilium ions rather than the glycosyl sulfonates [103–108].

6.6 Applications in S-Glycoside Synthesis

Glycosyl triflates generated from glycosyl sulfoxides with triflic anhydride are excellent acceptors for the glycosylation of thiols. In the 4,6-O-benzylidene-protected mannopyranose series, this chemistry affords the β-thiomannosides in high yield and selectivity [109]. A further example, taken from the work of the Vasella group, illustrates application to the formation of lincomycin analogs (Scheme 6.16) [110].

6.7 Applications in C-Glycoside Synthesis

The preactivation of 4,6-O-benzylidene-protected gluco- and mannopyranosyl donors leading to the formation of intermediate mannosyl triflates followed by reaction with sp^2-C carbon nucleophiles follows the pattern established for O-glycoside formation. Thus, with ether protection at the 2- and 3-O-positions, the mannosyl donors are highly β-selective while their glucosyl counterparts are α-selective [111]. This pattern of selectivity is maintained even in the more reactive 2-O-benzyl-4,6-O-benzylidene-3-deoxy gluco- and mannosyl donors prompting the suggestion that, with the weakly nucleophilic C-nucleophiles, reaction occurs via the S_N1 manifold rather than via S_N2-type attack on the covalent triflates, which simply serve as a reservoir for the highly reactive and short-lived oxocarbenium ion [112]. Selectivity is explained by face-selective attack on the opposite face of the oxocarbenium ion (in the $^4H_3$ and $B_{2,5}$ for the gluco- and manno-isomers, respectively) to the pseudoaxial C2—H bond [112]. In the absence of the 4,6-O-benzylidene acetal, the galacto-, gluco-, and manno-isomers are all highly α-selective (Scheme 6.17) [113].

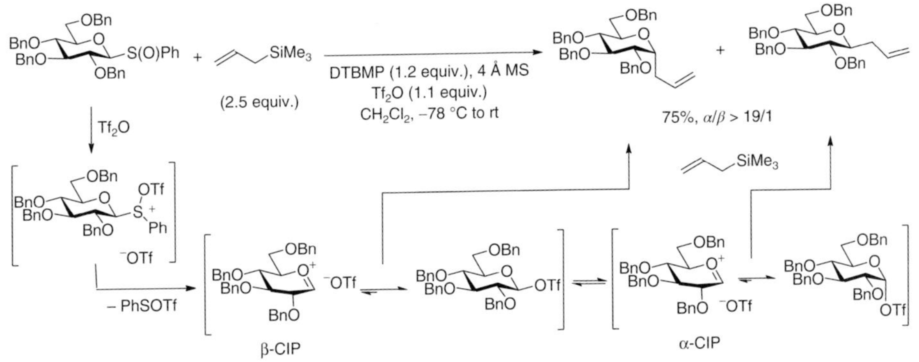

Scheme 6.17 Application of glycosyl sulfonates to the synthesis of *C*-glycosides.

Returning to the 4,6-*O*-benzylidene-protected mannosyl donors, and continuing the parallel with *O*-glycoside formation (Scheme 6.13) [96], replacement of the 3-*O*-benzyl group by an ester results in the reversal of selectivity and formation of the α-*C*-glycoside [114]. A satisfactory explanation has yet to be furnished for the dramatic influence of the 3-*O*-protecting group in these reactions.

6.8 Polymer-Supported Glycosylation with Sulfonates

Triflic-anhydride-activated glycosyl sulfoxides were first applied to the polymer-supported synthesis of oligosaccharides and the parallel synthesis of oligosaccharide libraries, using an acceptor-bound approach, by the Kahne laboratory [115, 116]. The polymer-supported synthesis of β-mannosides was described by Crich and Smith using donor-bound strategy in which the 4,6-*O*-polystyrylboronate group serves both as a surrogate for the more common 4,6-*O*-benzylidene acetal and as a linker to the resin [117]. Seeberger and coworkers subsequently reported the automated synthesis of β-mannosides containing oligosaccharides making use of 4,6-*O*-benzylidene-protected mannosyl 2-carboxybenzyl glycoside donors, resin-bound acceptors, and activation by triflic

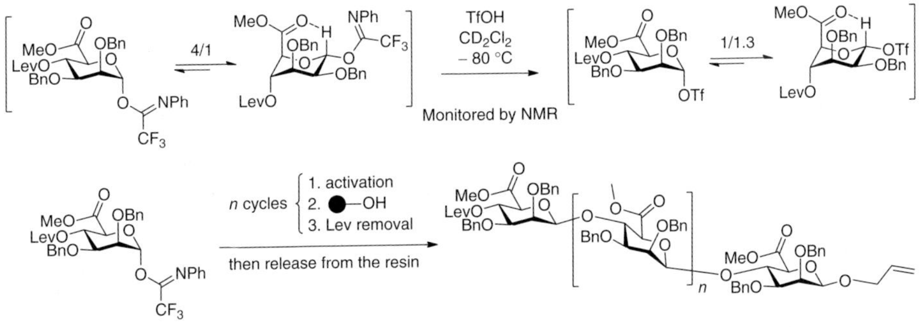

Scheme 6.18 Application of glycosyl sulfonates in the polymer-supported synthesis of oligosaccharides.

anhydride [118]. To date, the most impressive use of glycosyl sulfonates in polymer-supported oligosaccharide synthesis, however, has been the β-selective synthesis of oligomannuronates by Codée and coworkers employing an *N*-phenyl trifluoroacetimidate donor, activated by TMS triflate, in the repeated glycosylation of the polymer-supported acceptor (Scheme 6.18).

6.9 Conclusion

From the mechanistic standpoint, the glycosyl sulfonates comprise one of the most extensively studied classes of glycosyl donor and serve as a paradigm for the general mechanism of glycosylation. Glycosyl sulfonate chemistry has afforded tried and tested solutions to several of what were previously considered "difficult" classes of glycosidic bond.

References

1 Helferich, B. and Gootz, R. (1929) *Ber. Dtsch. Chem. Ges.*, 62, 2788–2792.
2 Kronzer, F.J. and Schuerch, C. (1973) *Carbohydr. Res.*, 27, 379–390.
3 Eby, R. and Schuerch, C. (1974) *Carbohydr. Res.*, 34, 79–90.
4 Marousek, V., Lucas, T.J., Wheat, P.E., and Schuerch, C. (1978) *Carbohydr. Res.*, 60, 85–96.
5 Lucas, T.J. and Schuerch, C. (1975) *Carbohydr. Res.*, 39, 39–45.
6 Srivastava, V.K. and Schuerch, C. (1980) *Carbohydr. Res.*, 79, C13–C16.
7 Srivastava, V.K. and Schuerch, C. (1981) *J. Org. Chem.*, 46, 1121–1126.
8 Crich, D. (2001) in Glycochemistry: Principles, Synthesis, and Applications (eds P.G. Wang and C.R. Bertozzi), Dekker, New York, pp. 53–75.
9 Crich, D. (2002) *J. Carbohydr. Chem.*, 21, 663–686.
10 Garcia, B.A. and Gin, D.Y. (2000) *J. Am. Chem. Soc.*, 122, 4269–4279.
11 Leroux, J. and Perlin, A.S. (1976) *Carbohydr. Res.*, 47, C8–C10.
12 Leroux, J. and Perlin, A.S. (1978) *Carbohydr. Res.*, 67, 163–178.
13 Szeja, W. (1988) *Synthesis*, 223–224.
14 Szeja, W. and Bogusiak, J. (1988) *Synthesis*, 1988, 224–225.
15 Morishima, N., Koto, S., and Zen, S. (1982) *Chem. Lett.*, 1039–1040.
16 Koto, S., Sato, T., Morishima, N., and Zen, S. (1980) *Bull. Chem. Soc. Jpn.*, 53, 1761–1762.
17 Koto, S., Morishima, N., and Zen, S. (1984) *Carbohydr. Res.*, 130, 73–83.
18 Pavia, A.A., Rocheville, J.M., and Ung, S.N. (1980) *Carbohydr. Res.*, 79, 79–89.
19 Lacombe, J.M., Pavia, A.A., and Rocheville, J.M. (1981) *Can. J. Chem.*, 59, 473–481.
20 Pavia, A.A. and Ung-Chhun, S.N. (1981) *Can. J. Chem.*, 59, 482–489.
21 Issa, J.P., Lloyd, D., Steliotes, E., and Bennett, C.S. (2013) *Org. Lett.*, 15, 4170–4173.
22 Issa, J.P. and Bennett, C.S. (2014) *J. Am. Chem. Soc.*, 136, 5740–5744.
23 Kim, K.S., Fulse, D.B., Baek, J.Y., Lee, B.-Y., and Jeon, H.B. (2008) *J. Am. Chem. Soc.*, 130, 8537–8547.
24 Crich, D. and Sun, S. (1997) *J. Am. Chem. Soc.*, 119, 11217–11223.
25 Crich, D. and Lim, L.B.L. (2004) *Org. React.*, 64, 115–251.

26 Gildersleeve, J., Pascal, R.A., and Kahne, D. (1998) *J. Am. Chem. Soc.*, 120, 5961–5969.

27 Crich, D. and Sun, S. (1998) *J. Am. Chem. Soc.*, 120, 435–436.

28 Huang, X., Huang, L., Wang, H., and Ye, X.-S. (2004) *Angew. Chem. Int. Ed.*, 43, 5221–5224.

29 Yang, L., Qin, Q., and Ye, X.-S. (2013) *Chem. Asian J.*, 2, 30–49.

30 Verma, V.P. and Wang, C.-C. (2013) *Chem. Eur. J.*, 19, 846–851.

31 Crich, D. and Smith, M. (2000) *Org. Lett.*, 2, 4067–4069.

32 Crich, D. and Smith, M. (2001) *J. Am. Chem. Soc.*, 123, 9015–9020.

33 Crich, D., Banerjee, A., Li, W., and Yao, Q. (2005) *J. Carbohydr. Chem.*, 24, 415–424.

34 Codée, J.D.C., Litjens, R.E.J.N., den Heeten, R., Overkleeft, H.S., van Boom, J.H., and van der Marel, G.A. (2003) *Org. Lett.*, 5, 1519–1522.

35 Codée, J.D.C., van den Bos, L.J., Litjens, R.E.J.N., Overkleeft, H.S., van Boeckel, C.A.A., van Boom, J.H., and van der Marel, G.A. (2004) *Tetrahedron*, 60, 1057–1064.

36 Durón, S.G., Polat, T., and Wong, C.-H. (2004) *Org. Lett.*, 6, 839–841.

37 Wang, C., Wang, H., Huang, X., Zhang, L.-H., and Ye, X.-S. (2006) *Synlett*, 2846–2850.

38 Tatai, J. and Fügedi, P. (2007) *Org. Lett.*, 9, 4647–4650.

39 Nokami, T., Shibuya, A., Tsuyama, H., Suga, S., Bowers, A.A., Crich, D., and Yoshida, J.-i. (2007) *J. Am. Chem. Soc.*, 129, 10922–10928.

40 Rencurosi, A., Lay, L., Russo, G., Caneva, E., and Poletti, L. (2006) *Carbohydr. Res.*, 341, 903–908.

41 Walvoort, M.T.C., van den Elst, H., Plante, O.J., Kröck, L., Seeberger, P.H., Overkleeft, H.S., van der Marel, G.A., and Codée, J.D.C. (2012) *Angew. Chem. Int. Ed.*, 51, 4393–4396.

42 Peng, P., Xiong, D.-C., and Ye, X.-S. (2014) *Carbohydr. Res.*, 384, 1–8.

43 Aubry, S., Sasaki, K., Sharma, I., and Crich, D. (2011) *Top. Curr. Chem.*, 301, 141–188.

44 Frihed, T.G., Bols, M., and Pedersen, C.M. (2015) *Chem. Rev.*, 115, 4963–5013.

45 Walvoort, M.T.C., Lodder, G., Mazurek, J., Overkleeft, H.S., Codée, J.D.C., and van der Marel, G.A. (2009) *J. Am. Chem. Soc.*, 131, 12080–12081.

46 Walvoort, M.T.C., de Witte, W., van Dijk, J., Jasper Dinkelaar, J., Lodder, G., Overkleeft, H.S., Codée, J.D.C., and van der Marel, G.A. (2011) *Org. Lett.*, 13, 4360–4363.

47 Rönnols, J., Walvoort, M.T.C., van der Marel, G.A., Codée, J.D.C., and Widmalm, G. (2013) *Org. Biomol. Chem.*, 11, 8127–8134.

48 Crich, D., Dai, Z., and Gastaldi, S. (1999) *J. Org. Chem.*, 64, 5224–5229.

49 Zeng, Y., Wang, Z., Whitfield, D., and Huang, X. (2008) *J. Org. Chem.*, 73, 7952–7962.

50 Callam, C.S., Gadikota, R.R., Krein, D.M., and Lowary, T.L. (2003) *J. Am. Chem. Soc.*, 125, 13112–13119.

51 Bohé, L. and Crich, D. (2011) *C.R. Chim.*, 14, 3–16.

52 Martin, A., Arda, A., Désiré, J., Martin-Mingot, A., Probst N., Sinaÿ, P., Jiménez-Barbero, J., Thibaudeau, S., Blériot, Y. (2016) *Nature Chem.* 8, 186–191.

53 Akien, G.R. and Subramaniam, B. (2014) 247th ACS National Meeting & Exhibition Dallas, 2014, p. CARB-96.

54 Winstein, S., Clippinger, E., Fainberg, A.H., Heck, R., and Robinson, G.C. (1956) *J. Am. Chem. Soc.*, 78, 328–335.

55 Rhind-Tutt, A.J. and Vernon, C.A. (1960) *J. Chem. Soc.*, 4637–4644.

56 Lemieux, R.U., Hendriks, K.B., Stick, R.V., and James, K. (1975) *J. Am. Chem. Soc.*, 97, 4056–4062.

57 Bohé, L. and Crich, D. (2015) *Carbohydr. Res.*, 403, 48–59.

58 Bohé, L. and Crich, D. (2014) Comprehensive Organic Synthesis, 2nd edn, vol. 6, Elsevier, Oxford, pp. 1–33.
59 Smith, D.M. and Woerpel, K.A. (2006) *Org. Biomol. Chem.*, 4, 1195–1201.
60 Walvoort, M.T.C., Dinkelar, J., van den Bos, L.J., Lodder, G., Overkleeft, H.S., Codée, J.D.C., and van der Marel, G.A. (2010) *Carbohydr. Res.*, 345, 1252–1263.
61 Cumpstey, I. (2012) *Org. Biomol. Chem.*, 10, 2503–2508.
62 Fraser-Reid, B. and Lopez, C. (2011) *Top. Curr. Chem.*, 301, 1–29.
63 Gomez, A.M. (2011) *Top. Curr. Chem.*, 301, 31–68.
64 Wu, C.-Y. and Wong, C.-H. (2011) *Top. Curr. Chem.*, 301, 223–252.
65 Premathilake, H.D. and Demchenko, A.V. (2011) *Top. Curr. Chem.*, 301, 189–222.
66 Pedersen, C.M., Marinescu, L.G., and Bols, M. (2011) *C.R. Chim.*, 14, 17–43.
67 Jensen, H.H. and Bols, M. (2006) *Acc. Chem. Res.*, 39, 259–265.
68 Heuckendorff, M., Pedersen, C.M., and Bols, M. (2010) *Chem. Eur. J.*, 16, 13982–13994.
69 Igarashi, K., Honma, T., and Irisawa, J. (1970) *Carbohydr. Res.*, 15, 329–337.
70 Hosoya, T., Takano, T., Kosma, P., and Rosenau, T. (2014) *J. Org. Chem.*, 79, 7889–7894.
71 Hosoya, T., Kosma, P., and Rosenau, T. (2015) *Carbohydr. Res.*, 401, 127–131.
72 Hosoya, T., Kosma, P., and Rosenau, T. (2015) *Carbohydr. Res.*, 411, 64–69.
73 Huang, M., Garrett, G.E., Birlirakis, N., Bohé, L., Pratt, D.A., and Crich, D. (2012) *Nat. Chem.*, 4, 663–667.
74 Huang, M., Retailleau, P., Bohé, L., and Crich, D. (2012) *J. Am. Chem. Soc.*, 134, 14746–14749.
75 Adero, P.O., Furukawa, T., Huang, M., Mukherjee, D., Retailleau, P., Bohé, L., and Crich, D. (2015) *J. Am. Chem. Soc.*, 137, 10336–10345.
76 Crich, D. and Bowers, A.A. (2008) in Handbook of Chemical Glycosylation: Advances in Stereoselectivity and Therapeutic Relevance (ed. A.V. Demchenko), Wiley-VCH Verlag GmbH, Weinheim, pp. 303–329.
77 Crich, D. and Sun, S. (1997) *J. Org. Chem.*, 62, 1198–1199.
78 Crich, D., Li, H., Yao, Q., Wink, D.J., Sommer, R.D., and Rheingold, A.L. (2001) *J. Am. Chem. Soc.*, 123, 5826–5828.
79 Cai, F., Wu, B., and Crich, D. (2009) *Adv. Carbohydr. Chem. Biochem.*, 62, 251–309.
80 Rahkila, J., Ekholm, F.S., Panchadhayee, R., Ardá, A., Cañada, F.J., Jiménez-Barbero, J., and Leino, R. (2014) *Carbohydr. Res.*, 383, 58–68.
81 Ohara, K., Lin, C.-C., Yang, P.-J., Hung, W.-T., Yang, W.-B., Cheng, T.-J.R., Fang, J.-M., and Wong, C.-H. (2013) *J. Org. Chem.*, 78, 6390–6411.
82 Crich, D. and Picione, J. (2003) *Org. Lett.*, 5, 781–784.
83 Crich, D., Hutton, T.K., Banerjee, A., Jayalath, P., and Picione, J. (2005) *Tetrahedron: Asymmetry*, 16, 105–119.
84 Crich, D. and Li, L. (2009) *J. Org. Chem.*, 74, 773–781.
85 Gagarinov, I.A., Fang, T., Liu, L., Srivastava, A.D., and Boons, G.-J. (2015) *Org. Lett.*, 17, 928–931.
86 Crich, D. and Xu, H. (2007) *J. Org. Chem.*, 72, 5183–5192.
87 Bousquet, E., Khitri, M., Lay, L., Nicotra, F., Panza, L., and Russo, G. (1998) *Carbohydr. Res.*, 311, 171–181.
88 Crich, D. and Cai, W. (1999) *J. Org. Chem.*, 64, 4926–4930.
89 Crich, D. and Li, L. (2007) *J. Org. Chem.*, 72, 1681–1690.
90 Lourenço, E.C., Maycock, C.D., and Ventura, M.R. (2009) *Carbohydr. Res.*, 344, 2073–2078.

91 Chu, A.-H.A., Nguyen, S.H., Sisel, J.A., Minciunescu, A., and Bennett, C.S. (2013) *Org. Lett.*, 15, 2566–2569.
92 Matsushita, T., Chen, W., Juskeviciene, R., Teo, Y., Shcherbakov, D., Vasella, A., Böttger, E.C., and Crich, D. (2015) *J. Am. Chem. Soc.*, 137, 7706–7717.
93 Michalska, M. and Borowiecka, J. (1983) *J. Carbohydr. Chem.*, 2, 99–103.
94 Komarova, B.S., Ustyuzhania, N.E., Tsvetkov, Y.E., and Nifantiev, N.E. (2014) in Modern Synthetic Methods in Carbohydrate Chemistry: From Monosaccharides to Complex Glycoconjugates (eds D.B. Werz and S. Vidal), Wiley-VCH Verlag GmbH, Weinheim, pp. 125–160.
95 Crich, D., Hu, T., and Cai, F. (2008) *J. Org. Chem.*, 73, 8942–8953.
96 Crich, D., Cai, W., and Dai, Z. (2000) *J. Org. Chem.*, 65, 1291–1297.
97 Crich, D. and Yao, Q. (2004) *J. Am. Chem. Soc.*, 126, 8232–8236.
98 van den Bos, L.J., Dinkelaar, J., Overkleeft, H.S., and van der Marel, G.A. (2006) *J. Am. Chem. Soc.*, 128, 13066–13067.
99 Dinkelaar, J., de Jong, A.R., van Meer, R., Somers, M., Lodder, G., Overkleeft, H.S., Codée, J.D.C., and van der Marel, G.A. (2009) *J. Org. Chem.*, 74, 4982–4991.
100 Codée, J.D.C., Christina, A.E., Walvoort, M.T.C., Overkleeft, H.S., and van der Marel, G.A. (2011) *Top. Curr. Chem.*, 301, 253–290.
101 Tang, S.-L. and Pohl, N.L.B. (2015) *Org. Lett.*, 17, 2642–2645.
102 Taylor, J.G., Li, X., Oberthür, M., Zhu, W., and Kahne, D.E. (2006) *J. Am. Chem. Soc.*, 128, 15084–15085.
103 Crich, D. and Li, W. (2006) *Org. Lett.*, 8, 959–962.
104 Sun, B. and Jiang, H. (2011) *Tetrahedron Lett.*, 52, 6035–6038.
105 Wang, Y.-J., Jia, J., Gu, Z.-Y., Liang, F.-F., Li, R.-C., Huang, M.-H., Xu, C.-S., Zhang, J.-X., Men, Y., and Xing, G.-W. (2011) *Carbohydr. Res.*, 346, 1271–1276.
106 Gu, Z.-y., Zhang, J.-x., and Xing, G.-w. (2012) *Chem. Asian J.*, 7, 1524–1528.
107 Wang, Y., Xu, F.-F., and Ye, X.-S. (2012) *Tetrahedron Lett.*, 53, 3658–3662.
108 Gu, Z.-y., Zhang, X.-t., Zhang, J.-x., and Xing, G.-w. (2013) *Org. Biomol. Chem.*, 11, 5017–5022.
109 Crich, D. and Li, H. (2000) *J. Org. Chem.*, 65, 801–805.
110 Collin, M.-P., Hobbie, S.N., Böttger, E.C., and Vasella, A. (2009) *Helv. Chim. Acta*, 92, 230–266.
111 Crich, D. and Sharma, I. (2008) *Org. Lett.*, 10, 4731–4734.
112 Moumé-Pymbock, M. and Crich, D. (2012) *J. Org. Chem.*, 77, 8905–8912.
113 McGarvey, G.J., LeClair, C.A., and Schmidtmann, B.A. (2008) *Org. Lett.*, 10, 4727–4730.
114 Crich, D. and Sharma, I. (2010) *J. Org. Chem.*, 75, 8383–8391.
115 Yan, L., Taylor, C.M., Goodnow, R., and Kahne, D. (1994) *J. Am. Chem. Soc.*, 116, 6953–6954.
116 Liang, R., Yan, L., Loebach, J., Ge, M., Uozumi, Y., Sekanina, K., Horan, N., Gildersleeve, J., Thompson, C., Smith, A., Biswas, K., Still, W.C., and Kahne, D. (1996) *Science*, 274, 1520–1522.
117 Crich, D. and Smith, M. (2002) *J. Am. Chem. Soc.*, 124, 8867–8869.
118 Codée, J.D.C., Kröck, L., Castagner, B., and Seeberger, P.H. (2008) *Chem. Eur. J.*, 14, 3987–3994.

Part III

Catalytic Activation of Glycosides

7

Stereoselective C-Glycosylation from Glycal Scaffolds

Kim Le Mai Hoang, Wei-Lin Leng, Yu-Jia Tan, and Xue-Wei Liu

7.1 Introduction

Complex carbohydrate structures display very diverse and deep roles in many vital biological pathways, from extensive network of hyaluronans forming the backbone of extracellular matrix in the interstitial space to the dense arrays of sugars coating the cell surfaces, the so-called glycocalyx, and even deep within the nucleus and cytoplasm serving as various regulatory switches [1]. This effectively puts down the initial assumption that glycans were just part of the structural support and energy reservoirs for the organisms but delegate them to the front row of many critical interactions between a cell and its environment.

Among the important glycans discovered so far, *C*-glycosides have attracted considerable interest as the more chemically and enzymatically stable analogs of their *O*- and *N*-glycoconjugates such as glycolipids and glycoproteins. It has been observed that *C*-glycosides possessing similar 3D structures display comparable biological activities but with longer lifetime *in vivo* [2]. One example is depicted in Figure 7.1. The discovery of agelasphines [2, 3], novel anticancer sphingolipids isolated from the marine sponge *Agelas mauritianus*, has led to the development of KRN7000, a synthetic sphingolipid. Remarkably, the *C*-glycoside analog of KRN7000 was reported to exhibit 1000 times more effectiveness against mouse malaria and 100 times increase in activity against mouse melanoma, compared to KRN7000. Since then, many natural occurring *C*-glycosides have been isolated and studied (Figure 7.2), showing promising pharmacological and biological properties.

Extensive research on stereoselective synthesis of *C*-glycosides employed a variety of starting materials, including saturated sugars as well as 1,2-unsaturated glycal donors. Even though there are numerous methods available for making anomeric carbon–carbon bonds from fully saturated structures, through transformation of C-1 into an electrophilic, nucleophilic, or radical moiety, the selectivity in most cases originated from substrate control [4]. Catalytic cross-coupling reactions from saturated sugars have seen relatively moderate success due to the susceptibility of anomeric-substituted metal complex to undergo β-hydride or alkoxy elimination. Hence, glycal donors are

Selective Glycosylations: Synthetic Methods and Catalysts, First Edition. Edited by Clay S. Bennett.

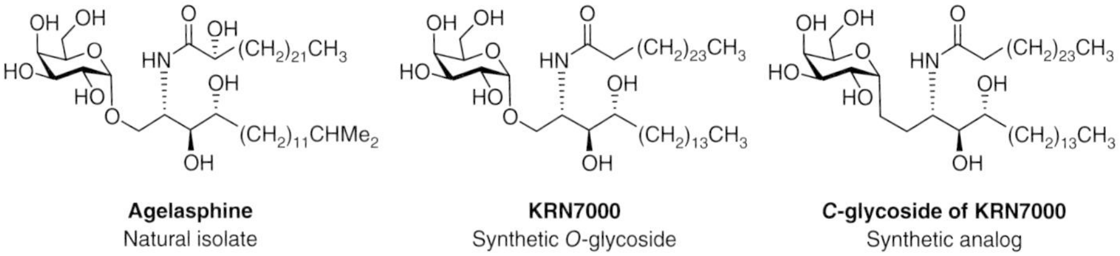

Figure 7.1 Agelasphines, novel anticancer sphingolipids, and their derived compounds.

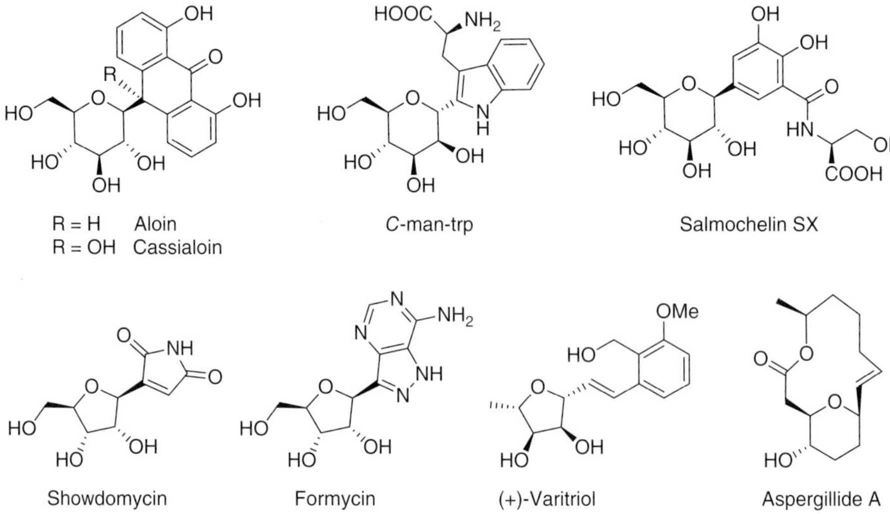

Figure 7.2 Examples of naturally occurring C-glycosides.

regarded as suitable scaffolds for a metal-catalyzed approach. Herein, stereoselective C-glycosylations from glycals are reviewed with an emphasis on the underlying mechanisms.

7.2 Classification of *C*-Glycosylation Reactions

The majority of C-glycosylation reactions employing glycals fall into one of the following three categories: Ferrier-type rearrangement, Pd-catalyzed Heck-type, or Pd-catalyzed Tsuji–Trost-type glycosylation (Scheme 7.1, Eqs. (1)–(3), respectively). Other methods include [3,3]-sigmatropic rearrangement and *N*-heterocyclic carbene (NHC)-catalyzed glycosylations.

7.3 Ferrier-Type Rearrangement

Among the reported methods, Lewis-acid-promoted Ferrier rearrangement [5] is undoubtedly receiving the most attention. First observed by Fischer in 1914 [6], the synthetic utility of this transformation was fully realized by Ferrier [7] much later.

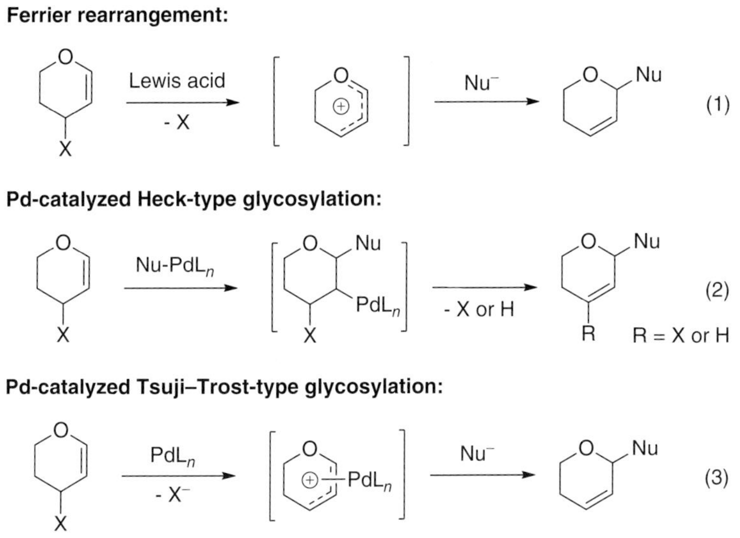

Scheme 7.1 General strategies of 2,3-unsaturated sugar synthesis.

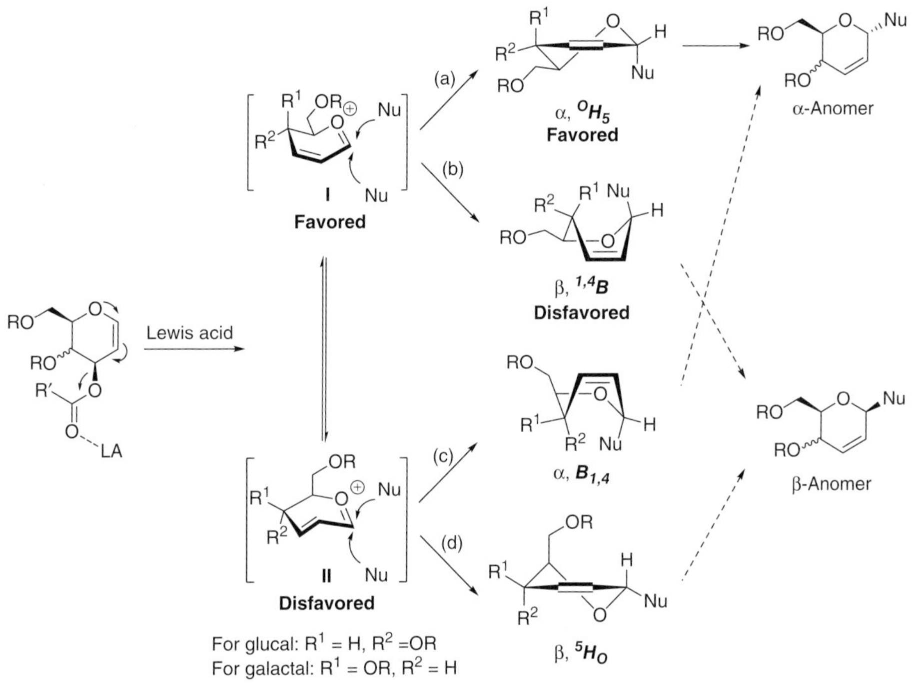

Scheme 7.2 Stereoselectivity of Ferrier-type rearrangement.

As seen from Scheme 7.2, the Lewis acid activates the acyl group at C-3 position to generate the oxocarbenium ion, which is attacked by a carbon nucleophile. The stereochemistry is mainly governed by the conformation of oxocarbenium intermediate as well as the attack pathway of the lower energy conformation **I** over **II**. Bottom face attack of **I** leads to favored half-chair conformation ${}^{O}H_5$, whereas top face attack of **I** would give rise to the disfavored boat conformation ${}^{1,4}B$. Consequently, α-anomer is the major product, but it is often difficult to completely suppress the formation of β-anomer.

Table 7.1 illustrates the typical Lewis acids and nucleophiles involved in Ferrier reactions with glycals as starting materials. The stereochemical outcome of the reactions generally favors the α-isomer, which is consistent with the underlying mechanism described *vide-supra*. The versatility of allyltrimethylsilanes as acceptors is evident in the good yields and selectivities obtained (entries 1–7). Strong Lewis acids such as $TiCl_4$ and $BF_3{\cdot}OEt_2$ were employed first [8, 9]. Since then, milder reagents [10–14] such as $Yb(OTf)_3$ and DDQ were found to be efficient in promoting the Ferrier rearrangement. Alkynyl [15] (entry 8) and propargyl silanes [16] (entry 9) were suitable nucleophiles to produce acetylenic and allenic glycosides, respectively. Similarly, silylallene nucleophiles [17] reacted smoothly to give propargylic glycosides with good yield and selectivity (entry 10). The scope of nucleophiles was further extended to silyl enol ethers [18] (entries 11–13), with $Yb(OTf)_3$ notably being carried out at ambient temperature in ionic liquid [18c]. Silylated alkyne iodide, in the presence of indium catalyst [19], could afford propargylic glycosides with retention of the silylated group (entry 14). Potassium alkynyltrifluoroborates [20], being air and moisture stable, were suitable nucleophiles in the $BF_3{\cdot}OEt_2$-catalyzed Ferrier reaction (entry 15). It was postulated that $BF_3{\cdot}OEt_2$ first reacted with alkynyltrifluoroborate to generate a reactive intermediate, which activated the acetate group in glucal. Compared to the aforementioned nucleophiles, C(sp^3)-hybridized examples were relatively rare, with reports of diethylmalonate [14] (entry 16) and diethyl zinc [21] (entry 17) as suitable glycoside acceptors in the Ferrier reaction.

As discussed earlier, hexopyranose glycal donors favored α-anomers through the half-chair conformation ${}^{O}H_5$. On the other hand, pentopyranose glycal donors were found to preferentially produce the 1,4-*anti*-adduct due to favorable axial orientation of C-4 substituent. Table 7.2 summarizes the reaction on xylal and arabinal substrates [22–24]. In the case of xylal donors, the oxocarbenium intermediate adopted conformation **II** to minimize the allylic strain. Bottom face attack of **II** led to favored half-chair conformation ${}^{5}H_O$, whereas top face attack of **II** would give rise to the disfavored boat conformation $B_{1,4}$. Therefore, 1,4-anti-anomer is the major product.

Both xylal and arabinal reacted efficiently with allylsilane in the presence of $BF_3{\cdot}OEt_2$ as the Lewis acid (entries 1 and 4), to afford allylic glycosides in good yield and 1,4-anti-stereoselectivities. Reactions with bis(trimethylsilyl)acetylene similarly proceeded smoothly, and high yields and selectivities were observed (entries 2 and 5). Low temperatures (−40 to −15 °C) were generally sufficient for the reactions with alkynyl- and allylsilanes. Reaction with vinylsilyl ether (entry 3) was comparatively slower and, hence, was conducted at 0 °C. Indium-catalyzed cyanation of arabinal was demonstrated to be efficient (entry 6), producing the desired glycoside in high yield and moderate α-selectivity. The reaction was essentially complete within 30 s under microwave conditions.

7.4 Pd-Catalyzed Heck-Type

The 2010 Nobel-winning Mizoroki–Heck reaction is commonly used to construct C—C bond in organic syntheses [25]. Given the vinyl feature of glycals, the Heck reaction is

Table 7.1 Lewis-acid-catalyzed Ferrier rearrangements with *C*-nucleophiles.

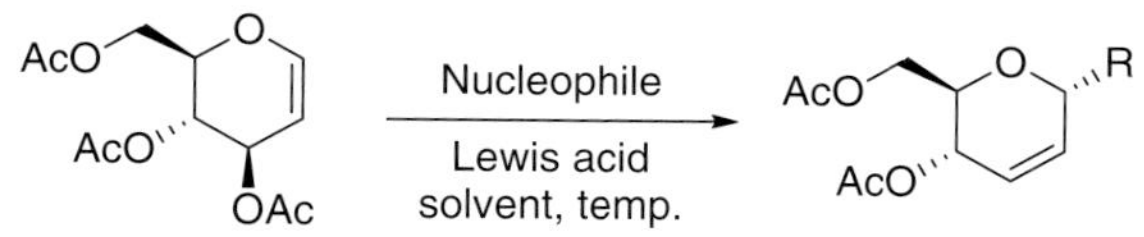

Entry	Nucleophile	Lewis acid	Temperature	Products: R	Products: Yield ($\alpha:\beta$)	References
1[a]		$TiCl_4$	−78 °C	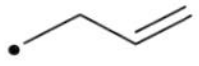	85% (16:1)	[8]
2[a]		$BF_3{\cdot}OEt_2$	−50 °C		99% (16:1)	[9]
3[a]		$Yb(OTf)_3$	rt		94% (α only)	[10]
4[b]		$Zn(OTf)_2$	40 °C		97% (99:1)	[11]
5[c]		DDQ	−50 °C		85% (16:1)	[12]
6[d]		TMSOTf	−78 °C		63% (37:1)	[13]
7[a]		$AuCl_3$	rt		81% (3:1)	[14]
8[a]	 R′ = TMS, Me, Ph, etc...	$InBr_3$ or I_2	rt		95% (α only)	[15]
9[a]	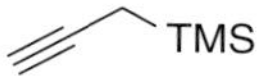	$SnCl_4$ or $TiCl_4$	−20 °C		89% (α only)	[16]

(Continued)

Table 7.1 (Continued)

Entry	Nucleophile	Lewis acid	Temperature	Products: R	Products: Yield (α:β)	References
10[c]	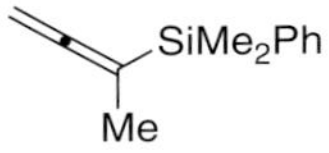	TMSOTf	−40 °C		93% (20 : 1)	[17]
11[a]	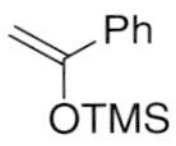	$BF_3 \cdot OEt_2$	−40 °C	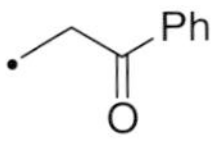	99% (4 : 1)	[18a]
12[a]		$Yb(OTf)_3$	rt		90% (8 : 1)	[18b]
13		$Yb(OTf)_3$ in [bmim][NTf_2]	rt		65% (19 : 1)	[18c]
14[a]	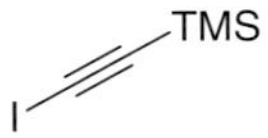	In^0	Reflux		82% (6 : 1)	[19]
15[c]	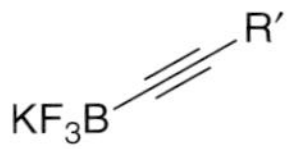	$BF_3 \cdot OEt_2$	−45 °C		85% (24 : 1)	[20]
16[a]	$CH_2(CO_2Et)_2$	$AuCl_3$	rt	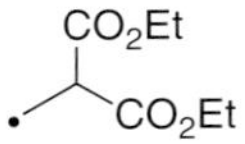	80% (4 : 1)	[14]
17[a]	Et_2Zn	$BF_3 \cdot OEt_2$	rt	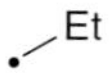	95% (3 : 1)	[21]

Solvent used:

a) CH_2Cl_2.

b) 1,2-Dichloroethane.

c) MeCN.

d) CH_2Cl_2 : MeCN 2 : 1.

rt: room temperature.

Table 7.2 Lewis-acid-catalyzed Ferrier rearrangements with pentopyranose glycal donors.

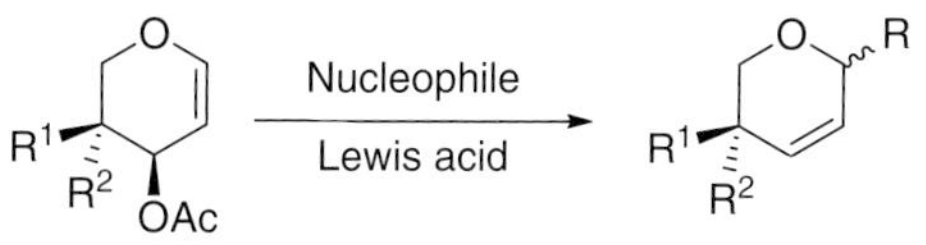

for xylal: R¹ = H, R² = OAc
for arabinal: R¹ = OAc, R² = H

Entry	Donor	Nucleophile	Lewis acid	Products: R	Products: Yield (*anti*:*syn*)	References
1	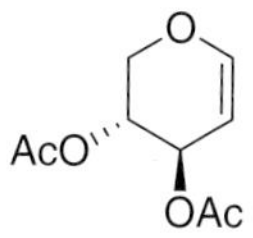 Xylal	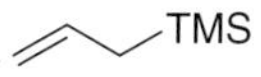	$BF_3·OEt_2$	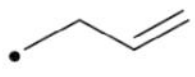	99% (>19:1)	[22]
2		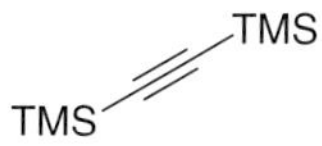	$TiCl_4$		73% (>19:1)	[22]
3		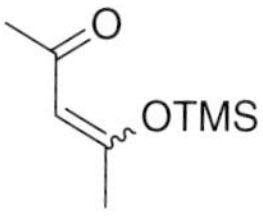	$BF_3·OEt_2$	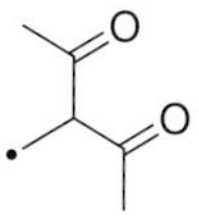	65% (6:1)	[22]
4	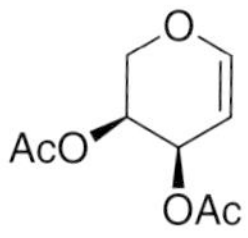 Arabinal	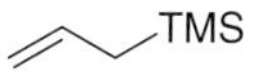	$BF_3·OEt_2$	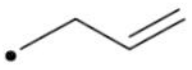	95% (19:1)	[22]
5			$TiCl_4$	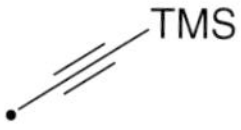	97% (>19:1)	[22]
6		TMSCN	$InCl_3$	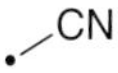	72% (10:1)	[23]

expected to be a reliable synthetic method toward *C*-glycosides. A general mechanism of Heck-type C-glycosylation reaction is depicted in Scheme 7.3.

The reactive organopalladium complex **R′PdX** can be generated through oxidative addition or transmetalation reaction, followed by syn-addition to the glycal double bond to form complex **I**. From here, the intermediate can undergo elimination through either β-hydride or β-heteroatom. The released Pd(II) can then be reduced to Pd(0) to continue the next catalytic cycle. Aryl iodides are the most commonly used coupling partners in Heck reaction. Both furanose [26] and pyranose glycals [27] were introduced to this reaction. The stereoselectivity was controlled by stereoconfiguration at

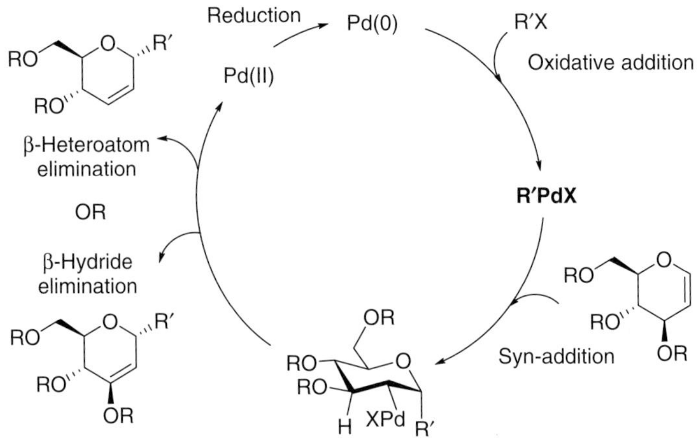

I

Scheme 7.3 Mechanism of a Heck-type C-glycosylation reactions.

C-3, which dictated the syn-addition from the opposite face of the leaving group (Scheme 7.4). Interestingly, silyl protecting groups seemed to facilitate the reaction, whereas other protective groups such as acetyl or benzyl failed to deliver any product under the same conditions.

Other aromatic coupling partners such as aryl boronic acids [28], electron-rich benzoic acids [29], aryl hydrazines [30], and aryl sulfonates [31] were reported. A Heck-type coupling of arylboronic acid [32] to glycals was found to provide three structural motifs based on the choice of oxidants (Table 7.3). No product was observed when peracetylated glucal was used (entry 1). Tri-*O*-benzyl glucal also returned low yields (32%, entry 2). TBS-protected glucal was found to be the most effective donor, giving

Table 7.3 Heck-type C-glycosylation with glycal donors.

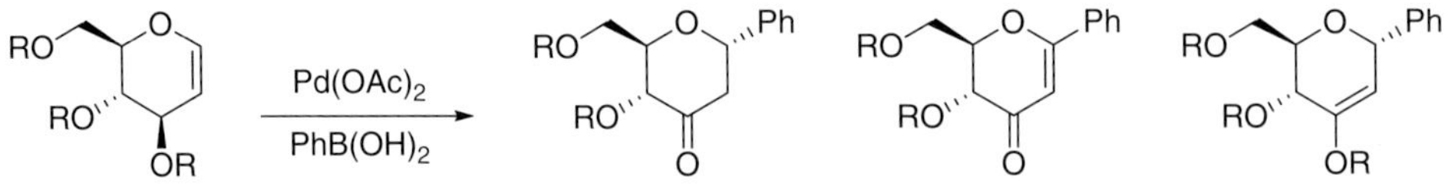

A B C

Entry	R	Oxidants	Products (yield)		
			A (%)	**B (%)**	**C (%)**
1	Ac	BQ	—	—	—
2	Bn	BQ	32	—	—
3	TBS	BQ	84	—	—
4	TBS	$Cu(OAc)_2/O_2$	—	—	94
5	TBS	DDQ	—	69	—

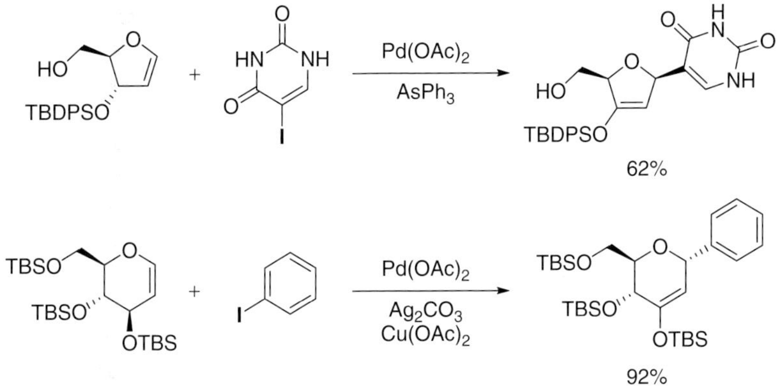

Scheme 7.4 Example of Heck-type C-glycosylation of glycals and aryl iodides.

pyranone glycoside **A** in the presence of benzoquinone (BQ) (entry 3). The combination of $Cu(OAc)_2$ and O_2 provided the enol ether **C** (entry 4) with exclusive α-selectivity. Finally, addition of DDQ as oxidant yielded the Heck-type glycoside **B** (entry 5). A variety of arylboronic acids were found to be compatible with the three conditions; nevertheless, acceptable yields were limited to glycals with TBS ethers at C-3. It was proposed that conformational rigidity imposed by the TBS protective group facilitated the *syn*-β-hydride elimination of the intermediate.

Liu's group recently reported a stereoselective C-glycosylation of peracetylated glycals with enol triflates [33]. It was proposed that the inclusion of multiequivalent chloride ion (from the catalyst and the additive) as an electron-rich ligand significantly increased the electron density at the Pd center. An electron-rich Pd would favor the β-heteroatom elimination pathway since the leaving group anion was less electron-rich compared to hydride (Scheme 7.5).

7.5 Tsuji–Trost-Type *C*-Glycosylation

Compared to Ferrier-type and Heck-type reactions, Tsuji–Trost-type C-glycosylation has seen development at a slower pace. This was partly due to the difficulty in generating Pd π-allyl complex on glycal system [34]. Early works by RajanBabu [35] substituted the 3-*O*-acetyl for 3-*O*-trifluoroacetyl to improve the affinity of allylic system on glycal to palladium (Scheme 7.6, Eq. (1)). Regarding the nucleophile, only strong *C*-nucleophiles such as potassium and sodium malonate were able to afford the *C*-glycoside in

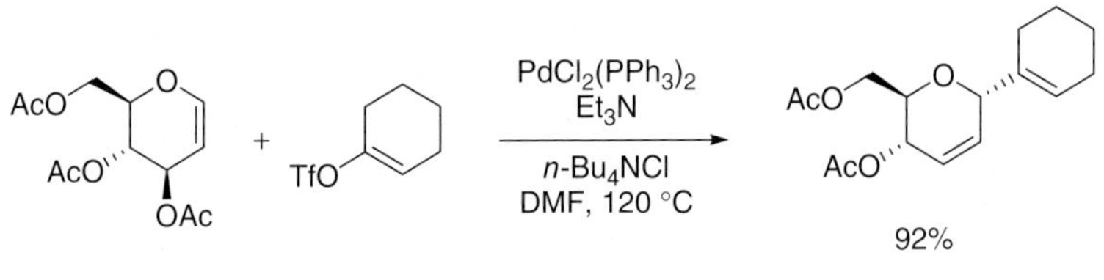

Scheme 7.5 Heck-type C-glycosylation of glycals and enol triflate.

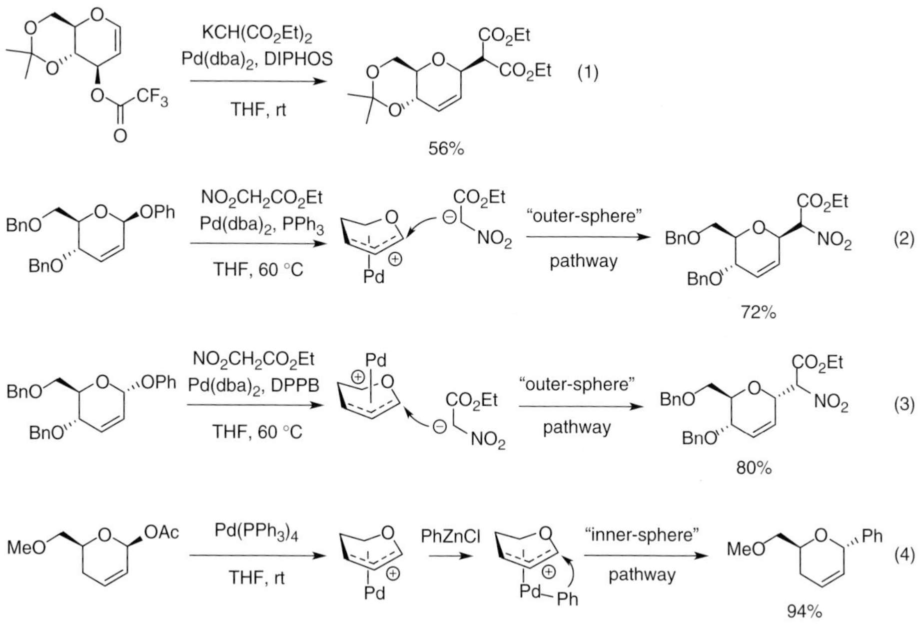

Scheme 7.6 Early attempts at generating the Pd π-allyl complex on glycal.

acceptable yield. Attempts to alleviate the problem included starting from the preactivated pseudo-glycal [34b, 36] (Eqs. (2)–(4)) or addition of activators such as Et_2Zn [37], the latter being unfortunately confined to O-glycosylation.

Similar to the original Tsuji–Trost-type reaction, the mechanism involves formation of Pd π-allyl complex via S_N2 attack of Pd(0) to the allylic system from the opposite face of the leaving group. From here, the stereoselectivity of the reaction depends on the nature of carbon nucleophile. Softer nucleophiles such as malonate ester prefer to attack the anomeric carbon from the opposite face of the palladium complex to form the retention product (double inversion) (Eqs. (2) and (3)). This is the so-called

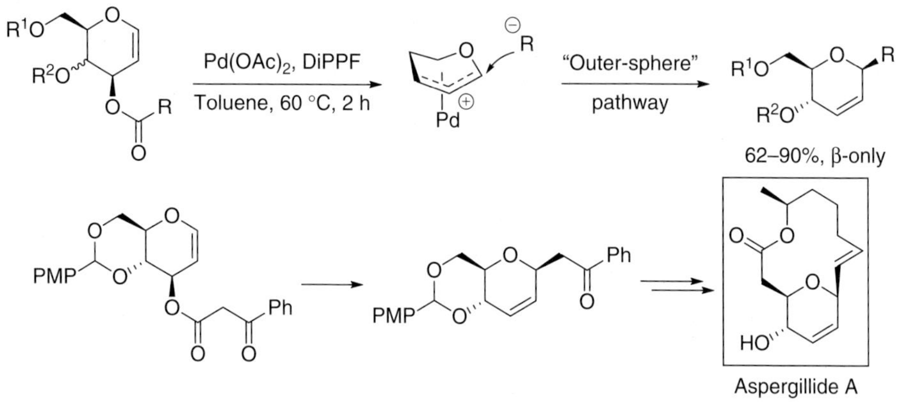

Scheme 7.7 Pd-catalyzed decarboxylative C-glycosylation from glycal scaffolds.

outer-sphere pathway. On the other hand, harder nucleophiles such as organozinc reagent in Eq. (4) would prefer to attack the more electrophilic Pd center, followed by intramolecular delivery of the *C*-nucleophile to the C-1 position via a so-called inner-sphere pathway. The result is an inverted product. One distinction of Tsuji–Trost-type C-glycosylation is the ability to obtain β-anomers without the need to employ the more cost-prohibitive allal donors (C-3 epimer of glucal).

Despite these early efforts, the Tsuji–Trost-type C-glycosylation remained underutilized until a breakthrough was made in 2013. Liu's group reported a decarboxylative allylation reaction [36] from readily available glycal scaffolds to synthesize the elusive β-*C*-glycosides in high yields with exclusive regioselectivity and diastereoselectivity (Scheme 7.7). An extensive array of substrate scopes was examined, and the utility of this method was demonstrated by the concise total synthesis of various natural products including (±)-centrolobine [38], decytospolide A, B [38], and aspergillide A [39].

The reaction mechanism was further studied by experimental and DFT calculations. It was found that the reaction proceeded through an ionization–allylation–decarboxylation sequence via an "outer-sphere" mechanism. Subsequent works from the same group successfully extended the method to N- and O-glycosylation [40]. Interestingly, glycals equipped with 3-*O*-picoloyl were found to reverse the stereochemical outcome of the reaction [41]. It was proposed that the Pd π-allyl complex was generated from the same face of the leaving group via Pd—N coordination. In addition, fine-tuning of acceptor strength allows the reaction to proceed via an "inner-sphere" or "outer-sphere" pathway.

7.6 Sigmatropic Rearrangement

Apart from the Lewis-acid-catalyzed and transition-metal-catalyzed C-glycosylations, one of the earlier methodologies for C-glycosylation is [3,3]-sigmatropic rearrangement, in particular the Claisen rearrangement, which often does not require a catalyst since the reaction is driven thermally. When using glycal ester as substrate, Claisen rearrangement occurs via a concerted mechanism through a six-membered transition state that resembles boat conformation, as opposed to the chair transition state in typical Claisen reactions. This leads to the chirality of C-1 position being determined by C-3 stereo-configuration in starting glycal (Scheme 7.8).

Fraser-Reid reported an Eschenmoser–Claisen rearrangement for C-glycosylation [42]. As seen from Scheme 7.9, an allyl ketene aminal intermediate was generated *in situ* by heating glycal and dimethylacetamide dimethyl acetal, which underwent sigmatropic rearrangement to obtain the 2,3-unsaturated *C*-glycosides.

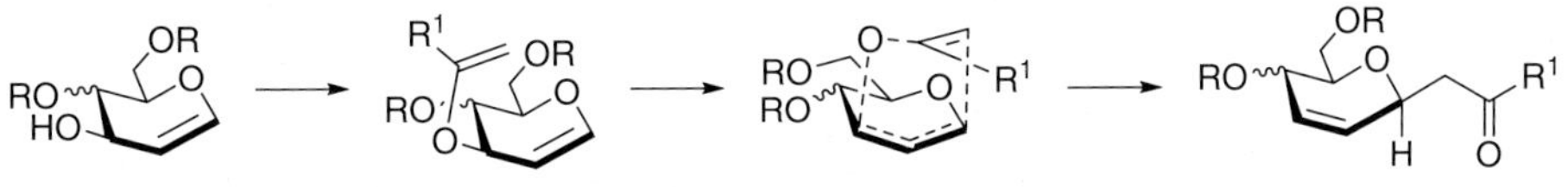

Scheme 7.8 Mechanism of Claisen rearrangement.

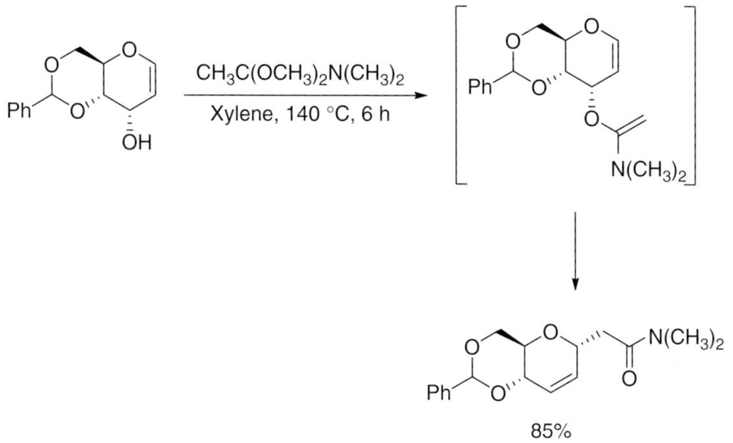

Scheme 7.9 Eschenmoser–Claisen rearrangement.

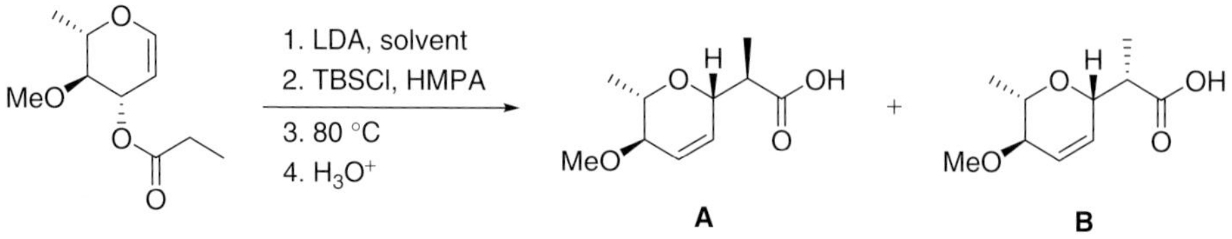

Solvent	Yield % (A : B)
23% HMPA-THF	71% (82 : 18)
100% THF	73% (19 : 81)

Scheme 7.10 Ireland–Claisen rearrangement.

In the same year, Ireland reported C-glycosylation using ketene silyl acetals [43] (Scheme 7.10). Both furanoid and pyranoid glycals could be applied in this Ireland–Claisen rearrangement. When enolizable an enolizable ester was used as the substrate, stereoisomers varying at the α-position were obtained. The ratio of stereoisomers could be altered by addition of HMPA as additive, as it could coordinate to the lithium ion and favor the formation of (*Z*)-enolate.

Later, Fairbanks developed tandem Tebbe methylenation–Claisen rearrangement [44], starting from esters of 3-hydroxyglycal. As seen in Scheme 7.11, Tebbe methylenation of the glycal ester afforded the enol intermediate, followed by Claisen rearrangement to furnish *C*-glycosides. The reaction was applied in the preparation of various disaccharides.

The versatility of these [3,3]-sigmatropic rearrangements in C-glycosylations was further demonstrated in various total syntheses including Lasalocid A by Ireland [45] and *C*-mannosyl-alanine by Colombo [46].

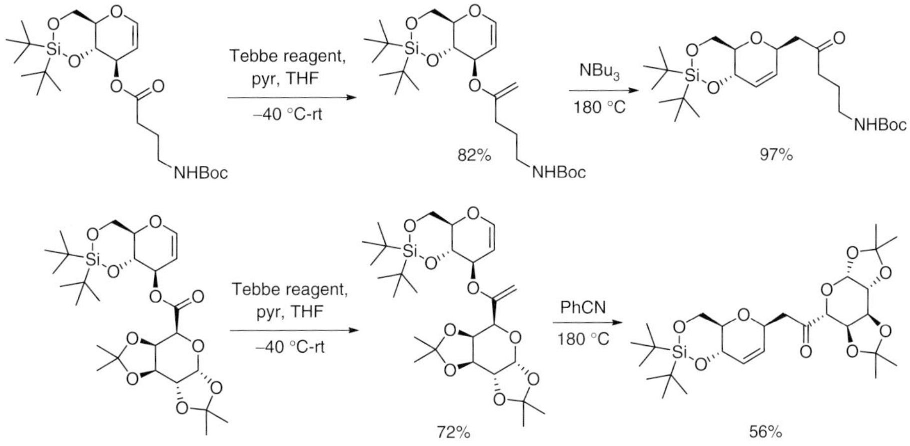

Scheme 7.11 Tandem Tebbe methylenation–Claisen rearrangement.

7.7 NHC-Catalyzed C-Glycosylations

The use of NHC is gaining popularity as it overcomes many shortcomings of other heterogeneous catalysts, such as air and moisture sensitivity, higher cost, and environmental toxicity. Although the field of organocatalysis is well developed, we have observed limited applications to asymmetric glycosylations using glycal substrates.

In the presence of NHC and a base, the umpolung effect transformed an aldehyde from an electrophile into a nucleophile. This nucleophilic acylanion, namely the Breslow intermediate, could potentially react with other electrophiles, such as α,β-unsaturated ketones (Stetter reaction) or another aldehyde (benzoin condensation). To date, there have been few reports on the application of NHCs for glycosylations, probably attributed to the inherent preference of Breslow intermediate for electrophiles and the difficult generation of suitable electrophiles with the glycal system.

Liu's group successfully applied Stetter reaction for C-glycosylation by choosing 2-nitroglucal as the electrophile, which was demonstrated as suitable for 1,4-addition [47] (Scheme 7.12). Varying the reaction conditions could afford β-selective or nitro-eliminated products. As depicted in Scheme 7.13, the reaction commenced with the nucleophilic addition of the carbine to aldehyde **I** to form Breslow intermediate **II**. 2-Nitroglucal **1a** existed predominantly in the preferred $^{5}H_{4}$ conformation, which minimized steric interactions arising from the unfavorable 1,3-allylic strain. Hydrogen bonding with the hydroxyl group in the Breslow intermediate then allowed the acyl anion to approach from the β-face of **III**, forming adduct **IV**. Subsequent proton shift resulted in the release of NHC catalyst to form **1b**, which, upon deprotonation, gave the observed product **1c**.

The same group later extended to dual-catalyzed C-glycosylation, employing palladium and NHC catalysts [48]. A dual catalytic system required the catalysts to be compatible with each other. However, NHC species could act as ligands and potentially coordinate to palladium, shutting off both catalytic cycles. The combination of the Tsuji–Trost reaction and the Stetter reaction was able to form the *C*-glycosides in good yields. The mechanism is depicted in Scheme 7.14. Palladium and NHC catalyzed the generation of cationic π-allyl Pd complex and the Breslow nucleophile,

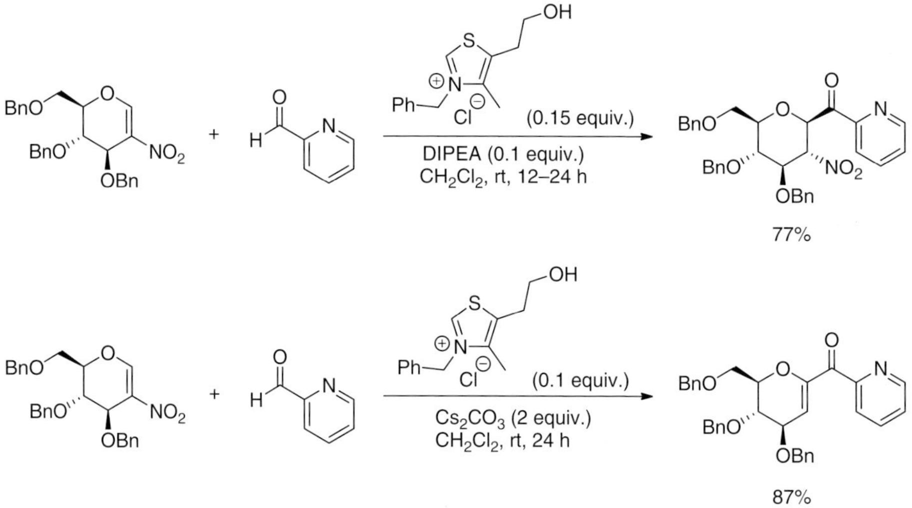

Scheme 7.12 NHC-catalyzed C-glycosylation reactions.

respectively. These intermediates were brought to close spatial proximity by Pd—N coordination, resulting in intermediate **I**. This Breslow intermediate underwent an intramolecular nucleophilic attack at C-1 position of the sugar to form intermediate **II**. The 2,3-unsaturated *C*-glycoside **III** was obtained after regeneration of the NHC catalyst. Under basic conditions, the reaction could further undergo proton transfer to obtain product **3**.

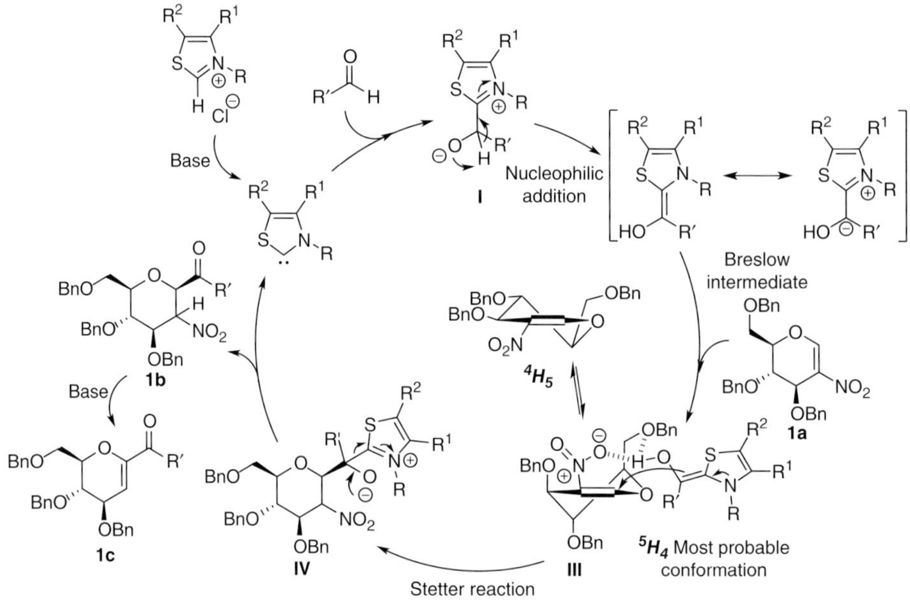

Scheme 7.13 Mechanism of NHC-catalyzed C-glycosylation.

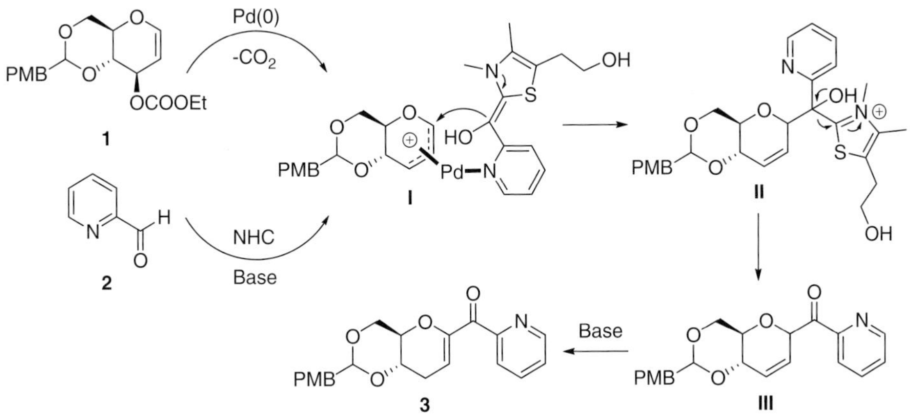

Scheme 7.14 Mechanism of dual-catalyzed C-glycosylation.

7.8 Conclusion

There have been significant advances in the area of C-glycosylation in recent years, especially in the three major areas: Ferrier-type rearrangement, Heck-type, and Tsuji–Trost-type glycosylation. Recent progress has enhanced our understanding of the underlying mechanisms behind the observed stereoselectivity, in particular the Tsuji–Trost-type glycosylation. Nevertheless, there remain challenges, which provide room for further exploration. On the other hand, areas such as NHC-organocatalyzed glycosylation and sigmatropic rearrangement are in their nascent stages and thus present opportunities for more in-depth research development. The discovery of new catalytic glycosylation reactions and their mechanisms will be beneficial in the synthesis of complex natural products and *C*-glycoside analogs of bioactive compounds.

References

1 For reviews on biological roles of glycoconjugates and syntheses, see: (a) Dwek, R.A. (1996) *Chem. Rev.*, 96, 683–720; (b) Varki, A., Cummings, R.D., Esko, J.D., Freeze, H.H., Stanley, P., Bertozzi, C.R., Hart, G.W., and Etzler, M.E. (2009) Essentials of Glycobiology, 2nd edn, Cold Spring Harbor Laboratory Press, Cold Spring Harbor, NY.

2 For reviews on biological activities of C-glycosides in comparison to O-glycosides analogues, see: (a) Härle, J., Günther, S., Lauinger, B., Weber, M., Kammerer, B., Zechel, D.L., Luzhetskyy, A., and Bechthold, A. (2011) *Chem. Biol.*, 18, 520–530; (b) Weatherman, R.V. and Kiessling, L.L. (1996) *J. Org. Chem.*, 61, 534–538; (c) Hultin, P.G. (2005) *Curr. Top. Med. Chem.*, 5, 1299–1331; (d) Zou, W. (2005) *Curr. Top. Med. Chem.*, 5, 1363–1391; (e) Compain, P. and Martin, O.R. (2001) *Bioorg. Med. Chem.*, 9, 3077–3092; (f) Yang, G., Schmieg, J., Tsuji, M., and Franck, R.W. (2004) *Angew. Chem. Int. Ed.*, 43, 3818–3822; (g) Levy, D.E. and Tang, C. (1995) The Chemistry of C-Glycosides, Pergamon, Tarrytown, NY.

3 Franck, R.W. and Tsuji, M. (2006) *Acc. Chem. Res.*, 39, 692–701.

4 McKay, M.J. and Nguyen, H.M. (2012) *ACS Catal.*, 2, 1563–1595.

5 Fraser-Reid, B. and Lopez, J.C. (2009) *Curr. Org. Chem.*, 13, 532–553.
6 Fischer, E. (1914) *Ber. Dtsch. Chem. Ges.*, 47, 196–210.
7 (a) Ferrier, R.J., Overend, W.G., and Ryan, A.E. (1962) *J. Chem. Soc.*, 3667–3670; (b) Ferrier, R.J. (1964) *J. Chem. Soc.*, 5443–5449; (c) Ferrier, R.J., Prasad, N., and Sankey, G.H. (1968) *J. Chem. Soc. C*, 974–977.
8 Danishefsky, S., and Kerwin , Jr. J.F. (1982) *J. Org. Chem.*, 47, 3803–3805.
9 Ichikawa, Y., Isobe, M., Konobe, M., and Goto, T. (1987) *Carbohydr. Res.*, 171, 193–199.
10 Takhi, M., Rahman, A.A.-H.A., and Schmidt, R.R. (2001) *Tetrahedron Lett.*, 42, 4053–4056.
11 Reddy, T.R., Rao, D.S., and Kashyap, S. (2015) *RSC Adv.*, 5, 28338–28343.
12 Toshima, K., Ishizuka, T., Matsuo, G., and Nakata, M. (1993) *Chem. Lett.*, 22, 2013–2016.
13 Toshima, K., Ishizuka, T., Matsuo, G., and Nakata, M. (1994) *Tetrahedron Lett.*, 35, 5673–5676.
14 Balamurugan, R. and Koppolu, S.R. (2009) *Tetrahedron*, 65, 8139–8142.
15 (a) Yadav, J.S., Reddy, B.V.S., Raju, A.K., and Rao, C.V. (2002) *Tetrahedron Lett.*, 43, 5437–5440; (b) Yadav, J.S., Reddy, B.V.S., Raju, A.K., and Rao, C.V. (2003) *Tetrahedron Lett.*, 44, 6211–6215.
16 (a) Zhu, Y.-H. and Vogel, P. (2001) *Synlett*, 2001, 0082–0086; (b) Huang, G. and Isobe, M. (2001) *Tetrahedron*, 57, 10241–10246.
17 Brawn, R.A. and Panek, J.S. (2010) *Org. Lett.*, 12, 4624–4627.
18 (a) Dawe, R.D. and Fraser-Reid, B. (1981) *J. Chem. Soc., Chem. Commun.*, 22, 1180–1181; (b) Sasaki, M., Tsubone, K., Shoji, M., Oikawa, M., Shimamoto, K., and Sakai, R. (2006) *Bioorg. Med. Chem. Lett.*, 16, 5784–5787; (c) Anjaiah, S., Chandrasekhar, S., and Grée, R. (2004) *J. Mol. Catal. A: Chem.*, 214, 133–136.
19 Lubin-Germain, N., Hallonet, A., Huguenot, F., Palmier, S., Uziel, J., and Augé, J. (2007) *Org. Lett.*, 9, 3679–3682.
20 Vieira, A.S., Fiorante, P.F., Hough, T.L.S., Ferreira, F.P., Lüdtke, D.S., and Stefani, H.A. (2008) *Org. Lett.*, 10, 5215–5218.
21 Thorn, S.N. and Gallagher, T. (1996) *Synlett*, 1996 (2), 185–186.
22 Hosokawa, S., Kirschbaum, B., and Isobe, M. (1998) *Tetrahedron Lett.*, 39, 1917–1920.
23 Das, S.K., Reddy, K.A., Abbineni, C., Roy, J., Rao, K.V.L.N., Sachwani, R.H., and Iqbal, J. (2003) *Tetrahedron Lett.*, 44, 4507–4509.
24 Nishikawa, T., Adachi, M., and Isobe, M. (2008) in Glycoscience (eds B. Fraser-Reid, K. Tatsuta, and J. Thiem), Springer, Berlin, Heidelberg, pp. 755–811.
25 For reviews on Mizoroki-Heck reaction, see: (a) Heck, R.F. (1979) *Acc. Chem. Res.*, 12, 146–151; (b) Daves, G.D. and Hallberg, A. (1989) *Chem. Rev.*, 89, 1433–1445; (c) Beletskaya, I.P. and Cheprakov, A.V. (2000) *Chem. Rev.*, 100, 3009–3066; (d) Zapf, A. and Beller, M. (2002) *Top. Catal.*, 19, 101–109; (e) Nicolaou, K.C., Bulger, P.G., and Sarlah, D. (2005) *Angew. Chem. Int. Ed.*, 44, 4442–4489; (f) Zeni, G. and Larock, R.C. (2006) *Chem. Rev.*, 106, 4644–4680; (g) Jutland, A. (2009) in The Mizoroki-Heck Reaction (ed. M. Oestreich), John Wiley & Sons, Ltd., pp. 1–50.
26 Zhang, H.C. and Daves, G.D. (1992) *J. Org. Chem.*, 57, 4690–4696.
27 Li, H.-H. and Ye, X.-S. (2009) *Org. Biomol. Chem.*, 7, 3855–3861.
28 Ramnauth, J., Poulin, O., Rakhit, S., and Maddaford, S.P. (2001) *Org. Lett.*, 3, 2013–2015.

29 Xiang, S., Cai, S., Zeng, J., and Liu, X.-W. (2011) *Org. Lett.*, 13, 4608–4611.
30 Bai, Y., Le Mai Hoang, K., Liao, H., and Liu, X.-W. (2013) *J. Org. Chem.*, 78, 8821–8825.
31 Ma, J., Xiang, S., Jiang, H., and Liu, X.-W. (2015) *Eur. J. Org. Chem.*, 2015 (5), 949–952.
32 Xiong, D.C., Zhang, L.H., and Ye, X.S. (2009) *Org. Lett.*, 11, 1709–1712.
33 Bai, Y., Leow, M., Zeng, J., and Liu, X.-W. (2011) *Org. Lett.*, 13, 5648–5651.
34 (a) Trost, B.M. and Gowland, F.W. (1979) *J. Org. Chem.*, 44, 3448–3450; (b) Dunkerton, L.V. and Serino, A.J. (1982) *J. Org. Chem.*, 47, 2812–2814.
35 RajanBabu, T.V. (1985) *J. Org. Chem.*, 50, 3642–3644.
36 (a) Brakta, M., Lhoste, P., and Sinou, D. (1989) *J. Org. Chem.*, 54, 1890–1896; (b) Moineau, C., Bolitt, V., and Sinou, D. (1998) *J. Org. Chem.*, 63, 582–591.
37 Kim, H., Men, H., and Lee, C. (2004) *J. Am. Chem. Soc.*, 126, 1336–1337.
38 Zeng, J., Tan, Y.J., Ma, J., Leow, M.L., Tirtorahardjo, D., and Liu, X.-W. (2014) *Chem. Eur. J.*, 20, 405–409.
39 Zeng, J., Ma, J., Xiang, S., Cai, S., and Liu, X.-W. (2013) *Angew. Chem. Int. Ed.*, 52, 5134–5137.
40 (a) Xiang, S., Lu, Z., He, J., Le Mai Hoang, K., Zeng, J., and Liu, X.-W. (2013) *Chem. Eur. J.*, 19, 14047–14051; (b) Xiang, S., He, J., Ma, J., and Liu, X.-W. (2014) *Chem. Commun.*, 50, 4222–4224; (c) Xiang, S., He, J., Tan, Y.J., and Liu, X.-W. (2014) *J. Org. Chem.*, 79, 11473–11482.
41 Xiang, S., Le Mai Hoang, K., He, J., Tan, Y.J., and Liu, X.-W. (2015) *Angew. Chem. Int. Ed.*, 54, 604–607.
42 Fraser-Reid, B., Dawe, R.D., and Tulshian, D.B. (1979) *Can. J. Chem.*, 57, 1746–1748.
43 Ireland, R.E., Wilcox, C.S., Thaisrivongs, S., and Vanier, N.R. (1979) *Can. J. Chem.*, 57, 1743–1745.
44 (a) Godage, H.Y. and Fairbanks, A.J. (2000) *Tetrahedron Lett.*, 41, 7589–7593; (b) Chambers, D.J., Evans, G.R., and Fairbanks, A.J. (2005) *Tetrahedron: Asymmetry*, 16, 45–55.
45 Ireland, R.E., Anderson, R.C., Badoud, R., Fitzsimmons, B.J., McGarvey, G.J., Thaisrivongs, S., and Wilcox, C.S. (1983) *J. Am. Chem. Soc.*, 105, 1988–2006.
46 Colombo, L., Di Giacomo, M., and Ciceri, P. (2002) *Tetrahedron*, 58, 9381–9386.
47 Vedachalam, S., Tan, S.M., Teo, H.P., Cai, S., and Liu, X.-W. (2012) *Org. Lett.*, 14, 174–177.
48 Bai, Y., Leng, W.L., Li, Y., and Liu, X.-W. (2014) *Chem. Commun.*, 50, 13391–13393.

8

Brønsted- and Lewis-Acid-Catalyzed Glycosylation

David Benito-Alifonso and M. Carmen Galan

8.1 Introduction

The stereoselective synthesis of glycosidic bonds still remains one of the most challenging areas within organic synthesis. Glycosylation reactions typically involve the coupling between two building blocks to form a glycosidic bond, the new bond is often an O-linkage, but examples of *S*-, *N*-, and *C*-glycosides are also possible. In a typical O-glycosylation reaction, a fully protected glycosyl donor, bearing an anomeric leaving group, is activated via reaction with a promoter/activator to facilitate the nucleophilic attack of the suitably functionalized glycosyl acceptor (which typically contains an OH group) and displaces the leaving group of the glycosyl donor. It is generally accepted that the glycosylation reaction follows an S_N1-like mechanism, although examples of S_N2-like displacements are also reported in the literature [1]. Typically, these reactions lead to a mixture of two diastereoisomers (α/β-anomers). Generally, 1,2-*trans* glycosidic bonds can be prepared by taking advantage of the neighboring group participation of an acyl group at C-2 position of the glycosyl donor, while the formation of 1,2-*cis*-linked glycosides remains a difficult task in many cases. Accordingly, the development of reagent/catalyst-controlled glycosylations offers many advantages over traditional methods, as these strategies have the potential to allow the use of less complex monosaccharide building blocks. Despite the broad use of organocatalysis within asymmetric synthesis, its application toward oligosaccharide synthesis is just emerging. Among recent methods developed for this purpose, the application of Brønsted and Lewis acid catalysis has been noteworthy. This chapter highlights the most relevant examples of Brønsted- and Lewis-acid-catalyzed glycosylations over the past decade, featuring their application in the stereoselective formation of glycosidic bonds.

8.2 Chiral Brønsted Acids

Chiral Brønsted acids have been successfully applied in different areas of organic chemistry, and carbohydrate chemistry is not an exception. Activation of the anomeric position with chiral acids is deemed to be a feasible way to control the stereo-outcome of the

Selective Glycosylations: Synthetic Methods and Catalysts, First Edition. Edited by Clay S. Bennett.

reaction; however, the induction of stereoselectivity in glycosylations with chiral acids has proven to be challenging. Brønsted-acid-mediated chirality induction is governed by a complex relationship between the acid, the glycosyl donor, and the glycosyl acceptor stereochemistry. Fairbanks and coworkers [2], inspired by Toste's activation of trichloroacetimidate leaving groups in the formation of episulfonium ions [3], showed for the first time the use of chiral phosphoric acids in glycosylations (Scheme 8.1). The team showed that activation of galactosyl trichloroacetimidate with both enantiomers of BINOL-derived acids ((***S***)-***Cat1*** *and* (***R***)-***Cat2***) yielded β-glycosides preferentially. Interestingly, reactions using the *S*-catalyst produced higher β-selectivity, proving the influence of the acid chirality in the stereo-outcome of the reaction and that glycosylation diastereocontrol might be achieved by the careful choice of a chiral promoter.

Indeed, expanding on Fairbanks' work, Toshima and coworkers [4] developed the first α,β-stereo- and diastereoselective glycosylation with glucosyl trichloroacetimidates. Exploiting the matched chirality between BINOL-derived chiral phosphoric acid (***S***)-***Cat1*** (Scheme 8.1), benzyl glucosyl donor, and chiral acceptors, the team were able to tune the reaction conditions to obtain only the β-*S*-glycoside product (Scheme 8.2).

This method was successfully extended to other secondary racemic alcohols and to the synthesis of a flavan glycoside natural product extracted from the *Cinnamomum cassia* bark. Starting from a racemic mixture of monohydroxylated (±) epicatechin and using (***S***)-***Cat1*** as the glycosylation promoter, only (−)epicatechin was glycosylated affording the (β,*R*)-glycoside in good yields (Scheme 8.3).

Similarly, the match/mismatch relationship between a glycosyl donor and a chiral acid can also be exploited for the diastereoselective synthesis of 2-deoxyglycosides. Bennett and coworkers [5] recently described the use of BINOL-based phosphoric acids to activate 2-deoxyglucosyl trichloroacetimidates. In this example, good stereoselectivities

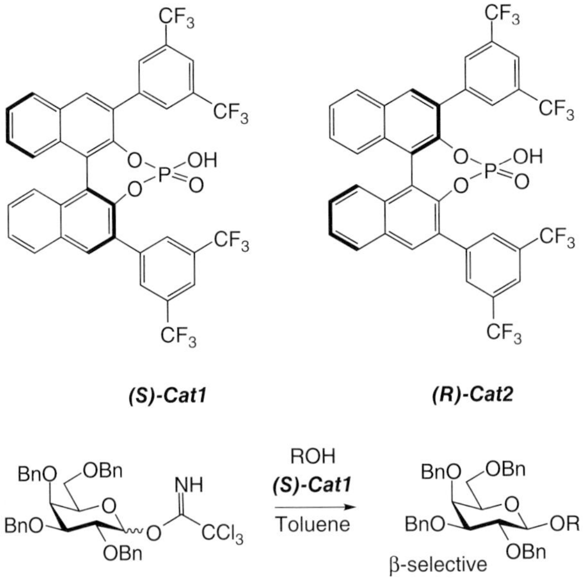

Scheme 8.1 Chiral BINOL-derived Brønsted-acid-catalyzed glycosylation of galactosyl trichloroacetimidates.

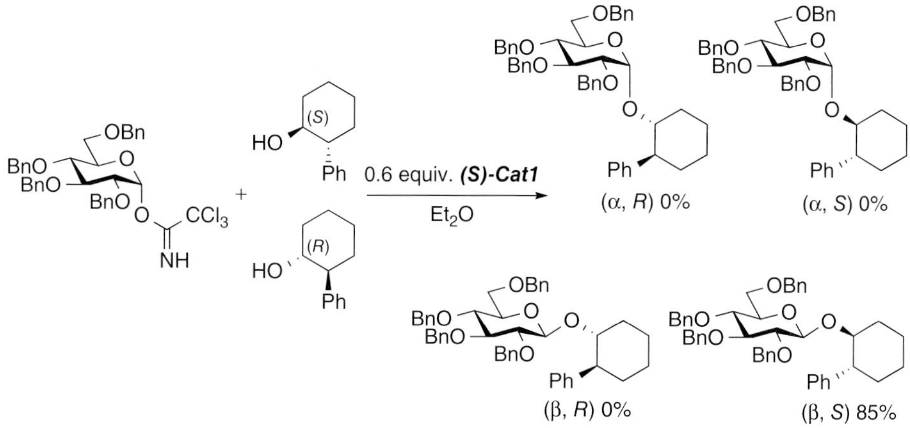

Scheme 8.2 Chiral BINOL-derived Brønsted-acid-catalyzed glycosylation of glucosyl trichloroacetimidates with chiral acceptors.

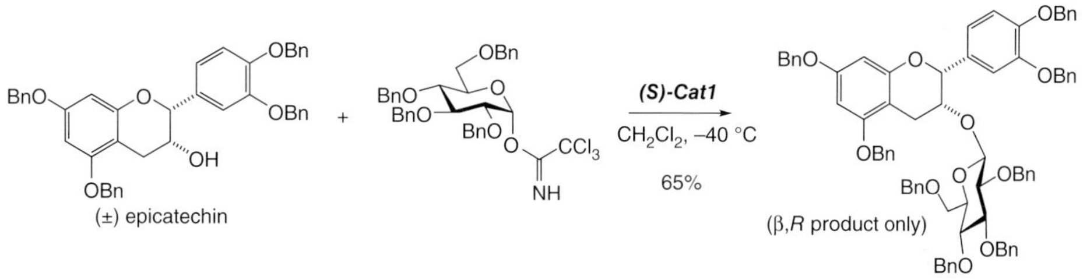

Scheme 8.3 Synthesis of flavan glycoside.

were only obtained when the stereochemistries of the glycosyl trichloroacetimidate (α)- and (*S*)-catalysts matched, affording the β-glycoside as the major isomer. Attempts to exploit a possible match relationship between the (*R*)-catalyst and the β-glycosyl donor were unsuccessful, affording only a loss of diastereomeric induction. The authors hypothesize that the outcome of the reaction is controlled by a different reaction pathway depending on the configuration of the glycosyl donor and the catalyst, which is in agreement with the explanation Toshima and coworkers had previously provided in their report [4]. In brief, the glycosylation between the α-glycosyl donor and the (*S*)-catalyst would proceed by hydrogen bond activation and follow an S_N2 pathway, while the β-glycosyl donor would follow an ionic S_N1 pathway under the same catalytic conditions, explaining thus the loss of stereoselectivity (Scheme 8.4).

Glycosyltransferases and hydrolases are enzymes that catalyze glycosidic bond formation and hydrolysis with exquisite regio- and stereocontrol. The mechanism for this transformation typically involves the action of two carboxylic acids (e.g., Glu and Asp); one amino acid would donate a proton to the oxygen of the glycosidic bond while the nucleophile, assisted by the other carboxylate acting as a general base, attacks the anomeric carbon. The reaction proceeds in a single step via an

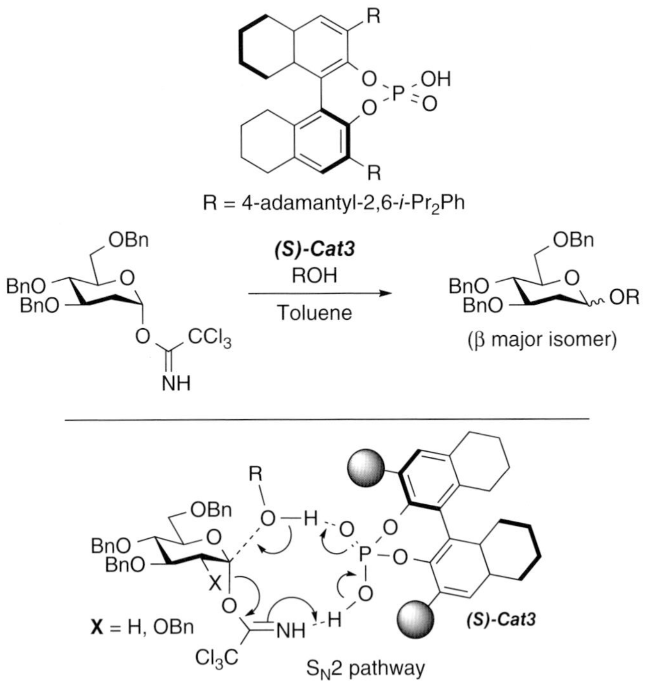

Scheme 8.4 Chiral BINOL-derived Brønsted-acid-catalyzed glycosylation of deoxyglycosides and proposed mechanism.

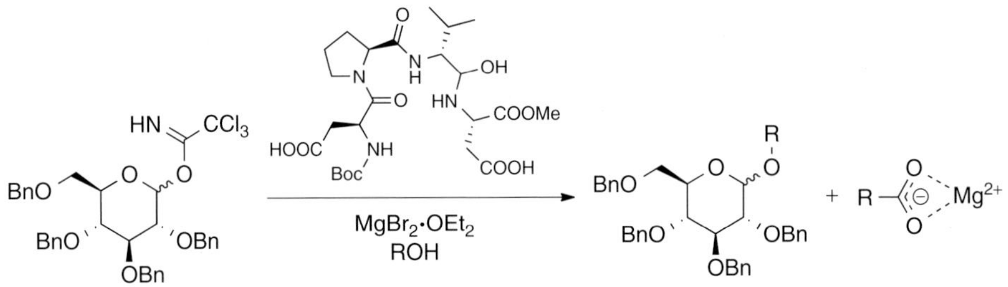

Scheme 8.5 Combined Lewis-acid- and Brønsted-acid-catalyzed glycosylation of glycosyl trichloroacetimidates.

oxocarbenium-ion-like transition state [6]. Inspired by nature, the Miller group endeavored to develop a catalytic glycosylation system using chiral Brønsted acids, including peptide-based acids, to emulate the structural environment of glycoenzymes [7]. Initial studies employing glucosyl trichloroacetimidate donors were unsatisfactory due to the irreversible reaction of the catalysts with the glycosyl donor, but addition of a Lewis acid such as $MgBr_2{\cdot}OEt_2$ as cocatalyst that can coordinate to the carboxylate afforded the glycosides and stopped the formation of the glycosyl ester side product (Scheme 8.5).

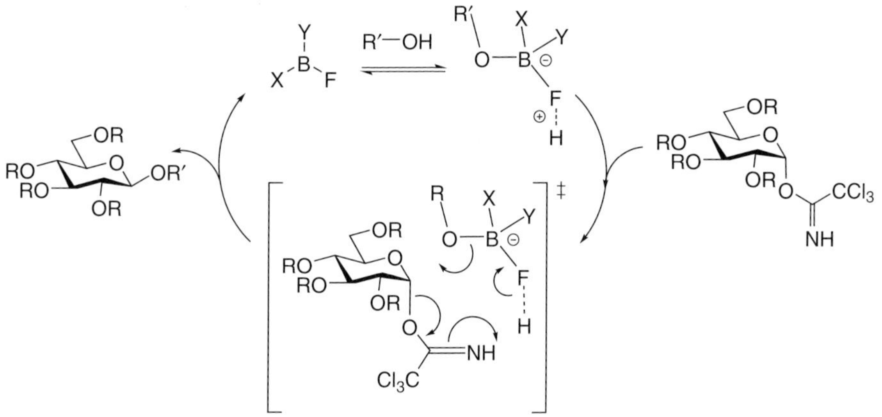

Scheme 8.6 Acid–base-catalyzed glycosylation.

8.3 Achiral Brønsted Acids

In the context of glycosylation reactions promoted by achiral catalysts, a wide variety of acids and conditions have been developed for the stereoselective synthesis of glycosides employing both homogeneous and heterogeneous catalyses.

8.3.1 Homogeneous Brønsted Acid Catalysis

A relevant case of Brønsted acid homogeneous catalysis is the system designed by Schmidt and coworkers [8], which combines acid and base in the same catalyst in the form of arylboron fluorides (Scheme 8.6). The acidic proton responsible for activating the basic imidate center is delivered by the tetragonal boron species upon the attack of the acceptor to the boronic center. Thus, the donor will only be activated in the presence of the acceptor. The concerted process by which the donor is protonated and the acceptor is delivered results in an S_N2-type mechanism affording preferentially β-glycosides in most examples. The applicability of this bifunctional system was tested with other primary and secondary alcohols affording a range of di- and trisaccharides with some of the best yields and stereoselectivities reported for Brønsted-acid-catalyzed glycosylations.

The use of ionic liquids (ILs) in organic synthesis has seen continuous growth during the past decade, and their application to glycosylation chemistry [9] has been no different, with ILs playing different roles in the reaction, for example, as solvents, as ionic tags for supported oligosaccharide synthesis, and as glycosylation promoters. In the context of Brønsted acid catalysis, ILs have been used in combination with protic acids such as trifluoromethane sulfonamide [10] or on their own as Brønsted acid promoters [11]. In particular, tetrazole-based Brønsted acid ILs were used in a Fisher-type glycosylation of unprotected monosaccharides in water to obtain furanosides and pyranosides in moderate-to-good yields and stereoselectivities [11].

Another area where Brønsted acid catalysis has been successfully applied is nucleoside glycosylation. The synthesis of nucleosides via Vorbrüggen glycosylation is usually performed using stoichiometric amounts of Lewis acid to activate the silylated nucleobases. This methodology has the problem of functional group compatibility, along with issues associated with the quenching and disposal of the excess acid, particularly when the reaction is performed at preparative scale. The development of flow chemistry has helped overcome some of the inherent problems associated with batch reactions. In this context, the system developed by the Jamison group for the synthesis of ribonucleosides is a very attractive alternative to traditional methods [12]. By telescoping the glycosylation and deprotection steps into a single, continuous, and uninterrupted reactor network, the streamlined synthesis of unprotected nucleosides in the presence of pyridinium-based Brønsted acids was achieved. Subsequently, the strategy was extended to the synthesis of 5′-deoxyribonucleosides with antiviral activity [13]. By performing the glycosylation under flow conditions, the amount of acid can be reduced and the desired *N*-glycosides were obtained in short times, good yields, and good stereoselectivities.

An interesting example of nonchiral Brønsted acid activation of 2-deoxyglycosides is reported by McDonald and Balthaser [14], whereby camphorsulfonic acid is used to catalyze the synthesis of trisaccharide fragments of Saccharomycins A and B (Scheme 8.7). Although in this report the use of 20-fold excess of camphorsulfonic acid is needed, activation of glycals does not always require such a large excess of acid. For instance, the McDonald team themselves showed that 6-deoxy-ribose glycal could be activated in the presence of 1% of PPh_3·HBr [15]. Similarly, the Galan team recently demonstrated that 1 mol% of toluene sulfonic acid was sufficient to catalyze the synthesis of a range of 2-deoxyglucosides [16], in a methodological study where the effects of ring fluxionality in the stereo-outcome of the reaction were addressed.

Inspired by the report from Galan *et al.* [17] on the use of *N,N′*-bis[3,5-bis-(trifluoromethyl)phenyl]thiourea for the efficient and stereoselective synthesis of 2-deoxygalactosides, the Schmidt group described the use of Schreiner's thiourea as a cocatalyst with a Brønsted acid in glycosylation reactions with *O*-glycosyl trichloroacetimidates as glycosyl donors [18]. The acid and thiourea exhibit a cooperative behavior during the reaction that had a strong effect on the reaction rate, yield, and β-selectivity of glycosidations. The authors proposed that the thiourea enables hydrogen-bond-mediated complex formation between *O*-glycosyl trichloroacetimidate donors, acceptors, and acid catalysts, which facilitates acid–base-catalyzed S_N2-type glycoside bond formation even at room temperature and in the absence of a chimeric assistance.

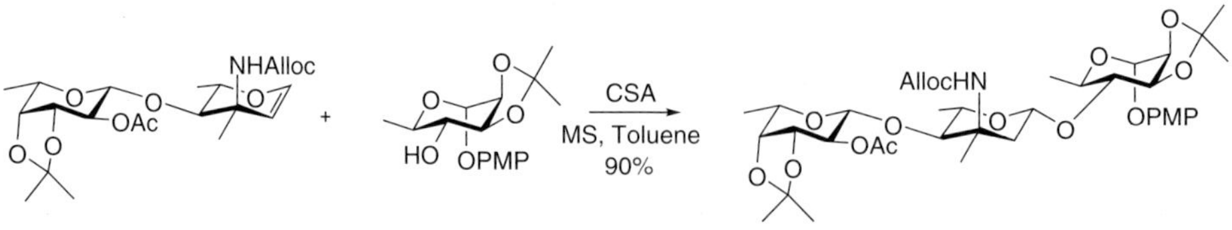

Scheme 8.7 Brønsted-acid-catalyzed glycosylation of 2-deoxyglycosides.

8.3.2 Heterogeneous Brønsted Acid Catalysis

Due to their stability, Brønsted acids such as sulfuric [19] or perchloric [20] acids can be easily supported on solid matrices for the activation of glycosides under heterogeneous conditions. A relevant example is the activation of peracetylated glycosyl trichloroacetimidates with perchloric acid adsorbed on silica by Field and coworkers [20]. The efficiency of silica-supported $HClO_4$ was showcased in the one-pot sequential synthesis of Le^A and Le^X trisaccharides. The distinctive feature of this method lies in the fact that acidic silica was used to pack a conventional chromatography column and that glycosylation and subsequent product isolation were performed "on-column," yielding a range of 1, 2-*trans* glycosylated amino acids in good yields.

Other examples of heterogeneous Brønsted acid catalysts such as β-Zeolites for the activation of 1,2,3,4,6-penta-*O*-acetyl-β-D-galactopyranose [21], montmorillonite K-10 [22], or acidic resins [23] as catalysts for Fischer-type glycosylations under microwave irradiation have been described. In all these reports, the ability of the acid to activate the glycosyl donor is demonstrated, although the acceptor scope is in most cases limited to primary alcohols.

8.4 Lewis-Acid-Catalyzed Glycosylations

8.4.1 Synthesis of O-Glycosides

Lewis acid catalysis is perhaps one of the most common modes of activation in glycoside-forming reactions. Among the many reported examples, glycosylations with lipid alcohols to form biologically relevant glycoconjugates are particularly challenging due to the lower nucleophilicity of the alcohol nucleophile (e.g., ceramide and long-chain fatty alcohols) [24]. Most available methods rely on traditional approaches [25] whereby peracetylated glycosyl bromides, glycosyl trichloroacetimidates, thioglycosides, or glycosyl fluorides are used as glycosyl donors since these compounds are easy to prepare and to activate with standard Lewis acids (e.g., $SnCl_4$, $BF_3{\cdot}Et_2O$, $ZnCl_2$, or TMSOTf), and the presence of a C-2 ester group preferentially yields 1,2-*trans* glycosides. Unfortunately, when lipid alcohols are employed as nucleophiles, reactions are often low yielding (30–40%) and lead to orthoester formation as a side product. To circumvent this problem, the Zuilhof group [26] reported the direct and high-yielding (77–99%) β-glycosylation of ω-functionalized alcohols using per-*O*-pivaloylated derivatives of lactose, galactose, and glucose with $ZnCl_2$ (1.5 equiv.) as the Lewis acid in toluene at 65–70 °C (Scheme 8.8). The conditions are practical, and although anhydrous conditions are required, the use of molecular sieves is not recommended. This work highlights the potential use of anomeric pivaloylates as an alternative type of glycosyl donors in glycosylation reactions.

As an alternative to conventional reagents (e.g., TMSOTf, $BF_3{\cdot}OEt_2$, $ZnCl_2$) for the activation of trichloroacetimidate glycosyl donors, Misra and coworkers [27] demonstrated that nitrosyl tetrafluoroborate ($NOBF_4$) can also be used as an effective promoter for the stereoselective synthesis of glycosides, and a series of mono- and disaccharide derivatives were synthesized in excellent yields and with high stereoselectivity. It is important to note that the use of $NOBF_4$ as a glycosylation promoter for thioglycosides

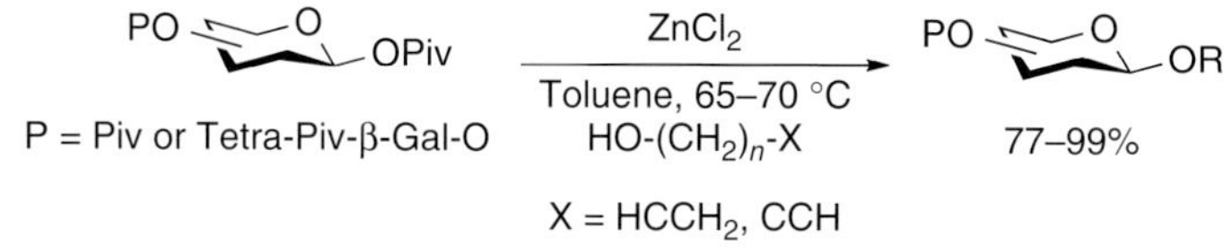

Scheme 8.8 $ZnCl_2$-catalyzed glycosylation of per-*O*-pivaloylated glycosides.

had already been previously reported in the 1980s by Pozsgay and Jennings [28]. Later on, the team further extended the use of this Lewis acid-activation to the preparation of 2,3-unsaturated sterically hindered alkyl, thio, and sulfonamido glycosides from D-glucal and L-rhamnal derivatives and 2-deoxy alkyl, thio, and sulfonamido galactosides from D-galactal derivatives [29].

Orthogonal glycosylation protocols, where a glycosyl donor can be selectively activated in the presence of another one are highly desirable as a more efficient alternative to traditional approaches, which often require additional number of protection and deprotection steps and give low overall yields [30]. Lowary and Chang [31] contributed to this area by describing the use of glycosyl 2-pyridyl sulfones as a novel class of hydrolytically stable glycoside donors that can be activated with samarium(III) triflate in refluxing dichloromethane to afford glycosides in moderate-to-excellent yields (Scheme 8.9). The authors hypothesized that sulfone activation is likely to take place via remote activation [32] by complexation of the metal ion with the pyridyl nitrogen and one of the sulfone oxygens. Subsequent cleavage of the C1—S bond in this activated complex leads to an oxonium ion, which is then trapped by the alcohol. The method was used to prepare a series of di- and trisaccharides containing both furanose and pyranose residues. Moreover, both sulfones and thioglycosides can be selectively activated in the presence of the others (anomeric sulfones are unreactive to thioglycoside activation with NIS/AgOTf), and this enabled rapid preparation of a series of oligosaccharides by the team. Although the method has several drawbacks, for example, requirement of high temperatures for sulfone activation and the fact that disarmed (acylated) sulfones cannot be used as glycosylation partners, the orthogonality of sulfones with regard to activation of other glycoside donors makes this method a significant addition to the carbohydrate chemist toolbox.

As mentioned earlier, ILs offer an interesting alternative to traditional reagents in organic synthesis, including recent applications in the area of oligosaccharide synthesis [9, 10, 33–36]. Galan *et al.* reported the first application of 1-butyl-3-methylimidazolium triflate [bmim][OTf] as a mild and versatile IL cosolvent and recyclable promoter for the room-temperature glycosylation of both thiophenyl and trichloroacetimidate

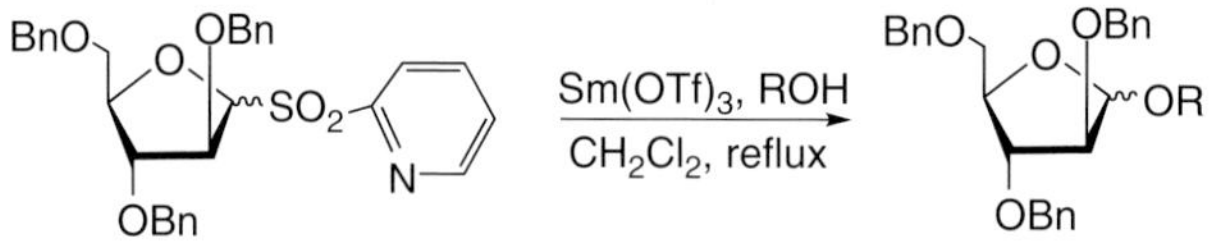

Scheme 8.9 Sm(III) triflate-catalyzed glycosylation of glycosyl 2-pyridyl sulfones.

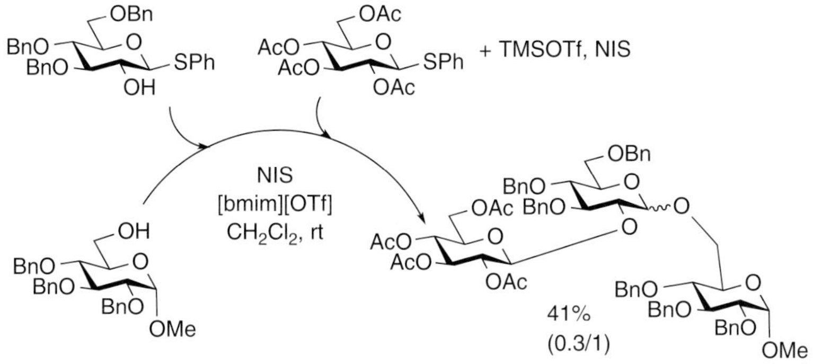

Scheme 8.10 IL-catalyzed one-pot-reactivity-based glycosylation.

glycoside donors (Scheme 8.10) [37]. The conditions are mild and compatible with a range of hydroxyl- and amino-protecting groups, such as acetates, benzyl ethers, acetals, phthalimido (Phth), and trichloroethoxycarbonyl (Troc). Initial mechanistic studies suggested that [bmim][OTf] can facilitate glycosylation reactions by the slow release of catalytic amounts of triflic acid and that the IL also protects the newly formed glycosidic linkage from hydrolysis [37, 38]. The team also showed that the triflated IL could selectively promote activated (armed) thiophenyl and trichloroacetimidate glycosyl donors, while less active (disarmed) donors required the addition of catalytic triflic acid. The group showcased the versatility of the IL/NIS promoter in a series of regio- and chemoselective-reactivity-based one-pot glycosylation reactions at room temperature, where both donor and acceptor bear a free OH of distinct reactivity, to access branched and linear trisaccharides (Scheme 8.10) [39].

8.4.2 Conformationally Constraint Glycosyl Donors

Lewis acids can also be used as α-directing additives in glycosylation reactions of 2,3-*O*-carbonate-protected thioglycosides using preactivation protocols [40]. The preactivation glycosylation strategy developed within the Ye group for the iterative one-pot synthesis of oligosaccharides, whereby the glycosyl donor is completely activated prior to the addition of the glycosyl acceptor, is an effective strategy that has been shown to have a significant effect on the stereoselectivity of the reaction. The same group and others also demonstrated that the use of glycosylation additives can help direct the glycosylation outcome [16, 41]. Taking into consideration that conformation-constraining protecting groups are also known to have an effect on the stereoselectivity of the reaction [42–45], the team screened a range of Lewis acids as additives in their preactivation glycosylation protocol and were able to achieve excellent α-stereoselectivities with very unreactive substrates, such as the 4,6-di-*O* acetyl-2,3-*O*-carbonate-protected thioglycoside donors, by the addition of 0.2 equiv. of $BF_3{\cdot}OEt_2$ (Scheme 8.11). On the other hand, in the presence of 1 equiv. of $SnCl_4$, the selectivity of the reaction with 4,6-di-*O* benzyl-2,3-*O*-carbonate-protected thioglucoside donors was completely reversed from β to α. Furthermore, the poor stereoselectivities typically observed for 4,6-di-*O*-benzyl-2,3-*O*-carbonate-protected thiogalactoside donor in glycosylations were also improved by using $SnCl_4$ as additive. The authors proposed, based on their results and previous mechanistic studies on other

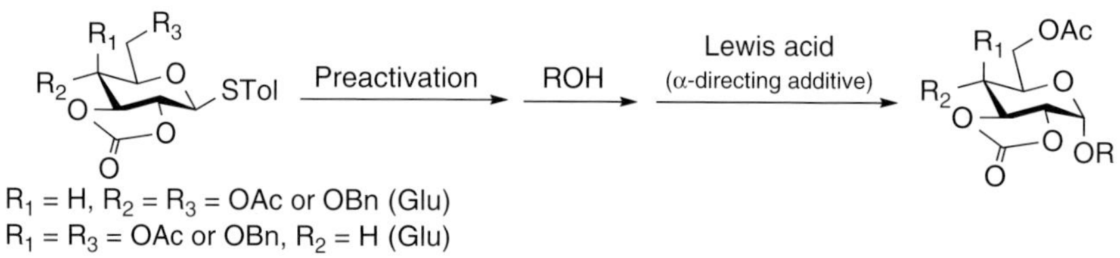

Scheme 8.11 Lewis-acid-catalyzed α-glycosylation of 2,3-*O*-carbonate-protected preactivated thioglycosides.

protecting group-constraint glycosylation systems [46], that in situ Lewis-acid-mediated anomerization of the β-glycosidic bond to the most thermodynamically stable anomer is mainly responsible for the high α-selectivity observed.

The stereoselective synthesis of α-*O*-sialosides is one of the most difficult challenges in glycosylation chemistry [47]. To address this issue, several groups have developed a number of practical strategies based on the use of 4-*O*-5-*N*-oxazolidinone-protected sialyl donors [48, 49] and their *N*-acetyl derivatives [50–53]. To expand on this excellent work, the Crich group used the same oxazolidinone-based glycoside donors in the highly stereoselective synthesis of α-*S*-sialosides [54]. Thiosialosides, which contain a thiol group instead of an anomeric oxygen, are particularly useful compounds as nonhydrolyzable mimics of *O*-glycosides for use in enzymatic studies and as potential inhibitors and stable substrate analogs that can facilitate crystallographic studies [55]. To that end, the team demonstrated that *N*-acetyl-5-*N*,4-*O*-oxazolidinone-protected sialyl phosphates are excellent donors under TMSOTf-catalyzed conditions for the formation of α-*S*-sialosides at −78 °C in dichloromethane with a range of thiols (e.g., including galactose 3-, 4-, and 6-thiols) in good yields and high stereocontrol (Scheme 8.12).

8.4.3 Synthesis of O-Glycoside Mimics

A Lewis acid such as TMSOTf can also be used to catalyze the diastereoselective opening of carbohydrate 1,2-lactones with nucleophile alcohols to access 2-*C*-malonyl glycosides in good-to-excellent yields and with high α-selectivity [56]. Interestingly, the methyl ester group remains intact in the products, whereas the former lactone is converted into a new ester group, giving access to unsymmetrical 2-*C*-malonyl carbohydrates (Scheme 8.13). The high diastereoselectivity can be explained by the selective attack of TMSOTf at the lactone carbonyl group and subsequent esterification and concomitant trapping of the transient anomeric oxocarbenium ion with the corresponding alcohol. However, the

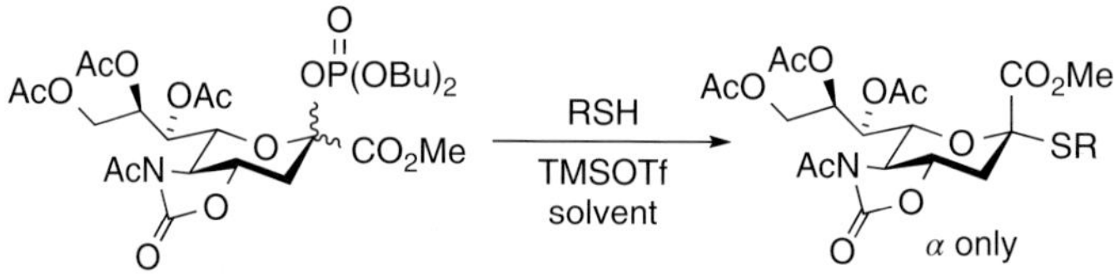

Scheme 8.12 Lewis-acid-catalyzed α-*S*-sialoside formation.

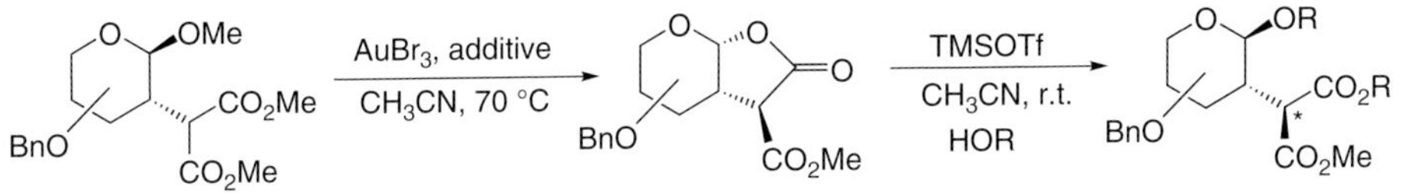

Scheme 8.13 Lewis-acid-catalyzed stereoselective lactonization and subsequent glycosylation.

formation of a transient β-anomeric triflate and subsequent S_N2 reaction with alcohols could also account for the selectivity observed and should also be considered plausible. The anomeric lactones used in the study are easily accessible by reaction of 2-*C*-malonyl glycosides in the presence of catalytic gold(III) bromide and afford only one stereoisomer due to steric interactions of the ester group with the carbohydrate ring, which is minimized in the observed configuration as suggested by the authors. Thus, these types of bicyclic materials offer an attractive alternative as chiral building blocks in carbohydrate chemistry.

Another interesting example of a Lewis-acid-catalyzed glycosylation comes from the Shao and Hu group [57]. The team studied the glycosylation reaction of 1,2-cyclopropaneacetylated glycosyl donors in the presence of a Lewis acid to produce oligosaccharides and glycoconjugates bearing a 2-*C*-acetylmethyl substituent. These types of glycosides are very useful 2-*N*-acetamido sugar mimics that can be used as probes in glycobiology research [58]. Interestingly, the diastereo-outcome of the glycosylation reactions with the 2-*C*-acetylmethyl group was dependent on the reactivity of the glycosyl donor and the type of Lewis acid used as the catalyst (Scheme 8.14). For instance, galactosyl donors could undergo S_N1 pathway when a strong Lewis acid such as TMSOTf was used as the catalyst, while reactions under $BF_3{\cdot}OEt_2$-catalyzed conditions undergo an S_N2 pathway. On the other hand, less reactive glucosyl donors appear to follow an S_N2-type glycosylation. The stereoselectivity was also dependent on neighboring group participation of the 2-*C*-acetylmethyl group during the glycosylation, whereby a C2-acetal intermediate could form and favor β-products.

8.5 Metals as Lewis Acids

8.5.1 Gold

Cationic gold complexes have been shown to be highly carbophilic Lewis acids that can activate C—C multiples bonds toward nucleophilic attack, while Au(I) complexes, in

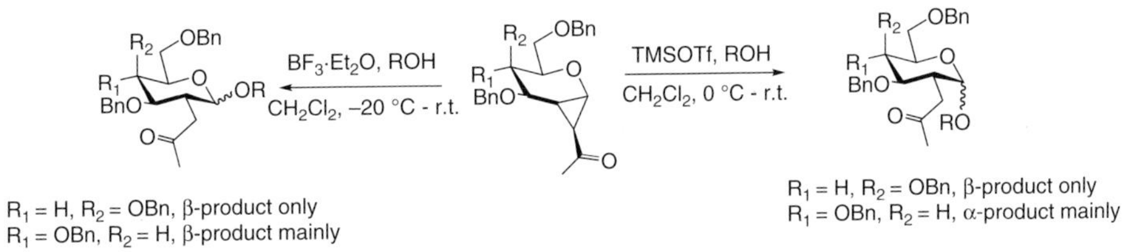

Scheme 8.14 Lewis acid stereoselective synthesis of 2-acetylmethyl glycosides.

particular, possess little oxophilic character, therefore displaying good functional oxo-group compatibility and good air and moisture stability [59]. Yu and coworkers demonstrated the use of triphenylphosphine gold(I) trifluoromethanesulfonate ($Ph_3PAuOTf$) as an effective catalyst for the activation of exceedingly reactive sugar 1,2-epoxides in glycosylation reactions [60]. The team envisioned that a Lewis acid weaker than current promoters (e.g., $ZnCl_2$, $BF_3{\cdot}OEt_2$, and AgOTf, which are often used in multiequivalent quantities), and devoid of a nucleophilic counterion, would be a milder and thus a better promoter for the nucleophilic opening of the 1,2-oxirane ring. Indeed, reactions of 1,2-anhydrosugars using only 0.1–0.2 equiv. of the Au(I) complex yielded the desired products with high-to-good β-stereocontrol and in high yields, which were >20% higher than those obtained with other Lewis acids (Scheme 8.15). The reaction shows that soft Lewis acid such as the Au(I) complex can efficiently substitute hard Lewis acid promoters in glycosylation reactions.

Another application of Au(I) complexes as soft Lewis acids for glycosylation promoters comes from the Kunz group [61]. In this example, gold(I) chloride was used to catalyze glycosylation reactions with common glycosyl donors such as glycosyl halides and trichloroacetimidates. Although reactions with glycosyl halides were sluggish and require high temperatures (110 °C) due to catalyst inhibition by halide anions, glycosylations with trichloroacetimidate glycosyl donors gave products with high diastereoselectivity and in good yields (Scheme 8.16).

The combination of Gold(III) chloride with phenylacetylene has recently been devised as a new relay catalytic system to promote the Ferrier rearrangement of acetyl-protected glycals and 2-acetoxymethylglycals, as well as the O-glycosylation of 1-*O*-acetyl sugars with different nucleophiles [62]. Reaction times are short, and the desired products were obtained in good-to-excellent yields and with high anomeric selectivity in the case of glycals, while moderate-to-good selectivities were observed for the 1-*O*-acetyl glycosyl donors (Scheme 8.17). Subsequently, the same group expanded the scope of the Au(III)-based relay catalytic system to glycosylations using glycosyl trichloroacetimidates, which was shown to be more efficient than $AuCl_3$ alone [63]. Reactions proceeded efficiently at room temperature and in good yields and with diastereoselectivity for

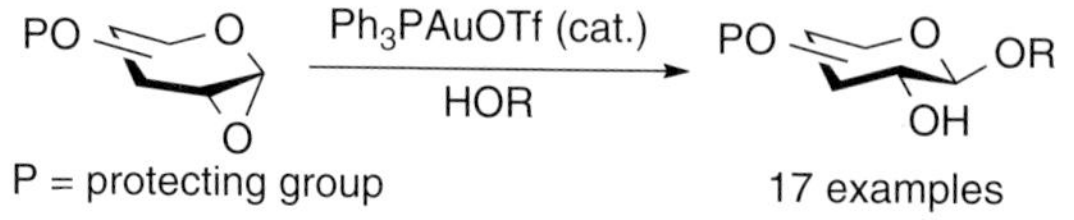

Scheme 8.15 Au(I)-catalyzed glycosylation of 1,2-anhydro sugars.

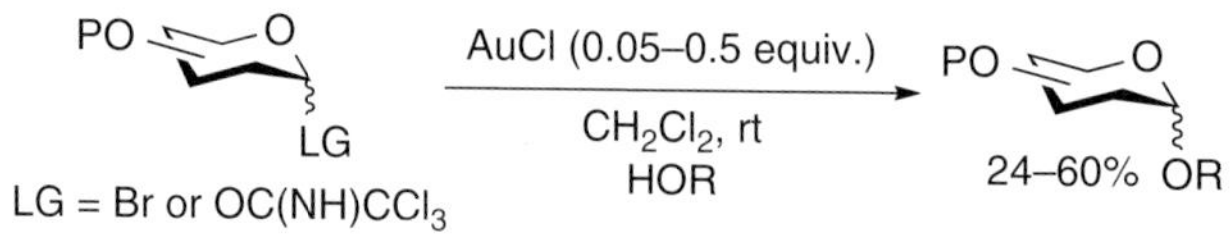

Scheme 8.16 Au(I)-catalyzed glycosylation of glycosyl halides and trichloroacetimidates.

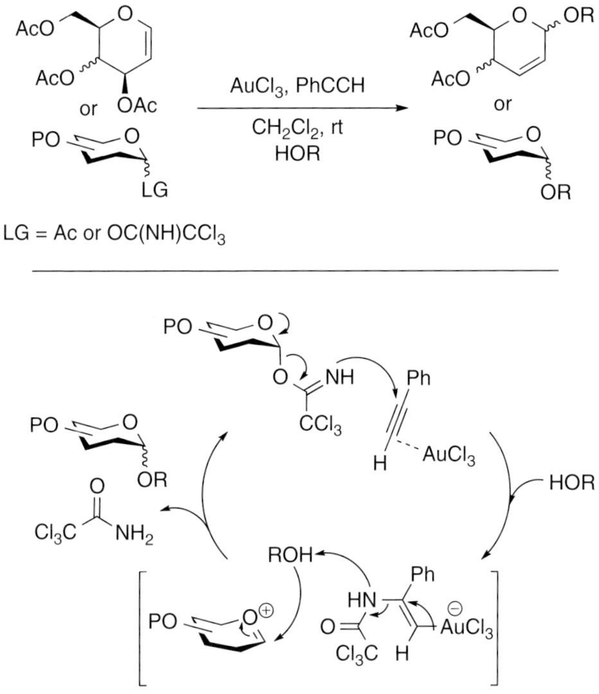

Scheme 8.17 Au(III)/phenylacetylene-catalyzed glycosylation of glycals, glycosyl acetates, and trichloroacetimidates.

2-acetyl-protected disarmed donors, whereas anomeric mixtures were obtained for armed glycosides and the use of acid-sensitive nucleophiles afforded moderate yields. The Vankar group proposed that coordination of $AuCl_3$ with phenylacetylene makes the phenylacetylene somewhat electron-deficient, so that it becomes susceptible to nucleophilic attack by the lone pair of the trichloroacetimidate nitrogen and thus effecting its departure. This makes the gold salt/alkyne combination a more powerful catalyst compared to $AuCl_3$ alone. Protonation by the alcohol nucleophile to form the amide as a by-product regenerates the catalytic system for the next cycle. However, the team does not exclude that direct activation of the trichloroacetimidate moiety on the glycosyl donors with $AuCl_3$ as the Lewis acid may also take place.

8.5.2 Cobalt

More recently, the use of (salen)Co catalysts as a new class of bench-stable stereoselective glycosylation promoters of glycosyl trichloroacetimidates has been reported by the Galan group [64] (Scheme 8.18). The reaction conditions are practical and proceed in good-to-excellent yields at room temperature and without the need for molecular sieves, demonstrating the potential of this class of catalysts as promising synthetic tools for diastereoselective acetal chemistry.

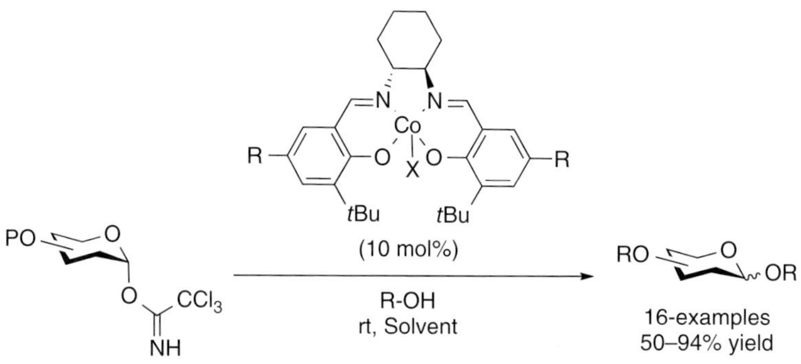

Scheme 8.18 (Salen)Co-catalyzed glycosylation of glycosyl trichloroacetimidates.

8.5.3 Nickel

The stereoselective synthesis of 1,2-*cis*-2-amino-2-deoxy glycosides of D-glucosamine and D-galactosamine, which are essential moieties of oligosaccharide and glycoconjugates with important biological roles, is particularly challenging as it requires a nonparticipating group at the C(2)-amino functionality on the glycosyl donor to direct the α-selectivity [65–67]. To address this issue, a very elegant and practical method for the stereoselective synthesis of α-2-deoxy-2-amino glycosides has been developed. Nguyen and Mensah [68] report the use of C(2)-*para*-methoxybenzylideneamino trichloroacetimidates as glycosyl donors, which, upon activation with catalytic amounts of Nickel (II) at room temperature, afforded the desired glycosides in excellent yields and α-selectivity. It was found that the diastereoselectivity of the reaction was dependent on the nature of the cationic nickel catalyst, while the nature of the nucleophile or the protecting groups had little effect (Scheme 8.19).

8.5.4 Iron

The use of anhydrous ferric chloride to catalyze the α-glycosylation of glycosamine penta-*O*-acetate glycosyl donors with a variety of nucleophile acceptors in refluxing 1,2-dichloroethane was reported by Du and coworkers [69]. The conditions are practical and cost-effective with yields ranging from moderate to good and with high α-stereocontrol.

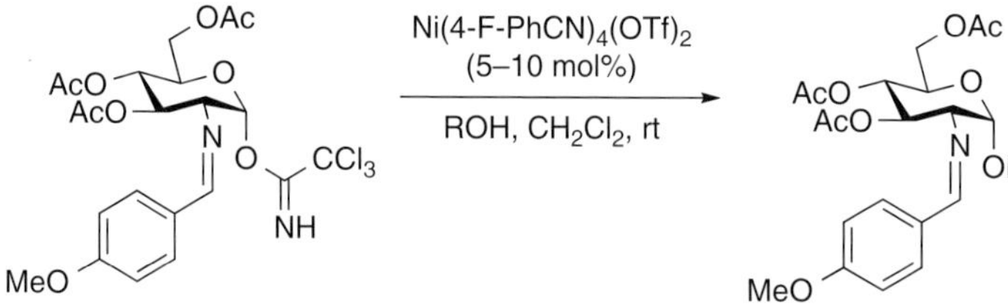

Scheme 8.19 Ni-catalyzed stereoselective formation of α-2-deoxy-2-amino glycosides.

8.6 Synthesis of C-Glycosides

C-glycosides are carbohydrates in which the anomeric oxygen has been replaced by a carbon atom and represent a very important class of molecules with interesting biological and pharmaceutical activities. The lack of anomeric oxygen imparts an increased stability on the glycosides, which makes them very useful analogs to study the role of common O-linked sugars in biological processes. Therefore, there is a lot of interest in the development of efficient methodologies to access these types of glycosides in a stereoselective manner. Lewis-acid-catalyzed C-glycosylation has been used to prepare a wide range of C-linked glycosides, albeit most methods tend to preferentially afford the α-glycosides, while the β-products are much more difficult to access. To circumvent this problem, Shuto and coworkers took advantage of the inherent conformational restrictions of the pyranose ring and developed an anomeric-effect-dependent, highly α- or β-stereoselective Lewis-acid-catalyzed C-glycosylation protocol [70]. The team hypothesized that the conformation of the transition state and that of the reactive intermediate are strongly influenced by the conformational effects that stabilize the ground-state conformation. Thus, by the careful choice of protecting groups, the team prepared xylosyl fluorides with either a $^4C_1$-restricted conformation (by employing a 3,4-diketal) or a $^1C_4$-restricted conformation (by using bulky *O*-silyl groups) and compared the outcome of the anomeric allylation reaction with allyltrimethylsilane and $BF_3{\cdot}Et_2O$ in dichloromethane at room temperature with that of an unrestricted glycosyl donor (tri-*O*-benzylated 1-fluoro-D-xylose). Indeed, high α-selectivity was observed for the $^4C_1$-restricted, while complete β-selectivity was observed for the $^1C_4$-restricted starting material and only moderate α-selectivity for the unrestricted glycosyl donor (Scheme 8.20). These results demonstrate that controlling the conformation of putative

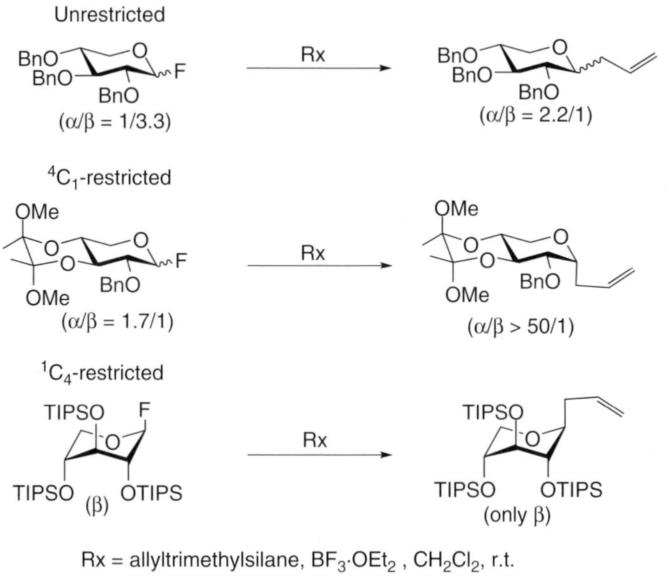

Scheme 8.20 Lewis-acid-catalyzed anomeric allylation of conformationally restricted xyloxyl fluorides.

reaction intermediates can be used as an effective approach to achieve high stereocontrol in glycosylation reactions catalyzed by common Lewis acids.

8.7 Conclusions and Outlook

Much progress has been made in the field of chemical glycosylation over the past decade by the development of novel catalysts or catalyst combinations in conjunction with highly efficient strategies and a better understanding of the glycosylation mechanism. However, there are still no general methods for the routine stereoselective synthesis of all types of complex oligosaccharides. The work reported in this chapter highlights the capriciousness of the glycosylation process, which is not only dependent on the reaction media and conditions (e.g., temperature, solvent, reagent concentration), but also dictated by the type of glycoside donor, acceptor nucleophile, and nature/chirality of the catalytic system employed to instigate the acetal formation step. As our fundamental understanding of the mechanism of this simple and yet very complex reaction improves, better and more efficient methodologies to streamline the synthesis of oligosaccharides are likely to be developed. Given the heterogeneous nature of complex oligosaccharides, the ultimate goal of finding a universal glycosylation catalyst might be elusive, and perhaps the development of a toolbox of glycosylation promoters specific for different types of glycoside building blocks that can give us stereoselective access to any glycoside required will be a more realistic and achievable target.

References

1 Ranade, S.C. and Demchenko, A.V. (2013) *J. Carbohydr. Chem.*, 32, 1.
2 Cox, D.J., Smith, M.D., and Fairbanks, A.J. (2010) *Org. Lett.*, 12, 1452.
3 Hamilton, G.L., Kanai, T., and Toste, F.D. (2008) *J. Am. Chem. Soc.*, 130, 14984.
4 Kimura, T., Sekine, M., Takahashi, D., and Toshima, K. (2013) *Angew. Chem. Int. Ed.*, 52, 12131.
5 Liu, D., Sarrafpour, S., Guo, W., Goulart, B., and Bennett, C.S. (2014) *J. Carbohydr. Chem.*, 33, 423–434.
6 Lairson, L.L., Henrissat, B., Davies, G.J., and Withers, S.G. (2008) *Annu. Rev. Biochem.*, 77, 521.
7 Gould, N.D., Liana Allen, C., Nam, B.C., Schepartz, A., and Miller, S.J. (2013) *Carbohydr. Res.*, 382, 36.
8 Kumar, A., Kumar, V., Dere, R.T., and Schmidt, R.R. (2011) *Org. Lett.*, 13, 3612.
9 Galan, M.C., Jones, R.A., and Tran, A.T. (2013) *Carbohydr. Res.*, 375, 35.
10 Sasaki, K., Nagai, H., Matsumura, S., and Toshima, K. (2003) *Tetrahedron Lett.*, 44, 5605.
11 Delacroix, S., Bonnet, J.-P., Courty, M., Postel, D., and Van Nhien, A.N. (2013) *Carbohydr. Res.*, 381, 12.
12 Sniady, A., Bedore, M.W., and Jamison, T.F. (2011) *Angew. Chem. Int. Ed.*, 50, 2155.
13 Shen, B. and Jamison, T.F. (2012) *Org. Lett.*, 14, 3348.
14 Balthaser, B.R. and McDonald, F.E. (2009) *Org. Lett.*, 11, 4850.
15 McDonald, F.E. and Subba Reddy, K. (2001) *Angew. Chem. Int. Ed.*, 40, 3653.

16 Balmond, E.I., Benito-Alifonso, D., Coe, D.M., Alder, R.W., McGarrigle, E.M., and Galan, M.C. (2014) *Angew. Chem. Int. Ed.*, 53, 8190.
17 Balmond, E.I., Coe, D.M., Galan, M.C., and McGarrigle, E.M. (2012) *Angew. Chem. Int. Ed.*, 51, 9152.
18 Geng, Y.Q., Kumar, A., Faidallah, H.M., Albar, H.A., Mhkalid, I.A., and Schmidt, R.R. (2013) *Angew. Chem. Int. Ed.*, 52, 10089.
19 Roy, B. and Mukhopadhyay, B. (2007) *Tetrahedron Lett.*, 48, 3783.
20 Mukhopadhyay, B., Maurer, S.V., Rudolph, N., Van Well, R.M., Russell, D.A., and Field, R.A. (2005) *J. Org. Chem.*, 70, 9059.
21 Aich, U. and Loganathan, D. (2007) *Carbohydr. Res.*, 342, 704.
22 Roy, D.K. and Bordoloi, M. (2008) *J. Carbohydr. Chem.*, 27, 300.
23 Bornaghi, L.F. and Poulsen, S.A. (2005) *Tetrahedron Lett.*, 46, 3485.
24 Dwek, R.A. (1996) *Chem. Rev.*, 96, 683.
25 Demchenko, A.P. (ed.) (2008) Handbook of Chemical Glycosylation, Wiley-VCH Verlag GmbH & Co. KGaA.
26 Pukin, A.V. and Zuilhof, H. (2009) *Synlett*, 2009, 3267.
27 Sau, A., Santra, A., and Misra, A.K. (2012) *Synlett*, 23, 2341.
28 Pozsgay, V. and Jennings, H.J. (1987) *J. Org. Chem.*, 52, 4635.
29 Santra, A., Guchhait, G., and Misra, A.K. (2013) *Synlett*, 24, 581.
30 Kanie, O., Ito, Y., and Ogawa, T. (1994) *J. Am. Chem. Soc.*, 116, 12073.
31 Chang, G.X. and Lowary, T.L. (2000) *Org. Lett.*, 11, 1505.
32 Hanessian, S. (1997) In Preparative Carbohydrate Chemistry, Dekker, New York.
33 Forsyth, S.A., MacFarlane, D.R., Thomson, R.J., and von Itzstein, M. (2002) *Chem. Commun.*, 714.
34 Murugesan, S., Karst, N., Islam, T., Wiencek, J.M., and Linhardt, R.J. (2003) *Synlett*, 1283.
35 Rencurosi, A., Lay, L., Russo, G., Caneva, E., and Poletti, L. (2005) *J. Org. Chem.*, 70, 7765.
36 Diaz, G., Ponzinibbio, A., and Bravo, R.D. (2012) *Top. Catal.*, 55, 644.
37 Galan, M.C., Brunet, C., and Fuensanta, M. (2009) *Tetrahedron Lett.*, 50, 442.
38 Galan, M.C., Tran, A.T., Boisson, J., Benito, D., Butts, C., Eastoe, J., and Brown, P. (2011) *J. Carbohydr. Chem.*, 30, 486.
39 Galan, M.C., Tran, A.T., and Whitaker, S. (2010) *Chem. Commun.*, 46, 2106.
40 Geng, Y.Q., Qin, Q., and Ye, X.S. (2012) *J. Org. Chem.*, 77, 5255.
41 Geng, Y.Q., Zhang, L.H., and Ye, X.S. (2008) *Chem. Commun.*, 597.
42 Lu, S.R., Lai, Y.H., Chen, J.H., Liu, C.Y., and Mong, K.K.T. (2011) *Angew. Chem. Int. Ed.*, 50, 7315.
43 Zhu, T. and Boons, G.J. (2001) *Org. Lett.*, 3, 4201.
44 Crich, D. and Jayalath, P. (2005) *J. Org. Chem.*, 70, 7252.
45 Heuckendorff, M., Pedersen, C.M., and Bols, M. (2013) *J. Org. Chem.*, 78, 7234.
46 Olsson, J.D.M., Eriksson, L., Lahmann, M., and Oscarson, S. (2008) *J. Org. Chem.*, 73, 7181.
47 Boons, G.J. and Demchenko, A.V. (2000) *Chem. Rev.*, 100, 4539.
48 Harris, B.N., Patel, P.P., Gobble, C.P., Stark, M.J., and De Meo, C. (2011) *Eur. J. Org. Chem.*, 2011, 4023.
49 a) Tanaka, H., Nishiura, Y., and Takahashi, T. (2006) *J. Am. Chem. Soc.*, 128, 7124; b) Farris, M. D., De Meo, C. (2007) *Tet. Lett.* 48, 1225 and c) Harris, B. N., Patel, P. P.; Gobble, C. P.; Stark, M. J.; De Meo, C. (2011) *Eur. J. Org. Chem.* 4023.
50 Crich, D. and Li, W.J. (2007) *J. Org. Chem.*, 72, 2387.

51 Chu, K.C., Ren, C.T., Lu, C.P., Hsu, C.H., Sun, T.H., Han, J.L., Pal, B., Chao, T.A., Lin, Y.F., Wu, S.H., Wong, C.H., and Wu, C.Y. (2011) *Angew. Chem. Int. Ed.*, 50, 9391.
52 Hsu, C.H., Chu, K.C., Lin, Y.S., Han, J.L., Peng, Y.S., Ren, C.T., Wu, C.Y., and Wong, C.H. (2010) *Chem. Eur. J.*, 16, 1754.
53 Wang, Y.J., Jia, J., Gu, Z.Y., Liang, F.F., Li, R.C., Huang, M.H., Xu, C.S., Zhang, J.X., Men, Y., and Xing, G.W. (2011) *Carbohydr. Res.*, 346, 1271.
54 Noel, A., Delpech, B., and Crich, D. (2012) *Org. Lett.*, 14, 4138.
55 Davies, G.J., Planas, A., and Rovira, C. (2012) *Acc. Chem. Res.*, 45, 308.
56 Pimpalpalle, T.M., Vidadala, S.R., Hotha, S., and Linker, T. (2011) *Chem. Commun.*, 47, 10434.
57 Tian, Q.A., Dong, L.A., Ma, X.F., Xu, L.Y., Hu, C.W., Zou, W., and Shao, H.W. (2011) *J. Org. Chem.*, 76, 1045.
58 Hang, H.C. and Bertozzi, C.R. (2001) *J. Am. Chem. Soc.*, 123, 1242.
59 Hashmi, A.S.K. (2007) *Chem. Rev.*, 107, 3180.
60 Li, Y., Tang, P.P., Chen, Y.X., and Yu, B. (2008) *J. Org. Chem.*, 73, 4323.
61 Gotze, S., Fitzner, R., and Kunz, H. (2009) *Synlett*, 2009, 3346.
62 Roy, R., Rajasekaran, P., Mallick, A., and Vankar, Y.D. (2014) *Eur. J. Org. Chem.*, 2014, 5564.
63 Roy, R., Palanivel, A.K., Mallick, A., and Vankar, Y.D. (2015) *Eur. J. Org. Chem.*, 2015, 4000.
64 Medina, S., Henderson, A.S., Bower, J.F., and Galan, M.C. (2015) *Chem. Commun.*, 51, 8939.
65 Mydock, L.K. and Demchenko, A.V. (2010) *Org. Biomol. Chem.*, 8, 497.
66 Zhu, X.M. and Schmidt, R.R. (2009) *Angew. Chem. Int. Ed.*, 48, 1900.
67 Galan, M.C., Benito-Alifonso, D., and Watt, G.M. (2011) *Org. Biomol. Chem.*, 9, 3598.
68 Mensah, E.A. and Nguyen, H.M. (2009) *J. Am. Chem. Soc.*, 131, 8778.
69 Wei, G.H., Lv, X., and Du, Y.U. (2008) *Carbohydr. Res.*, 343, 3096.
70 Tamura, S., Abe, H., Matsuda, A., and Shuto, S. (2003) *Angew. Chem. Int. Ed.*, 42, 1021.

9

Nickel-Catalyzed Stereoselective Formation of 1,2-*cis*-2-Aminoglycosides

Eric T. Sletten, Ravi S. Loka, Alisa E. R. Fairweather, and Hien M. Nguyen

9.1 Introduction

Recent advances in the field of glycoscience have revealed carbohydrates as an essential component of many biologically significant molecules found in nature [1–3]. Once seen only as an energy source, carbohydrates have been shown to govern important biological processes such as immune response [4], viral transfection [5], cancer metastasis [6], coagulation processes [7], cell–cell recognition, and cellular transport [8]. This increase in knowledge of carbohydrates has attracted many researchers. The major obstacle in the glycobiology field in deciphering the specific function of bioactive carbohydrate molecules is the lack of sufficient quantity of pure and well-defined oligosaccharides and glycoconjugates. Obtaining an adequate supply of well-defined glycan targets from natural sources can be quite challenging, because they are found in low concentrations and heterogeneous forms. Chemical synthesis is considered as one of the most efficient and indispensable routes to obtain sufficient quantities of the defined and desired biologically relevant glycans [9–11].

9.2 Biological Importance of 1,2-*cis*-Aminoglycosides

1,2-*cis*-2-Aminosugars **1** (Figure 9.1) are commonly found as key components in a wide variety of naturally occurring oligosaccharides and glycoconjugates [12, 13]. Many of these C(2)-aminosugars are attached to proteins on the cell surface to help mediate the interaction between cells by acting as ligands for lectins, antibodies, and enzymes [14–16]. C(2)-aminosugars are coupled to amino acids or other carbohydrate residues in one of two ways: either 1,2-*cis*- or 1,2-*trans*-glycosidic bonds (Figure 9.1) [13, 17, 18]. Control over stereoselectivity is the important issue in the synthesis of these bioactive glycan motifs bearing C(2)-aminosugar components.

1,2-*cis*-Aminoglycosides **1** (Figure 9.1) are ubiquitous constituents in many bioactive glycoconjugates. Examples of natural products bearing an α-1,2-*cis*-glycosidic bond are illustrated in Figure 9.2. Glycosylphosphatidylinositol (GPI) anchor **3** (Figure 9.2) serves to bridge a wide variety of signaling proteins to a fatty acid anchored in the cell

Selective Glycosylations: Synthetic Methods and Catalysts, First Edition. Edited by Clay S. Bennett.

1 — 1,2-*cis*-2-aminosugars

2 — 1,2-*trans*-2-aminosugars

R^1 = H, Ac, SO_3^-; R^2 = Amino acids or sugar residues

Figure 9.1 Structure of α- and β-C(2)-aminoglycosides.

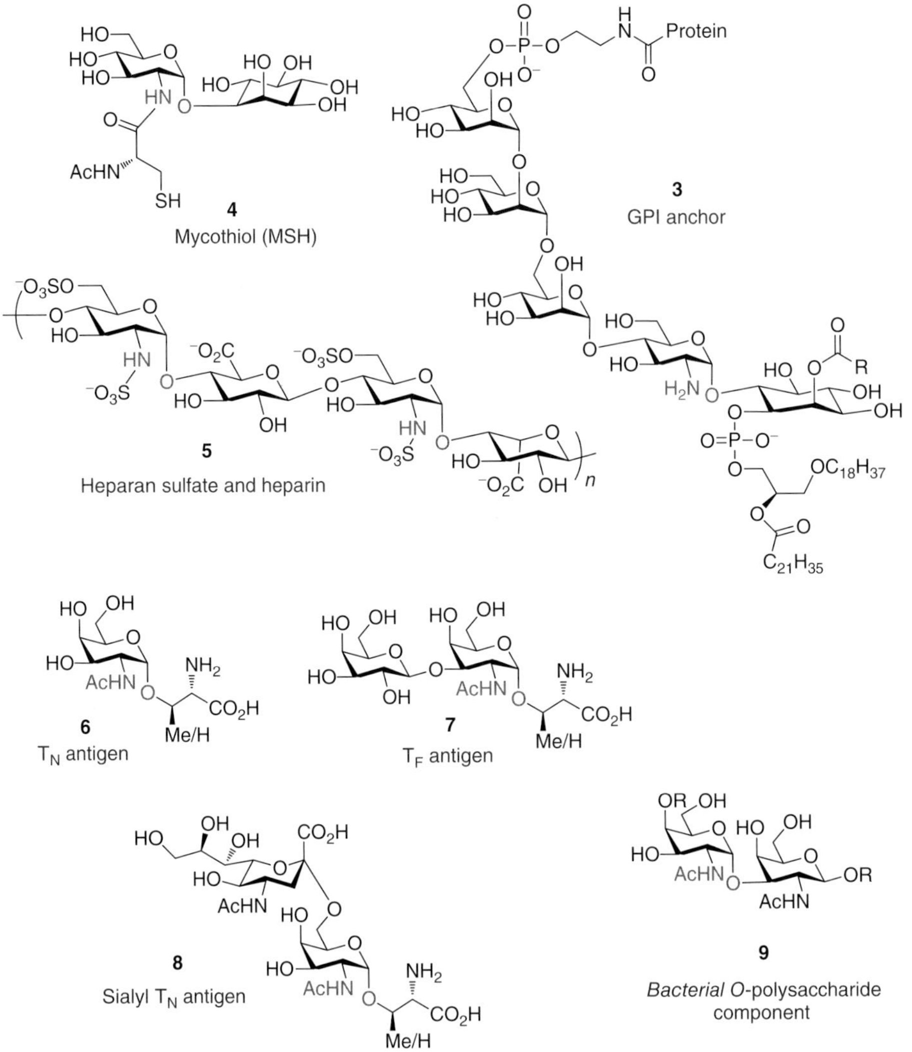

Figure 9.2 Biologically important 1,2-*cis*-aminoglycoside targets.

membrane [19]. This unique 1,2-*cis*-2-aminoglycosidic bond is also in mycothiol (MSH, **4**), a glucosamine–inositol motif with a pendant cysteine residue, which is found intracellularly in most actinomycetes as a defense mechanism. Mycothiol **4** has also been recognized as a potential antituberculosis drug target [20]. Both heparan sulfate and heparin (**5**, Figure 9.2) are anionic polysaccharides made up of repeating glucosamine–uronic acid disaccharides. Heparan sulfate is found on the surface of nearly all mammalian cell types, whereas heparin is just stored in mast cells. While heparin is well known for its anticoagulant activity [14], heparan sulfate is involved in adhesion, signaling, aiding in maintaining extracellular matrix (ECM) structural integrity, and storage of growth factors. In addition, heparan sulfate is in control of normal processes such as morphogenesis, tissue repair, inflammation, and vascularization [21]. There are several tumor-associated carbohydrate antigens (TACAs) linked to aberrant glycosylation by tumor cells, with their core structure containing galactosamine α-linked to either serine or threonine (**6–8**, Figure 9.2) [15]. These mucin antigens have been explored as potential cancer vaccine candidates [16]. Furthermore, disaccharide **9** (Figure 9.2) with 1,2-*cis*-aminosugar-linkage is the major component of the *O*-polysaccharides present on the outer membrane of Gram-negative bacteria. This could potentially be used to serve as immunogenic protection against *Salmonella enterica* and *Providencia rustigianii*. These bacteria are pathogenic to humans and are typically contracted through the digestion of contaminated food and water, which can lead to short-term gastronomical and urinary problems and eventually to complications such as Reiter's syndrome and typhoid fever [22–24].

The preparation of 1,2-*trans*-aminoglycosides (Scheme 9.1a), which are most commonly found in nature as chitin and chitosan, as well as many O-linked GlcNAc targets and all N-linked glycans, is readily established with the utilization of a C(2)-amide neighboring group to provide anchimeric assistance giving the β-linked anomeric product **13**, (Scheme 9.1a) [3, 25]. In contrast, the synthesis of the 1,2-*cis*-aminoglycosides **17** (Scheme 9.1b) is not trivial because the neighboring ester group would produce the undesired product **16**. With the use of a C(2)-nonparticipatory group, the anomeric selectivity at the newly formed glycosidic bond is unpredictable with the cis-anomer **17** as the major product due to thermodynamic stability (Scheme 9.1b) [12].

Since there are several elegant reviews on the stereoselective construction of 1,2-*cis*-2-aminoglycosides [13, 17, 18], the following chapter will first briefly summarize the synthesis of this carbohydrate motif using C(2)-nonparticipatory protecting groups and then focus mainly on the recent advances in this regard from the Nguyen group.

9.3 Use of Nonparticipatory Groups to Form 1,2-*cis*-Aminoglycosides

Conventional synthesis of 1,2-*cis*-aminoglycosides is accomplished with the use of a nonparticipatory protecting group placed on the C(2)-amine and has been thoroughly discussed in several previously published review articles [13, 17, 18]. This nonparticipatory strategy is predominantly accomplished through the use of Lemieux and Paulsen's C(2)-azido donor [26–28]. The C(2)-azido functionality serves as an excellent amino mask that is stable under both acidic and basic conditions, affording a range of

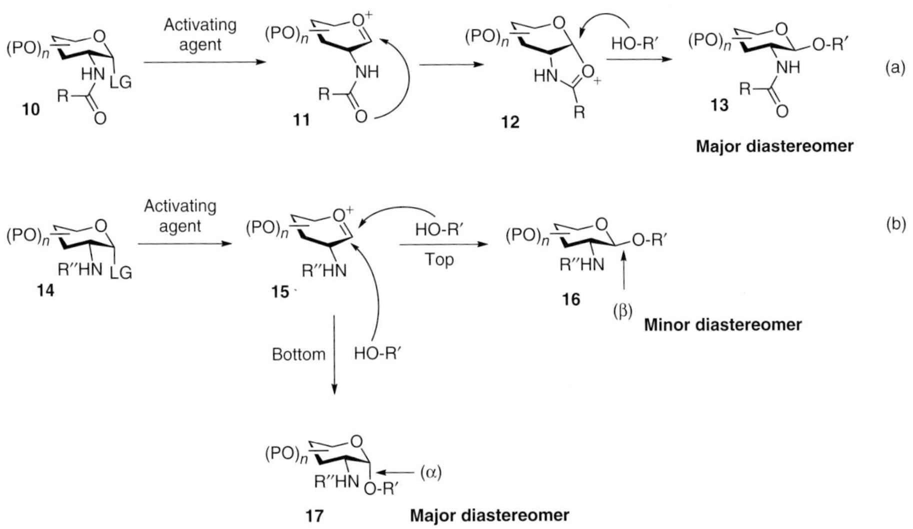

Scheme 9.1 General strategy for the formation of C(2)-aminoglycosidic bonds.

subsequent orthogonal-protecting group manipulations. Due to the robustness of C(2)-azido functionality, it has been shown to be very compatible with many different types of glycosyl donors including glycosyl halides, trichloroacetimidates, and thioglycosides. In the late stage of the synthetic sequence, the C(2)-azido functionality can then be selectively transformed back into the amino group. Its versatility can be seen in the countless number of total syntheses of biologically relevant oligosaccharide and glycoconjugate targets. One of these representative examples is for the synthesis of heparin disaccharide **24** (Scheme 9.3a) [29].

No universal formula exists when using C(2)-azido donors; solvent, temperature, and leaving group all need to be carefully controlled and screened in order to provide α-linkage in good-to-high anomeric selectivity, rarely generating only the α-product. In addition, formation of the C(2)-azide is not trivial. There are several methods to access C(2)-azido donors including Lemieux's azidonitration of glycal **18** with ceric ammonium nitrate and NaN_3 in acetonitrile (Scheme 9.2). Lemieux's azidonitration method is

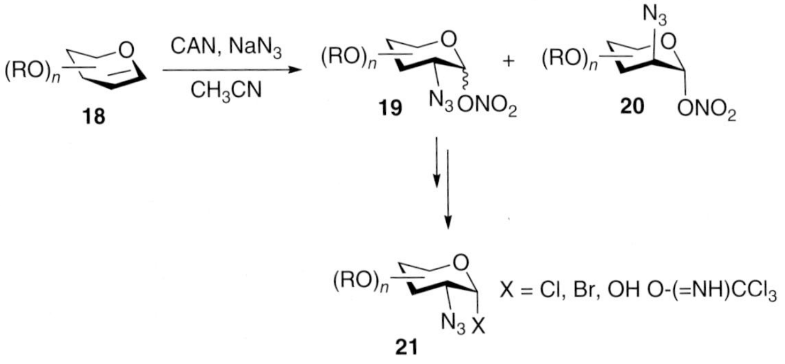

Scheme 9.2 Lemieux's azidonitration method.

not stereoselective and often provides a mixture of 2-azido-glucoside **19** and mannoside **20**, which is atomically inefficient and requires chromatographic purification to separate the diastereomers (Scheme 9.2) [28]. The ratio of **19** and **20** formed in the reaction is dependent on the glycal structure and the nature of protecting groups on glycal **18**. Alternatively, a diazotransfer method can be utilized to generate C(2)-azido derivatives by the action of trifluoromethanesulfonyl azide or imidazole-1-sulfonyl azide on glycosamine starting materials; however, these azide reagents are potentially explosive, which limits their scalability [30, 31].

Since the introduction of the C(2)-azido glycosyl donor, there have been some efficient improvements toward the synthesis of 1,2-*cis*-aminoglycoside over the past 30 years. Kerns demonstrated high α-selectivity through the utilization of 2,3-*trans*-carbamate to distort the ring structure of the glycosyl donor, resulting in an enriched cis-product (e.g., **27**, Scheme 9.3b) [32]. Both Schmidt and Gin were able to prepare α-linked GalNAc-Ser/Thr glycosides (Scheme 9.3c and d) using a protected 2-nitrogalactal **32** (Scheme 9.3d) as the glycosyl donor or through the opening of aziridines with C(1)-hemiacetals **28** (Scheme 9.3c), respectively [33, 34].

These few examples, as well the many more reported, all are constrained to several limitations. Prior methods rely on the anomeric effect for selectivity, confining them to

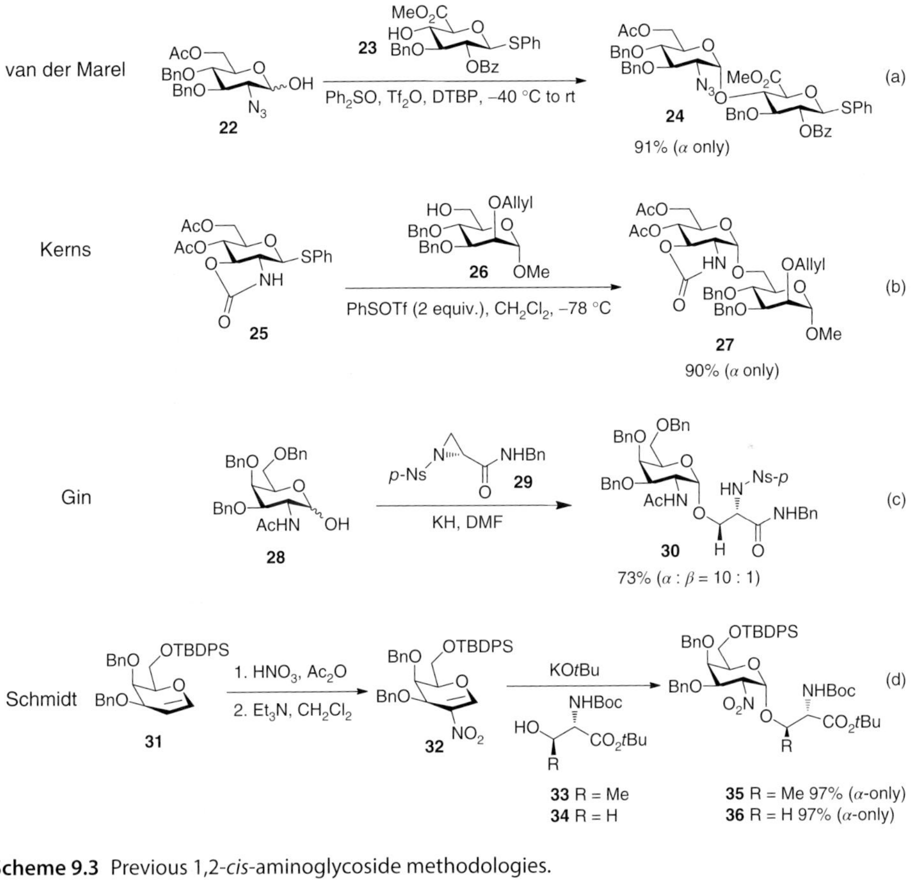

Scheme 9.3 Previous 1,2-*cis*-aminoglycoside methodologies.

the nature of the glycosyl donors resulting in unpredictable α-anomeric stereoselectivity. This unpredictability requires chromatographic purification to separate the desired 1,2-*cis*-2-amino product from the 1,2-*trans*. In some cases, this purification can be very tedious. In addition, many of these approaches require stoichiometric amount of activating reagents, resulting in the unnecessary waste of materials. Several of these activating reagents are air- and moisture-sensitive and not compatible with all protecting groups. Therefore, they require low temperatures, anhydrous conditions, and advanced technical laboratory skills for ideal selectivity and results. In general, these previous methods are highly dependent on the nature of the protecting groups on both coupling partners to influence the selective formation of 1,2-*cis*-2-amino linkage, and thus, they cannot be easily generalized [13]. These requirements remain at odds with principles laid out by Schmidt for efficient glycosylation methodologies in a recent review [9]: (i) use of catalytic amount of reagent, (ii) stereoselective and high yielding, and (iii) applicable on a large scale.

9.4 Nickel-Catalyzed Formation of 1,2-*cis*-Aminoglycosides

Over the past few years, the Nguyen group has been working on the development of a novel, scalable method to address many of the significant synthetic limitations previously associated with the preparation of 1,2-*cis*-2-aminoglycosides [13, 17, 18]. Trihaloacetimidates possess chemical and structural properties that have been utilized to provide varied and novel reactions. While carbohydrate chemists view trihaloacetimidate as a good "leaving group," [35, 36] the trihaloacetimidate nitrogen can also be viewed as a good "directing group" in transition-metal-catalyzed reactions [35]. On the other hand, the benzylidene functionality plays a dual role: (i) a "protecting group" for the C(2)-amino group of carbohydrate substrates and (ii) a "directing group" for transition-metal catalysts. We hypothesize that the combination of the benzylidene group and the transition-metal catalyst would guide the glycosyl acceptor to selectively attack the α-face of the activated donor forming the desired 1,2-*cis*-2-aminoglycoside product. With these concepts in mind, we designed and then synthesized donors **37** and **38** that contain the benzylidene functionality at C(2) and the trihaloacetimidate at C(1). We hypothesize that these donors would be effective at coordinating to the transition-metal catalyst to form complex **39** (Figure 9.3). Subsequent ionization/coupling events would generate the corresponding 1,2-*cis*-2-aminosugars **40** in a stereoselective manner [37–42].

The success of this strategy depends on the nature of a transition-metal catalyst that can simultaneously coordinate to both the trihaloacetimidate nitrogen and the arylamine group of donors **37** or **38** (Figure 9.3). We successfully identified cationic Ni(4-F-PhCN)$_4$(OTf)$_2$, generated *in situ* from Ni(4-F-PhCN)$_4$Cl$_2$ and AgOTf, as the most effective catalyst at promoting the glycosylation event. *The detailed mechanism of cationic nickel-catalyzed 1,2-cis-2-aminoglycosylation reaction is under investigation and will be reported in due course.* In this chapter, we only discuss the reaction, discovery, substrate scope, and application to the synthesis of biologically relevant oligosaccharides and glycoconjugates containing 1,2-*cis*-2-aminosugar substituents. In 2009, we reported a novel approach for the preparation of 1,2-*cis*-2-aminoglycosides **45** (Scheme 9.4a) via nickel-catalyzed stereoselective glycosylation with C(2)-*N*-*para*-(methoxy)benzylideneamino

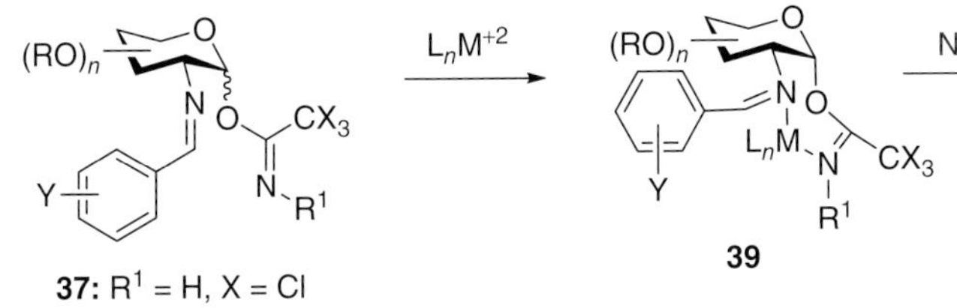

Figure 9.3 Proposed transition-metal-catalyzed formation of 1,2-*cis*-2-aminosugars.

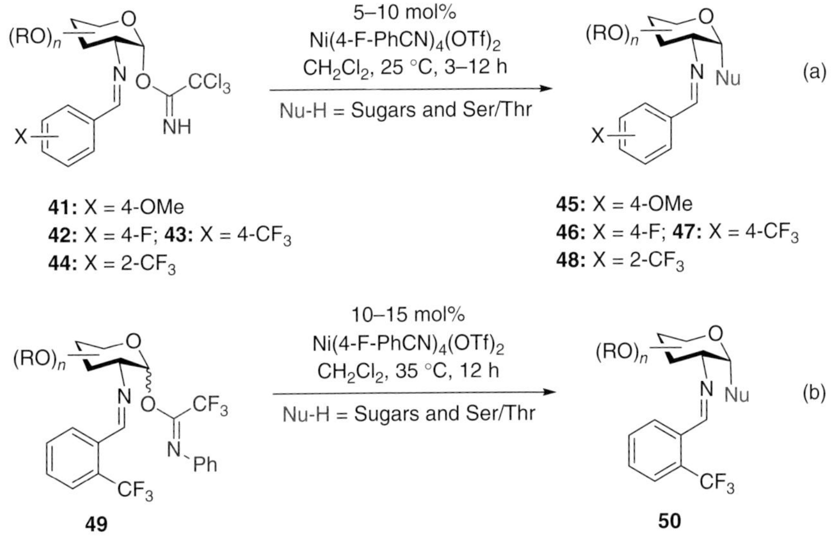

Scheme 9.4 Nickel-catalyzed stereoselective 1,2-*cis*-2-amino glycosylation.

trichloroacetimidate donors **41** [39]. We later discovered an improvement on donor **41**; the electron-withdrawing C(2)-*N*-*para*-fluoro- and trifluoromethyl-benzylidenamino donors **42** and **43** are more stable compared to the methoxy counterpart **41** for handling and purification [39]. In our approach, glycosylation can be carried at 25 °C with 5–10 mol% of nickel catalyst, Ni(4-F-PhCN)$_4$(OTf)$_2$. The products **45–48** (Scheme 9.4a) were formed with high anomeric selectivity ($\alpha:\beta = 12:1 \rightarrow 25:1$) and high yield (70–98%) [37–42].

During this process, we discovered that the C(2)-*N*-*ortho*-(trifluormethyl)benzylidene functionality is not only the most stable protecting group but also the most effective directing group at promoting the glycosylation under nickel conditions (Scheme 9.4a). The major drawback of using this *ortho*-(trifluormethyl)benzylidene group is that only the α-isomer of glycosyl trichloroacetimidate donor **44** (Scheme 9.4a) reacted under cationic nickel conditions. The β-trichloroacetimidate donor-bearing C(2)-*N*-*ortho*-(trifluoromethyl)benzylidenamino group was, however, the major anomer ($\alpha:\beta = 1:3$) from the reaction of hemiacetal with Cl$_3$CCN. *We subsequently found that both α- and β-anomers of N-phenyl trifluoroacetimidates* **49** *(Scheme 9.4b) reacted to provide the products with excellent α-selectivity* [42], *although the β-isomer reacted more slowly*

[38]. Although trichloroacetimidate derivatives are widely employed donors in carbohydrate synthesis, *the advantages of using donors **49** are that no separation of anomers is required and that they are more stable for handling and storage compared to trichloroacetimidates.* Their efficacy has been demonstrated in the chemoselective coupling to hydroxyl groups of thioglycoside acceptors to afford the desired disaccharides, which were then used as the donors in another coupling iteration to generate oligosaccharides [42]. The utility of these donors was also highlighted in the syntheses of T_N-antigen [41], mycothiol [38], and GPI pseudosaccharides [37].

While the first part of this chapter focuses on the utility of the α-trichloroacetimidate donors bearing C(2)-N-substituted benzylideneamino group in the stereoselective formation of 1,2-*cis*-2-aminoglycosides via nickel catalysis, the second part focuses on the broad utility of the α/β mixture of *N*-phenyl trifluoroacetimidate donors bearing C(2)-*N*-*ortho*-trifluoromethyl-benzylideneamino group.

9.5 C(2)-*N*-Substituted Benzylidene Glycosyl Trichloroacetimidate Donors[1]

In preliminary studies, $Pd(PhCN)_2(OTf)_2$ was the initial catalyst of choice to facilitate the glycosyl coupling of trichloroacetimidate **51** with diacetone galactose acceptor **52** (Table 9.1, entry 1). This catalyst was previously discovered by the Nguyen group to the efficiently catalyze a β-selective glycosylation with glycosyl trichloroacetimidate donors in the absence of a C(2)-anchimeric-protecting group [43]. Although the disaccharide **56** was isolated in a moderate yield of 60%, an enrichment of the desired α-anomer (4 : 1 $\alpha:\beta$) proved promising. This result supports our hypothesis that the use of a cationic transition-metal catalyst in partnership with the C(2)-benzylidene group could concurrently activate the donor and direct the anomeric selectivity. Encouraged by these results, the activating reagent was switched to a more azaphilic, cationic nickel catalyst, $Ni(PhCN)_4(OTf)_2$, which can be generated *in situ* from $Ni(PhCN)_4Cl_2$ and AgOTf in CH_2Cl_2. With just 5 mol% catalytic loading (entry 2), a drastic increase in the yield (60 → 95%) and anomeric stereoselectivity ($\alpha:\beta = 4:1 \rightarrow 8:1$) with a decrease in the reaction time (10 → 4 h) was observed in the coupling event. Modulation of the catalyst's electronic nature by the addition of para-substituents onto the benzonitrile ligands with an electron-donating group (entry 4) showed an increase in selectivity (8 : 1 → 10 : 1 $\alpha:\beta$) but a drop in yield (95 → 76%) and also extended the reaction time. Installation of an electron-withdrawing fluorine substituent to the para position of the benzylidene group resulted in a similar yield; however, the selectivity increased to 10 : 1 $\alpha:\beta$ (entry 3). To further demonstrate that $Ni(4\text{-}F\text{-}PhCN)_4(OTf)_2$ is the actual activating reagent required for the glycosylation to occur, not the initial catalyst precursors, use of $Ni(4\text{-}F\text{-}PhCN)_4Cl_2$, AgOTf, or triflic acid in control experiments was carried out with the same donor and acceptor (entries 5–7). The use of neutral $Ni(4\text{-}F\text{-}PhCN)_4Cl_2$ yielded none of the desired disaccharide **56**, proving that the less coordinating triflate anion was

1 The Nguyen group has envisioned that a donor incorporated with a C(2)-benzylidene-protected imine and an anomeric trichloroacetimidate leaving group can be activated through the use of a substoichiometric amount of transition metal, by simultaneous metal coordination to the C(2)-benzylidene of the activated donor, and the acceptor will enhance the α-stereoselectivity.

Table 9.1 Optimized conditions for selective formation of 1,2-*cis*-2-aminoglycosides.

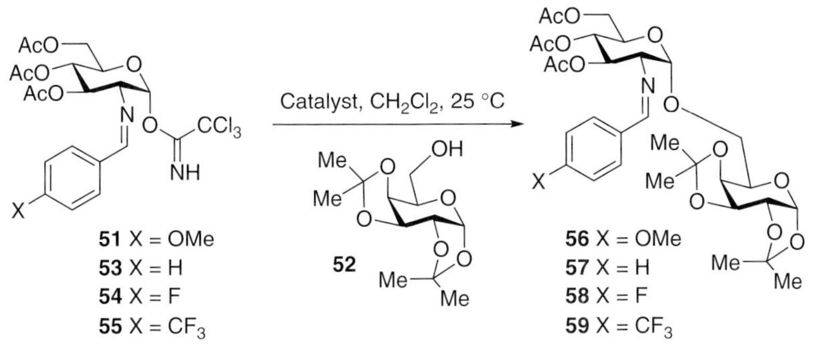

Entry	Donor	Catalyst	Loading (mol%)	Time (h)	Yield (%)	α:β
1	**51**	$Pd(PhCN)_2(OTf)_2$	5	10	60	4:1
2	**51**	$Ni(PhCN)_4(OTf)_2$	5	4	95	8:1
3	**51**	$Ni(4-F-PhCN)_4(OTf)_2$	5	3	93	10:1
4	**51**	$Ni(4-MeO-PhCN)_4(OTf)_2$	5	6	76	10:1
5	**51**	$Ni(4-F-PhCN)_4Cl_2$	5	10	—	—
6	**51**	AgOTf	10	10	—	—
7	**51**	TfOH	10	5	10	3:1
8	**53**	$Ni(4-F-PhCN)_4(OTf)_2$	5	3	92	10:1
9	**54**	$Ni(4-F-PhCN)_4(OTf)_2$	5	1	96	9:1
10	**55**	$Ni(4-F-PhCN)_4(OTf)_2$	5	1	87	9:1

necessary for the reaction to occur (entry 5). Stoichiometric amounts of AgOTf were previously capable of activating trichloroacetimidate donors [44], and an experiment performed with 10 mol% of AgOTf revealed less than 1% conversion (entry 7). Potential generation of triflic acid, from the cationic nickel catalyst, may act as the catalytic source to promote the glycosylation reaction. In order to eliminate this possibility, the coupling of **51** and **52** was performed in the presence of 10 mol% of triflic acid (entry 8). The desired disaccharide **56** was isolated in 10% yield with $\alpha:\beta=3:1$. The results obtained from these control experiments conclude that the presence of the cationic nickel catalyst is important for the stereoselective coupling event, and it is not just simply acting as a Lewis acid.

Several other donors **53–55**, which incorporated para-electron-withdrawing substituents on the C(2)-benzylidene group, were also investigated (Table 9.1, entries 8–10); the results illustrate that the reaction was faster with use of donors **54** and **55,** which bear the C(2)-*N*-electron-withdrawing benzylidene functionality (entries 9 and 10). To further investigate the unique features of the cationic nickel catalyst, glycosylation of galactosyl acceptor **52** with C(2)-*N*-4-methoxybenzylidene trichloroacetimidate donor **51** was conducted with other mild Lewis acids ($Zn(OTf)_2$ and $Cu(OTf)_2$) and transition-metal catalysts, ($Au(PPh_3)(OTf)$ and $Rh(cod)_2OTf$). Under these conditions, the desired disaccharide **56** was obtained in poor yields (20 → 26%) and as a mixture of α- and β-isomers (1 : 1 → 2 : 1) [45].

9.5.1 Comparison to Previous Methodologies

To prove the utility and uniqueness of this system, which merges nickel catalysis with carbohydrate synthesis, comparisons were performed against conventional trichloroacetimidate donor activators and established 1,2-*cis*-aminosugar methodologies [46]. Glycosylation of glucoside acceptor **60** with donor **51** was attempted by Mara and Sinay in 1990 under Lewis acid conditions [46]. The reaction did not occur in the presence of TMSOTf as the promoter, and the use of $BF_3{\cdot}OEt_2$ resulted in a trace amount of the desired disaccharide **61** (Scheme 9.5a). Encouraged by the results obtained in Table 9.1, we decided to explore the feasibility of the cationic nickel method with both coupling partners **51** and **60** (Scheme 9.5b). The reaction proceeded smoothly to provide disaccharide **61** in 77% yield with excellent diastereoselectivity ($\alpha:\beta = 20:1$). Compared to other glycosylation methods [46, 47], this nickel chemistry is more α-selective. For instance, coupling of **60** with glycosyl bromide derivative of **51** in the presence of AgOTf (1.5 equiv.) as a promoter provided the β-isomer of **61** as the major product ($\alpha:\beta = 1:9$) [46]. On the other hand, glycosylation of **60** with a C(2)-oxazolidinone thioglycoside donor afforded the glycoconjugate in 81% yield with $\alpha:\beta = 3:1$ [47].

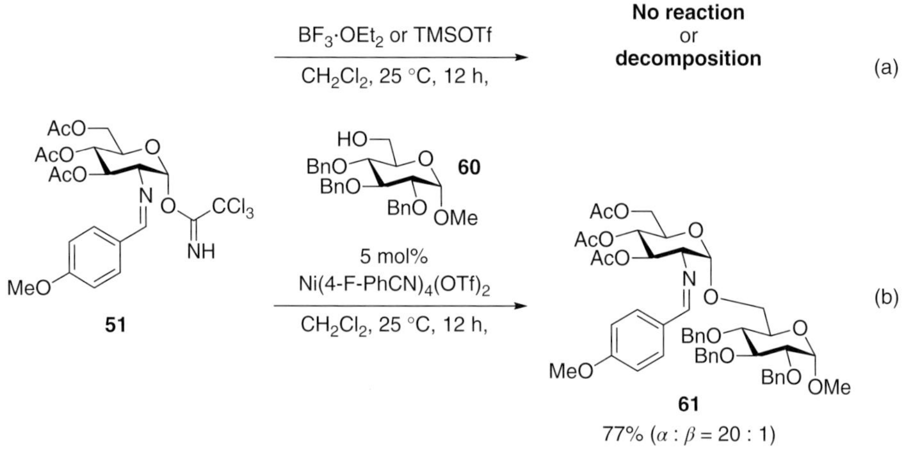

Scheme 9.5 Comparison between traditional Lewis acids and nickel catalyst.

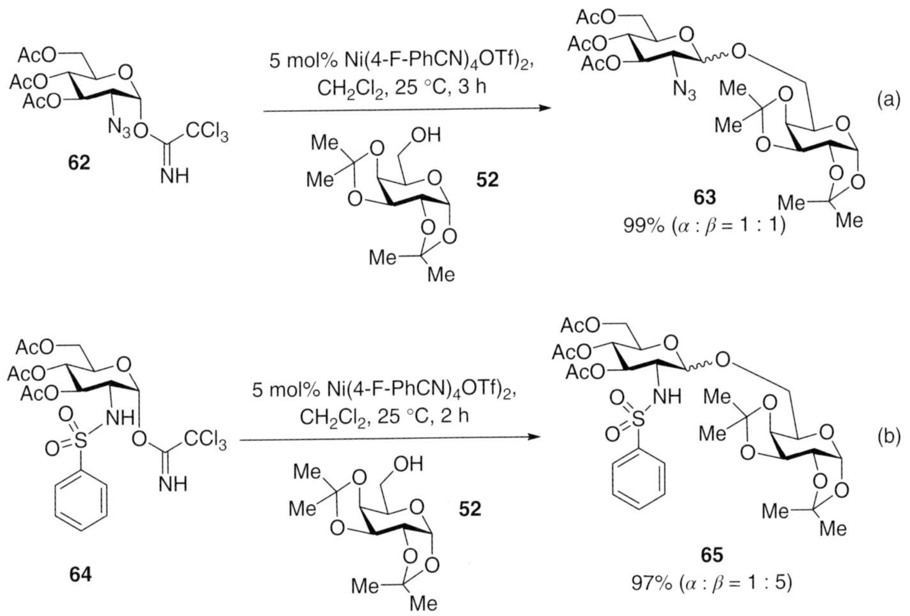

Scheme 9.6 Investigation of nickel catalyst with other C(2)-functional groups.

When the C(2)-*N*-benzylidene functionality was substituted with Lemieux's popular C(2)-azido group in the reaction of galactose acceptor **52** with C(2)-azido trichloroacetimidate **62** (Scheme 9.6a) in the presence of 5 mol% Ni(4-F-PhCN)$_4$(OTf)$_2$, the desired disaccharide **63** was obtained as a 1 : 1 mixture of α- and β-isomers. Phenylsulfonamide has been commonly used as the directing group for palladium-catalyzed C—H activation reactions [48, 49], so we attempted to explore a C(2)-*N*-phenylsulfonamide trichloroacetimidate **64** as a potential glycosyl donor (Scheme 9.6b). Contradictory to what was hypothesized [48, 49], the β-isomer **65** was isolated as the major product (α:β = 1 : 5) under nickel conditions (Scheme 9.6b). These two control experiments highlight the critical role of the C(2)-N-substituted benzylidene group in the nickel-catalyzed α-selective glycosylation reaction.

9.5.2 Expansion of Substrate Scope

With optimal conditions established, the generality of the nickel method was examined with a variety of glycosyl acceptors and glucosamine donors **51–54** (Table 9.2). All the glycosylations provided the desired α-products **66–80** in high yield and with excellent anomeric selectivity. Even the coupling to the challenging, sterically hindered, tertiary adamantol **79** (entry 6) provided the desired glycoconjugate **80** in 96% yield and with 17 : 1 α-selectivity. Alternative approaches involving sterically hindered secondary acceptor **76** have produced poor yields or low selectivity [47, 50]. For example, the use of the C(2)-oxazolidinone thioglycoside or the C(2)-azido donors provided the expected product with α:β = 2 : 1 [47] or only 40% yield of α-isomer, respectively [50]. Different iterations of the glycosamine donors have a dramatic effect over the nature of the

Table 9.2 Substrate scope of nickel-catalyzed 1,2-*cis*-amino glycosylation.

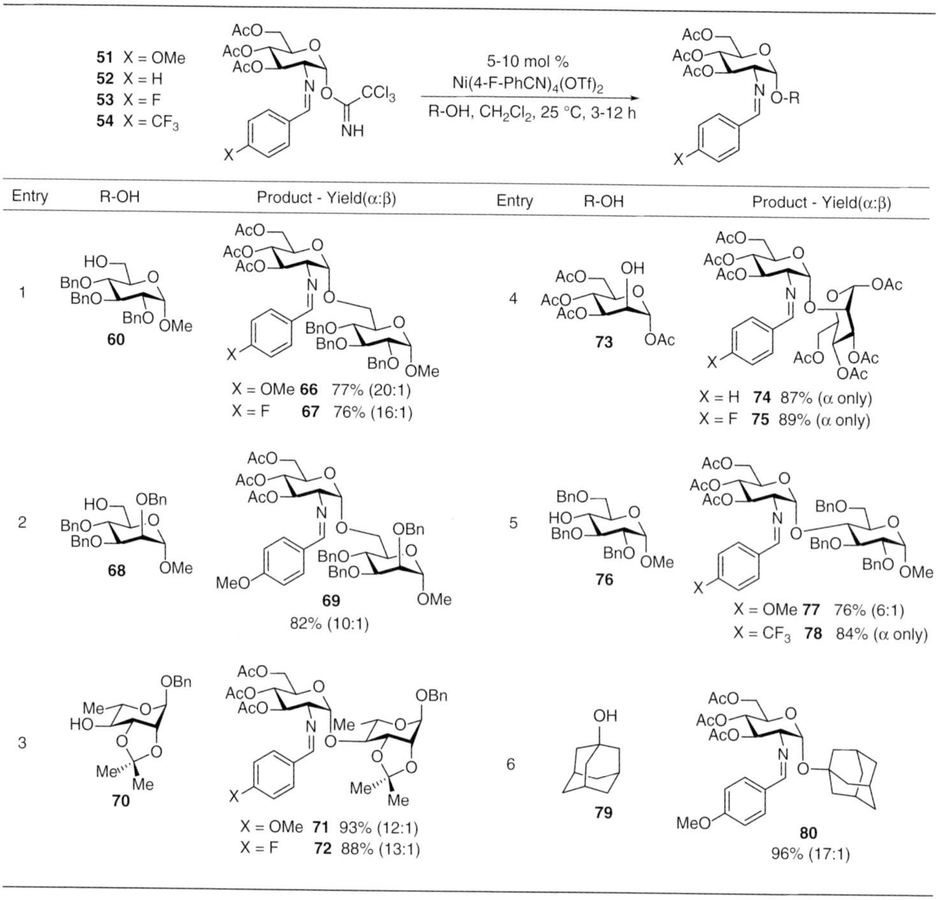

α-selectivity of the glycosidic bond formation as illustrated in entry 5 of Table 9.2. For instance, simple modification of the para-substituent on the benzylidene group in the glycosylation of carbohydrate acceptor **76** showed increased yield and α-selectivity with the *para*-CF_3-benzylidene over the *para*-OMe derivative.

With the successful results obtained from the glucosamine donor reactions under cationic nickel conditions, galactosamine trichloroacetimidate **81** was then employed as a potential donor (Table 9.3) in the glycosylation reaction with both primary and secondary alcohol acceptors. Once again, these coupling reactions were found to be highly α-selective and provided 1,2-*cis*-2-aminoglycosides **84–86** in high yield regardless of the position of the hydroxyl moiety and the nature of the protecting groups on the nucleophilic acceptors.

The generality of nickel-catalyzed α-selective glycosylation was extended to the synthesis of oligosaccharides, as illustrated in Scheme 9.7. The catalytic system was capable of linking the unprotected secondary alcohol of disaccharide acceptor **87** (Scheme 9.7a) with the C(2)-*N*-*para*-trifluoromethyl benzylidene glucosamine donor **54** yielding trisaccharide **88** (76%) with a high stereoselectivity ($\alpha:\beta = 11:1$). In a [2+2] approach,

Table 9.3 α-Selective coupling with D-galactosamine trichloroacetimidate.

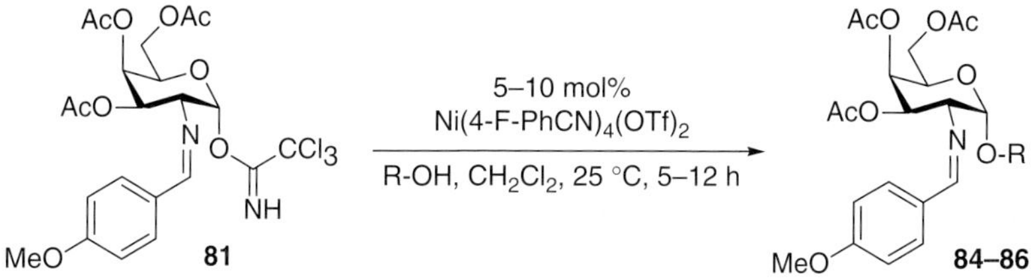

Entry	R-OH	Product	Yield (α : β)
1	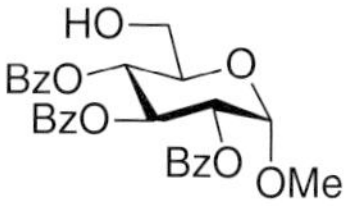 **82**	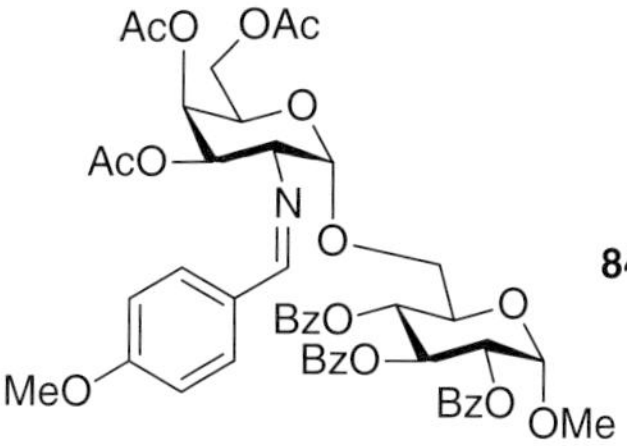 **84**	74% (14 : 1)
2	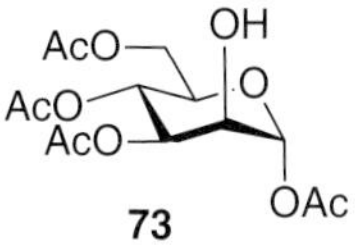 **73**	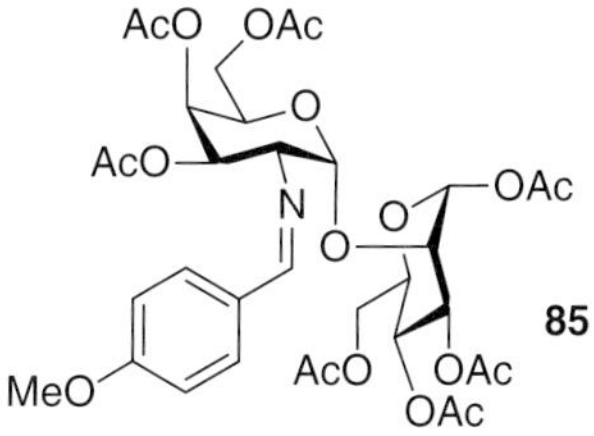 **85**	93% (α only)
3	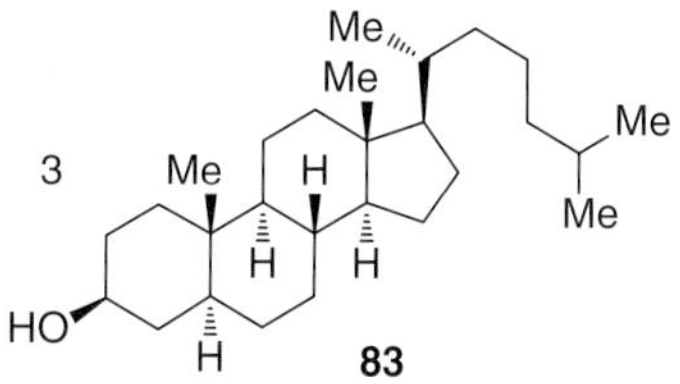 **83**	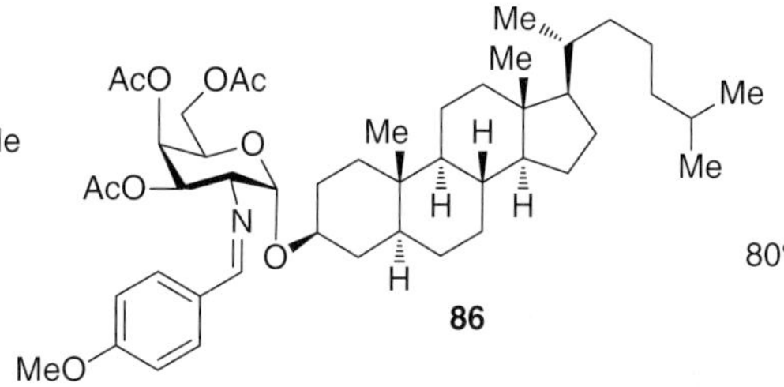 **86**	80% (10 : 1)

coupling of disaccharide acceptor **90** with disaccharide donor **89** proceeded smoothly to provide tetrasaccharide **91** in 72% yield with $\alpha:\beta = 11:1$ (Scheme 9.7b).

After the formation of the desired 1,2-*cis*-2-aminoglycosidic bond, the benzylidene group can be orthogonally removed and subsequently functionalized. Optimal condition for the removal of the benzylidene-protecting group is treatment with 1.1 equiv. of 2N or 5N HCl (depending on the acid sensitivity of the rest of the molecule), acetone/CH_2Cl_2, at 25 °C, resulting in the free amine product in less than 5 min (Scheme 9.8). Subsequent acylation of the amine salt intermediate **93** afforded the fully protected α-*O*-GalNAc-Ser **94** in 81% yield over two steps (Scheme 9.8a). Similarly, N-sulfation is possible upon treatment of SO_3/pyridine to the amine salt intermediate as illustrated in Scheme 9.8b.

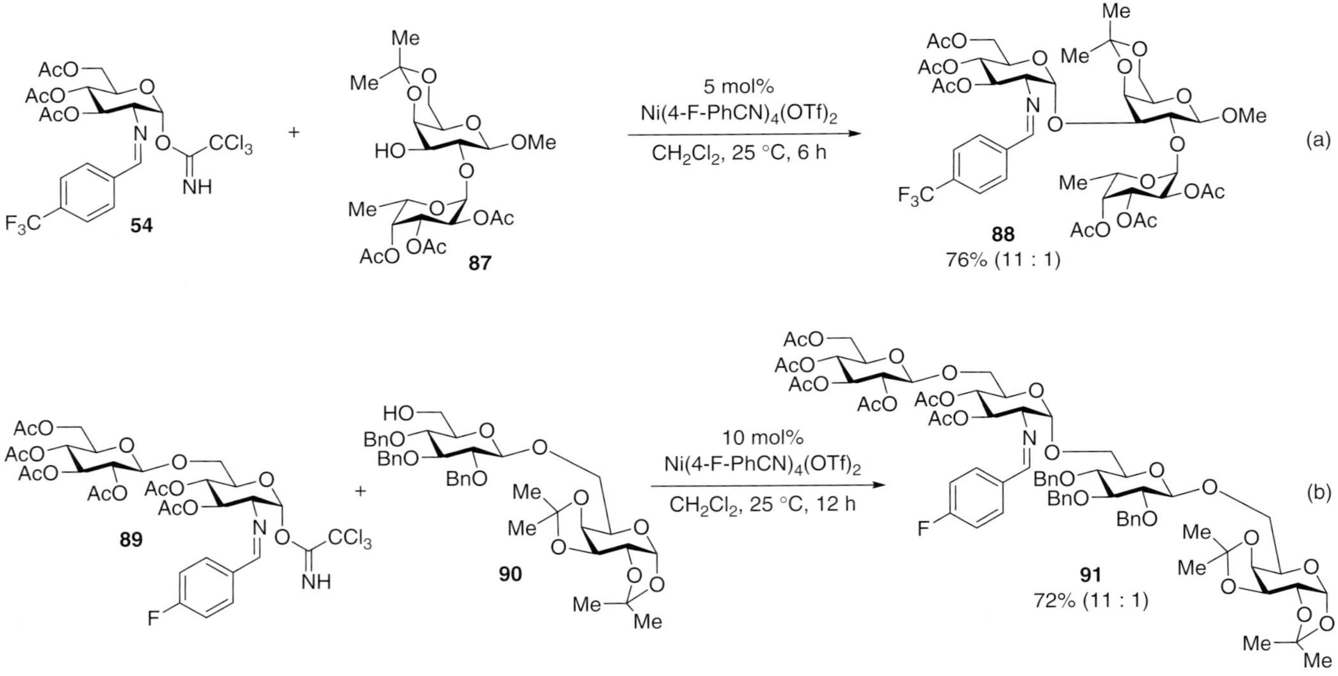

Scheme 9.7 Nickel-catalyzed α-oligosaccharide formation.

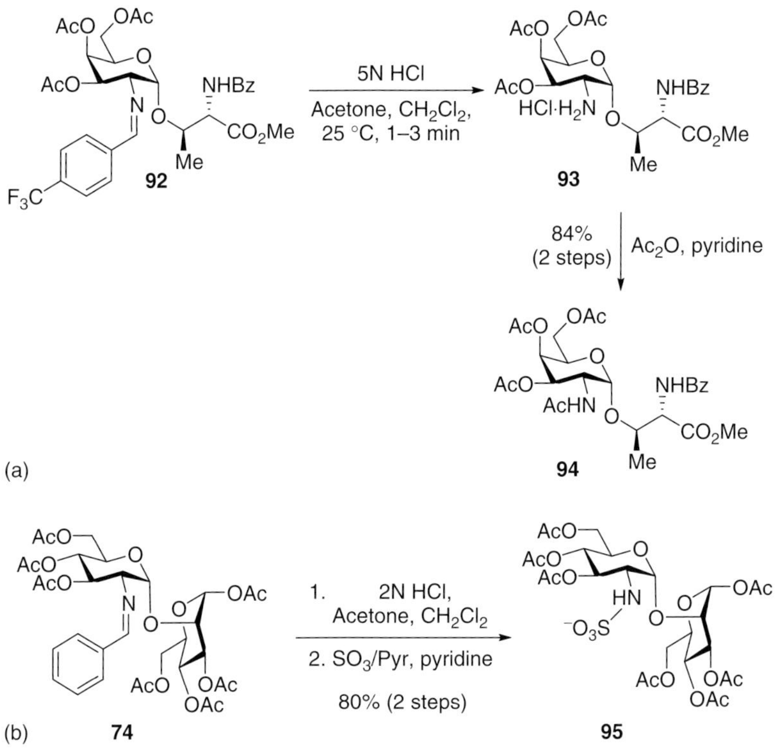

Scheme 9.8 Removal of *N*-benzylidene followed by subsequent derivatization.

9.6 Studies of C(2)-*N*-Substituted Benzylideneamino Glycosyl *N*-Phenyl Trifluoroacetimidate Donors

In an effort to expand the capabilities of this nickel methodology, focus was switched to the optimization of the anomeric leaving group. This was done in an effort to alleviate unwanted by-products **96** and **97** (Scheme 9.9). Due to their stability and ease of handling, *N*-phenyl trifluoroacetimidate donors were explored *in lieu* of the traditional trichloroacetimidate [51–53]. Additionally, their attenuated activity compared to trichloroacetimidates slowed the rate of the [1,3]-rearrangement of the imidate donors when sterically hindered alcohol acceptors were utilized in the glycosylation process. This modification also suppressed the elimination process to the undesired glycal by-product **97**.

Unfortunately, the activation by the cationic nickel catalyst of para-substituted C(2)-*N*-benzylidene *N*-phenyl trifluoroacetimidate donors, similar to the trichloroacetimidates donors, only proceeded through the α-anomer, whereas the β-anomer was unreactive (Scheme 9.10). Although the α-anomer was the more effective donor, it was the minor isomer ($\alpha:\beta = 1:3$) and therefore required tedious chromatographic separation, resulting in low atomic efficiency. Gratifyingly, with the modification of the leaving group to the *N*-phenyl trifluoroacetimidate in conjunction with the use of

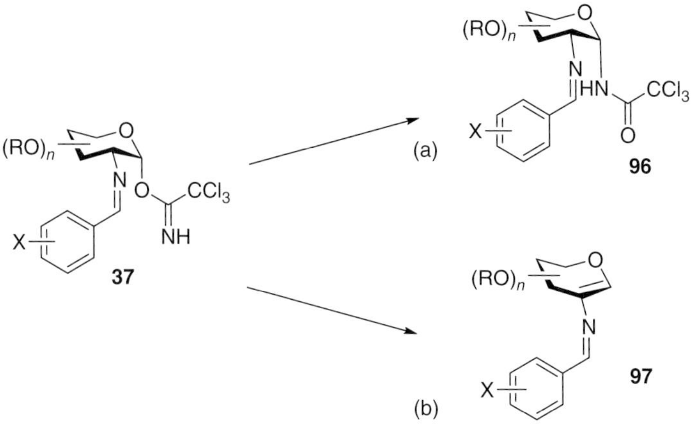

Scheme 9.9 Undesired by-products observed from the use of trichloroacetimidate donors.

ortho-trifluoromethyl-*N*-benzylidene **49**, both the α- and β-anomers were found to be reactive and with no loss in α-selectivity during the coupling process. This new class of donors increases both synthetic efficiency and atom economy (Scheme 9.10c).

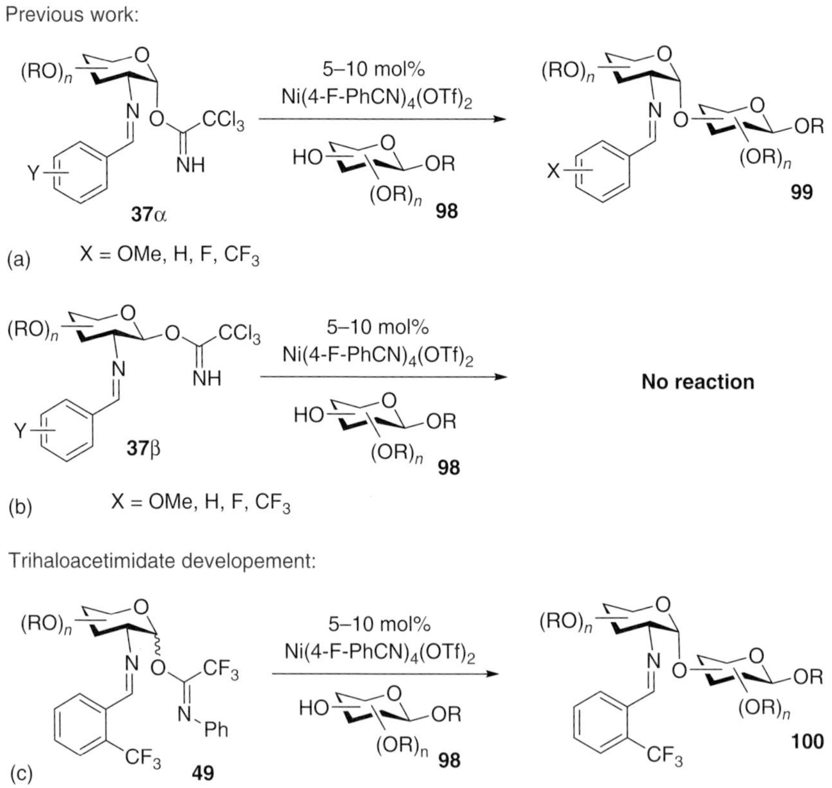

Scheme 9.10 Trihaloacetimidate modification easing synthesis of glycosyl donor.

9.6.1 Synthetic Advantage of *N*-Phenyltrifluoroacetimidate Donors

The synthetic merit of the *N*-phenyltrifluoroacetimidate donor was demonstrated over the previous methodology during the construction of the key 1,2-*cis*-amino pseudodisaccharide unit of mycothiol (**4**, Figure 9.2). The initial α-coupling of C(1)-OH inositol **101** to glucosamine attempted with first-generation α-trichloroacetimidate donors, **54** and **55**, resulted in the formation of **102** and **103** with high selectivity ($\alpha:\beta = 14:1$ and 20 : 1), but the yields were modest, 54% and 66%, respectively (Table 9.4, entries 1 and 2). Switching from the 4-CF_3 to the 2-CF_3-*N*-benzylidene **104**, entry 3, provided the core mycothiol component with excellent yield (66% → 94%) [38]. This type of benzylidene trichloroacetimidate donor was also previously reported to provide high yields of other coupling products [40]. Disappointingly, during the synthesis of the 2-CF_3-*N*-benzylidene trichloroacetimidates, it was found that although α-trichloroacetimidate **104** was the most effective donor, it was the minor isomer resulting from the reaction of hemiacetal with trichloroacetonitrile ($\alpha:\beta = 1:3$). As previously mentioned, the major product, the β-anomer, was unreactive.

Surprisingly, when *N*-phenyl trifluoroacetimidate version of donor was investigated, both α- and β-anomers were found to be reactive in the coupling event. Due to its lesser activity compared to the trichloroactitimidate version, this reaction was heated to 35 °C to facilitate the coupling process. Accordingly, coupling of C(1)-hydroxyl inositol **101** to the α-*N*-phenyl trifluoroacetimidate **106** proceeded to completion in 4 h, exclusively affording the α-pseudodisaccharide **105** in 77% yield (Table 9.5, entry 1). While its counterpart, the β-anomeric donor, proved to be sluggish, after 4 h, it provided just 48% of **105** (entry 2). It was observed that prolonging the reaction time to 12 h gave similar yields (81%, entry 3) while maintaining the α-1,2-*cis* selectivity. Based on these results, direct subjection of the 1 : 3 $\alpha:\beta$ mixture of *N*-phenyltrifluoroacetimidate donor **106** to the nickel-catalyzed conditions yielded **105** in 78%, *α*-only. These results suggest that a potential different mechanistic pathway could be followed as compared to trichloroacetimidate donors under nickel conditions.

Table 9.4 Initial studies of the formation of mycothiol with trichloroacetimidates.

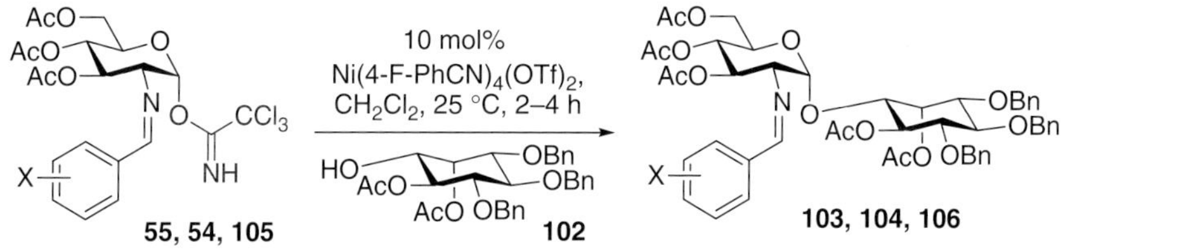

Entry	Donors	Products	Yield (%)	$\alpha:\beta$
1	**54:** X = 4 − F	**103**	54	14:1
2	**55:** X = 4 − CF_3	**104**	66	20:1
3	**104:** X = 2 − CF_3	**106**	94	20:1

Table 9.5 Study of *N*-phenyl trifluoroacetimidates donors.

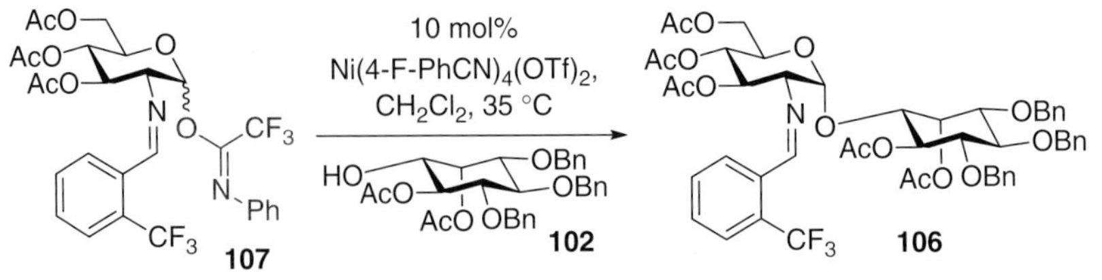

Entry	Imidate donor	Time (h)	Yield (%)	α:β
1	**107**α	4	77	α only
2	**107**β	4	48	α only
3	**107**β	12	81	α only
4	**107**/β mixture	18	78	α only

Furthermore, the results illustrated in Table 9.5 indicate that there could be a potential anomerization of the unreactive β-anomer to the more reactive α-species prior to the coupling event. To determine if this was in fact occurring, a control experiment was performed by monitoring the treatment of pure β-*N*-phenyltrifluoroacetimidate donor **106** with Ni(4-F-PhCN)$_4$(OTf)$_2$ in the absence of an acceptor with ^{1}H NMR, during which no isomerization of the β-anomer to the anticipated reactive α-anomer species was observed. The β-anomer instead proceeded through direct activation to glycal **107**. As illustrated in Figure 9.4, the resonance signal of the anomeric proton of **106β** (δ 6.05 ppm) over the course of the reaction (seen in gray) did not shift to δ 6.74 ppm (anomeric proton resonance of **106α**, pure spectrum shown in black), and it is simply reduced in size while the elimination product **107** proton peak resonance at δ 7.28 ppm increased as the reaction progresses [37].

9.7 1,2-*cis*-Amino Glycosylation of Thioglycoside Acceptors

The capabilities of this methodology were further investigated by varying the glycosyl acceptor: one being the chemoselective coupling to thioglycoside acceptors without losing α-stereoselectivity, as depicted in Scheme 9.11. This would allow for two successive and chemoselective glycosylations without the manipulation of the protecting groups. First, the selective formation of the 1,2-*cis*-aminoglycoside bond catalyzed by the nickel complex would be then followed by orthogonal activation of the thioglycoside, all occurring in a domino-type reaction. Glycosylation reactions with thioglycoside acceptors are typically difficult due to aglycon transfer of the anomeric sulfide to the donor, forming undesired thioglycoside **110** (Scheme 9.11). Another obstacle in this type of reaction is the nucleophilic nature of sulfur, which could potentially serve as a good ligand for transition-metal catalysts. Coordination of sulfur to the cationic nickel catalyst could

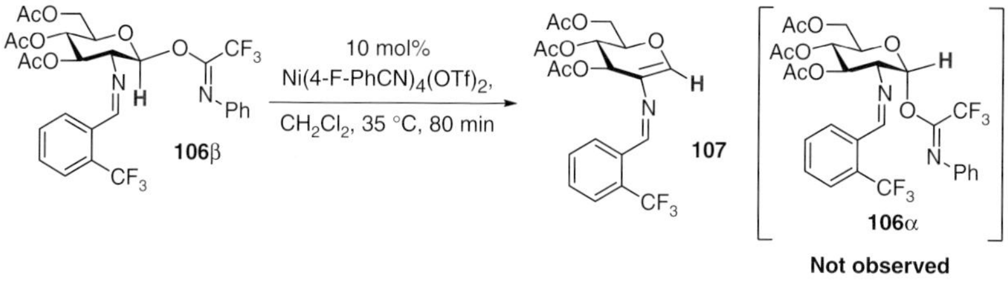

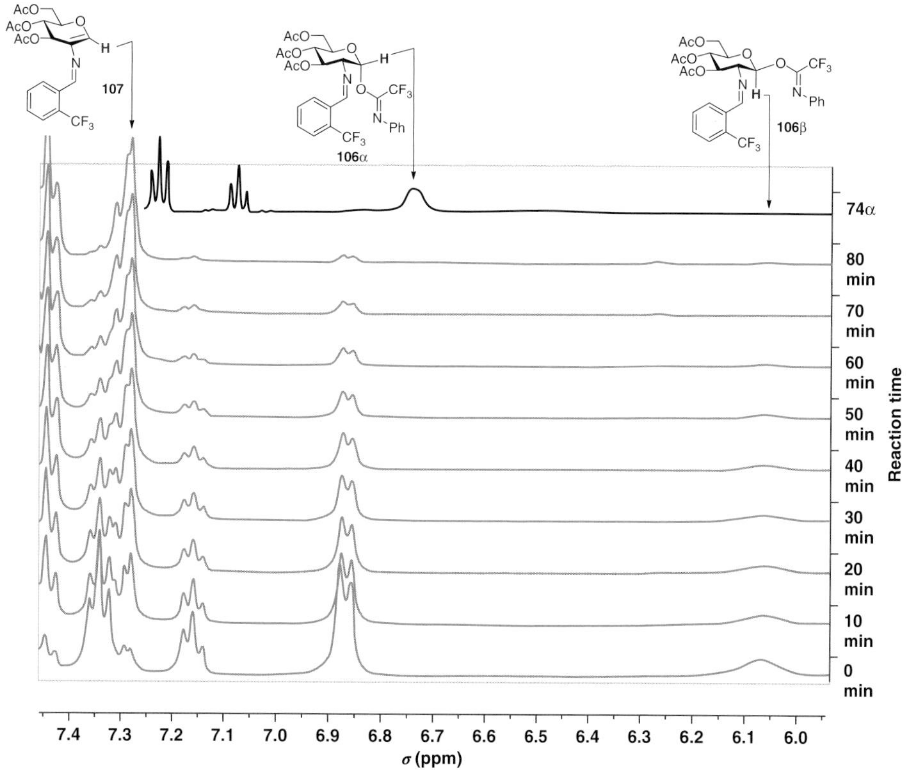

Figure 9.4 NMR studies on the potential anomerization of *N*-phenyl trifluoroacetimidate donors by cationic nickel catalyst.

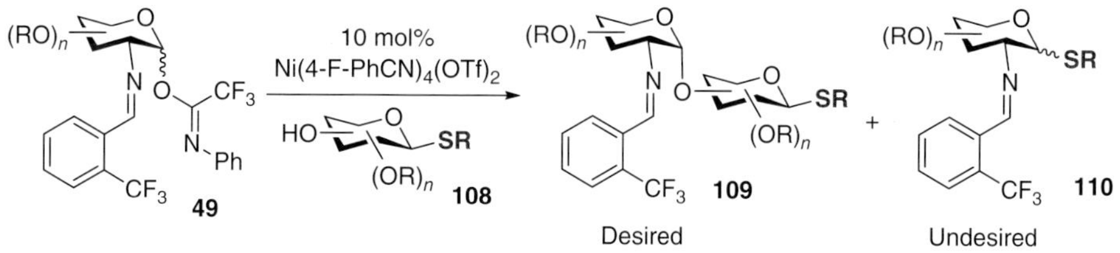

Scheme 9.11 1,2-*cis*-Amino nickel-catalyzed glycosylation to thioglycoside acceptors.

potentially slow/shut down the catalytic turnover, rendering the nickel catalyst inactive. The ability to overcome these inherent challenges would pave the way for streamlined synthesis of complex carbohydrates bearing 1,2-*cis*-2-aminoglycosidic linkages.

To initiate our studies, we first investigated the coupling of *N*-Troc-protected thioglycoside acceptor **111** with both C(2)-*N*-*ortho*- and *para*-trifluoromethyl-benzylidene α-trichloroacetimidate donors under the previously established nickel-catalyst conditions (Table 9.6). The *ortho*-trifluoromethyl-benzylideneamino donor **106** (entry 1) provided the best coupling to acceptor **111**, producing the desired disaccharide **116** in moderate yield (57%) along with low production of undesired thioglycoside **121** (14%). Alternatively, when trichloroacetimidate donor **112** was utilized in the reaction, a significant amount of the undesired [1,3]-rearrangement product (*vide supra*, Scheme 9.9a) was also observed. To prevent this rearrangement, the *N*-phenyl trifluoroacetimidate donor **106** was then employed. Upon initial implementation of this donor during the coupling of *ortho*-trifluoromethyl-benzylideneamino donor **106** with thioglycoside **111**, no [1,3]-rearrangement product was observed in the glycosylation reaction. Another advantage highlighted during this experiment was that both the α- and β-isomers were reactive. Notably, the formation of **114** showed that the sulfur did not render the nickel catalyst inactive.

Suppression of aglycon transfer by modulation of the sterics and electronics of the sulfide group is crucial for optimization of this reaction (Table 9.6). The addition of an

Table 9.6 Optimization with various thiophenyl derivatives.

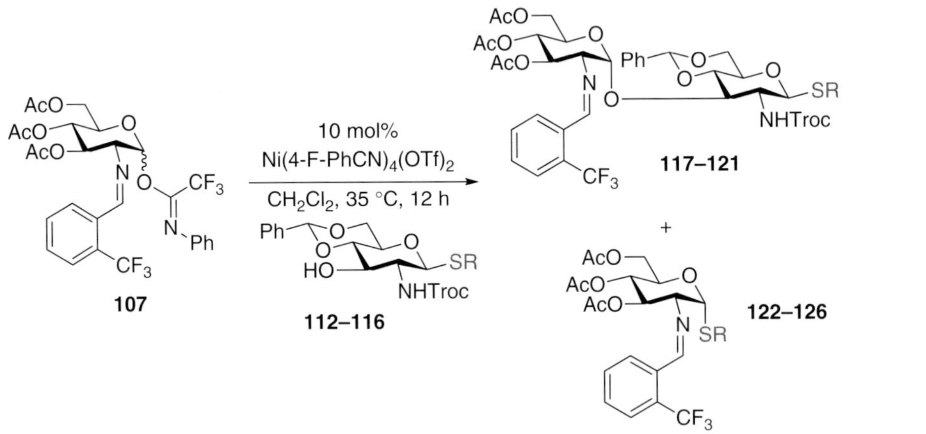

Entry	Acceptors	Donor/ Acceptor ratio	Disaccharides yield (%) ($\alpha:\beta$)	Thioglycosides yield (%)
1	**112**: R = Ph	1/2	**117**: 58 (α only)	**122**: 14
2	**113**: R = 4 − MeO − Ph	1/2	**118**: 37 (11 : 1)	**123**: 30
3	**114**: R = 4 − F − Ph	1/2	**119**: 58 (20 : 1)	**124**: 9
4	**115**: R = 2 − F − Ph	1/2	**120**: 60 (α only)	**125**: 4
5	**116**: R = 2 − CF_3 − Ph	1/2	**121**: 61 (α only)	**126**: 0
6	**116**: R = 2 − CF_3 − Ph	2/1	**121**: 58 (α only)	**126**: 0

electron-donating *para*-OMe substituent onto the thiophenyl (entry 2) of glycosyl acceptor **112** caused an increase of the aglycon transfer (14% → 30%), whereas the use of electron-withdrawing groups (entry 3–5) virtually alleviated all thiol transfer. Employment of electron-withdrawing trifluoromethyl on the ortho position of glycosyl acceptor **115** completely blocked aglycon transfer while exclusively producing the α-isomer of the desired disaccharide **120** in 61% yield (entry 5). Even after the addition of 2 equiv. of donor **106** to the reaction, no aglycon transfer product **125** was observed in the coupling event (entry 6). With these promising results, the 2-trifluoromethylthiophenyl became thioglycoside acceptor of choice.

Several other Lewis acids were evaluated to compare if they could provide the desired disaccharide with the same results as does the nickel catalyst. Traditional Lewis acids such as TMSOTf as well as other metal triflates, $Cu(OTf)_2$, $Rh(cod)_2OTf$, $Ir(cod)_2OTf$, $Fe(OTf)_2$, $Pd(4\text{-}F\text{-}PhCN)_2(OTf)_2$, $Ph_3PAuOTf$, were all tested with donor **106** and thiophenyl acceptor **111** (Table 9.7). Even though the desired disaccharide **116** was produced, the yields were low (11–45%) with $\alpha:\beta = 1:0$ to $6:1$. In some cases, the thiol transfer product **121** was produced in 30–40% yield (entries 1 and 2). In comparison to

Table 9.7 Activation of *N*-Phenyl trifluoroacetimidate with various catalysts.

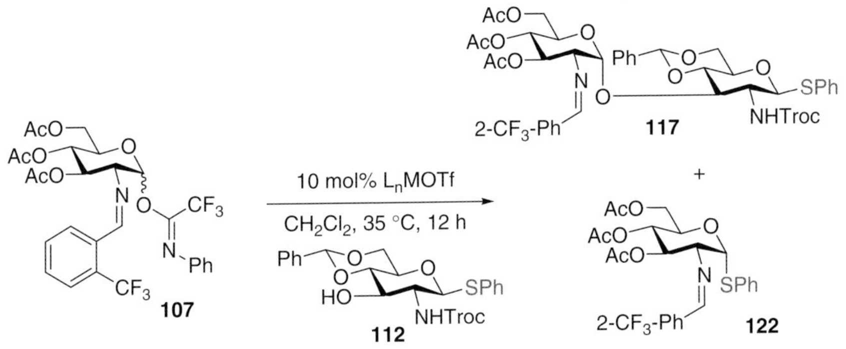

Entry	Catalysts	Disaccharide 116 yield (%) (α:β)	Thioglycoside 121 yield (%)
1	TMSOTf	13 (10:1)	40
2	$Cu(OTf)_2$	26 (10:1)	30
3	$Rh(COD)_2OTf$	43 (6:1)	20
4	$Ir(COD)_2OTf$	45 (8:1)	26
5	$Pd(4-F-PhCN)_2(OTf)$	36 (20:1)	23
6	$Ph_3PAuOTf$	43 (α only)	24
7	$Fe(OTf)_2$	11 (11:1)	1

$Ni(4\text{-}F\text{-}PhCN)_4(OTf)_2$, none of these promoters were able to perform well in terms of yield, α-selectivity, and suppression of aglycon transfer.

The scope of this chemoselective coupling reaction was subsequently expanded to the glycosylation of a variety of thioglycosides with both armed and disarmed donors. Several representative examples are shown in Table 9.8. Overall, the desired glycosylation occurred with both armed and disarmed donors to various glycoside acceptors with good yield and α-selectivity. The effectiveness of this methodology can be directly identified in entries 1–3. Three l-rhamnose acceptors with different anomeric sulfides (**127**–**129**) were evaluated in the coupling to donor **106**. With phenylthioglycoside **127** (entry 1), disaccharide **131** was obtained in high yield (81%, *α*-only), although, as anticipated, a significant amount of algycon transfer product **122** was observed (15%). On the other hand, the use of electron-withdrawing 2-trifluoromethylphenyl thioglycoside **128** (entry 2) further improved the yield (81% → 98%), but most importantly, the sulfide transfer was completely curbed (15% → 0%). The bulky 2,6-dimethylphenyl (DMP) thioglycoside functionality was also applied to reduce the aglycon transfer [54]. Compared to the use of **128**, coupling of this DMP acceptor **129** (entry 3) with donor **106** produced disaccharide **133** in slightly lower yield (98% → 78%) while maintaining the *α*-only selectivity and completely prevented the formation of the undesired thioglycoside **136**. The utility of the 2-trifluoromethylphenyl aglycon was also illustrated in the coupling of galactosamine acceptor **130** with donors **106** and **126** (entries 4 and 5). These reactions efficiently provided α-disaccharides **134** and **135** at 71 and 62% yield, respectively, and thiol transfer products were not observed.

Overall, the 2-trifluoromethylphenyl group (Table 9.8) has proven to be an alternative aglycon of thioglycoside acceptors in the glycosylation with C(2)-*N*-*ortho*-trifluoromethylbenzylideneamino donors for the stereoselective construction of oligosaccharide thioglycosides containing 1,2-*cis*-2-amino glycosidic bonds. In comparison, when the C(2)-azido version of *N*-phenyltrifluoroacetimidate donor **138** was activated by 10 mol% $Ni(4\text{-}F\text{-}PhCN)_4(OTf)_2$, the yield suffered significantly for both thioglucosamine acceptors **112** and **115** (18% and 13%), albeit provided only α-linked disaccharide products (Scheme 9.12).

To illustrate the utility of this method, chemo- and stereoselective sequential formation of a trisaccharide was performed (Scheme 9.13). Under the aforementioned nickel-catalyzed conditions, selective formation of the 1,2-*cis*-amino bond of disaccharide **124** was accomplished without any undesired aglycon transfer. Seamlessly, the disaccharide product **115** obtained under nickel conditions now becomes disaccharide thioglycoside donor **127**. After subsequent activation of thioglycoside by NIS and AgOTf, the addition of galactosyl acceptor **52** at 35 °C provided the corresponding trisaccharide **142** in good yield (81%) without any protecting group manipulation. This result also highlights the robustness of the benzylidene functionality. It serves not only as a directing group for 1,2-*cis*-amino bond formation but also as a stable protecting group under a variety of conditions [42].

9.8 Application to the Synthesis of Biologically Active Glycans

The synthetic potential and efficacy of the nickel methodology were applied to several synthetically challenging biologically relevant glycans, which have been stereoselectively constructed with minimal additional effort [37–42].

Table 9.8 Coupling of various thioglycosides with *N*-phenyl trifluoroacetimidate donors.

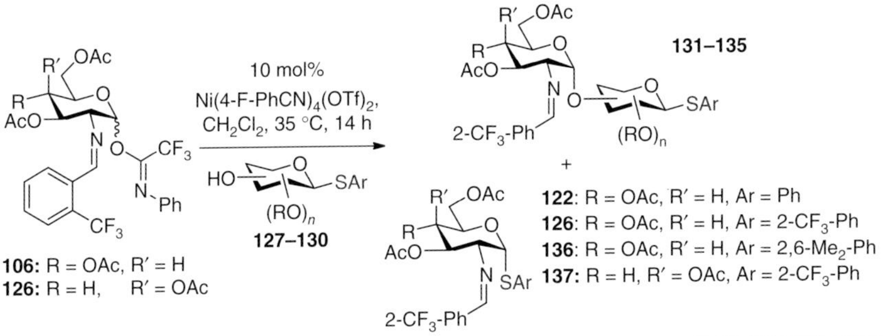

106: R = OAc, R′ = H
126: R = H, R′ = OAc

122: R = OAc, R′ = H, Ar = Ph
126: R = OAc, R′ = H, Ar = 2-CF_3-Ph
136: R = OAc, R′ = H, Ar = 2,6-Me_2-Ph
137: R = H, R′ = OAc, Ar = 2-CF_3-Ph

Entry	Donors	Acceptors	Disaccharides yield (α : β)	Thioglycosides yield (%)
1	**107**	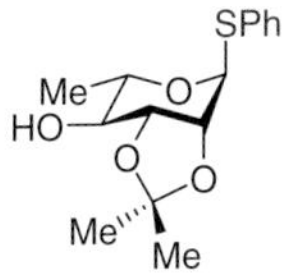 **127**	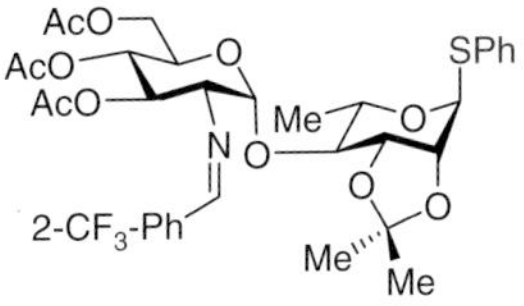 **131:** 81% (α only)	**123** (15%)
2	**107**	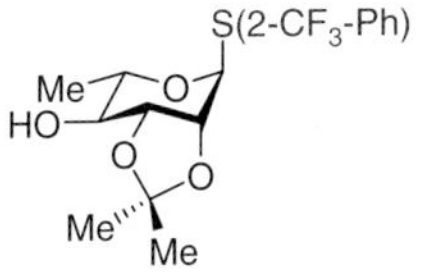 **128**	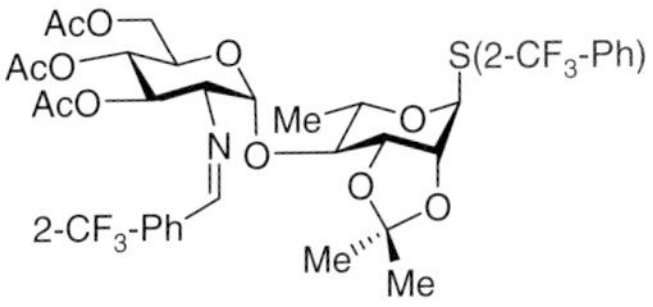 **132:** 98% (α only)	**127** (0%)
3	**107**	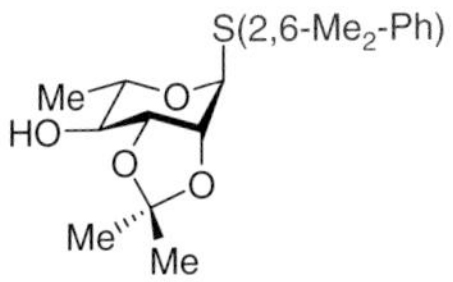 **129**	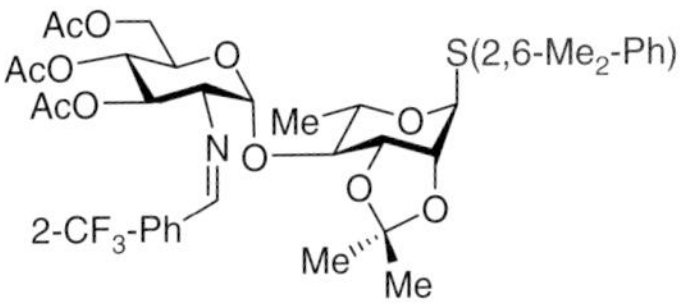 **133:** 74% (α only)	**137** (0%)
4	**107**	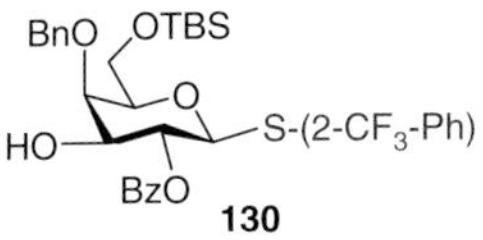 **130**	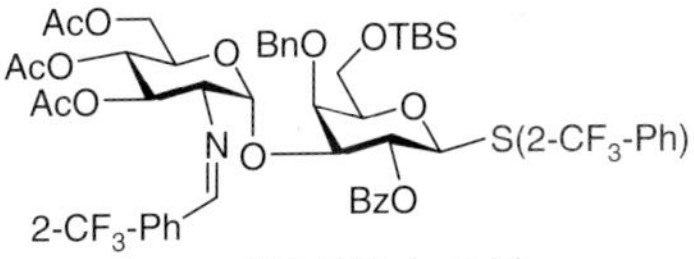 **134:** 71% (α only)	**127** (0%)
5	**127**	**130**	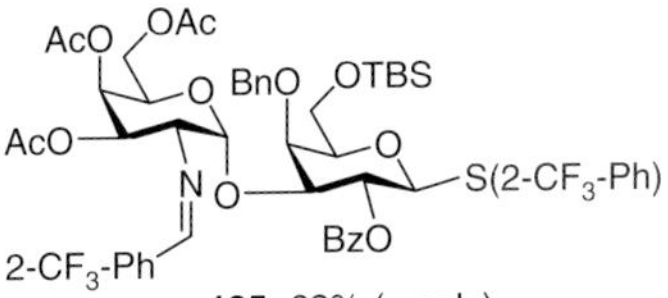 **135:** 62% (α only)	**138** (0%)

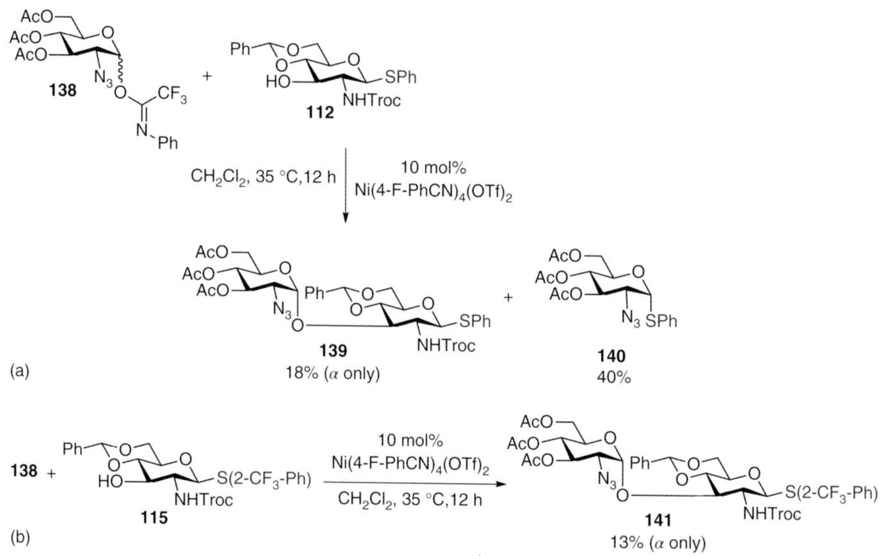

Scheme 9.12 Glycosylation with C(2)-azido imidate donor.

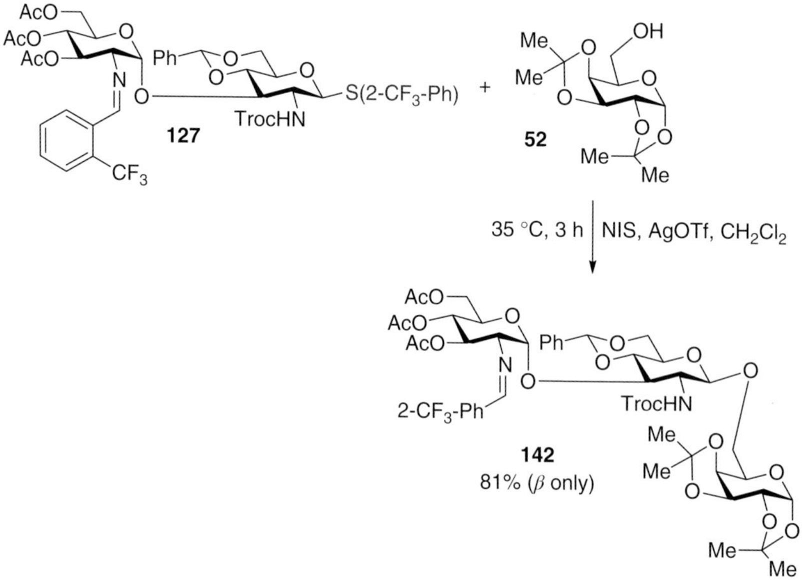

Scheme 9.13 Activation of thioglycoside after formation of 1,2-*cis*-amino glyosidic bond.

9.8.1 Mycothiol

Mycothiol (MSH) (**4**, Figure 9.2) is the major thiol found in Actinobacteria and is used to defend against foreign electrophilic agents such as oxidants, radical, and drugs [55, 56]. Disruption of the MSH metabolic pathway is fatal to bacteria, including the tuberculosis causing *Mycobacterium tuberculosis*. From numerous studies, the cysteine-linked pseudodisaccharide motif of MSH has been established as a potential tuberculosis drug target [57–61]. Current enzymatic synthesis of this target requires 1 l of *Mycobacterium smegmatis* to produce less than 1.5 mg of mycothiol **4** [20]. Existing chemical synthetic methodologies have afforded the pseudodisaccharide of **4** with modest selectivity ($\alpha:\beta = 1:1$ to $6:1$) [62–66]. Only a complex sequence of transformations involving intramolecular α-glucosaminidation [65] and desymmetrization of *meso*-inositols were able to afford an α-pseudodisaccharide product [66]. With the hypothesis that the nickel methods could be potentially more efficient and selective compared to the existing 1,2-*cis*-2-amino glycosylation methods, the use of a mixture of α- and β-isomers of C(2)-N-substituted benzylidenamino donors was attempted with C(1)-hydroxyl inositol acceptor **101** (Table 9.9) to form mycothiol core pseudosaccharides.

Accordingly, coupling of C(1)-OH inositol **101** with glycosyl donor **106** in the presence of 10% $Ni(4\text{-}F\text{-}PhCN)_4(OTf)_2$ produced the desired mycothiol pseudodisaccharide **105** (Table 9.9, entry 1) in 78% yield exclusively as a single α-isomer. Under optimized conditions, mycothiol core analogs were generated effortlessly (entries 2–5) in high yield (55–85%) and with α-only selectivity. Notably, the anomeric selectivity differs in the presence of the tribenzoylated substrate donor **143** to afford **145** with $\alpha:\beta = 13:1$ (entry 2). Encouraged by these results, expansion of the substrate tolerance for α-pseudosaccharide formation was then investigated. Switching to galactosamine donor **126** still solely provided the α-pseudodisaccharide **146** in 85% yield (entry 3). Even with the more challenging 1,6-linked disaccharide donor **144**, the desired pseudotrisaccharide **147** was still obtained in 64% yield and exclusively as α-isomer (entry 4). These results show that an array of mycothiol analogs is able to be rapidly generated by the nickel-catalyzed methodology in good yield and with minimal additional effort [38].

From the synthetic assembly of pseudodisaccharide **105**, subsequent derivatization was needed to formally generate mycothiol (Scheme 9.14). The 2-trifluoromethyl-benzylidene group of **105** was first removed under acidic conditions, followed by hydrogenolysis with Pearlman's catalyst in buffered *t*-BuOH. Finally, Zemplén deacetylation yielded fully deprotected **148** in 55% yield over two steps. The coupling of the ammonium salt on **149** with the cysteine residue **150** can be accomplished following the methods previously reported to furnish mycothiol **4** [65, 66].

9.8.2 GPI Anchor

GPI anchor (**5**, Figure 9.2) is a glycolipid, which tethers proteins to the cell membrane through hydrophobic fatty acids connected to a carbohydrate linker [67]. These anchored proteins play a biochemically vital role in processes such as cancer metastasis, immune response, and signal transduction [68, 69]. Although structurally diverse, all GPI-anchored proteins share the same basic core of a phosphoethanolamine linker, a tetrasaccharide attached to the C(6)-position of *myo*-inositol, and a phospholipid tail. While it is known that the main function of GPI is binding proteins to a cell surface, synthetic analogs of the GPI anchor can provide insight into the relationship between structural diversity and biological function [70–75].

Even though several synthetic routes have been reported, there is still the challenge to stereoselectively construct the 1,2-*cis*-glycosidic bond between the glucosamine unit and the *myo*-inositol component [76–78]. The inositol containing 1,2-*cis*-amino

Table 9.9 Formation of mycothiol core pseudosaccharides.

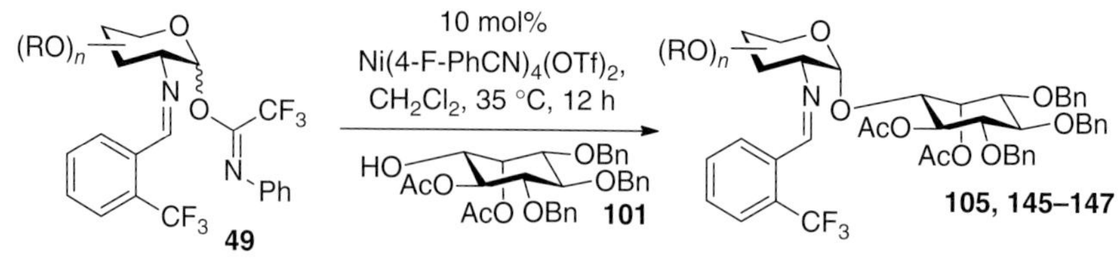

Entry	Donors	Pseudosaccharides yield (α : β)
1	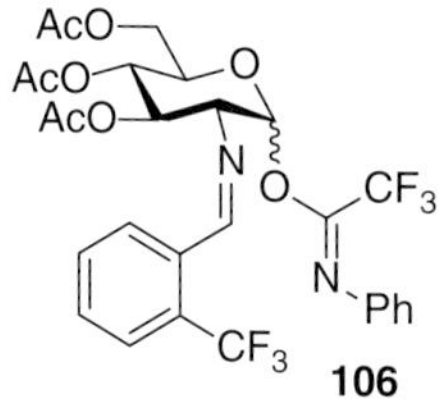 **106**	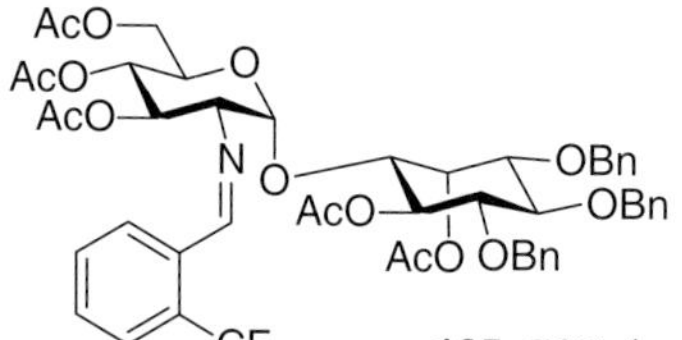 **105:** 78% (α only)
2	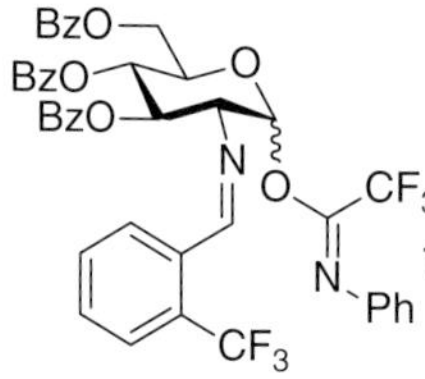 **143**	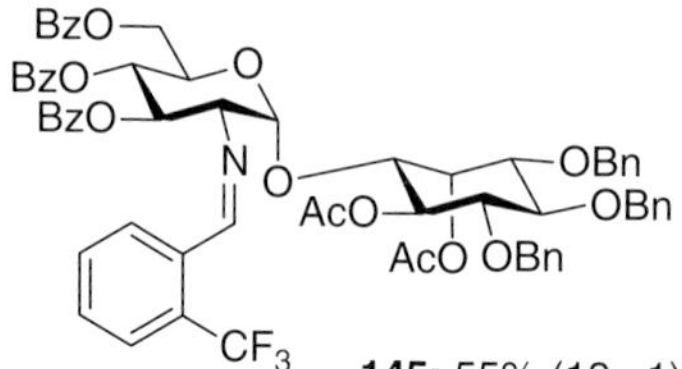 **145:** 55% (13 : 1)
3	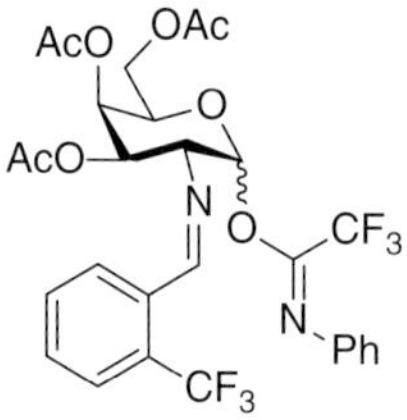 **126**	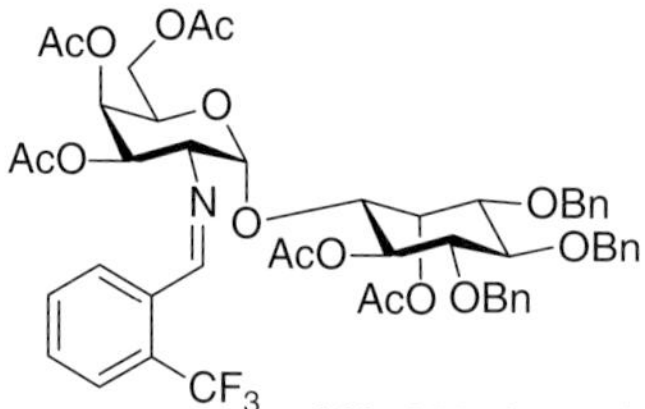 **146:** 85% (α only)
4	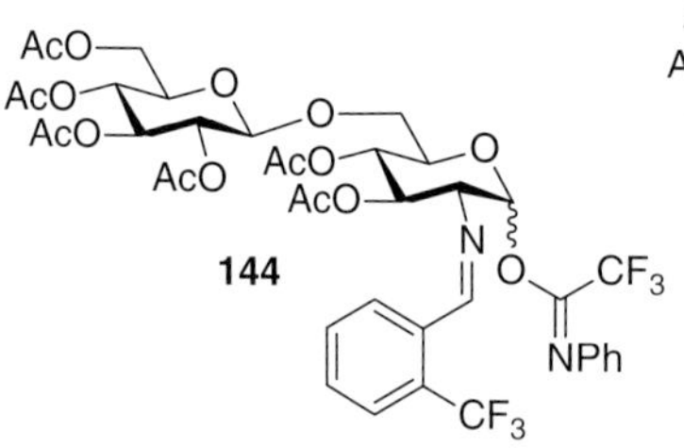 **144**	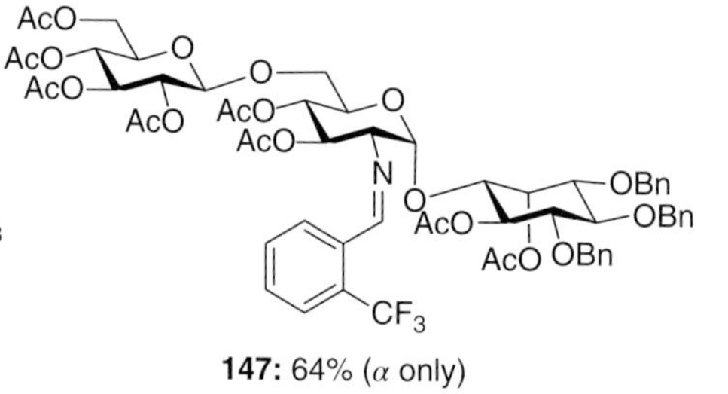 **147:** 64% (α only)

pseudosaccharide of the GPI core being similar to that of mycothiol (*vide supra*, Section 9.8.1), it was inferred that construction could proceeded efficiently by nickel catalysis (Scheme 9.15).

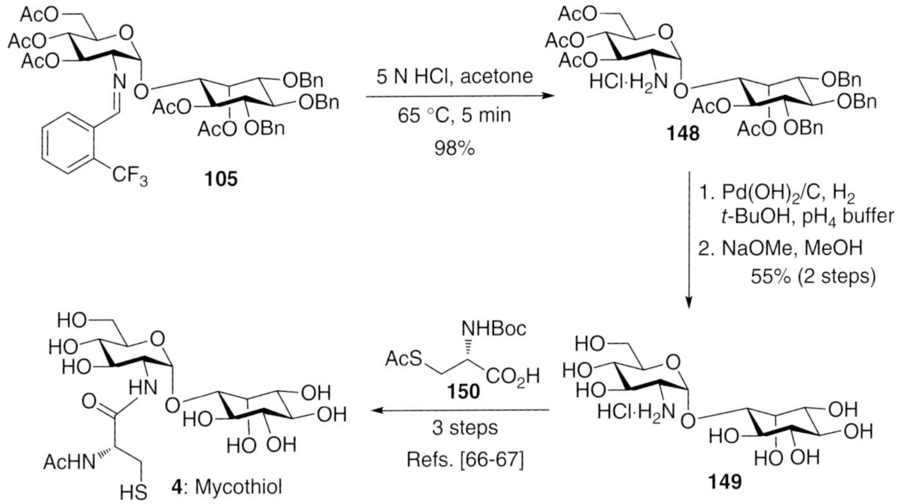

Scheme 9.14 Formal synthesis of mycothiol.

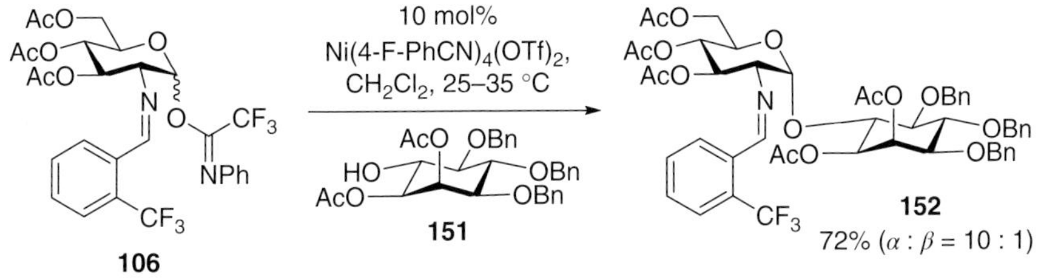

Scheme 9.15 Preliminary glycosylation testing for the synthesis of the GPI core.

The coupling of inositol **151** with an α/β-mixture of the *N*-phenyl trifluoroacetimidate donor **106**, mediated by Ni(4-F-PhCN)$_4$(OTf)$_2$, produced the desired pseudodisaccharide **152** in 72% yield with $\alpha{:}\beta = 10{:}1$ (Scheme 9.15). Demonstrating that this system was suitable for the construction of the GPI glycan core, it was also found that if bulky protecting groups such as benzyl, PMB, and allyl were placed on C(1) and C(2) of the inositol acceptor, the reaction did not proceed [37].

Further investigation established a formal synthesis of the pseudotrisaccharide core, including the problematic α-glucosamine–inositol glycosidic bond (Scheme 9.16). The C(4)-TES C(2)-*N*-benzylidene donor **153** was coupled to C(6)-hydroxyl *myo*-inositol **151** in the presence of 10 mol% of Ni(4-F-PhCN)$_4$(OTf)$_2$, producing the corresponding pseudodisaccharide **153** in 61% yield and $\alpha{:}\beta = 11{:}1$. The TES group was then selectively removed with TBAF to generate disaccharide acceptor **154**. Subsequent glycosylation of **154** with tetrabenzylated d-mannose trichloroacetimidate **155** afforded the desired pseudotrisaccharide **156** in 47% yield over two steps, completing the core. This synthetic route lays the foundation for further expansion to the entire GPI anchor. This synthetic scheme validates the dual role that the N-substituted benzylidene functionality can play; it possesses the capability of serving as a glycosidic

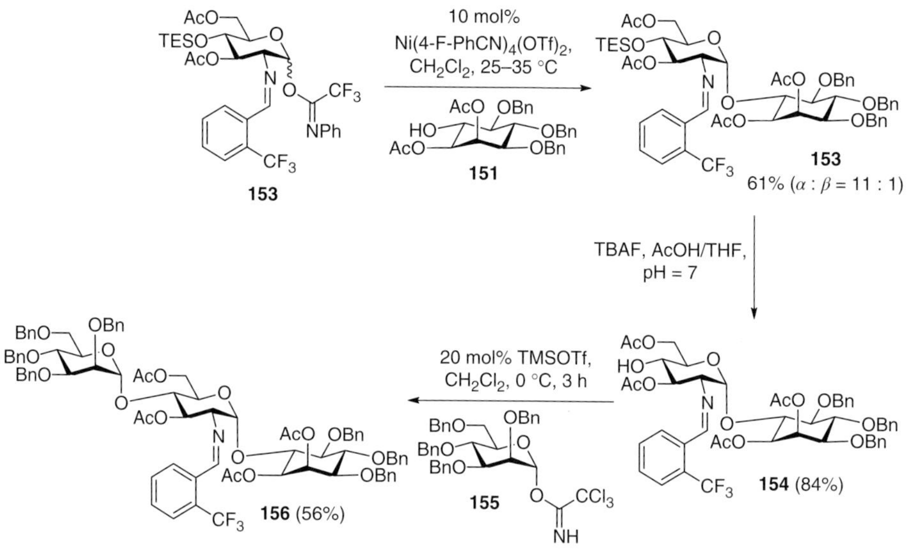

Scheme 9.16 Formal synthesis of GPI anchor core.

directing group and also as a robust protecting group under a variety of coupling conditions [37].

9.8.3 *O*-Polysaccharide Component of Gram-Negative Bacteria *S. enterica* and *P. rustigianii*

Upon optimization of the nickel-catalyzed chemoselective synthesis of oligosaccharide thioglycosides (*vide supra*, Section 9.7), it was concluded that 2-trifluoromethtylthiol-phenyl acceptors were capable of being utilized in the construction of crucial GalNAc-α-(1→3)-GlcNAc repeating unit of *O*-polysaccharides (**9**, Figure 9.2). These polysaccharides are present on the outer membrane of Gram-negative bacteria such as *S. enterica* and *P. rustigianii* (Scheme 9.17). These bacteria are major pathogens to both humans and animals that commonly contracted through contaminated water and food. They have been linked to urinary and gastronomical complications particularly in children. These bacteria are also connected to much more severe problems including Reiter's syndrome and typhoid fever [22–24]. The advantage of utilizing the previously described chemoselective nickel-catalyzed approach with thioglycoside acceptors is that, from these results, subsequent extension to the synthesis of the entire Gram-negative bacteria oligosaccharides can be readily achieved by the activation of the thioglycoside.

Accordingly, coupling of C(3)-hydroxyl group of glucosamine thioglycoside acceptor **115** with C(2)-N-ortho-substituted benzylidene-galactosamine-derived donor **126** (Scheme 9.17) proceeded smoothly in 10 mol% of $Ni(4\text{-}F\text{-}PhCN)_4(OTf)_2$ at 35 °C to generate the desired disaccharide **157** in 73% yield exclusively as the α-isomer. The undesired aglycon transfer product **137** (Scheme 9.17) was not observed in the coupling event. This result validates the efficacy of the nickel-catalyzed α-chemoselective glycosylation method for the synthesis of oligosaccharide repeating unit of the polysaccharides of *S. enterica* and *P. rustigianii.*

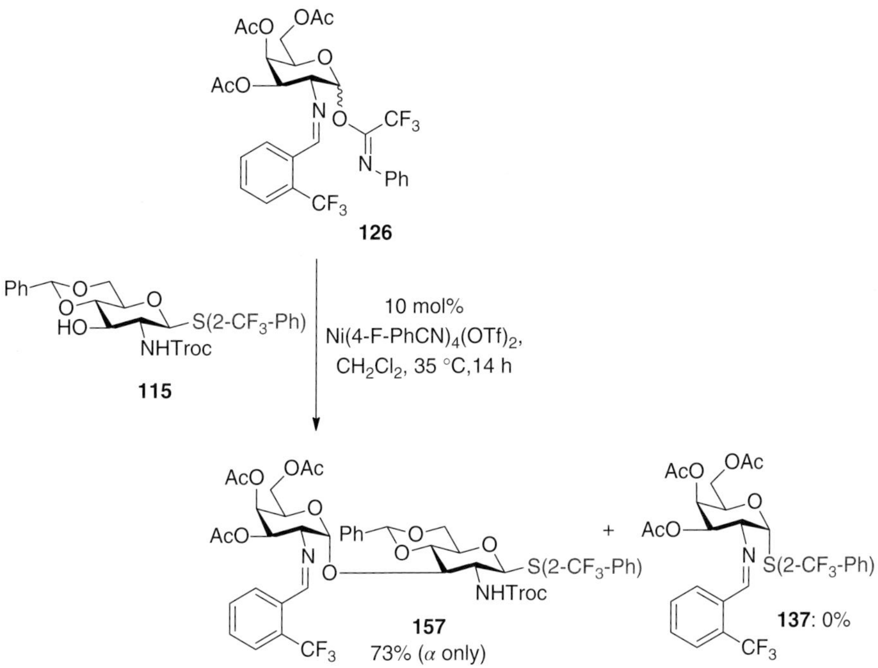

Scheme 9.17 Nickel-catalyzed formation of GalNAc-α-(1,3)-GlcNAc linkage of *O*-polysaccharides present on the outer membrane of Gram-negative bacteria.

9.8.4 T_N Antigen

Aminoglycans are often attached to the oxygen atom of either serine/threonine amino acid residues, which makes them part of a large group known as O-linked proteoglycans. The α-GalNAc-Ser/Thr unit of these glycoproteins makes up the core of a unique mucin family that has been identified as TACAs. Upon further expansion from the core GalNAc motif, T_N antigen (**6**, Figure 9.2) off the C(3) and/or C(6) hydroxyls gives rise to several other TACAs including TF antigen and ST_N antigen (**7** and **8**, Figure 9.2). These antigens are broadly expressed in tumors as a result of aberrant glycosylation [15]. With the fact that they are highly expressed in carcinomas, they have potential as a biomarker or as a vaccine against cancer [16, 79–83].

Existing methodologies such as the C(2)-oxazolidinone and the C(2)-azido donor methods, when used for this glycopeptide coupling, afforded $\alpha:\beta$ mixtures of 1:1 and 4:1, respectively [84, 85]. With this being a potential high-impact prophylactic, the ability for reproducible large-scale production of a well-defined and pure T_N antigen would be highly beneficial. Having previously illustrated that 5% $Ni(4\text{-}F\text{-}PhCN)_4(OTf)_2$ is capable of producing this challenging 1,2-*cis*-2-amino linkage more efficiently, a preliminary reaction was carried out under the Nguyen system, which saw the selectivity increase to 15:1 $\alpha:\beta$ and also a high yield (81%, Scheme 9.18a) [39]. However, this coupling was accomplished with Cbz-protected threonine **159**, which is not compatible with the two standard solid-phase peptide synthesis (SPPS) methods. The SPPS methods employ either Fmoc- or Boc-based chemistry [86]. In line with these guidelines, Fmoc-protected threonine **161** substituted into the glycosylation with donor **162**

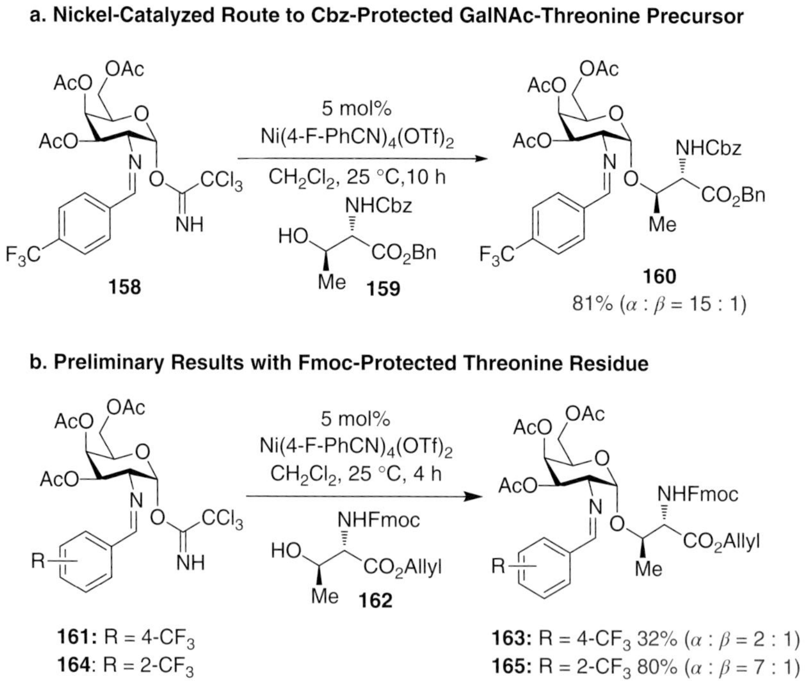

Scheme 9.18 Route to GalNAc–threonine residue by nickel catalysis.

resulted in only 32% of the desired glycopeptide **163** with $\alpha:\beta = 2:1$ (Scheme 9.18b). On the other hand, the use of C(2)-*N*-*ortho*-trifluoromethylbenzylidenamino trichloroacetimidate **164** increased both the yield (32% → 80%) and the α-selectivity ($\alpha:\beta = 2:1 \rightarrow 7:1$). Unfortunately, as previous described throughout this chapter, with *ortho*-(trifluoromethyl)benzylideneamino donors, only the minor α-isomer was reactive and the major β-isomer was not. Thus, glycosyl trichloroacetimidate donors are not suitable substrates for generating T_N antigen core under nickel conditions.

This was readily solved by the use of *N*-phenyl trifluoroacetimidates in which both the α- and β-isomers are effective donors. Small-scale employment of this imidate was first attempted with *o*-(trifluoromethyl)benzylidenamino donor **126** combining with threonine acceptor **166** and 10% Ni(4-F-PhCN)$_4$(OTf)$_2$ as the catalyst of choice (Scheme 9.19a). The reaction proceeded smoothly to exclusively produce protected T_N-antigen core **167** in 74% yield (Scheme 9.19a). With the desire to see if the nickel-catalyzed glycosylation system is suitable for a large-scale preparation of T_N-antigen core **167**, two more scale-up iterations were attempted (Scheme 9.19b and c). The final trial was able to produce 3.77 g of **167** in the same yield as the small-scale one and with exclusive α-selectivity (Scheme 9.19c). This scale-up with no loss in yield and selectivity showed that the nickel-catalyzed formation of 1,2-*cis*-aminoglycosides is applicable in large-scale industrial practices [41].

In two steps, glycan–amino acid residue **167** could then be prepared to be sequenced into a large peptide using a solid-phase peptide synthesizer. Fully protected T_N-antigen core **167** from Scheme 9.19 batch C was treated to acetyl chloride in methanol followed

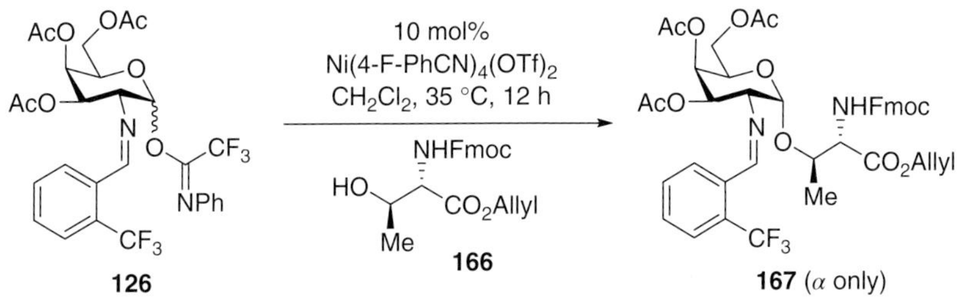

a.	0.808 g (1.0 equiv.)	0.585 g (1.2 equiv.)	74%, 0.782 g
b.	0.947 g (1.0 equiv.)	0.685 g (1.2 equiv.)	63%, 0.779 g
c.	5.730 g (1.28 equiv.)	2.710 g (1.0 equiv.)	66%, 3.77 g

Scheme 9.19 Large-scale synthesis of T_N antigen core.

by the addition of Ac_2O to transform the *N*-benzylidene to 2.25 g of N-acetylated **168** (Scheme 9.20). The batch was then split into two equal parts. Selective removal of the allyl-protecting group by palladium-tetrakis(triphenylphosphine) in THF and NMA produced 1.02 g of glycan–amino acid residue **169** in 97% yield, a valuable precursor for use in generating a variety of glycopeptides via SPPS (Scheme 9.20). A global deprotection of **168** with 0.2 N NaOH in methanol resulted in 0.4 g of fully deprotected T_N-antigen **6** at 82% yield. Alternatively, the addition of trimethylamine to 0.2N NaOH in methanol produced a nearly quantitative yield of **6** (Scheme 9.20).

Further investigation showed that the nickel–benzylidene methodology was tolerant of many substrates in the synthesis of T_N-antigen core structures **169–173** (Figure 9.5). Both armed and disarmed glycosyl donors were able to effectively construct this challenging 1,2-*cis*-2-amino glycosidic bond. Several different carboxylic-acid-protecting groups on the threonine acceptor were also suitable under nickel conditions. Notably, compound **170** is protected by a propargyl alkyne, which could then undergo a

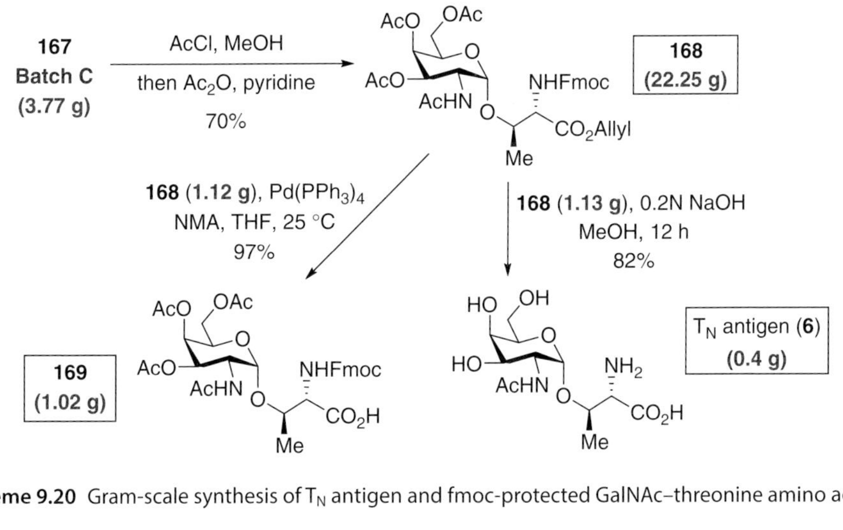

Scheme 9.20 Gram-scale synthesis of T_N antigen and fmoc-protected GalNAc–threonine amino acid.

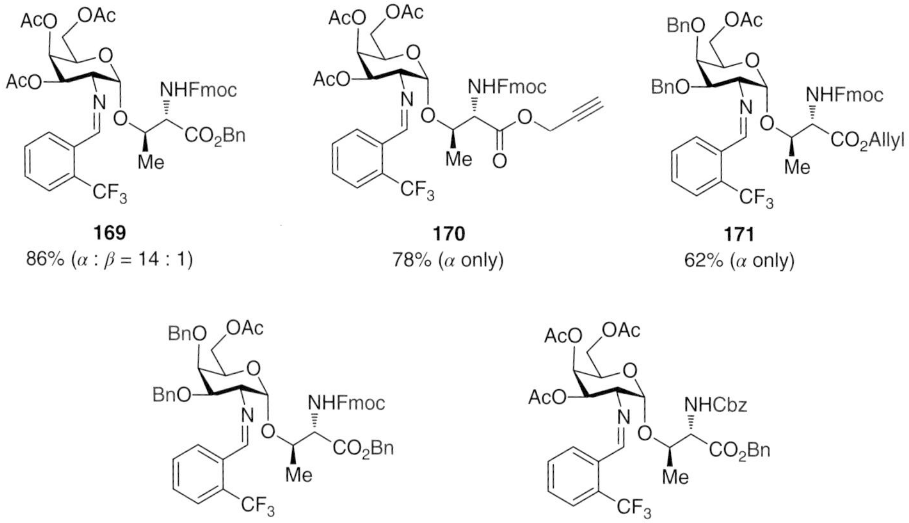

169 86% ($\alpha:\beta = 14:1$)
170 78% (α only)
171 62% (α only)
172 69% (α only)
173 67% (α only)

Figure 9.5 Scope of nickel-catalyzed formation of glycan–threonine products.

copper-mediated azide–alkyne cycloaddition reaction to be orthogonally linked to an azide functionality later in the synthetic sequence.

9.8.5 Heparin

Heparin (**5**, Figure 9.2) is a highly sulfated, anionic glycosaminoglycan polysaccharide with alternating α-(1→4)-linked glucosamine (GlcN) and uronic acids: d-glucuronic acid (GlcA) and its C-5 epimer and l-iduronic acid (IdoA) disaccharide units [14, 87–89]. Structurally, heparin closely resembles heparan sulfate proteoglycans (HSPGs). HSPGs are found on the cell surface and aid in maintaining the structural integrity of the ECM [21]. After heparin's discovery in 1916, its medicinal prowess as an anticoagulant was realized, and it became a commercial drug [90]. Heparin also has the ability to control a variety of other biological functions through binding to many different proteins [91–94]. Construction of the well-defined oligosaccharide repeating unit of heparin would be a powerful tool to understand its biological processes [29, 95, 96].

Given the previous success with the formation of these challenging 1,2-*cis*-2-amino bonds mediated by cationic nickel activation of C(2)-benzylidene imidate donors, an effort was made to synthesize heparin disaccharides with this methodology. Furthermore, the glucosamine-α-(1→4)–glucuronic acid disaccharides **179–183** (Table 9.10) can also serve as glycosyl donors for iterative glycosylation sequence to prepare a series of well-defined heparin oligosaccharides.

Systematic investigation of both armed and disarmed donors with various thioglycoside-protected glucuronic acid acceptors to produce heparin disaccharide thioglcyosides **179–183** is illustrated is Table 9.10. In stark contrast to our previous results, coupling of phenyl thioglycoside acceptor **175** with disarmed *N*-phenyl trifluoroacetimidate donor **106** resulted in no aglycon transfer of the sulfide (entry 1). Even

Table 9.10 Coupling of disarmed glucuronic acid thioglycosides with disarmed and armed glucosamine donors.

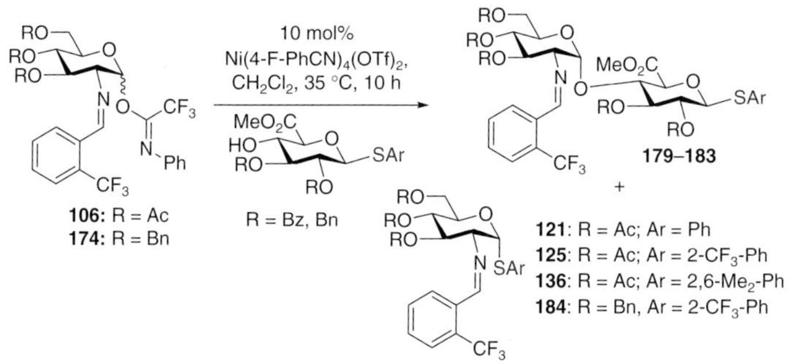

106: R = Ac
174: R = Bn

179–183

121: R = Ac; Ar = Ph
125: R = Ac; Ar = 2-CF$_3$-Ph
136: R = Ac; Ar = 2,6-Me$_2$-Ph
184: R = Bn, Ar = 2-CF$_3$-Ph

Entry	Donor	Acceptor	Disaccharides yield (α : β)	Thioglycosides yield (%)
1	**107**	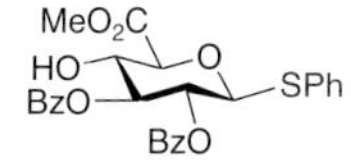 **175**	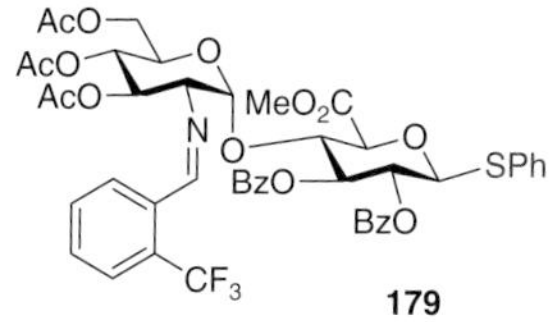 **179** 73% (α only)	**121**: 0
2	**107**	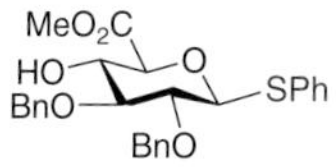 **176**	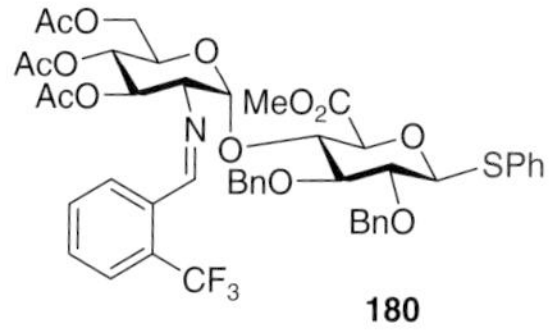 **180** 50% (α only)	**121**: 30
3	**107**	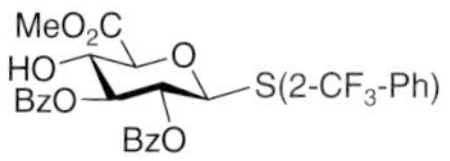 **177**	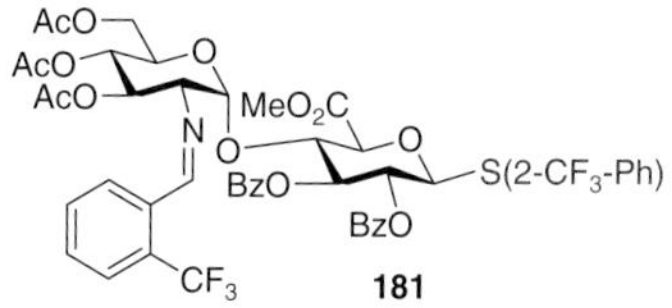 **181** 59% (α only)	**125**: 0
4	**175**	**177**	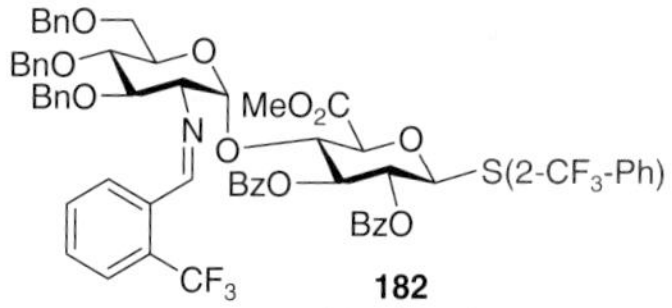 **182** 61% (7 : 1)	**184**: 0
5	**107**	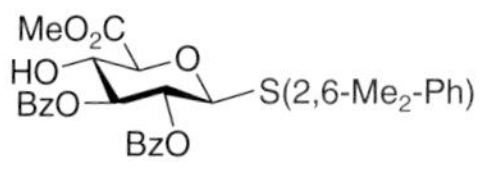 **178**	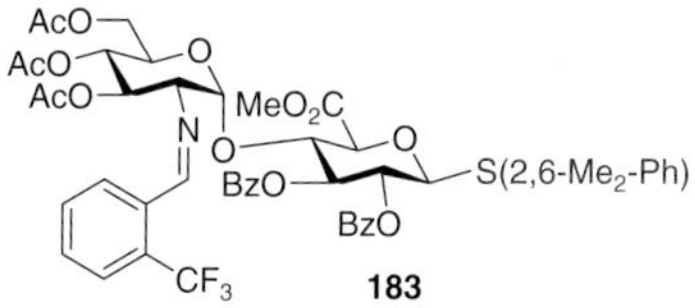 **183** 68% (α only)	**136:** 0

when 2 equiv. of glycosyl donor **106** were used, no thioglycoside **121** was observed, and only disaccharide **179** was attained in 73% yield and exclusively as the α-isomer (entry 1). Comparatively, when the partially armed phenyl aglycon acceptor **176** was utilized in the coupling with donor **106**, 30% yield of undesired thioglycoside **121** was obtained along with a diminished yield (73% → 50%) of the desired α-exclusive disaccharide **180** (entry 2). Modifications of thioglycoside **175** by replacing the phenyl with either the electron-withdrawing 2-trifluoromethylphenyl to generate **177** or the sterically hindered DMP group to produce **178** were then explored with our nickel system (entries 3–5). Coupling of these acceptors to donor **106** by means of 10 mol% Ni(4-F-PhCN)$_4$(OTf)$_2$ activation resulted in production of α-only disaccharides **181** and **183** in 59% and 68%, respectively. In all cases, aglycon transfer products were not observed in the reaction (entries 3 and 5). Finally, use of the tribenzylated armed donor **174** in the reaction with glycosyl acceptor **177** resulted in a decrease in α-selectively ($\alpha:\beta =$ 1 : 0 → 7 : 1), but a similar yield of its respective disaccharide **182** was still observed (entry 4). Overall, the results obtained in Table 9.10 suggest that the use of the 2-trifluoromethylphenyl aglycon might not be necessary in the case of glucuronic acids because the phenyl thioglycoside **175** is the most effective acceptor in the preparation of glucosamine-α-(1 → 4)–glucuronic acid unit of heparin oligosaccharides [97].

Although natural heparin does consist of 75% l-iduronic acid, glucuronic acid was substituted because iduronic acid is a commercially unavailable starting material and high cost to synthesize. This can still be seen as a promising preliminary result in the formation of heparin [39]. In comparison, Seeberger reported a synthesis of the heparin disaccharide through the coupling of a d-glucuronic acid acceptor to a C(2)-azido trichloroacetimidate donor where the desired unit was only produced in a yield of 57% with $\alpha:\beta$ of 3 : 1 [29].

9.9 Conclusion

There are a remarkable number of novel and efficient methods reported to access 1,2-*cis*-aminoglycosides. Nevertheless, these methods suffer with some limitations such as unpredictable anomeric selectivity, low yields, operational difficulty, stoichiometric use of activating reagents, low functional group tolerance, and limited substrate scope. The novel, efficient, and robust nickel-catalyzed method recently developed by Nguyen group to stereoselectively construct 1,2-*cis*-2-aminoglycosidic bonds may have addressed most of these limitations.

1) The nickel methodology is predictable, in that the α-stereoselective outcome of the newly formed glycosidic bond is not influenced by the protecting groups on either of the coupling partners. Thus, this nickel method provides a platform to access a wide variety of 1,2-*cis*-2-aminoglycosides with minimal effort.
2) In contrast to existing methods, the nickel method utilizes a directing group at the C(2) position of the glycosyl donors and substoichiometric amount of nickel–ligand complex to provide α-anomeric selectivity.
3) A simple condensation of arylaldehydes with aminosugar substrates yields the *ortho*-(trifluoromethyl)benzylidenamino donors. This procedure is much "safer" than that for the synthesis of the commonly used C(2)-azido glycosyl donors via the potentially explosive diazotransfer reaction.

4) This well-established nickel-catalyzed glycosylation protocol has been successfully applied to synthesize a number of 1,2-*cis*-2-aminoglycosides including T_N-antigen [41], heparin disaccharides [40], anti-TB mycothiol [38], GPI anchor pseudo-oligosaccharides [37], and the disaccharide repeating unit of Gram-negative bacterial polysaccharides [40]. In all cases, the challenging 1,2-*cis*-2-aminoglycosidic bonds are formed in high yields and α-selectivity, overcoming many of the shortcomings previously associated with existing methods.
5) Operationally simple procedure, mild conditions, high α-selectivity, scalability, and use of moisture and air-stable nickel catalyst all render this method to be adopted to automation.

Efforts to understand the mechanism of this glycosylation method are underway in Nguyen laboratory. The insights into the α-selectivity of this simple and novel protocol for generating 1,2-*cis*-2-amino linkage can open new avenues, further expanding the frontiers of the transition-metal-catalyzed stereoselective glycosylation reactions.

References

1 Fraser-Reid, B.O., Tatsuta, K., and Thiem, J. (2001) Glycoscience – Chemistry and Chemical Biology, 1st edn, Springer, New York.
2 Bertozzi, C.R. and Kiessling, L.,.L. (2001) *Science*, 291, 2357.
3 Dwek, R.A. (1996) *Chem. Rev.*, 96, 683.
4 Kooyk, Y.V. and Rabinovich, G.A. (2008) *Nat. Immunol.*, 9, 593.
5 Kwon, D.S., Gregorio, G., Bitton, N., Hendrickson, W.A., and Littman, D.R. (2002) *Immunity*, 16, 135.
6 de Silva, R.A., Appulage, D.K., Pietraszkiewicz, H., Bobbit, K.R., Media, J., Shaw, J., Valeriote, F.A., and Andreana, P.R. (2012) *Cancer Immunol. Immunother.*, 61, 581.
7 Capila, I. and Lindhardt, R.J. (2002) *Angew. Chem. Int. Ed.*, 41, 390.
8 Varky, A.R.C., Esko, J., Freeze, H., Hart, G., and Marth, J. (1999) Essentials of Glycobiology, 2nd edn, Cold Spring Harbor Laboratory Press, Cold Harbor, NY.
9 Zhu, X. and Schmidt, R.R. (2009) *Angew. Chem. Int. Ed.*, 48, 1900.
10 Boltje, T.J., Buskus, T., and Boons, G.-J. (2009) *Nat. Chem.*, 1, 611.
11 Stallforth, P., Lepenies, B., Adibekian, A., and Seeberger, P.H. (2009) *J. Med. Chem.*, 52, 5561.
12 Manabe, S. (2010) *Methods Enzymol.*, 478, 413.
13 Bongat, A.F. and Demchenko, A.V. (2007) *Carbohydr. Res.*, 342, 374.
14 Petitou, M. and van Boeckel, C.A.A. (2004) *Angew. Chem. Int. Ed.*, 43, 3118.
15 Hakomori, S. (1984) *Annu. Rev. Immunol.*, 2, 103–126.
16 Danishefsky, S.J. and Allen, J.R. (2000) *Angew. Chem. Int. Ed.*, 39, 836.
17 Banoub, J., Boullanger, P., and Lafont, D. (1992) *Chem. Rev.*, 92, 1167.
18 Kerns, R.J. and Wei, P. (2006) *ACS Symp. Ser.*, 932, 205.
19 Ferguson, M.A.J. and Williams, A.F. (1988) *Annu. Rev. Biochem.*, 57, 285.
20 Newton, G.L., Arnold, K., Price, M.S., Sherrill, C., Delcardayre, S.B., Aharonowitz, Y., Cohen, G., Davies, J., Fahey, R.C., and Davis, C. (1996) *J. Bacteriol.*, 178, 1990.
21 Vlodavsky, I., Ilan, N., Naggi, A., and Casu, B. (2007) *Curr. Med. Chem.*, 13, 2057.
22 Bindschadler, P., Dialer, L.O., and Seeberger, P.H. (2009) *J. Carbohydr. Chem.*, 28, 395.

23 Perepelov, A.V., Liu, B., Senchenkoya, S.N., Shashkov, A.S., Feng, L., Knirel, Y.A., and Wang, L. (2010) *Carbohydr. Res.*, 345, 971.
24 Fitzgerald, C., Collins, M., van Duyne, S., Mikoleit, M., Brown, T., and Fields, P. (2007) *J. Clin. Microbiol.*, 45, 3323.
25 Goodman, L. (1967) *Adv. Carbohydr. Chem. Biochem.*, 22, 109.
26 Paulsen, H., Kolar, C., and Stenzel, W. (1978) *Chem. Ber. Recl.*, 111, 2358.
27 Paulsen, H. and Stenzel, W. (1978) *Chem. Ber. Recl.*, 111, 2348.
28 Lemieux, R.U. and Ratcliffe, R.M. (1979) *Can. J. Chem.*, 57, 1244.
29 Orgueira, H.A., Bartolozzi, A., Schell, P., Litjens, R.E.J.N., Palmacci, E.R., and Seeberger, P.H. (2003) *Chem. Eur. J.*, 9, 140.
30 Vasella, A., Witzig, C., Chiara, J.-L., and Martin-Lomas, M. (1991) *Helv. Chim. Acta*, 74, 2073.
31 Nyffeler, P.T., Liang, C.-H., Koeller, K.M., and Wong, C.-H. (2002) *J. Am. Chem. Soc.*, 124, 10773.
32 Benakli, K., Zha, C., and Kerns, R.J. (2001) *J. Am. Chem. Soc.*, 123, 9461.
33 Ryan, D.A. and Gin, D.Y. (2008) *J. Am. Chem. Soc.*, 130, 15228.
34 Winterfeld, G.A. and Schmidt, R.R. (2001) *Angew. Chem. Int. Ed.*, 40, 2654.
35 Arnold, J.S., Zhang, Q., and Nguyen, H.M. (2014) *Eur. J. Org. Chem.*, 2014, 4925.
36 Schmidt, R.R. and Kinzy, W. (1994) in Advances in Carbohydrate Chemistry and Biochemistry, vol. 50 (ed. H. Derek), Academic Press, p. 21.
37 McConnell, M.S., Mensah, E.A., and Nguyen, H.M. (2013) *Carbohydr. Res.*, 381, 146.
38 McConnell, M.S., Yu, F., and Nguyen, H.M. (2013) *Chem. Commun.*, 49, 4313.
39 Mensah, E.A. and Nguyen, H.M. (2009) *J. Am. Chem. Soc.*, 131, 8778.
40 Mensah, E.A., Yu, F., and Nguyen, H.M. (2010) *J. Am. Chem. Soc.*, 132, 14288.
41 Yu, F., McConnell, M.S., and Nguyen, H.M. (2015) *Org. Lett.*, 17, 2018.
42 Yu, F. and Nguyen, H.M. (2012) *J. Org. Chem.*, 77, 7330.
43 Mensah, E.A., Azzarelli, J.M., and Nguyen, H.M. (2009) *J. Org. Chem.*, 74, 1650.
44 Douglas, S.P., Whitfield, D.M., and Krepinsky, J.J. (1993) *J. Carbohydr. Chem.*, 12, 131.
45 Mensah, E.A. (2011) Palladium and nickel catalyzed stereoselective formation of glycosides. PhD dissertation. University of Iowa, Iowa City, IA.
46 Marra, A. and Sinay, P. (1990) *Carbohydr. Res.*, 200, 319.
47 Geng, Y.Q., Zhang, L.H., and Ye, X.S. (2008) *Tetrahedron*, 64, 4949.
48 Li, J.J., Mei, T.S., and Yu, J.Q. (2008) *Angew. Chem. Int. Ed.*, 47, 6452.
49 Miura, M., Tsuda, T., Satoh, T., Pivsa-Art, S., and Nomura, M. (1998) *J. Org. Chem.*, 63, 5211.
50 Park, J., Kawatkar, S., Kim, J.H., and Boons, G.J. (2007) *Org. Lett.*, 9, 1959.
51 Yu, B. and Tao, H. (2001) *Tetrahedron Lett.*, 42, 2405.
52 Cai, S. and Yu, B. (2003) *Org. Lett.*, 5, 3827.
53 Tanaka, H., Iwata, Y., Takahashi, D., Adachi, M., and Takahashi, T. (2005) *J. Am. Chem. Soc.*, 127, 1630.
54 Li, Z. and Gildersleeve, J.C. (2006) *J. Am. Chem. Soc.*, 128, 11612.
55 Newton, G.L., Av-Gay, Y., and Fahey, R.C. (2000) *Biochemistry*, 39, 10739.
56 Newton, G.L. and Fahey, R.C. (2002) *Arch. Microbiol.*, 178, 388.
57 Payne, D.J., Gwynn, M.N., Holmes, D.J., and Pompliano, D.L. (2007) *Nat. Rev. Drug Discovery*, 6, 29.
58 Rawat, M., Johnson, C., Cadiz, V., and Av-Gay, Y. (2007) *Biochem. Biophys. Res. Commun.*, 363, 71.

59 Metaferia, B.B., Fetterolf, B.J., Shazad-ul-Hussan, S., Moravec, M., Smith, J.A., Ray, S., Gutierrez-Lugo, M.-T., and Bewley, C.A. (2007) *J. Med. Chem.*, 50, 6326.

60 Slattergard, R., Gammon, D.W., and Oscarson, S. (2007) *Carbohydr. Res.*, 342, 1943.

61 Knapp, S., Gonzalez, S., Myers, D.S., Eckman, L.L., and Bewley, C.A. (2002) *Org. Lett.*, 4, 4337.

62 Lee, S. and Rosazza, J.P.N. (2004) *Org. Lett.*, 6, 365.

63 Nicholas, G.M., Kovac, P., and Bewley, C.A. (2002) *J. Am. Chem. Soc.*, 124, 3492.

64 Jardine, A., Spies, H.S.C., Nkambule, C.M., Gammon, D.W., and Steenkamp, D.J. (2002) *Bioorg. Med. Chem. Lett.*, 10, 875.

65 Ajayi, K., Thakur, V.V., Lapo, R.C., and Knapp, S. (2010) *Org. Lett.*, 14, 2630.

66 Chung, C.-C., Zulueta, M.M.L., Padiyar, L.T., and Hung, S.-C. (2011) *Org. Lett.*, 2011, 5496.

67 Goel, M., Azev, V.N., and d'Alarcao, M. (2009) *Future Med. Chem.*, 1, 95.

68 Nosjean, O., Briolay, A., and Roux, B. (1997) *Biochim. Biophys. Acta*, 1331, 153.

69 Chesebro, B., Triflilo, M., Race, R., Meade-White, K., Teng, C., LaCasse, R., Raymond, L., Favara, C., Baron, G., Priola, S., Caughey, B., Masliah, E., and Oldstone, M. (2005) *Science*, 308, 1435.

70 van den Berg, C.W., Cinek, T., Hallett, M.B., Horejsi, V., and Morgan, B.P. (1995) *J. Cell Biol.*, 131, 669.

71 Zhang, F., Schmidt, W.G., Hou, T., Williams, A.F., and Jacobson, K. (1992) *Proc. Natl. Acad. Sci. U.S.A.*, 89, 5231.

72 Paulick, M.G., Forstner, M.B., Groves, J.T., and Bertozzi, C.R. (2007) *Proc. Natl. Acad. Sci. U.S.A.*, 104, 20332.

73 Paulick, M.G., Wise, A.R., Forstner, M.B., Groves, J.T., and Bertozzi, C.R. (2007) *J. Am. Chem. Soc.*, 129, 11543.

74 Tsai, Y.-H., Liu, X., and Seeberger, P.H. (2012) *Angew. Chem. Int. Ed.*, 51, 11438.

75 Swarts, B.M. and Guo, Z. (2011) *Chem. Sci.*, 2, 2342.

76 Guo, Z. and Bishop, L. (2004) *Eur. J. Org. Chem.*, 2004, 3585.

77 Swarts, B.M. and Guo, Z. (2012) *Adv. Carbohydr. Chem. Biochem.*, 67, 137.

78 Homans, S.W., Ferguson, M.A.J., Dwek, R.A., Rademacher, T.W., Anand, R., and Williams, A.F. (1988) *Nature*, 333, 269.

79 Singhal, A., Fohn, M., and Hakomori, S. (1991) *Cancer Res.*, 51, 1406.

80 Singhal, A. and Hakomori, S. (1990) *BioEssays*, 12, 223.

81 Henningsson, C.M., Selvaraj, S., MacLean, G.D., Suresh, M.R., Noujaim, A.A., and Longenecker, B.M. (1987) *Cancer Immunol. Immunother.*, 25, 231.

82 Samuel, J., Noujaim, A.A., MacLean, G.D., Suresh, M.R., and Longenecker, B.M. (1990) *Cancer Res.*, 50, 4801.

83 Xu, Y.F., Sette, A., Sidney, J., Gendler, S.J., and Franco, A. (2005) *Immunol. Cell Biol.*, 83, 440.

84 Kerns, R.J., Zha, C.X., Benakli, K., and Liang, Y.Z. (2003) *Tetrahedron Lett.*, 44, 8069.

85 Kuduk, S.D., Schwarz, J.B., Chen, X.-T., Glunz, P.W., Sames, D., Ragupathi, G., Livingston, P.O., and Danishefsky, S. (1998) *J. Am. Chem. Soc.*, 120, 12474.

86 Grogan, M.J., Pratt, M.R., Marcaurelle, L.A., and Bertozzi, C.R. (2002) *Annu. Rev. Biochem.*, 71, 593.

87 Avci, F.Y., Karst, N.A., and Linhardt, R.J. (2003) *Curr. Pharm. Des.*, 9, 2323.

88 Karst, N.A. and Linhardt, R.J. (2003) *Curr. Med. Chem.*, 10, 1993.

89 Liu, H., Zhang, Z., and Linhardt, R.J. (2009) *Nat. Prod. Rep.*, 26, 313.

90 Boysen, M., Gemma, E., Lahmann, M., and Oscarson, S. (2005) *Chem. Commun.*, 3044.
91 Lindahl, U., Kusche-Gullberg, M., and Kjellen, L. (1998) *J. Biol. Chem.*, 273, 24979.
92 Bernfield, M., Gotte, M., Park, P.W., Reizes, O., Fitzgerald, M.L., Lincecum, J., and Zako, M. (1999) *Annu. Rev. Biochem.*, 68, 729.
93 Capila, I. and Linhardt, R.J. (2002) *Angew. Chem.*, 41, 391.
94 Ori, A., Wilkinson, M.C., and Fernig, D.G. (2008) *Front. Biosci.*, 13, 4309.
95 Codee, J.D.C., Stubba, B., Schiattarella, M., Overkleeft, H.S., van Boeckel, C.A.A., van Boom, J.H., and van der Marel, G.A. (2005) *J. Am. Chem. Soc.*, 127, 3767.
96 Haller, M. and Boons, G.J. (2001) *J. Chem. Soc., Perkin Trans. 1*, 814.
97 Yu, F. and Nguyen, H.M. (2012) University of Iowa, Iowa City, IA, Unpublished work.

10

Photochemical Glycosylation

Justin Ragains

10.1 Introduction

The organic synthetic literature is replete with new examples of "mild" chemical transformations, and the rationale often provided for the development of new transformations is the replacement of "harsh" conditions with "mild" conditions. The field of chemical glycosylation is one that is often criticized for the development of harsh conditions. However, a certain caution should be exercised in using the terms "mild" and "harsh" with synthetic transformations. From the standpoint of the molecules undergoing conversion, a chemical glycosylation that proceeds in >90% isolated yield with >20 : 1 stereoselectivity could hardly be called harsh regardless of any highly reactive or toxic reagents. But despite the somewhat unfair designation of "harsh" ascribed to many chemical glycosylations, a commitment to developing high-yielding, stereoselective glycosylations that proceed quickly with high conversion and stable, inexpensive reagents is an important goal for advancing glycoscience. A recent report by the U.S. National Academy of Sciences has challenged the synthetic organic community to solve "long-standing and vexing problems of stereoselective, regioselective syntheses with *simple*, high-yielding reactions" [1].

The "simplicity" associated with easily handled and inert reagents that have low toxicity is often met with the development of synthetic photochemical reactions. In these cases, light provides the energy needed to activate otherwise moribund chemical species. With the development of photochemical flow reactors [2] and the explosion of publications in the area of visible-light photoredox catalysis [3], the synthetic organic community seems to be getting over the phobia once associated with photochemistry. Photochemical O- and C-glycosylations, in particular, are two areas that are benefitting from this simplicity with numerous publications reporting on high-yielding and stereoselective chemical glycosylations that employ relatively inert reagents. We identify the "perfect" photochemical glycosylation as one that proceeds quickly with visible-light irradiation, high yield and high stereoselectivity, and inexpensive, inert reagents with low toxicity. While such a glycosylation does not exist, we will identify the efforts by the synthetic community to attain this goal with a particular focus on the mechanism and problems that still need to be resolved.

Selective Glycosylations: Synthetic Methods and Catalysts, First Edition. Edited by Clay S. Bennett.

We will focus on recent publications (since ~2001); however, some older methods will be discussed especially to provide historical context or if they have pedagogical value. Unfortunately, space limitations preclude discussions of elegant methods involving singlet glycosylidene donors [4], transition metal glycosylidenes [5], and most of the work done on homolysis of anomeric C—X bonds to generate anomeric radicals for C-glycosylation [6]. Many of these topics have been the subject of excellent reviews cited herein. Photochemical modification of carbohydrates not involving the formation of glycosidic linkages will similarly not be discussed.

10.2 Photochemistry Basics

Photochemistry is defined as the "branch of chemistry concerned with the chemical effects of ultraviolet, visible, and infrared radiation" [7]. We will discuss examples of both visible-light and ultraviolet photochemical glycosylations. The photochemistry equivalent of the organic functional group is the *chromophore*, a group of atoms that absorbs a photon to access an *excited state*, generally an *electronically excited state* (designated with "*" throughout the chapter) that results from the promotion of an electron from a bonding or nonbonding to an antibonding orbital. The energy associated with promotion to the electronic excited state matches the energy of the photon. Thus, shorter wavelengths promote more energetic transitions and, as a result, more energetic photochemistry. Because most organic chromophores cannot absorb visible light, visible-light photochemistry was once viewed as a curiosity by the majority of synthetic organic chemists. Nevertheless, visible light is preferable to UV given that it is less hazardous and easily accessed.

Photochemistry generally occurs via the chemical reactions of electronically excited states. A *singlet excited state* occurs when electrons in the resulting singly occupied orbitals have paired spins. However, one of these electrons may undergo a change in spin (*intersystem crossing*) resulting in a *triplet excited state* where the two electrons have matched spins. Photon emission from singlet excited states is known as fluorescence, while emission from triplet states is known as phosphorescence. Because these two processes do not involve chemical reaction, they are photophysical (rather than photochemical); however, they receive frequent mention in the photochemical literature, and measuring them can provide valuable mechanistic insight. All of the previously mentioned processes are illustrated with the simplified Jablonski diagram in Figure 10.1.

Because of the facility of intersystem crossing in many systems and because triplets are generally long-lived (~1 μs or more), a great deal of synthetically useful bimolecular photochemistry involves the intervention of triplet states and enables the use of substrate/reagent concentrations typically employed by synthetic chemists.

Photochemical glycosylations involving singlet glycosylidene donors (e.g., **1**, Scheme 10.1) result from the irradiation of substrates including cyclobutanones and glycosyldiazirines [4]. In these cases, the donor substrate contains the chromophore. On the other hand, a number of photochemical glycosylations involve the use of chromophore-containing additives known as photosensitizers (examples of the direct irradiation of a chromophore-containing glycosyl donor and a photosensitized glycosylation are shown in Scheme 10.1) [4a], [8]. All of the photosensitized glycosylations

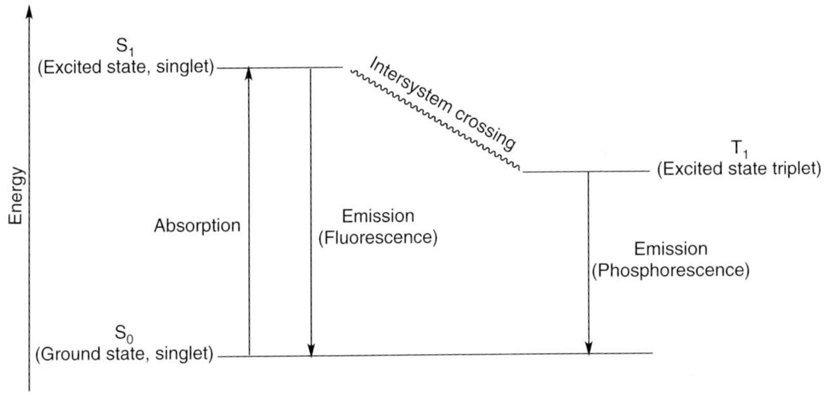

Figure 10.1 Simplified Jablonski diagram.

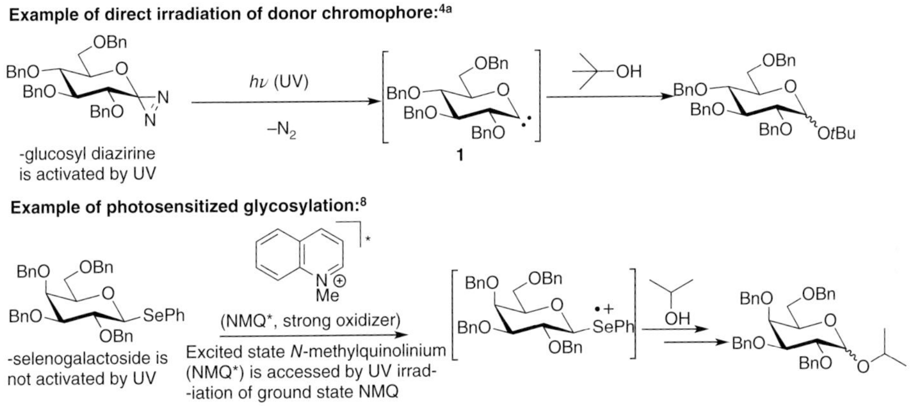

Scheme 10.1 Glycosylation involving direct irradiation of donor and photosensitization.

that we are aware of involve *electron transfer photosensitization* (rather than *energy transfer* where the excited state of a chromophore induces the excited state of another species). Thus, photosensitizers (or "cosensitizers," *vide infra*) can donate electrons to or accept electrons from nonabsorbing glycosyl donors. The favorability of electron transfer to and from excited-state photosensitizers can be estimated if the redox potentials of the electron donors and acceptors in addition to the excited-state energy of the photosensitizer (be it an electron donor or acceptor) are known. Reaction planning can be done with a rudimentary understanding of spectroscopy, electrochemistry, and the kinetics associated with intermediates resulting from electron transfer. Finally, the term "photosensitization" is generally restricted to cases in which the photosensitizer is not consumed. Many of the glycosylations that are discussed herein will involve excitation of photosensitizers by absorption of a photon, a series of electron transfers to and from the excited-state photosensitizer, and regeneration of the photosensitizer. As a result, the terms photosensitization, photocatalysis, and photoredox catalysis [3] can often be used interchangeably in this chapter.

10.3 Photosensitized O-Glycosylation with Chalcogenoglycoside Donors

Due to their easy chemical synthesis and handling in addition to their configurational stability, chalcogenoglycosides including thioglycosides and selenoglycosides are popular glycosyl donors among synthetic chemists [9]. Unfortunately, the stability of these species necessitates strongly electrophilic thio-/selenophiles for activation. Energy from photons may obviate this high reactivity. *O*-Aryl glycosides, on the other hand, have received considerably less attention as glycosyl donors; however, we will discuss two examples involving the photosensitized activation of this substrate class.

In 2013, Bowers and coworkers demonstrated the glycosylation of alcohol acceptors with *p*-methoxyphenylthioglycosides with blue LED irradiation in the presence of $BrCCl_3$ (or CBr_4) and 5 mol% of the visible-light photocatalyst $[Ir(dF(CF_3)ppy)_2(dtbbpy)]PF_6$ ("IrdF," Scheme 10.2) [10] in CH_3CN. The authors proposed a strategy involving single electron oxidation of substrates **2** to radical cations **3** (Scheme 10.3). Sulfur/selenium radical cations such as these have been targeted as activated intermediates with a number of photochemical [8, 11], chemical [12], and electrochemical [13] methods due to the facility with which they extrude neutral sulfur/selenium radicals to generate

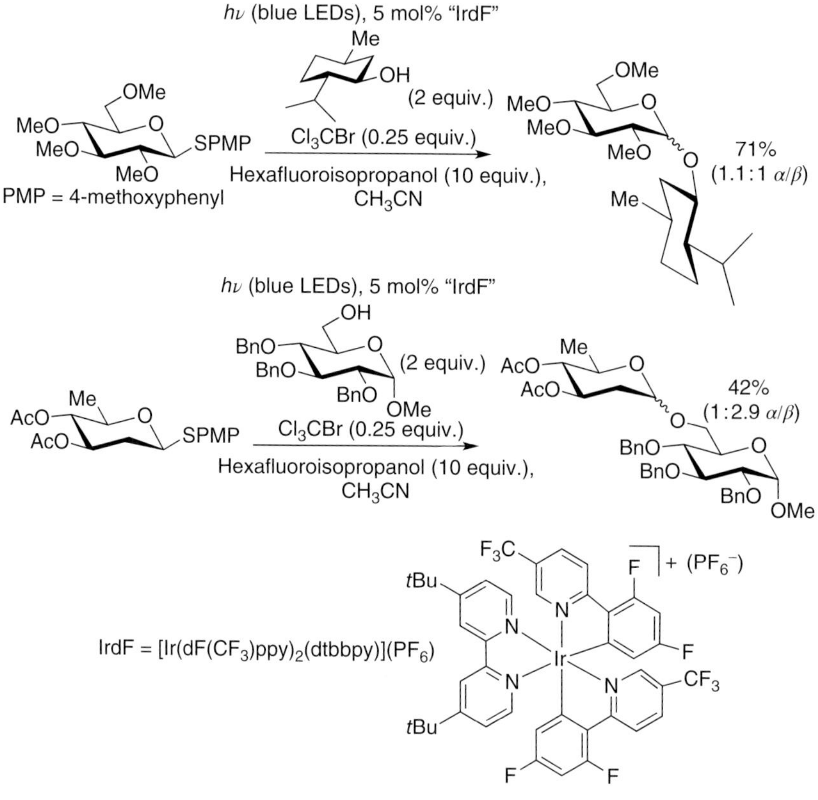

Scheme 10.2 Ir-photosensitized O-glycosylation with thioglycosides.

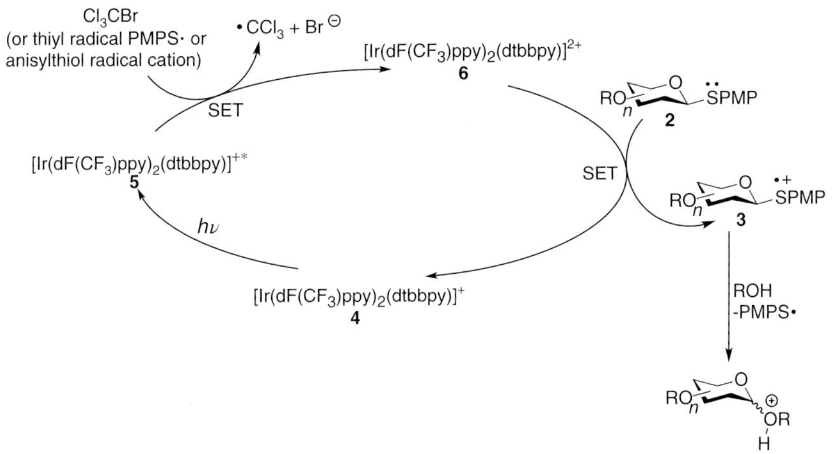

Scheme 10.3 Proposed mechanism for Ir-photosensitized glycosylation with thioglycosides.

oxocarbenium. A few examples of Bowers' procedure are depicted in Scheme 10.2. This procedure was effective for 1°, 2°, and 3° alcohol acceptors including a 1° sugar alcohol and an L-serine derivative with benzyl/methyl-protected D-glucosyl and acetylated 2-deoxy-D-rhamnosyl donors. Yields ranged from 23% to 96% and appear to be somewhat dependent on the steric bulk of the alcohol acceptors.

Bowers proposed the mechanism depicted in Scheme 10.3. Thus, irradiation of IrdF (**4**) [14] with blue LEDs (Ir III photocatalysts of this class have a metal-to-ligand charge transfer band (MLCT) that stretches well into the visible spectrum) [14, 15] results in a triplet excited state **5** that has enough reducing power ($E_{1/2}$,M*/M$^+$ = −0.89 V vs the saturated calomel electrode (SCE)) [3a] to transfer an electron to bromotrichloromethane resulting in the formation of bromide anion and trichloromethyl radical. The authors demonstrated that the additive hexafluoroisopropanol (HFIP) was necessary and suggested that it facilitates the collapse of charge transfer complexes (e.g., resulting from the encounter of **5** with $BrCCl_3$) and prevention of energy-wasting back electron transfer. The resulting Ir(IV) complex **6** is a powerful oxidizing agent with a redox potential ($E_{1/2}$,M/M$^+$ = +1.69 V, SCE) that is significantly higher than the peak oxidation current of the substrate **2** (E_{pa} = +1.22 V, Ag/AgCl for the tetrabenzyl-protected system) [13a]. Single-electron transfer (SET) from **2** to Ir(IV) complex **6** is facile. By contrast, the authors noted that the electron-poor tetraacetyl analog of **2** was unreactive likely due to a substantially higher E_{pa}. Recalcitrance and unwanted side reactions are general problems with photosensitized glycosylation with tetraacetyl-protected thio-/selenoglycoside donors [8, 11a]. Rapid breakdown of radical cation **3** to oxocarbenium ensues and is followed by glycosylation of ROH. As is the case with many photochemical glycosylations of this class [8, 11], the stereoselectivity was low, and, in this case, the solvent nitrile effect did not predominate as the reactions proved to be slightly α-selective. The authors also noted that glycosylation works (albeit less efficiently) with the exclusion of $BrCCl_3$ while light and photocatalyst were absolute requirements. Anisylthiyl (PMPS·) radicals or anisylthiol radical cations resulting from decomposition of **3** may act in lieu of $BrCCl_3$ as electron acceptors for **5**. Dimerization of thiyl radicals results in the formation of disulfides that were isolated from reaction mixtures.

A mechanistically similar, UV-promoted glycosylation method that involved photosensitization with 10 mol% 1,4-dicyanonaphthalene (DCN) was reported in 1990 by Griffin and coworkers (Scheme 10.4) [11a]. Excited-state cyanoarenes such as DCN are powerful single-electron oxidants [16]. Reactions were purged with Ar before irradiation, and the authors have not commented on an electron acceptor that induces turnover of the photoreduced DCN radical anions (serving as an electron sink as did $BrCCl_3$ with Bowers' method). Appreciable decomposition of DCN was never observed. The detection of thiophenol in product mixtures suggests that thiyl radicals (as proposed by Bowers) or disulfides may act as electron acceptors that turn over DCN radical anions. This method was limited to phenylthioglucoside donors with free hydroxyl groups or methyl ether protection with methanol as the acceptor. A mixture of α and β anomers was afforded in this instance. In addition, 1,6-anhydroglucose was observed with the unprotected phenylthioglucoside donor in the absence of methanol. Benzyl protection did not lead to glycosylated products, and the authors suggest that the formation of encounter complexes between DCN and the benzyl groups may prevent thioglucoside activation.

An interesting combination of photochemical O-glycosylation with thioglycosides and reagent chemistry to engender selectivity was reported by Toshima and coworkers in 2013 (Scheme 10.5) [11b]. Stating the need for the development of glycosylations utilizing unprotected donors, they demonstrated that the UV irradiation (365 nm, 100 W lamp) of solutions of 2-deoxyphenylthioglycosides, protected *in situ* with boronic acid **9**, in the presence of excess DDQ (2,3-dichloro-5,6-dicyanobenzoquinone) and alcohol acceptors resulted in O-glycosidic products. The reaction required DDQ and UV irradiation. Irradiation of the substrate **7** in the presence of cyclohexylmethanol resulted, perhaps not surprisingly, in the formation of substantial amounts of anhydro-2-deoxygalactoside **8**. By contrast, precondensation of **7** with boronic acid **9** resulted in *in situ* protected intermediate **10** that underwent glycosylation with

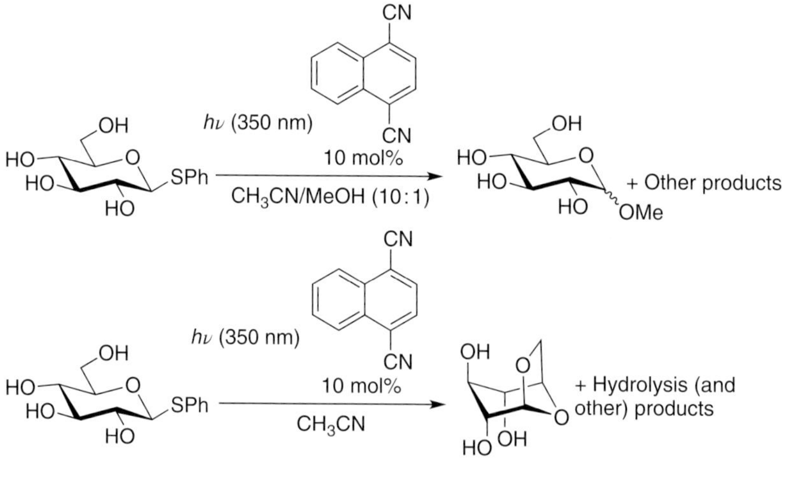

Scheme 10.4 DCN-photosensitized glycosylation with phenylglucoside donors.

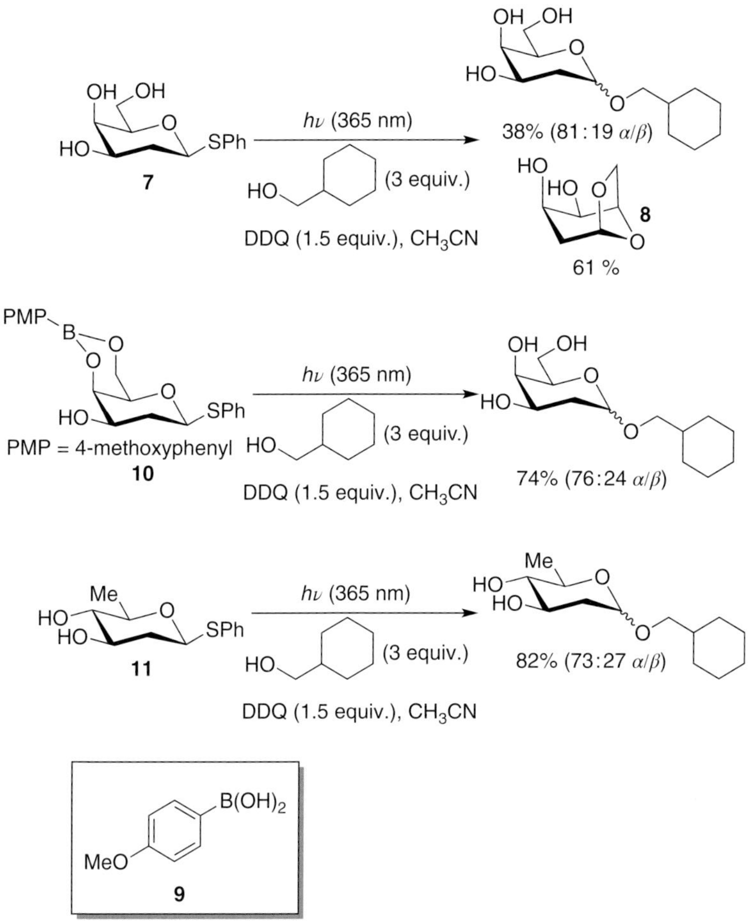

Scheme 10.5 UV/DDQ-promoted glycosylation with borinate-protected thioglycoside.

cyclohexylmethanol in 74% yield with modest α-selectivity (attributed by the authors to the kinetic anomeric effect) despite being run in CH_3CN. On the other hand, donor **11** performed best in the absence of boronic acid wherein the authors attribute this to a lack of any condensation between **11** and **9** in addition to interference of **9** with alcohol acceptor. Glycosylation was successful with 1° and 2° alcohols, and yields of glycosidic products ranged from 70% to 84% with the various monodeoxy- and dideoxy donors under optimized conditions.

While Toshima and coworkers did not propose a detailed mechanism, they cited a previous report in which deprotection of dithioacetals occurred with UV irradiation in the presence of DDQ and H_2O [17]. It is possible that excited-state DDQ accepts electrons in a manner that is analogous to excited-state cyanoarenes as proposed by Griffin and the Ir(IV) complex **6** as proposed by Bowers.

While phenylselenoglycosides have peak oxidation currents that are only slightly less positive than those of the analogous phenylthioglycosides [13a], they have proven somewhat more versatile in the photochemically driven formation of glycosidic linkages [8, 11c, d]. In 2012, Cumpstey and Crich demonstrated that the UV irradiation (350 nm, Rayonet lamps) of solutions of methyl- and benzyl-protected phenylselenoglucosides and -galactosides in the presence of 3 equiv. of alcohols and 0.3 equiv. of the sensitizer *N*-methylquinolinium hexafluorophosphate (NMQ-PF6) resulted in the formation of *O*-glycosides (Scheme 10.6) [8]. The authors used O_2 (air) as terminal oxidant and, as a consequence of H_2O being likely generated in the process of O_2 reduction, employed molecular sieves for best results. Nevertheless, hydrolysis products were frequently observed in addition to glycosylation. Interestingly, ball sieves were superior to crushed sieves due to decreased scattering of UV and the more efficient irradiation that resulted. Toluene was also used as a cosensitizer to increase the efficiency of the reaction [18]. The reaction was dependent on light, air (i.e., O_2), and NMQ-PF_6. The method was successful with 1° and 2° alcohol acceptors with yields ranging from 42% to 73% under optimized conditions. Primary alcohols of carbohydrate acceptors were reactive; however, 2° carbohydrate acceptors gave very low yields of product. Low levels of β-selectivity were attained, possibly because of the solvent nitrile effect. The authors also demonstrated that employment of a peracetylated phenylselenoglucoside resulted in unproductive hydrolysis, acyl migration, and Ritter reaction while thioglycoside donors also gave inferior results.

While Cumpstey and Crich did not propose a mechanism for this transformation, they discuss radical cation intermediates, the likes of which have been invoked by others as discussed herein [10, 11]. In addition, Floreancig has demonstrated that excited-state singlet *N*-methylquinolinium cation (NMQ+*) is a strong single-electron oxidant [18],

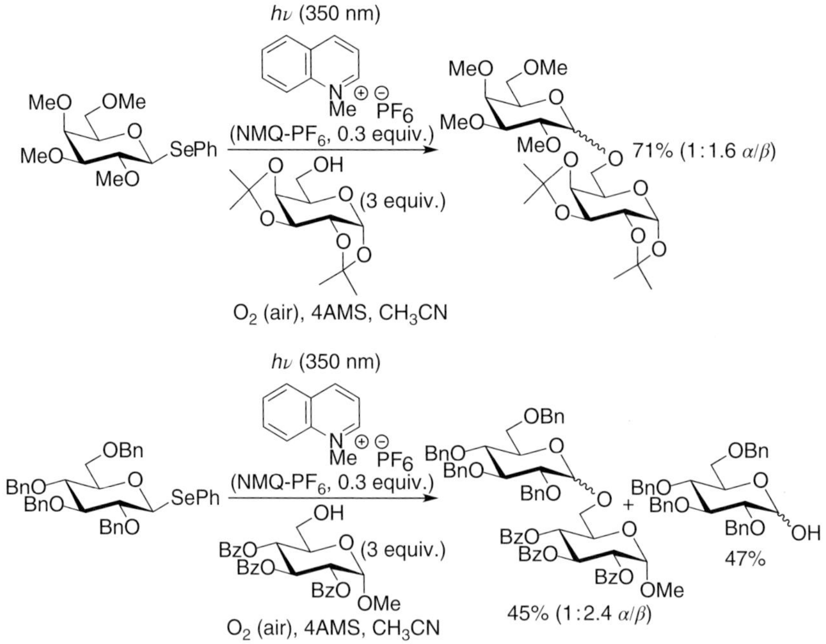

Scheme 10.6 NMQ-PF_6-photosensitized glycosylation with phenylselenoglycosides.

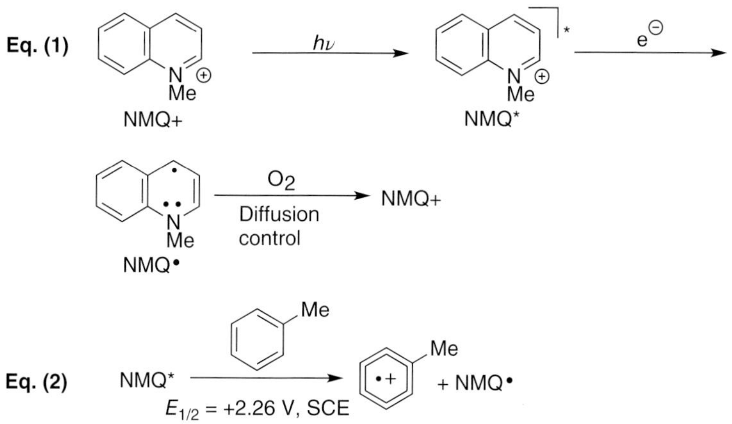

Scheme 10.7 Mechanistic considerations for NMQ-PF_6 and toluene cosensitizers.

while the photoreduced neutral radical NMQ· can be reoxidized by O_2 to cationic NMQ^+ at a diffusion-controlled rate (Scheme 10.7, Eq. (1)) [19]. Toluene, designated by Floreancig and Crich as a cosensitizer, is likely oxidized by NMQ+* to toluene radical cation ($E_{1/2}$ = +2.26 V, SCE) [20] that is also a strong SET oxidant and may act in lieu of NMQ+*, increasing the overall efficiency (Scheme 10.7, Eq. (2)) [18].

The method by Cumpstey and Crich is similar to the one reported by Furuta *et al.* in 1996 (Scheme 10.8) [11c]. This involved the UV irradiation (high-pressure Hg lamp through a Pyrex filter) of CH_3CN solutions of permethylated β-phenylselenoglucoside **12** with various cyanoaromatic photosensitizers or the strongly oxidizing ($E_{1/2}$, M^-/M^* = +2.28 V, SCE) [21] photosensitizer 2,4,6-triphenylpyrylium tetrafluoroborate (TPT) and excesses of alcohol acceptors. One equivalent of the photosensitizers was used, obviating the addition of stoichiometric oxidants. The authors settled on TPT for the development of reaction scope and demonstrated O-glycosylation with 1° (including a sugar alcohol) and 2° (including cholesterol) acceptors. Yields ranged from 59% to 75% except for the 8% of product isolated with cholesterol as the acceptor. A modest β-selectivity prevailed in most cases and was attributed to the nitrile effect. Furuta proposed a mechanism (Scheme 10.8) involving the UV-induced excitation of TPT, electron transfer from **12** to generate a radical cationic intermediate, breakdown to oxocarbenium, and interception by alcohol acceptor.

In 2013, shortly after Bowers' report was published, we reported two approaches to the visible-light-promoted O-glycosylation of alcohols with perbenzylated β-phenylselenoglucosides and -galactosides (Scheme 10.9) [11d]. We first employed 5 mol% of $Ru(bpy)_3(PF_6)_2$ (**13**) as photosensitizer with CBr_4 (a system that we believed would provide a similar approach to chalcogenoglycoside radical cations as that reported by Bowers, Scheme 10.2) and tetrabenzyl phenylselenoglycosides. Blue LED irradiation of the aforementioned species in the presence of 3 equiv. of 1° (including a sugar acceptor) and 2° alcohols with CH_3CN resulted in the formation of *O*-glycosides in moderate-to-high yields (43–81%) with a preponderance of α-anomers formed. Removal of alcohol acceptor resulted in the formation of a 1.4 : 1 mixture of α-bromoglucoside **14** and glycal **15** (respectively). The intermediacy of **14** and the presence of bromide

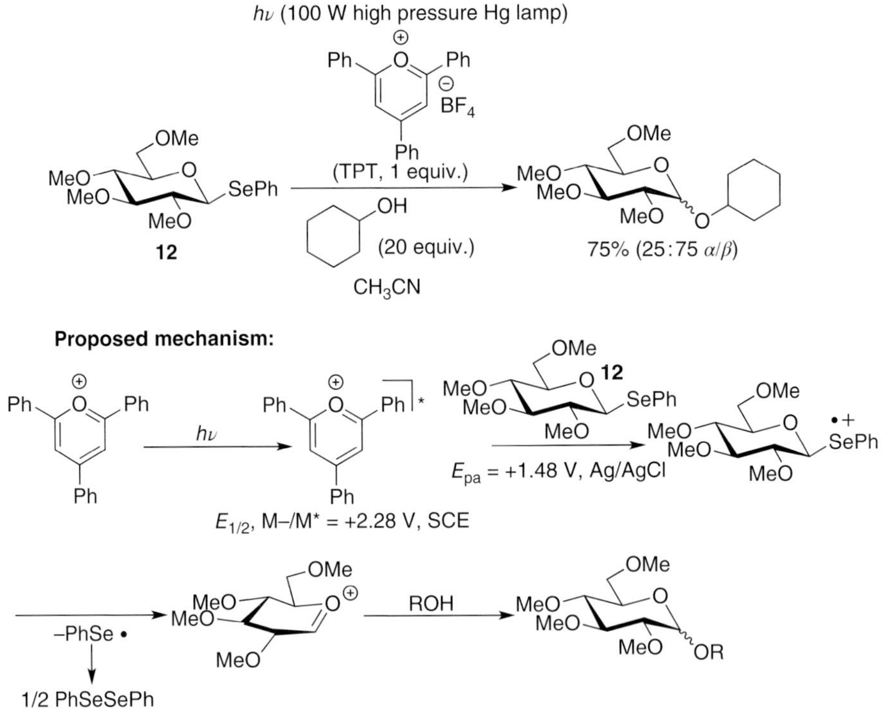

Scheme 10.8 2,4,6-Triphenylpyrylium-photosensitized glycosylation with phenylselenoglycoside.

anions may explain the α-selectivity of this process through previously described halide ion catalysis [22].

Most interestingly, the replacement of 5 mol% $Ru(bpy)_3(PF_6)_2$ with 10 mol% diphenyldiselenide resulted in similar reactivity (in CH_3CN) and yields (20–71%) and increased α-selectivity (but decreased reactivity) when CH_2Cl_2 was used as solvent (Scheme 10.9). ^{77}Se NMR experiments involving the blue LED irradiation of mixtures of diphenyldiselenide and CBr_4 resulted in the formation of PhSeBr, a species known to promote glycosylation with selenoglycoside donors [23]. While it is reasonable to suggest that initiation involves SET pathways and the formation of selenoglycoside radical cations in the presence of $Ru(bpy)_3(PF_6)_2$, a PhSeBr-promoted pathway is likely operative as well in these cases since diphenyldiselenide is a by-product of the $Ru(bpy)_3(PF_6)_2$-promoted reaction. We proposed a mechanism (Scheme 10.10) wherein the visible-light-promoted homolysis of the Se—Se bond of diphenyldiselenide to generate phenylselenyl radicals precedes reaction with CBr_4 to generate PhSeBr and tribromomethyl radical. The feasibility of this halogen atom transfer was later demonstrated with DFT calculations [24]. Reaction of PhSeBr with **16** [23] results in the formation of onium **17** that can then intercept bromide to generate α-bromoglycoside or alcohol to generate O-glycosidic products.

As alluded to earlier, photochemical glycosylations involving *O*-aryl glycosides as donors have not been studied as frequently as thio- and selenoglycosides. In 1985, Noyori and coworkers reported on the photosensitized tetrahydropyranylation and

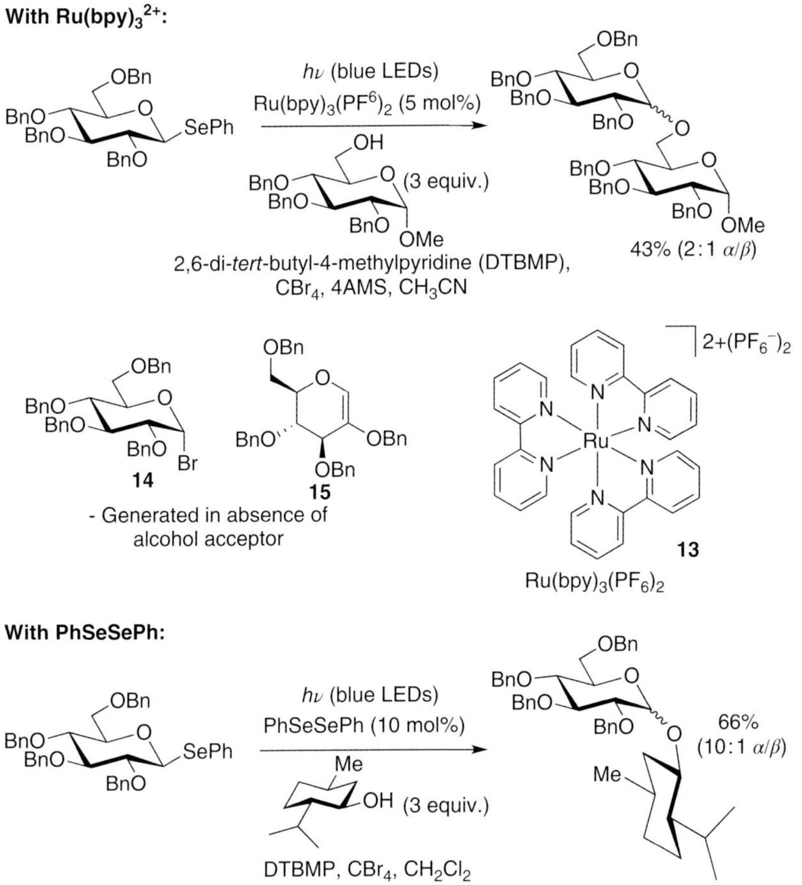

Scheme 10.9 Ru(bpy)$_3^{2+}$-photosensitized/PhSeSePh-promoted glycosylation with phenylselenoglycoside.

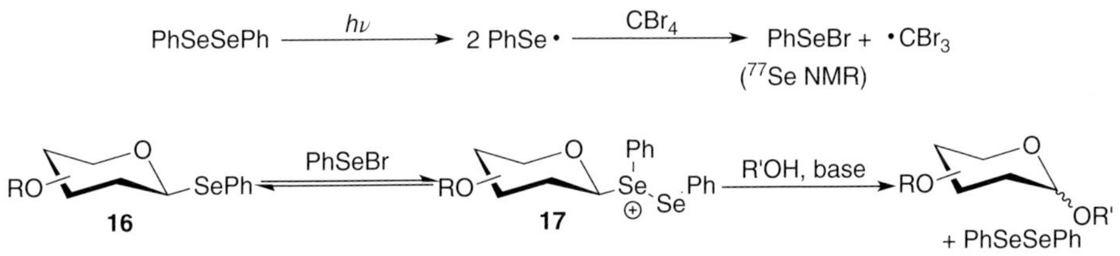

Scheme 10.10 Proposed mechanism for PhSeSePh-promoted photochemical glycosylation.

glycosylation of alcohols by UV irradiation (200 W high-pressure Hg lamp) of solutions of THP-protected phenols and *O*-aryl glycosides with 0.1 equiv. of aromatic hydrocarbons (photosensitizer), cyanoaromatics (also 0.1 equiv., cosensitizer), and alcohol acceptors [25]. Some of these results are depicted in Scheme 10.11. The most commonly used photosensitizer/cosensitizer combination was either phenanthrene or

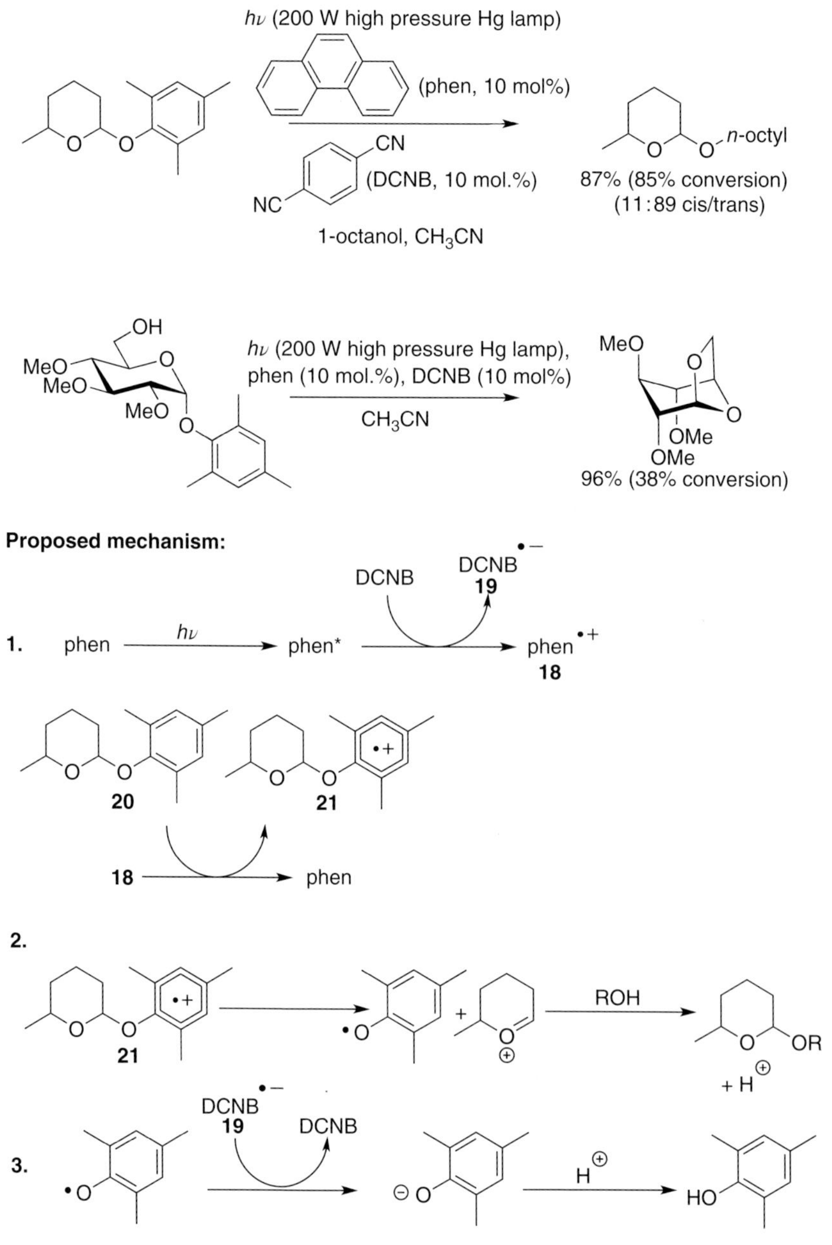

Scheme 10.11 Phenanthrene-photosensitized pyranylation/glycosylation with 2-aryloxypyrans.

triphenylene in combination with 1,4-dicyanobenzene (DCNB). The 2,4,6-trimethylphenylglycosides were more reactive compared to phenylglycosides due to the relatively low peak oxidation current ($E_{pa} = +1.64\,V$, Ag/AgCl) brought about by the electron-donating methyl groups. Pyranylation yields ranged from 16% to 88% at various levels of substrate conversion with phenanthrene as photosensitizer.

Based on a number of experiments including the quenching of the phenanthrene excited state by 1,4-dimethoxybenzene, Noyori proposed a detailed mechanism (Scheme 10.11). (i) Excitation of phenanthrene resulted in electron transfer to DCNB to generate phenanthrene radical cation **18** and DCNB radical anion **19**. Electron transfer from *O*-aryl donors such as **20** to **18** then generates radical cations **21** and turns over **18**. (ii) Breakdown of **21** to phenoxy radical and oxocarbenium precedes attack of alcohol to generate pyranylation/glycosylation products after loss of proton. (iii) The radical anion **19** is converted back to DCNB by transferring an electron to phenoxy radical. Proton transfer to the resulting phenoxide results in the formation of phenol product. Because the aryloxy moiety of **20** acts as electron donor to turn over **18** and, ultimately, as electron acceptor to turn over **19**, this constitutes a rare example of a redox neutral process in photosensitized glycosylation.

Griffin and coworkers reported seminal examples of photochemical O-glycosylation involving photosensitized activation of *O*-arylglycosyl donors in 1983 and 1984 (not pictured) [26]. These studies were initiated by interests in lignin degradation (unprotected β-phenylglucoside was designated as a model for lignin) rather than the construction of glycosidic linkages. Glycosylations were accomplished with the UV (350 nm) irradiation of solutions of *O*-arylglycoside donor in air- or O_2-saturated 10 : 1 CH_3CN/MeOH in the presence of DCN as photosensitizer. Methyl glycosides and 1,6-anhydrosugars were among the products of these conversions.

10.4 Photochemical *O*-Glycosylation with Other Donors

Photochemical *O*-glycosylations employing trichloroacetimidates and glycals have been reported. In 2014, a photoacid-promoted approach to the activation of benzyl-protected glycosyl trichloroacetimidates was reported once again by Toshima and coworkers (Scheme 10.12) [27]. This involved irradiation (Blackray 100 W lamp, 365 nm) of the organic photoacids **22** and **23**. While these species have a relatively high pK_a in their ground states, the excited states are highly acidic (OH group as proton donor) and were identified by the authors as potential promoters for glycosylation with trichloroacetimidate donors. Irradiation of Et_2O solutions of glucosyl, galactosyl, and mannosyl trichloroacetimidates, substoichiometric photoacid, and alcohol acceptor at 35 °C afforded high yields (generally >70%) of glycosidic products resulting from reaction with 1° and 2° alcohol acceptors. Anomeric ratios were often ~1 : 1; however, anchimeric assistance (2-O benzoylation) provided complete β-selectivity in some cases. Controls demonstrated that the poor selectivity was a consequence of kinetic control rather than isomerization of glycosidic products. To contrast with the use of acids such as TMSOTf or HOTf, this reaction did not require neutralization of acid since cessation of irradiation "removed" the acidic properties of **22**/**23**, which could furthermore be recovered in high yields and reused. The mechanism of this reaction (Scheme 10.12) likely involves photoexcitation of **22**/**23** to highly acidic excited states such as **22***. Proton transfer to trichloroacetimidate and subsequent oxocarbenium formation are prerequisites to the formation of glycosidic bond.

In 2001, Rojas and coworkers developed a highly β-selective glycosylation proceeding by nitrene insertion into glycals to generate β-2-amidoallopyranosides (Scheme 10.13) [28]. Irradiation (254 nm, Vycor filter) of allal azidoformate (**25**) in the presence of

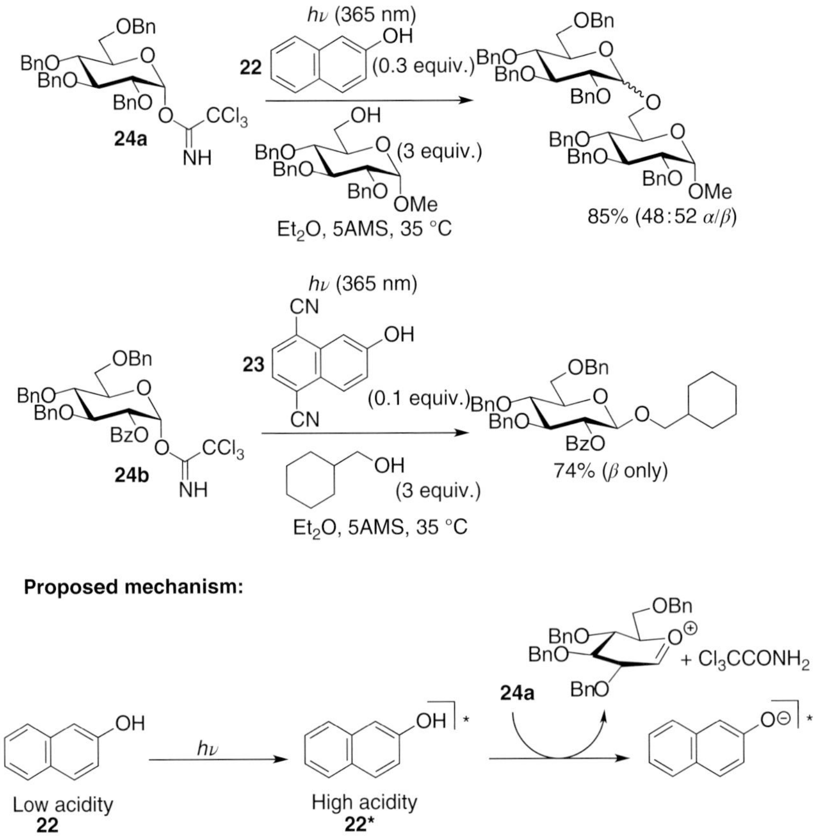

Scheme 10.12 Photoacid-promoted glycosylations with glycosyl trichloroacetimidates.

excess alcohol acceptors resulted in low-to-moderate (7–40%) yields of glycosidic products with exclusive β-selectivity. Products of the glycosylation of 1° and 2° (including sugar alcohols) alcohol acceptors were afforded. The authors suggested a mechanism (Scheme 10.13) involving the photochemical generation of nitrene that inserts into the glycal double bond. Reaction of alcohol with an aziridine-containing intermediate **26** results in the formation of glycosidic products and explains the high β-selectivity.

10.5 Photosensitized C-Glycosylation

The synthesis of *C*-glycosides using photosensitization, photoinitiation, or direct irradiation of the glycosyl donor has been the subject of fairly extensive development relative to photochemical O-glycosylation. Singlet glycosylidenes [4] and metalloglycosylidenes [5] have been the subjects of some of this development; however, the lion's share of investigation has involved the reaction of anomeric radicals with alkene acceptors (Scheme 10.14) [6]. Perhaps surprisingly to those uninitiated in the field of radical chemistry, these are among the most stereoselective of the photochemical glycosylations, providing very high

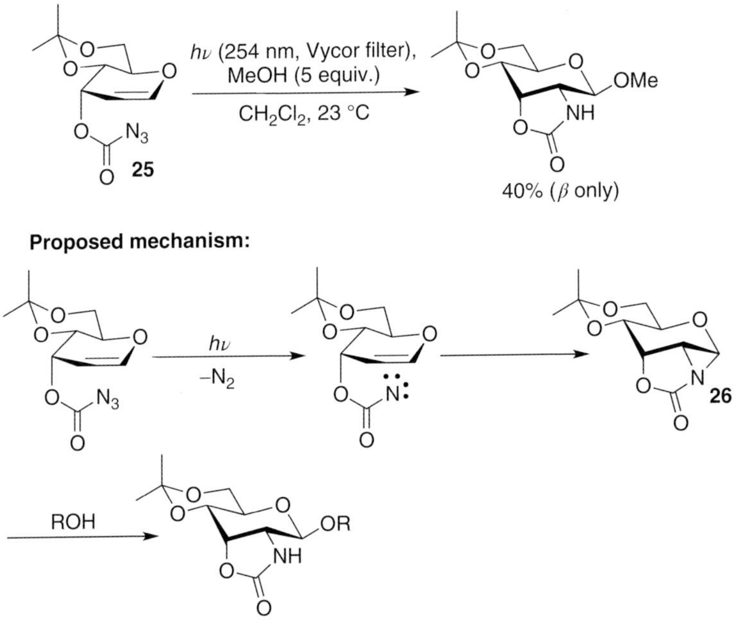

Scheme 10.13 Photochemical activation of allal azidoformates by nitrene insertion.

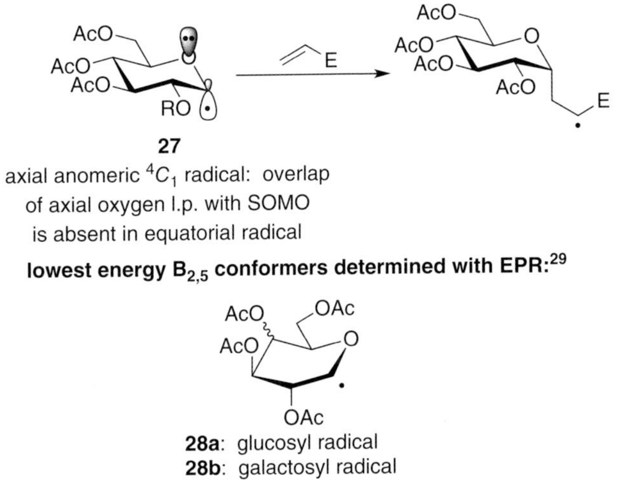

Scheme 10.14 Reaction of a glucosyl radical with an electron-poor alkene acceptor.

α-selectivity with, for example, glucosyl, mannosyl, and galactosyl donors with ester-protecting groups. This selectivity is often attributed to reaction of conformers with structures such as the $^4C_1$ conformer **27** with electron-deficient alkenes. The favorability of this mode of attack has been rationalized by the stabilizing overlap of the axial lone pair of the pyran oxygen with the axial C1 SOMO (singly occupied molecular orbital).

This interaction would be nonexistent with an equatorial C1 SOMO. The resulting highly nucleophilic conformer **27** thus undergoes facile reaction with electron-poor alkenes due to the polar nature of the attendant transition state [30]. On the other hand, it is important to note that extensive EPR studies demonstrate the near coplanarity of C1, C2, and the pyran oxygen of the *lowest energy* $B_{2,5}$ *conformers* of **28a** and **28b** [29]. Overlap between the C1 SOMO and the C2 C—O σ* antibonding orbital has been invoked as an explanation for this geometry.

The UV irradiation of solutions of AIBN (2,2′-azobisisobutyronitrile), glycosyl halide, tin hydride, and electron-deficient alkene or allyltin has been utilized on numerous occasions for C-glycosylation [6]. This involves the photochemical decomposition of AIBN to 2-cyanoprop-2-yl radicals that initiate a chain reaction. While this method has provided high yields of *C*-glycosides with high α-selectivity, the toxicity and often difficult purification associated with organotin reagents are serious drawbacks.

Starting in 2010, Gagné and coworkers published a series of papers on the visible-light-promoted C-glycosylation of electron-deficient alkenes [2a, 31]. Initial studies [31a] involved the visible-light irradiation (14 W fluorescent bulb) of solutions of acetyl-/benzoyl-protected glucosyl-, mannosyl-, and galactosyl bromides, *i*Pr_2NEt, Hantzsch ester (**29**), and visible-light-absorbing polypyridyl photosensitizer $Ru(bpy)_3(PF_6)_2$ in CH_2Cl_2. Some of the highest yields ever reported for such C-glycosylations (≥80% in all cases) marked these transformations (Scheme 10.15). Controls demonstrated that irradiation, photosensitizer, and *i*Pr_2NEt were critical for reaction to occur while **29** prevented the anomeric-radical-initiated telomerization of the alkene acceptor methyl acrylate. Preemptive reduction of anomeric radical with *t*BuSH, radical clock experiments, and time-resolved EPR provided evidence for radical intermediates.

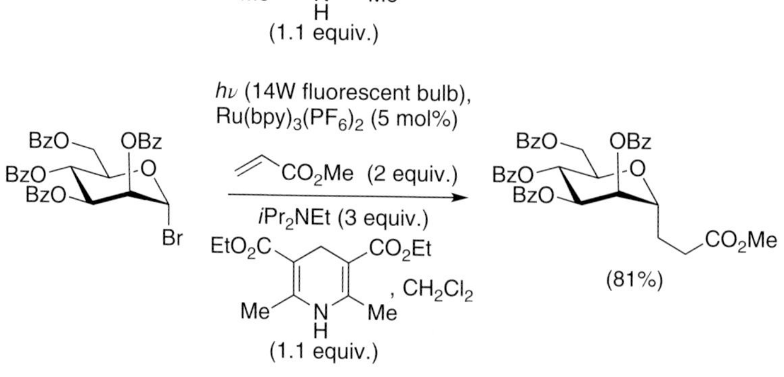

Scheme 10.15 Visible-light-photosensitized C-glycosylation of electron-deficient alkenes with glycosyl bromides.

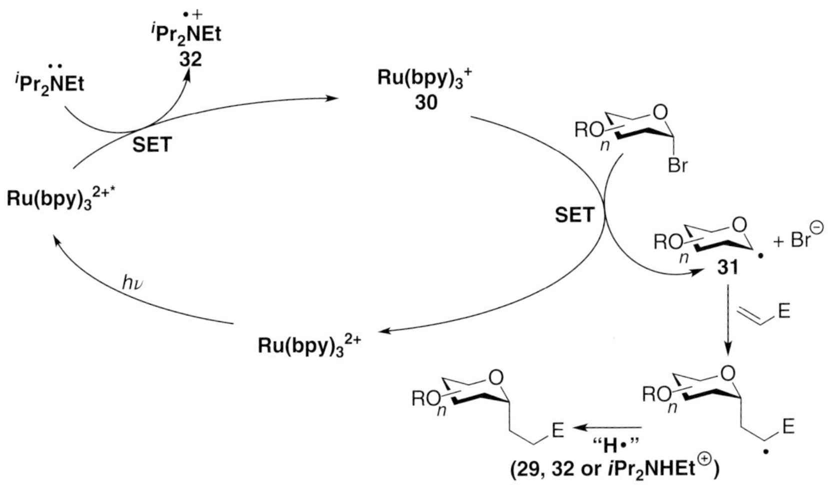

Scheme 10.16 Proposed mechanism for Ru-photosensitized C-glycosylation.

The authors proposed a mechanism (Scheme 10.16) involving the visible-light excitation of $Ru(bpy)_3^{2+}$ followed by single-electron reduction by iPr$_2$NEt to generate the strongly reducing ($E_{1/2}$, $M^-/M = -1.33$ V, SCE) [3a] Ru(I) complex Ru(bpy)$_3$+ (**30**). SET from **30** to glycosyl bromide results in the formation of nucleophilic anomeric radicals **31** that react with electron-deficient alkenes to generate α-*C*-glycosides after hydrogen atom transfer from **32**, iPr$_2$NHEt$^+$, or Hantzsch ester **29**.

Further contributions from the Gagné group have included the development of work aimed at further understanding the mechanism [31b] of their photochemical C-glycosylation and the use of flow reactors [2a] to increase efficiency. A recurring problem with transition metal polypyridyl-photosensitized reactions in general is low efficiency manifested by long reaction times with large-scale batch reactions. For example, Gagné observed a sluggish but nonetheless selective 85% conversion on a 2.43 mmol scale over a period of 24 h (turnover frequency, TOF = 3.5 h^{-1}) in a Schlenk flask for the reaction in Scheme 10.17. However, running the same reaction in an NMR tube on a 0.06 mmol scale resulted in a TOF of 70 h^{-1}. Further, employment of the Beer–Lambert law led the authors to conclude that, at a $Ru(dmb)_3^{2+}$ (another Ru polypyridyl photosensitizer identified as superior to $Ru(bpy)_3^{2+}$ [31b], dmb = 4,4′-dimethyl-2,2′-bipyridine) concentration of 1 mM, 98% of the light is absorbed within 1 mm of the vessel wall. To provide a solution to the low active volume of the photochemical reaction, Gagné and coworkers developed a flow reactor driven by a preparatory HPLC pump and consisting of clear fluorinated ethylene propylene (FEP) tubing with a small (0.8–1.6 mm) internal diameter. The FEP tubing was coiled several times around a Liebig condenser through which water could be passed to control the temperature and into which blue LED strips were inserted. A flow rate of 0.1 ml min^{-1} and a $Ru(dmb)_3^{2+}$ concentration of 0.5 mM with 0.8 mm FEP tubing gave a TOF of 120 h^{-1}. The small internal diameter and coiling of the FEP tubing around the LED strips enabled better light penetration and increased the active volume of the reaction, thereby dramatically increasing the efficiency. This and two other seminal contributions from the Seeberger and Stephenson groups [2] served to raise the profile of

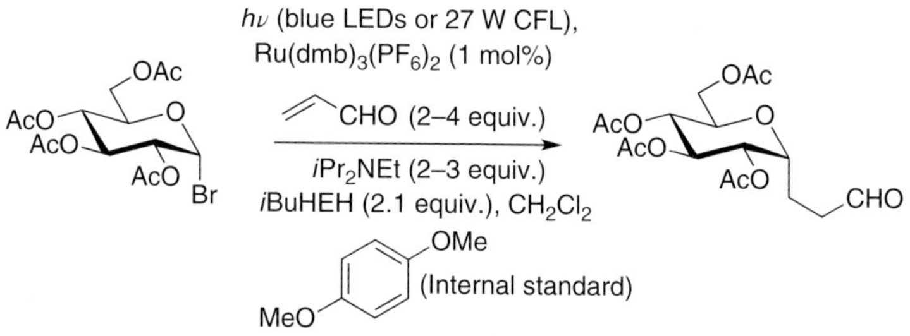

Reaction setup	TOF
2.43 mmol scale, 25 ml flask:	3.5 h^{-1}
0.06 mmol scale, NMR tube:	70 h^{-1}
flow reactor, 0.8 mm FEP tubing:	120 h^{-1}

Scheme 10.17 Turnover frequency (TOF) as a function of reactor design.

photochemical flow reactors, which are now commonly used in the development of large-scale visible-light-photosensitized transformations [32].

10.6 Conclusions

A number of problems remain in the realm of photochemical glycosylation. These include, but are not limited to, obtaining high yields and stereoselectivity with O-glycosylation, activating disarmed donors such as the tetraacetyl chalcogenoglycosides, and developing β-selective C-glycosylation. Importantly, these problems are not peculiar to photochemical glycosylation and constitute some of the "longstanding and vexing problems" [1] associated with chemical glycosylation that were mentioned earlier. We assert that photochemical glycosylation, especially within the context of O-glycosylation, is still a fledgling and underexplored field in need of further development. Nevertheless, there have been a number of noteworthy achievements in this area of study that include the development of mild, visible-light-promoted conditions, the employment of temporary protecting groups that can be installed and removed directly before and after a glycosylation procedure, the development of high-yielding and β-selective protocols with O-glycosylation and α-selective protocols with C-glycosylation, and the introduction of flow reactors that have been employed by the entire visible-light photocatalysis community to improve the efficiency of large-scale reactions. We also assert that, because of the mild conditions employed, there is no reason why photochemical glycosylations cannot provide solutions to some of the most difficult problems in glycosylation including the formation of 1,2-*cis* glycosides, α-sialosides, β-mannosides, and β-2-deoxyglycosides. Given enough time and investigation, it is entirely reasonable to think that mild, user-friendly approaches to these problems can be supplied by photochemistry.

References

1 (2012) Transforming Glycoscience: A Roadmap to the Future, The National Academies Press, Washington, DC.

2 (a) Andrews, R.S., Becker, J.J., and Gagné, M.R. (2012) *Angew. Chem. Int. Ed.*, 51, 4140–4143; (b) Tucker, J.W., Zhang, T.F., Jamison, T.F., and Stephenson, C.R.J. (2012) *Angew. Chem. Int. Ed.*, 51, 4144–4147; (c) Bou-Hamdan, F.R. and Seeberger, P.H. (2012) *Chem. Sci.*, 3, 1612–1616.

3 (a) Prier, C.K., Rankic, D.A., and MacMillan, D.W.C. (2013) *Chem. Rev.*, 113, 5322–5363; (b) Narayanam, J.M.R. and Stephenson, C.R.J. (2011) *Chem. Soc. Rev.*, 40, 102–113; (c) Hu, J., Wang, J., Nguyen, T.H., and Zheng, N. (2013) *Beilstein J. Org. Chem.*, 9, 1977–2001; (d) Teply, F. (2011) *Collect. Czech. Chem. Commun.*, 76, 859–917.

4 (a) Vasella, A., Bernet, B., Weber, M., and Wenger, W. (2000) in Carbohydrates in Chemistry and Biology (eds B. Ernst, G.W. Hart, and P. Sinaÿ), Wiley-VCH Verlag GmbH, Weinheim, pp. 155–175; (b) Vasella, A. (1991) *Pure Appl. Chem.*, 63, 507–518; (c) Lee-Ruff, E., Xi, F., and Qie, J.H. (1996) *J. Org. Chem.*, 61, 1547–1550; (d) Lee-Ruff, E., Wan, W.-Q., and Jiang, J.-L. (1994) *J. Org. Chem.*, 59, 2114–2118.

5 Dötz, K.H., Klumpe, M., and Nieger, M. (1999) *Chem. Eur. J.*, 5, 691–699.

6 (a) Giese, B. (1988) *Pure Appl. Chem.*, 60, 1655–1658; (b) Chen, G.-R. and Praly, J.-P. (2008) *C.R. Chim.*, 11, 19–28; (c) Togo, H., He, W., Waki, Y., and Yokoyama, M. (1998) *Synlett*, 1998, 700–717.

7 Braslavsky, S.E. (2007) *Pure Appl. Chem.*, 79, 293–465.

8 Cumpstey, I. and Crich, D. (2011) *J. Carbohydr. Chem.*, 30, 469–485.

9 (a) Codée, J.D.C., Litjens, R.E.J.N., van den Bos, L.J., Overkleeft, H.S., and van der Marel, G.A. (2005) *Chem. Soc. Rev.*, 34, 769–782; (b) Demchenko, A.V. (2003) *Synlett*, 15, 1225–1240; (c) Oscarson, S. (2000) in Carbohydrates in Chemistry and Biology (eds B. Ernst, G.W. Hart, and P. Sinaÿ), Wiley-VCH Verlag GmbH, Weinheim, pp. 93–116; (d) Mehta, S. and Pinto, B.M. (1993) *J. Org. Chem.*, 58, 3269–3276.

10 Wever, W.J., Cinelli, M.A., and Bowers, A.A. (2013) *Org. Lett.*, 15, 30–33.

11 (a) Griffin, G.W., Bandara, N.C., Clarke, M.A., Tsang, W.-S., Garegg, P.J., Oscarson, S., and Silwanis, B.A. (1990) *Heterocycles*, 30, 939–947; (b) Nakanishi, M., Takahashi, D., and Toshima, K. (2013) *Org. Biomol. Chem.*, 11, 5079–5082; (c) Furuta, T., Takeuchi, K., and Iwamura, M. (1996) *Chem. Commun.*, 157–158; (d) Spell, M., Wang, X., Wahba, A.E., Conner, E., and Ragains, J. (2013) *Carbohydr. Res.*, 369, 42–47.

12 Marra, A., Mallet, J.M., Amatore, C., and Sinaÿ, P. (1990) *Synlett*, 572–574.

13 (a) Yamago, S., Kokubo, K., Hara, O., Masuda, S., and Yoshida, J. (2002) *J. Org. Chem.*, 67, 8584–8592; (b) France, R.R., Rees, N.V., Wadhawan, J.D., Fairbanks, A.J., and Compton, R.G. (2004) *Org. Biomol. Chem.*, 2, 2188–2194.

14 Lowry, M.S., Goldsmith, J.I., Slinker, J.D., Rohl, R., Pascal, R.A. Jr., Malliaras, G.G., and Bernhard, S. (2005) *Chem. Mater.*, 17, 5712–5719.

15 Tamayo, A.B., Alleyne, B.D., Djurovich, P.I., Lamansky, S., Tsyba, I., Ho, N.N., Bau, R., and Thompson, M.E. (2003) *J. Am. Chem. Soc.*, 125, 7377–7387.

16 Turro, N.J., Ramamurthy, V., and Scaiano, J.C. (eds) (2010) Modern Molecular Photochemistry of Organic Molecules, University Science Books, Sausalito, CA, p. 774.

17 Mathew, L. and Sankararaman, S. (1993) *J. Org. Chem.*, 58, 7576–7577.

18 Kumar, V.S., Aubele, D.L., and Floreancig, P.E. (2001) *Org. Lett.*, 3, 4123–4125.

19 Dockery, K.P., Dinnocenzo, J.P., Farid, S., Goodman, J.L., and Gould, I.R. (1997) *J. Am. Chem. Soc.*, 119, 1876.

20 Merkel, P.B., Luo, P., Dinnocenzo, J.P., and Farid, S. (2009) *J. Org. Chem.*, 74, 5163–5173.

21 Gesmundo, N.J. and Nicewicz, D.A. (2014) *Beilstein J. Org. Chem.*, 10, 1272–1281.

22 Lemieux, R.U., Hendriks, K.B., Stick, R.V., and James, K. (1975) *J. Am. Chem. Soc.*, 97, 4056–4062.

23 Tingoli, M., Tiecco, M., Testaferri, L., and Temperini, A. (1994) *J. Chem. Soc., Chem. Commun.*, 1883–1884.

24 Conner, E.S., Crocker, K.E., Fernando, R.G., Fronczek, F.R., Stanley, G.G., and Ragains, J.R. (2013) *Org. Lett.*, 15, 5558–5561.

25 Hashimoto, S., Kurimoto, I., Fujii, Y., and Noyori, R. (1985) *J. Am. Chem. Soc.*, 107, 1427–1429.

26 (a) Timpa, J.D., Legendre, M.G., Griffin, G.W., and Das, P.K. (1983) *Carbohydr. Res.*, 117, 69–80; (b) Timpa, J.D. and Griffin, G.W. (1984) *Carbohydr. Res.*, 131, 185–196.

27 Iwata, R., Uda, K., Takahashi, D., and Toshima, K. (2014) *Chem. Commun.*, 50, 10695–10698.

28 Kan, C., Long, C.M., Paul, M., Ring, C.M., Tully, S.E., and Rojas, C.M. (2001) *Org. Lett.*, 3, 381–384.

29 (a) Dupuis, J., Giese, B., Ruegge, D., Fischer, H., Korth, H.-G., and Sustmann, R. (1984) *Angew. Chem. Int. Ed. Engl.*, 23, 896–898; (b) Praly, J.-P. (2001) *Adv. Carbohydr. Chem. Biochem.*, 56, 65–151.

30 Giese, B. (1983) *Angew. Chem. Int. Ed. Engl.*, 22, 753–764.

31 (a) Andrews, R.S., Becker, J.J., and Gagné, M.R. (2010) *Angew. Chem. Int. Ed.*, 49, 7274–7276; (b) Andrews, R.S., Becker, J.J., and Gagné, M.R. (2011) *Org. Lett.*, 13, 2406–2409.

32 Garlets, Z.J., Nguyen, J.D., and Stephenson, C.R.J. (2014) *Isr. J. Chem.*, 54, 351–360.

Part IV

Regioselective Functionalization of Monosaccharides

11

Regioselective Glycosylation Methods

Mark S. Taylor

11.1 Introduction

Glycosylation conventionally involves the reaction of a glycosyl donor with a protected acceptor having a single free hydroxy (OH) group. The prospect of streamlining oligosaccharide synthesis by minimizing or eliminating the protection of acceptor OH groups has motivated the development of protocols for regioselective glycosylation. This chapter provides an overview of methods for selective formation of glycosidic linkages from acceptors having two or more free OH groups, with a focus on results that have not been discussed in the previously published reviews on this and related subjects [1].

The topic will be divided into two major parts. The first (Sections 11.2–11.4) deals with regioselective glycosylations that result directly from structural features of the glycosyl acceptor and/or donor. These include transformations that exploit steric or electronic differences between OH groups in the acceptor substrate, either intrinsic or induced by variation of nearby acceptor protective groups. In certain instances, the preferred site of glycosylation is dependent on the donor employed. Examples of donor-based regiocontrol of this type will also be discussed. The second part (Sections 11.5–11.7) will be dedicated to the use of external reagents or catalysts to influence the regiochemical outcomes of glycosylations. After an overview of methods based on stoichiometric promoters, activators, and transient protective groups (Section 11.5), a brief, and by no means comprehensive, discussion of chemoenzymatic methods for selective formation of glycosidic linkages will be provided in Section 11.6. This will be followed by a more detailed description of processes that employ synthetic catalysts. Applications in oligosaccharide synthesis will be provided throughout, to illustrate the gains in efficiency that can be achieved using this approach.

Selective Glycosylations: Synthetic Methods and Catalysts, First Edition. Edited by Clay S. Bennett.

11.2 Substrate Control: "Intrinsic" Differences in OH Group Reactivity of Glycosyl Acceptors

Compounds having multiple OH groups often display some level of regioselectivity in their reactions with glycosyl donors. Representative examples and applications in synthesis are described as follows.

Glycosylations of primary OH groups in the presence of one or more secondary OH groups have been reported. An efficient synthesis of an α(2,9)-linked sialic acid trimer was achieved by sequential regioselective glycosylations of acceptors having free exocyclic 1,2-diol groups (Scheme 11.1) [2]. Both types of thioglycoside donor employed in this sequence (the *S*-benzoxazolyl (SBox) glycoside, activated by AgOTf, and the ethyl thioglycoside, activated by *N*-iodosuccinimide (NIS)), displayed a high level of selectivity for the less hindered 9-OH group of the diol acceptor, although the former resulted in a higher level of α-stereoselectivity compared to the latter.

Glycosylation of the free primary 6-OH group of a pyranoside acceptor having free secondary OH groups is possible. The presence of a single ester protective group at O-3 was sufficient to enable selective glycosylation at O-6 of a β-glucopyranoside (Scheme 11.2). This was followed by a second regioselective glycosylation at O-2, providing access to a branched flavonol glycoside [3]. Regioselectivity of this type has also been achieved in an intermolecular context, using glycosyl donors having a free OH group [4]. Scheme 11.3 depicts the glycosylation of bis(isopropylidene)galactose with a phosphate donor having an unprotected 2-OH group [5].

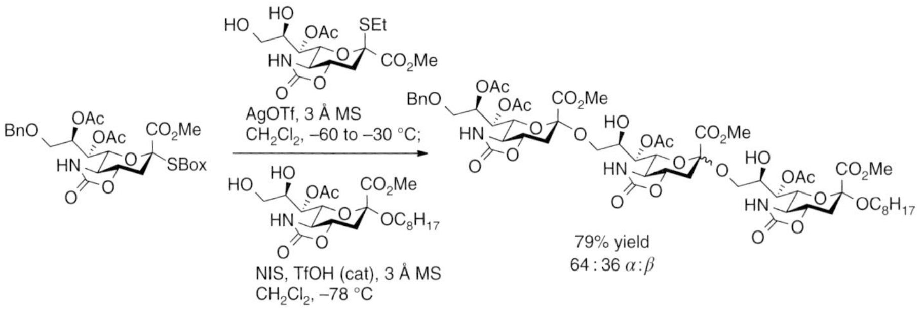

Scheme 11.1 Synthesis of a sialic acid trimer by selective glycosylation of primary OH groups.

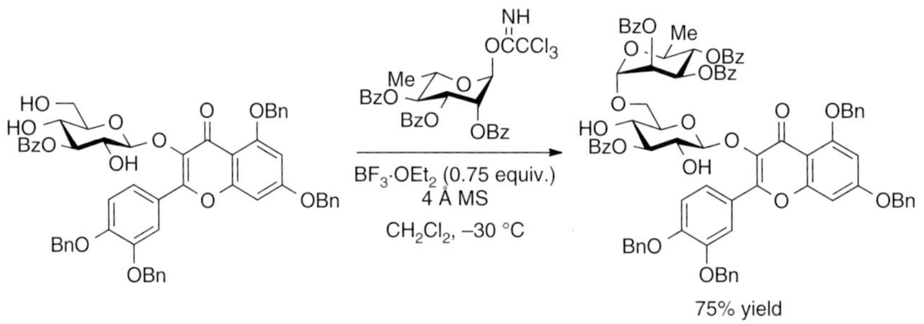

Scheme 11.2 Synthesis of a flavonol glycoside by selective 6-*O*-glycosylation.

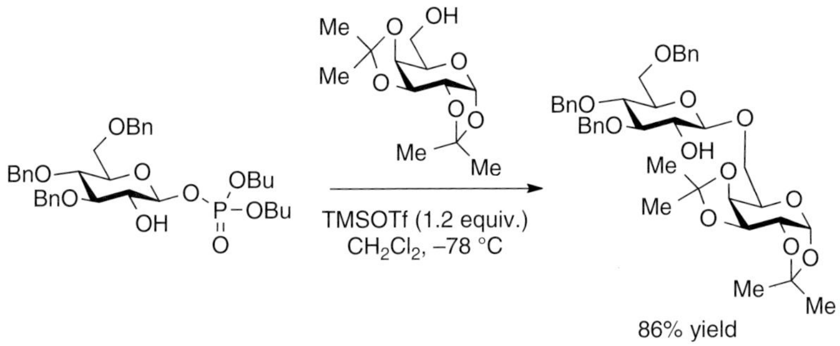

Scheme 11.3 6-*O*-Glycosylation using a donor having a free secondary OH group.

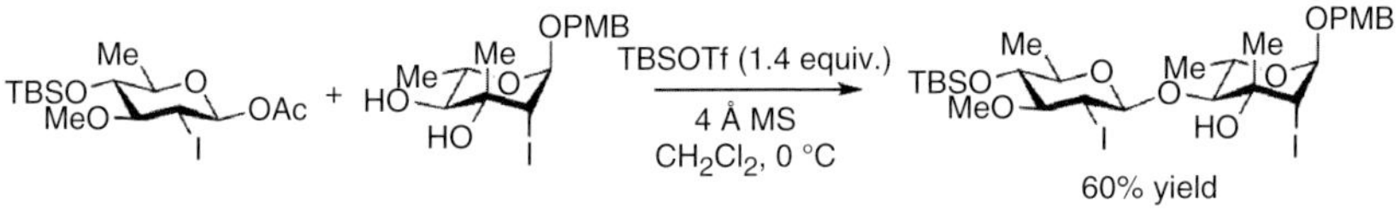

Scheme 11.4 Selective glycosylation of the secondary OH group of an olivomycose derivative.

Similarly, secondary OH groups may be glycosylated selectively in the presence of tertiary alcohols. This type of selectivity is at play in Roush and coworkers' synthesis of a precursor of the disaccharide moiety of apoptolidin A (Scheme 11.4) [6]. The olivomycose-derived acceptor underwent an O4-selective coupling, with the β-stereoselectivity resulting from the presence of an equatorially oriented iodo substituent on the glycosyl donor.

Differentiation of secondary OH groups through glycosylation is also possible. Equatorial OH groups often display higher glycosyl acceptor reactivity compared to their axial counterparts. Among equatorial OH groups, those flanked by equatorial substituents are usually less reactive than those having at least one axially oriented substituent at a neighboring position. Both of these types of selectivity correspond to the preferred reaction of the glycosyl donor with the least sterically hindered position of the acceptor. Sinaÿ and coworkers' synthesis of the heptasaccharide moiety of a glycolipid isolated from human erythrocyte stroma illustrates how these trends can be exploited to minimize protective group manipulations (Scheme 11.5) [7, 8]. The regioselective synthesis of tetrasaccharide **1** reflects the higher reactivity of the equatorial 3-OH group of the lactose-derived diol relative to both the adjacent axial 4-OH group and the sterically encumbered 3-OH group of the lactosamine-derived donor. Tetraol **2**, generated by ester hydrolysis and 4,6-*O*-benzylidene formation, was subjected to a second regioselective glycosylation. Again, it was the equatorial OH group having an adjacent axial OR substituent that was most reactive, leading to the β-1,3-linked heptasaccharide.

Selective glycosylations were employed in an efficient synthesis of the acetylated tetrasaccharide natural product cleistetroside-2 (Scheme 11.6) [9]. Both of the required α-1,3-linkages were constructed by taking advantage of the preferred reactivity of

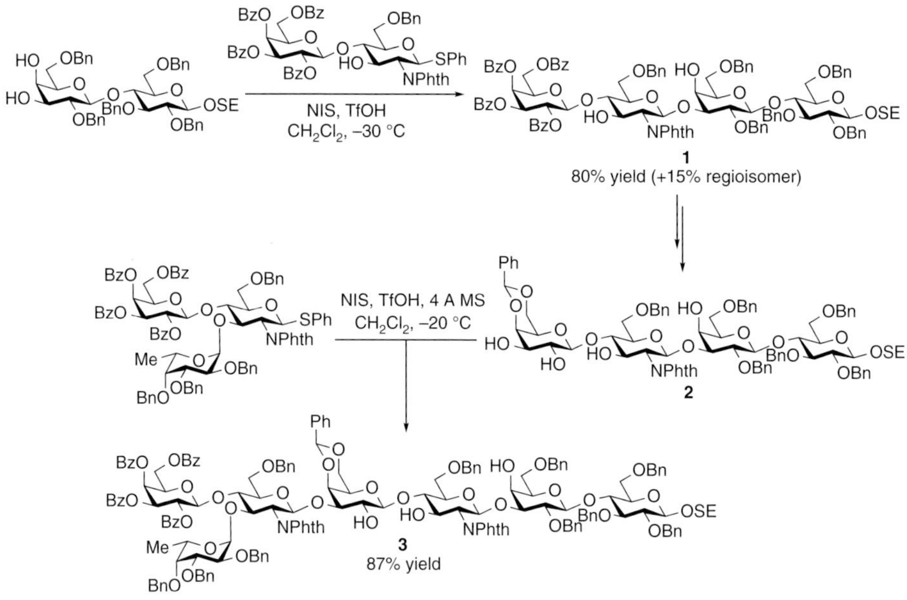

Scheme 11.5 Selective glycosylations of secondary OH groups in the synthesis of a heptasaccharide.

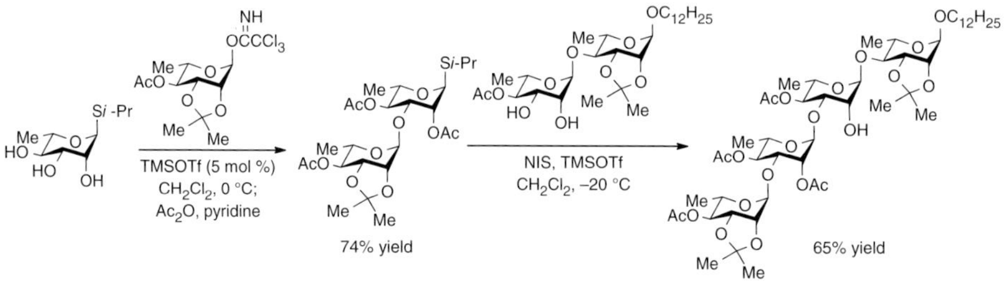

Scheme 11.6 Selective glycosylations of rhamnopyranosides in the synthesis of cleistetroside-2.

glycosyl donors toward the equatorial 3-OH group of rhamnopyranoside-derived diols and triols.

Access to branched oligosaccharides can be facilitated by the use of regioselective glycosylations. This approach was employed by the Kiso group in the synthesis of the tumor-associated α(2–3)/α(2–6) disialyl Lewis A heptasaccharide (Scheme 11.7) [10]. The *N*-acetylglucosamine-derived diol motif of tetrasaccharide **4** underwent glycosylation at O-3 using a disarmed trichloroacetimidate donor. While both of the free secondary OH groups of **4** occupy equatorial positions, the authors suggested that steric hindrance of the 4-OH group by the large sialyl residue at O-6 was responsible for this regiochemical outcome. The fucosyl moiety was then installed using an armed thioglycoside in the presence of NIS and TfOH.

Whereas the outcomes of the transformations discussed earlier were rationalized based on steric effects, there are numerous examples of regioselective glycosylations that appear to be under some level of electronic control. The inductive effects of

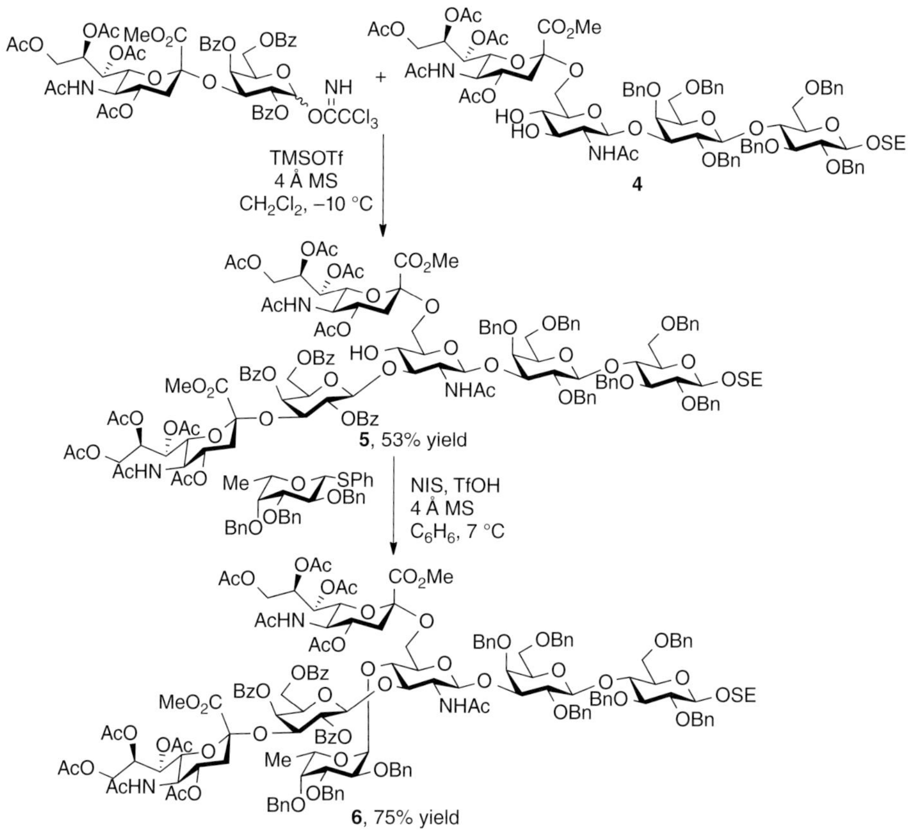

Scheme 11.7 Synthesis of a branched heptasaccharide by sequential glycosylations of a diol.

adjacent, electron-withdrawing substituents can reduce the glycosyl acceptor ability of a particular OH group. This type of trend can be exploited through variation of glycosyl acceptor protective groups, as will be discussed in the next section. Intramolecular hydrogen bonding interactions may enhance acceptor reactivity by stabilizing charge buildup in the glycosylation transition state: relatively high acceptor ability has been observed for OH groups that act as donors of intramolecular hydrogen bonds. Deactivation through intramolecular hydrogen bonding is also possible, with OH groups engaged as acceptors of intramolecular hydrogen bonds often showing lower reactivity toward glycosyl donors.

Wong and coworkers invoked intramolecular hydrogen bonding interactions to rationalize the outcomes of glycosylation reactions of the diastereomeric uronic-acid-based acceptors **7a** and **7b** (Scheme 11.8) [11]. Ethylidene-protected iduronic derivative **7a**, having two free axial OH groups, underwent O-4-selective coupling with a thioglycoside donor. Hydrogen bonding between the 4-OH and the ethylidene oxygen would enhance the nucleophilic reactivity of this group. In contrast, the corresponding glucuronic acid derivative **7b** displayed poor regioselectivity under the same conditions. The authors proposed that the axial orientation of the ester group of **7b** allowed for a geometrically favorable hydrogen bond with the 2-OH group. Because both of the free OH

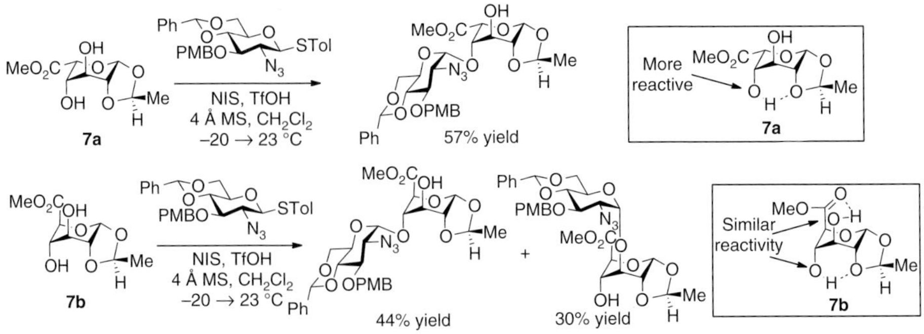

Scheme 11.8 Glycosylations of uronic acid-based diols.

groups of **7b** were engaged as donors of intramolecular hydrogen bonds, they displayed similar nucleophilicity, leading to a poorly selective glycosylation.

A distinct approach to glycoside synthesis has been developed by Vasella, wherein glycosylidene carbenes undergo O—H insertion reactions with glycosyl acceptors [12]. The insertion is likely a stepwise process, in which proton transfer from the hydroxyl group to the carbene generates an oxocarbenium ion and an alkoxide, which then combine to form the glycosidic linkage. The regiochemical outcomes of these glycosylations indicate that acceptor reactivity is influenced by the relative acidities of OH groups. For example, it was the 2-OH group of altropyranoside derivative **8** that underwent selective coupling with a glucose-derived diazirine (Scheme 11.9). Both the lower pK_a of the 2-OH group, due to the inductive electron-withdrawing effect of the adjacent anomeric position, and the lower kinetic acidity of the 3-OH group resulting from a relatively strong OH—OCH_3 hydrogen bond between cis-axial substituents, could contribute to this effect. The fact that the opposite regiochemical outcome was obtained from the coupling of **8** with a trichloroacetimidate donor – a reaction in which the two aforementioned effects would be expected to favor glycosylation of the more nucleophilic 3-OH group – highlights the mechanistic differences between the two approaches.

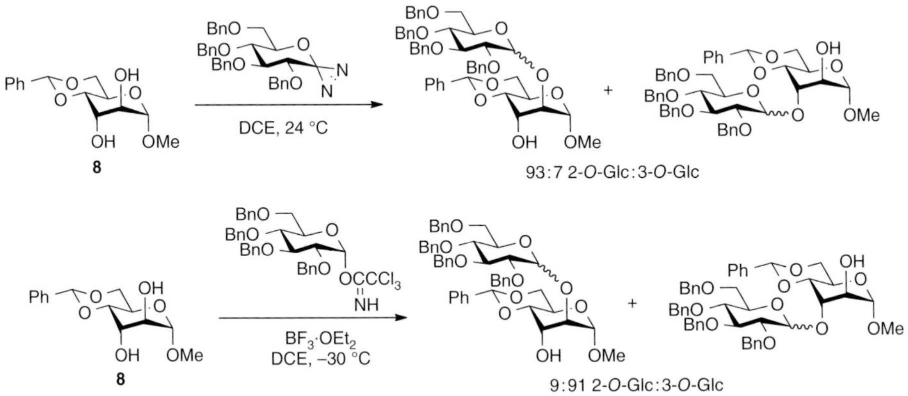

Scheme 11.9 Reactions of an altropyranoside-derived diol with a glycosylidene carbene (top) and a glycosyl trichloroacetimidate (bottom).

11.3 Substrate Control: Modulation of Acceptor OH Group Reactivity by Variation of Protective Groups

Regiocontrol in glycosylation reactions of partially protected acceptors can be influenced by the identity of nearby protective groups, due to steric or electronic effects or intramolecular hydrogen bonding interactions. For example, glycosylation reactions of methyl α-mannopyranosides having free OH groups at C2 and C3 displayed differences in regioselectivity depending on the protective groups used for the 4,6-diol group (Scheme 11.10) [13]. A significant amount of the β-(1→2)-linked regioisomer was obtained from the 4,6-di-*O*-acylated mannopyranoside substrate, whereas this isomer was not generated from the corresponding di-*O*-benzylated acceptor. The authors proposed that the inductive electron-withdrawing effect of the acetyl groups deactivated the 3-OH group to the point that glycosylation of the more sterically hindered 2-OH group became a competing pathway. Calculations of Fukui indices (a computational metric that has been correlated with site selectivity for nucleophilic or electrophilic functionalization) for the two protected acceptors were consistent with this hypothesis. The calculated Fukui indices for electrophilic attack of each OH group (f^-) are indicated in Scheme 11.10. In the case of the 4,6-di-*O*-benzylated derivative, the value of f^- was higher for the 3-OH group compared to the 2-OH group, suggesting that electronic effects (in addition to steric effects) would favor reaction at this position. Replacement of the benzyl protective groups by acetyl groups had a significant effect on the calculated Fukui indices, with the 2-OH having a slightly higher value compared to the 3-OH group. Thus, the lower regioselectivity obtained for this substrate may reflect the opposing effects of steric and electronic factors.

Glycosylation reactions of partially protected glucosamine derivatives are sensitive to the identity of the group used to protect the amine functionality. Intramolecular hydrogen bonding (e.g., with amide N—H groups serving as H-donors [14] or imide C=O groups as H-acceptors [15]) and steric effects have been invoked to explain these protective group effects on glycosyl acceptor reactivity and regioselectivity. An example of the latter type of effect is the use of a bulky *N*-phthaloyl [16] or *N*-tetrachlorophthaloyl (TCP) group [17] to deactivate the 3-OH group of glucosamine-derived diols, thus enabling the selective 4-*O*-galactosylation needed to prepare the core structure of

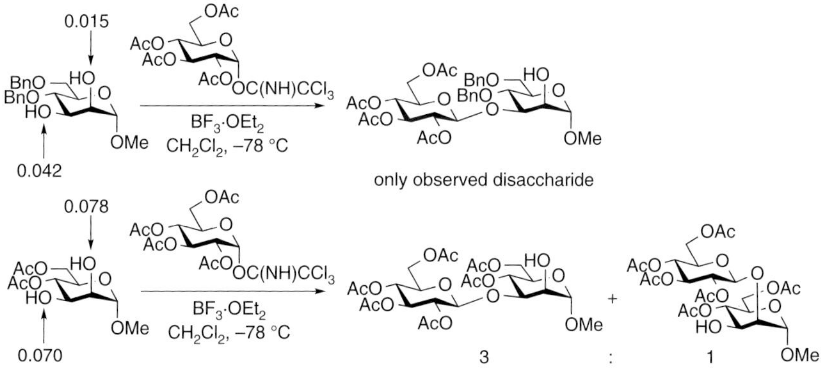

Scheme 11.10 Effects of ether versus ester protective groups on the selective glycosylation of a mannopyranoside-derived diol. Calculated Fukui indices for electrophilic attack are indicated for each structure.

oligosaccharides that are expressed on human cell surfaces (Scheme 11.11) [18]. Exchange of the TCP group for the less hindered acetyl group was conducted prior to installation of the 3-*O*-fucosyl moiety, thus accomplishing construction of the Lewis X trisaccharide core.

"Directing–protecting groups" (DPG) have been designed by Moitessier and coworkers to modulate the reactivity of nearby free OH groups through intramolecular hydrogen bonding interactions (Scheme 11.12) [19]. Glycosyl acceptor **9**, bearing a bis-(2-pyridyl)ethane-based DPG, underwent selective glycosylation at O-3 using a perbenzoylated trichloroacetimidate donor. The hydrogen bonding interactions that were proposed to give rise to this outcome are depicted in the structure of compound **9** (Scheme 11.12). The reactivity- and selectivity-enhancing effects of the DPG are illustrated by the result obtained using the TBDPS-protected control **10**, which

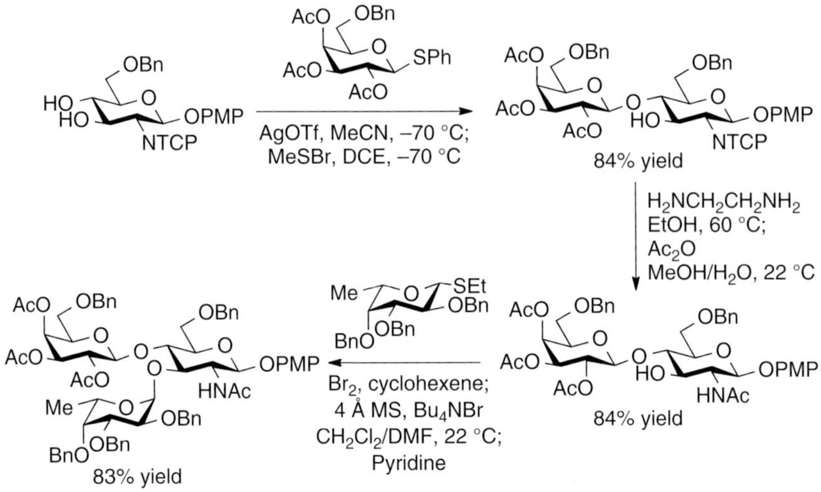

Scheme 11.11 Glycosylations of *N*-functionalized glucosamine derivatives.

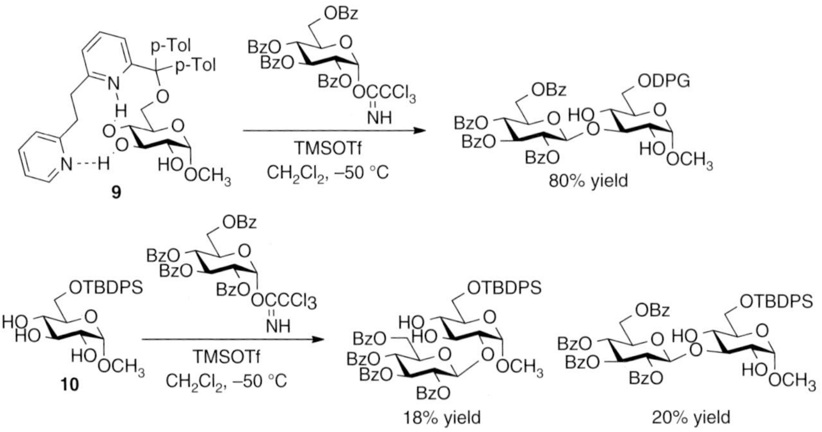

Scheme 11.12 Glycosylation using a directing–protecting group.

resulted in a roughly 1 : 1 ratio of 2-O-glycosylated and 3-O-glycosylated regioisomers, along with 62% recovery of unreacted glycosyl acceptor. The effect of the DPG was dependent on the reactivity of the glycosyl donor, with an armed [20], perbenzylated glycosyl donor giving rise to glycosylation at the most sterically accessible 2-OH group rather than the DPG-activated 3-OH group. Further examples of the effects of donor structure and reactivity on glycosylation regiocontrol are discussed in detail in the following section.

11.4 Substrate Control: Glycosyl Donor/Acceptor Matching in Regioselective Glycosylation

While the aforementioned discussion was focused on the effects of glycosyl acceptor structure on regioselectivity, the structure of the glycosyl donor also has a role to play in this regard. The relative reactivity of glycosyl acceptor OH groups may vary depending on the glycosylation mechanism (S_N2 versus S_N1, with or without anchimeric assistance), which in turn depends on structural features of the glycosyl donor such as the nature of the leaving group and the protective group pattern. Because glycosylation is a coupling of two chiral components, effects of the relative configuration of glycosyl donor and acceptor on regioselectivity can also be envisioned. In this section, representative examples of such matching/mismatching effects on glycosylation regioselectivity are discussed [21].

Ellervik and Magnusson investigated the effects of glycosyl donor protective groups on the regioselectivity of fucosylation reactions of a glucosamine-derived diol (Scheme 11.13) [18]. A mixture of isomers favoring the β-(1→3)-linked disaccharide (corresponding to reaction of the more sterically hindered OH group: see the discussion in Scheme 11.11) was obtained using the armed, perbenzylated donor. In contrast, glycosylation of the 4-OH group was observed using the peracetylated L-fucosyl donor. Although the authors did not pinpoint the specific mechanistic feature or features

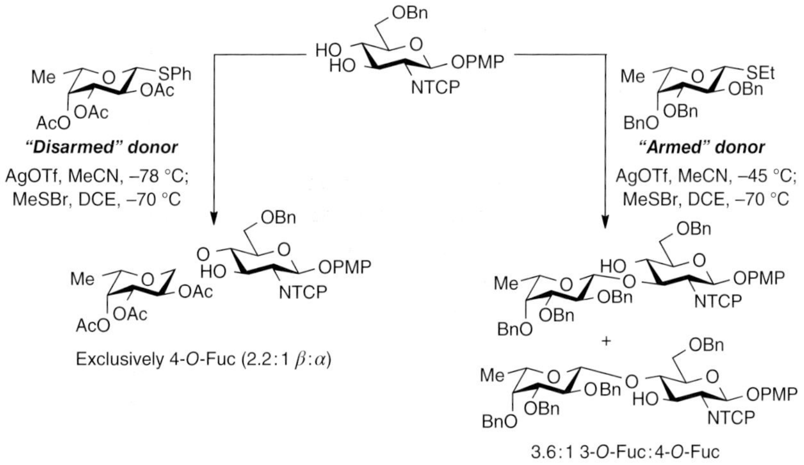

Scheme 11.13 Effects of glycosyl donor protective groups on regiochemical outcome.

responsible for this difference, the pattern of regiocontrol appears to be consistent with observations reported later by the Fraser-Reid group (see the following discussion).

Systematic investigations of the complementary regioselectivities of armed versus disarmed glycosyl donors have been undertaken by Fraser-Reid and coworkers, culminating in impressive synthetic applications [22, 23]. The initial report concerned the divergent reactivity of *myo*-inositol-derived diol **11** with mannose-derived perbenzylated pentenyl glycoside and pentenyl orthoester donors (Scheme 11.14) [23a]. The former displayed a moderate level of selectivity toward the axial 2-OH group, whereas the latter resulted exclusively with the equatorial 6-OH group. Reactions of other acceptors having two free OH groups revealed that the less sterically hindered group could be glycosylated selectively using the pentenyl orthoester, whereas varied levels of preference for coupling at the more hindered group were observed for the armed pentenyl glycoside or thioglycoside donors (Scheme 11.14).

The ability to generate a 1,3-dioxolan-2-ylium ion appears to be an important attribute for selective reaction with the less hindered OH group in these cases, as 2-*O*-benzoylated pentenyl glycoside donors also displayed this behavior, albeit with more modest levels of regiocontrol. The manno configuration of the pentenyl orthoester also plays a role, as the gluco- and galacto-configured derivatives displayed lower regioselectivity. The method is thus of particular utility for the construction of oligomannan components of pathogen-associated glycoproteins or glycolipids (e.g., glycolipids of Mycobacterium tuberculosis and the HIV-1 gp120 glycoprotein) [24].

Matching/mismatching effects on regioselectivity based on the relative configuration of donor and acceptor are also possible. The glycosylations of L- and D-*chiro*-inositol-derived diols L-**12** and D-**12** shown in Scheme 11.15 provide a clear-cut example [25]. The 2-azido-2-deoxyglucopyranosyl trichloroacetimidate donor was

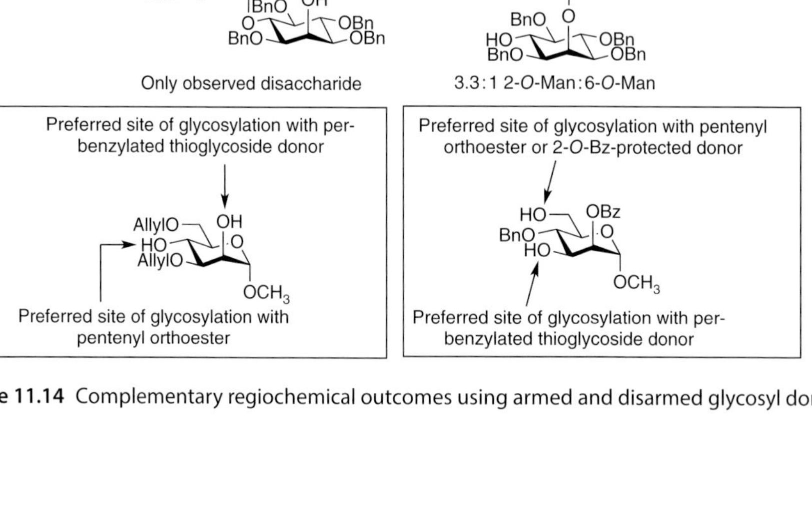

Scheme 11.14 Complementary regiochemical outcomes using armed and disarmed glycosyl donors.

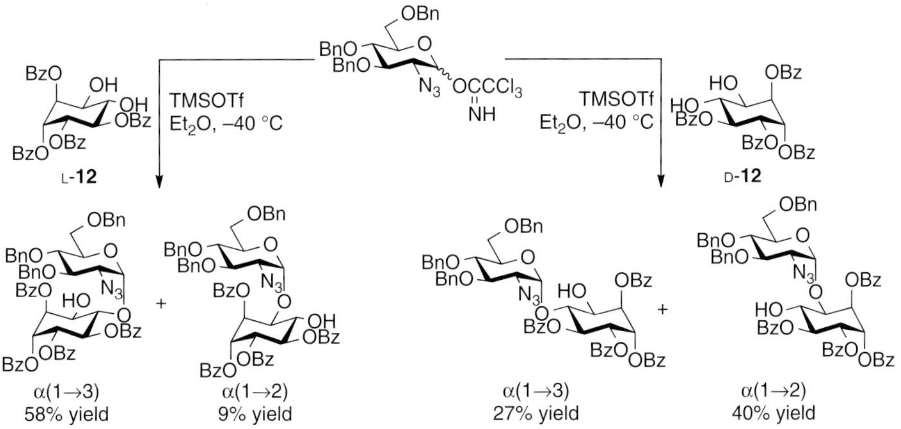

Scheme 11.15 Effects of donor/acceptor relative configuration on regiochemical outcome.

matched with the 3-OH group of the L-configured acceptor but with the 2-OH group of the D-enantiomer. Computational modeling was used to identify intra- and intermolecular hydrogen bonding interactions that were proposed to contribute to the matching/mismatching effect.

11.5 Reagent-Controlled, Regioselective Glycosylation

The possibility of influencing regioselectivity in glycosylation using an external reagent or promoter holds obvious appeal, in that such reagent-controlled methods could enhance or even override the types of bias described in the preceding sections. Approaches based on activation of the glycosyl acceptor (usually complexation-induced deprotonation using a Lewis acid) as well as the glycosyl donor (by variation of the promoter used to assist departure of the leaving group from the anomeric position) have been explored. Reagent-controlled glycosylation through transient protection of glycosyl acceptor OH groups constitutes another useful approach.

Activation of acceptor OH groups by complexation to organotin reagents has been used to achieve regioselective glycosylations. The foundations for this approach were set by the Moffatt group, who found that the 2′,3′-*O*-dibutylstannylene derivatives of ribonucleosides could be selectively functionalized by taking advantage of the enhanced reactivity of the Sn—O bonds toward acylating or alkylating agents [26]. Soon after, Ogawa and Matsui reported that tributylstannyl ethers derived from primary and secondary alcohols, including protected carbohydrate derivatives, were able to react with glycosyl halides under relatively mild conditions, generating either glycosides or orthoesters [27]. Regioselective glycosylation of a carbohydrate-derived stannylene acetal was achieved by Augé and Veyrières: coupling of perbenzylated α-galactopyranosyl chloride with a galactose-derived 3,4-*O*-stannylene acetal occurred selectively at O-3, yielding the α-(1→3)-linked disaccharide (Scheme 11.16, Eq. (1)) [28]. This reactivity has been extended to glycosyl acceptors having more than two free OH groups [29]. Free pyranosides generally undergo glycosylation at the 6-OH group (Scheme 11.16, Eq. (2)), while equatorial OH groups of *cis*-1,2-diols can be activated in substrates lacking a free

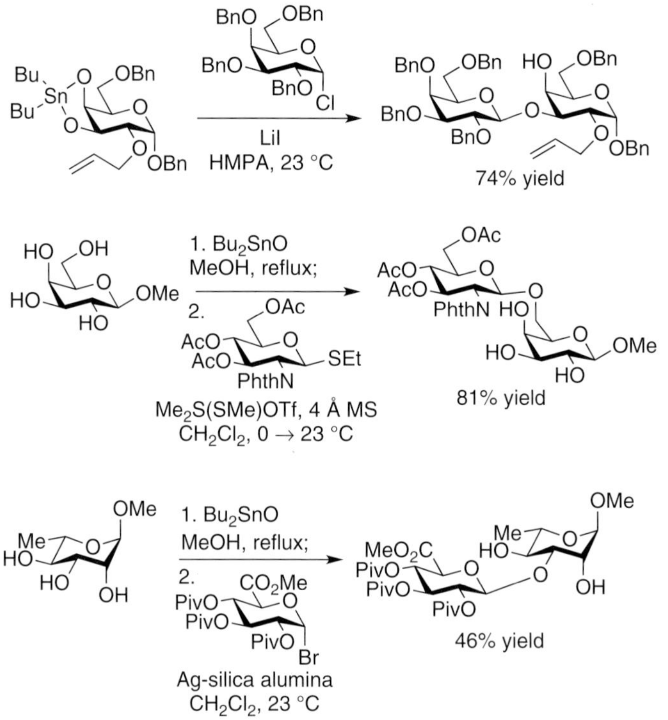

Scheme 11.16 Stannylene acetal-based glycosylations.

primary OH group (Scheme 11.16, Eq. (3)) or by addition of fluoride salts. O'Doherty and coworkers employed the Pd-catalyzed regioselective glycosylation of a stannylene acetal as a key step in the syntheses of cleistrioside and cleistetroside oligorhamnopyranoside natural products [30].

Paulsen and coworkers found that the regiochemical outcome of the Koenigs–Knorr-type glycosylation of a lactose-derived diol could be controlled by varying the Ag(I)-based activator of the glycosyl halide (Scheme 11.17) [31]. Under homogeneous conditions, using AgOTf as the activator in nitromethane solvent, the β-(1 → 3)-linked trisaccharide was obtained. However, the use of a heterogeneous promoter (silver silicate in dichloromethane) led to the selective formation of the β-(1 → 4) linkage. The former conditions are likely to favor an S_N1-type mechanism, whereas an S_N2-type mechanism has been implicated for the latter.

The Kochetkov group has exploited the glycosyl acceptor ability of triphenylmethyl (trityl) ethers to enable selective couplings at secondary positions over primary positions [32]. For example, bis-tritylated β-D-glucopyranoside **13** reacted selectively at O-4 versus O-6 in the presence of a thioglycoside donor (Scheme 11.18). The ability to glycosylate a secondary trityl ether in the presence of a primary one was found to be general for a variety of pyranoside derivatives and for several classes of glycosyl donors (cyanoethylidene acetals, glycosyl bromides, and thioglycosides). It has been suggested that these "contrasteric" outcomes are a reflection of a relatively early transition state for reactions of oxocarbenium ions with trityl ethers [33]. Both the steric strain present

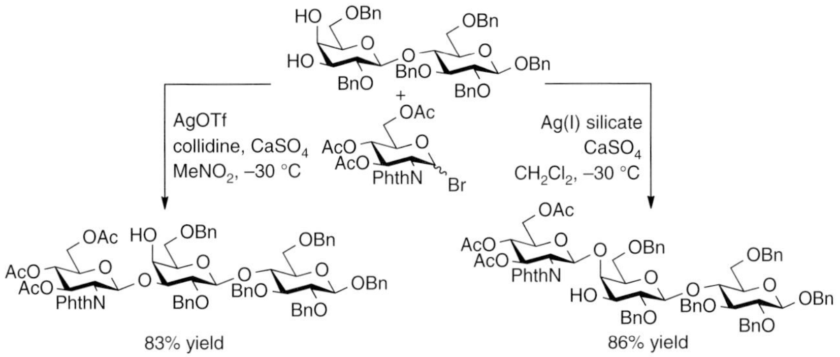

Scheme 11.17 Effects of Ag(I)-based activator on the regioselectivity of glycosyl bromide couplings.

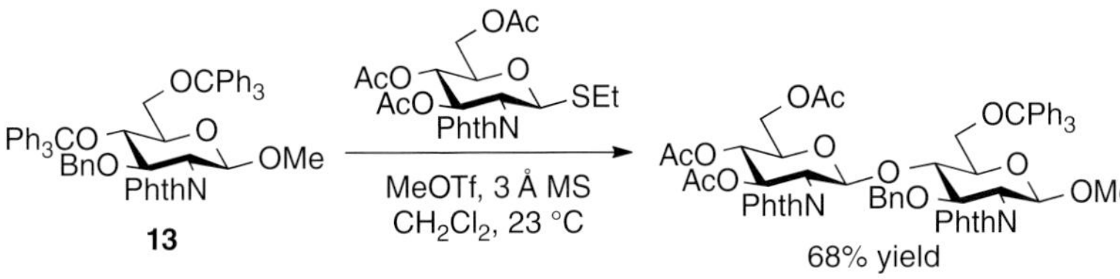

Scheme 11.18 Selective glycosylation via triphenylmethyl ethers.

in the tritylated alcohol starting material, and the stability of the triphenylmethyl carbocation by-product, would be expected to increase the thermodynamic driving force for the glycosylation step relative to that with a free alcohol acceptor. The more exergonic coupling would have an earlier transition state (lower degree of glycosidic bond formation) and thus be more sensitive to electronic effects compared to steric effects.

The transient protection of carbohydrate-derived diols as boronic esters has been used to achieve regioselective glycosylation. Treatment of methyl β-galactopyranoside with 4-methoxyphenylboronic acid, followed by addition of a thioglycoside donor in the presence of NIS and TMSOTf, generated the β-(1→3)-linked disaccharide after oxidative removal of the boronate group with sodium perborate (Scheme 11.19) [34]. Formation of a 4,6-*O*-boronate and glycosylation of the least sterically hindered free OH group is likely responsible for the observed regioselectivity. The scope of this approach was further explored by Madsen and coworkers, who used phenylboronic acid to transiently protect thioglycosides derived from glucose,

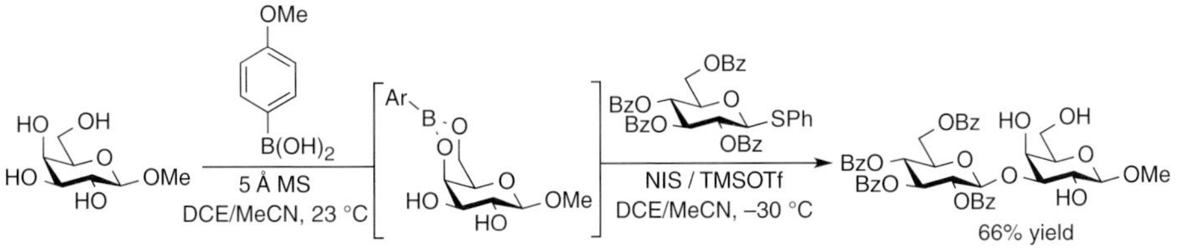

Scheme 11.19 Mono-glycosylation via transient formation of an acceptor-derived boronic ester.

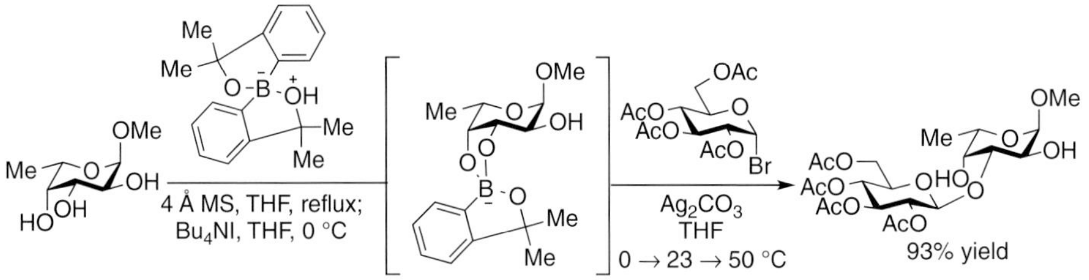

Scheme 11.20 Glycosyl acceptor activation by formation of a tetracoordinate organoboron adduct.

galactose, and glucosamine (4,6-*O*-boronate formation), as well as fucose (3,4-*O*-boronate) and rhamnose (2,3-*O*-boronate) [35].

Whereas formation of boronic esters from carbohydrates is a method for protection of diols, activation of diol groups by organoboron complexation is also possible. The key to achieving glycosyl acceptor activation is to generate a tetracoordinate rather than a tricoordinate boron complex. This was accomplished by the Aoyama group through the use of a boronic ester having a pendant, coordinating OH group, which served to activate *cis*-1,2-diol and 1,3-diol motifs toward coupling with glycosyl halides (Scheme 11.20) [36]. The *gem*-dimethyl substitution at the benzylic position of the arylboronate was installed to prevent undesired glycosylation of the coordinating OH group.

11.6 Enzyme-Catalyzed Regioselective Glycosylation

The regioselective biosynthesis of oligosaccharides is generally accomplished through catalyst control: glycosyltransferase enzymes show high levels of site selectivity in reactions of polyhydroxylated acceptors. These enzymes, along with other classes of naturally occurring or engineered enzymes that promote glycosidic bond formation, are powerful synthetic tools in their own right and have served to inspire the development of "small-molecule" catalysts that can perform analogous functions. The chemistry and biochemistry of glycosyltransferases and glycosidases have been the subject of detailed review articles [37]. The following discussion is intended to provide examples of the types of regioselective transformations that can be achieved using enzyme catalysis and will be followed by a more detailed discussion of synthetic catalysts for selective glycosylation.

Glycosyltransferases accelerate the displacement of a phosphate-derived leaving group from an activated glycosyl donor by a nucleophilic acceptor. In the case of the Leloir-type glycosyltransferases common in mammalian cells, the donors are sugar nucleosides. The family is further subdivided into inverting and retaining glycosyltransferases, depending on the stereochemical outcome of the glycosylation reaction. Both the stereospecificity of glycosyltransferase-mediated couplings and the ability of these enzymes to deliver glycosyl groups to a particular OH group of a specific mono- or disaccharide moiety are key advantages for the synthesis of complex oligosaccharides. To illustrate these features, a portion of Boons and coworkers' synthesis of a complex, asymmetrically substituted branched glycan is depicted in Scheme 11.21 [38]. Branched decasaccharide substrate **14** was prepared using a conventional "chemical synthesis"

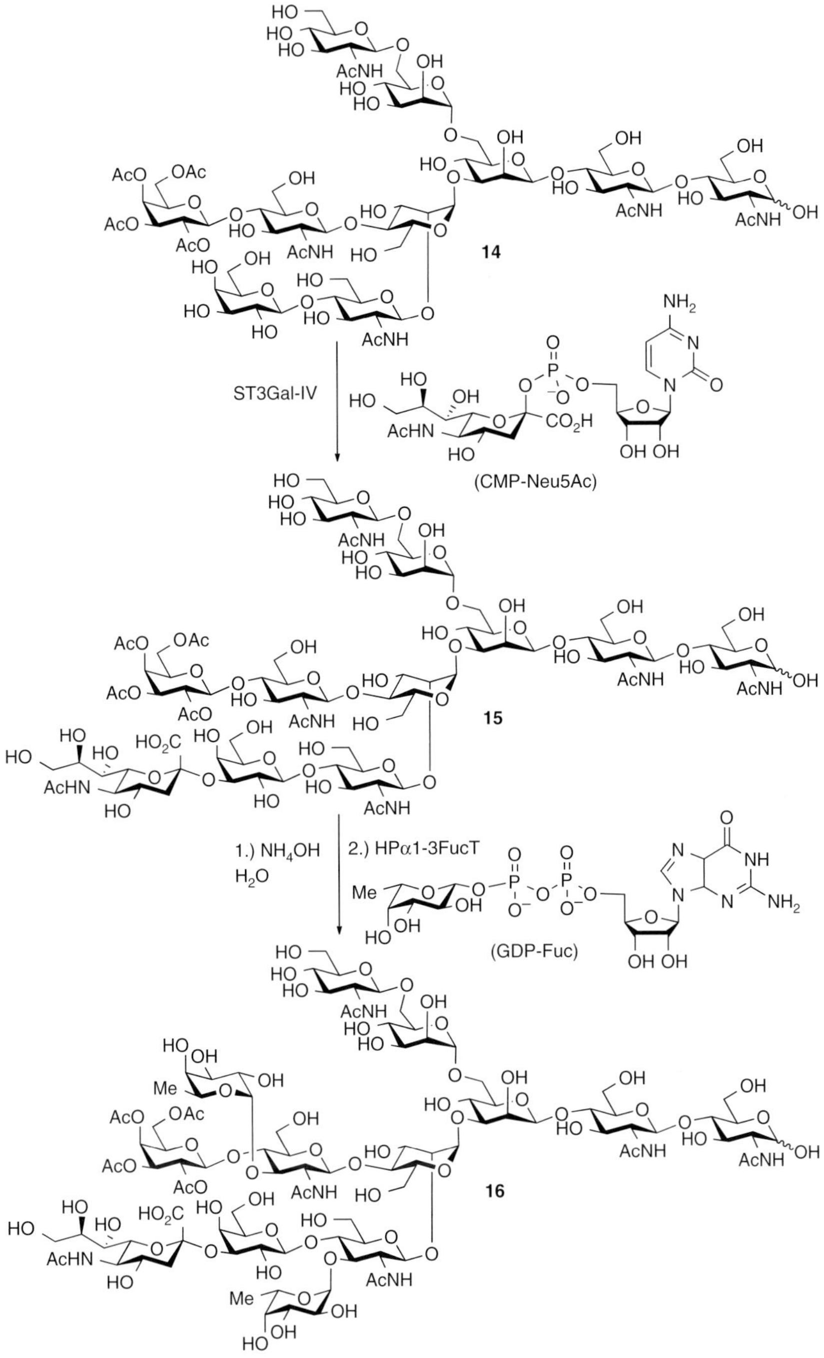

Scheme 11.21 Chemoenzymatic synthesis of an unsymmetrically substituted, branched glycan.

approach. A key structural feature of **14** is the presence of two *N*-acetyllactosamine (LacNAc) moieties, of which one is free and the other is partially protected by tetra-*O*-acetylation of the galactopyranoside portion. Only the free LacNAc group was recognized by the α2,3-sialyltransferase (ST3Gal-IV), thus enabling selective mono-sialylation in the presence of cytidine-5′-monophospho-*N*-acetylneuraminic acid (CMP-Neu5Ac). A phosphatase was included in this reaction mixture to minimize inhibition by the released uridine monophosphate. After saponification of the ester protective groups of **15**, the resulting free LacNAc group and the 3-sialylLacNAc were 2-*O*-fucosylated using guanidine-5′-diphosphoryl-L-fucose (GDP-Fuc) and an α1,3-fucosyltransferase (α3FucT). By taking advantage of the substrate specificities of the glycosyltransferases, the authors were able to further elongate the branches in an orthogonal fashion.

Glycosyl hydrolases, enzymes that catalyze the scission of glycosidic bonds to the corresponding free hemiacetals, have also been employed productively in the synthesis. The reaction stoichiometry and solvent, as well as the identity of the donor and acceptor, can be varied to favor glycoside formation over hydrolysis. Alternatively, the function of the enzymes can be modified rationally through mutation. Withers has pioneered the development of glycosynthases, mutant glycosidases lacking the nucleophilic amino acid residue that is responsible for hydrolysis activity [37b]. Glycosynthases efficiently promote glycosidic bond formation using readily available glycosyl fluorides as donors and are compatible with complex, highly functionalized substrates. For example, the synthesis of glycosphingolipids has been achieved using EGC II E351S, a glycosynthase based on the endoglycoceramidase II enzyme from *Rhodococcus* strain M-777 (Scheme 11.22) [39].

11.7 Synthetic Catalysts for Regioselective Glycosylation

Diarylboronic acids (Ar_2BOH) catalyze the regioselective couplings of pyranoside derivatives with glycosyl bromide or chloride donors, using Ag_2O as halide abstracting reagent and base [40]. This catalyst system can be employed to generate 1,2-*trans* glycosidic linkages at equatorial OH groups belonging to *cis*-vicinal diol motifs (e.g., Scheme 11.23, Eq. (1)). The mode of activation likely hinges on the formation of tetracoordinate borinate adducts similarly to the proposed intermediates in the stoichiometric, organoboron-mediated glycosylations by Aoyama as discussed in Section 11.5 [41]. Commercially available complex **17**, which displays improved stability toward oxidation in comparison to free Ph_2BOH, acts as a precatalyst by loss of the ethanolamine ligand under the glycosylation conditions. The regiochemical outcome apparently stems from the high nucleophilicity of tetracoordinate boron alkoxides relative to free OH groups, along with more subtle steric and/or electronic differences that favor the

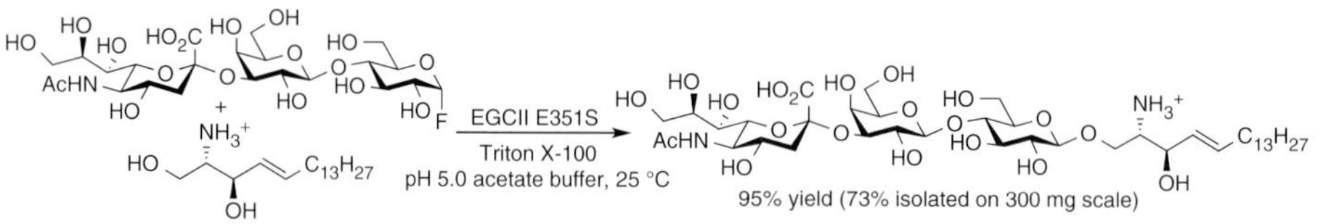

Scheme 11.22 Glycosynthase-catalyzed synthesis of a glycosphingolipid.

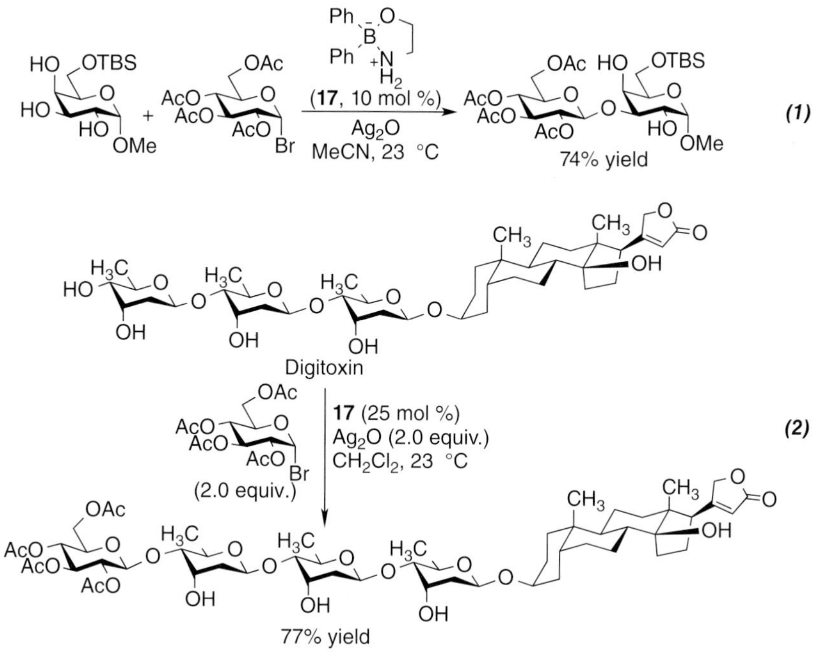

Scheme 11.23 Selective glycosylations catalyzed by borinic acid derivative **17**.

reaction of the equatorial over the axial position. This pattern of regiocontrol enabled mono-glycosylation of the steroidal glycoside digitoxin having four secondary and one tertiary OH groups, thus providing access to new analogs of this cardioprotective and cytotoxic natural product (Scheme 11.23, Eq. (2)) [42].

The 1,2-*trans* stereochemical outcome of glycosylations catalyzed by **17** was proposed to result from an S_N2-type inversion of configuration of the axially oriented halide leaving group (Scheme 11.24, Eq. (1)). This proposal is in line with the previous mechanistic studies of Koenigs–Knorr-type glycosylations using insoluble Ag(I)-based promoters and is supported by several experimental observations, including the following: (i) the formation of 1,2-*trans* glycosides from both peracylated and perbenzylated glycosyl halide donors; (ii) the dependence of the reaction outcome on the configuration of the starting peracetylated glycosyl halide (formation of an orthoester from the β-glycosyl halide versus the disaccharide from the α-glycosyl halide); and (iii) S_N2-type reaction kinetics (first-order dependence on glycosyl donor, acceptor, and catalyst concentrations). Consistent with the proposed S_N2-type mechanism, catalyst **17** was able to facilitate regio- and stereoselective couplings of α-2-deoxyglycosyl chlorides to generate β-linked products (Scheme 11.24, Eq. (2)) [43]. Control experiments revealed that both yield and β-stereoselectivity were improved in the presence of catalyst **17**, thus enabling the formation of this challenging class of glycosidic linkages.

The O'Doherty group has developed a dual organoboron/palladium catalyst system for regioselective couplings of glycosyl acceptors with pyranone-derived allylic carbonates (Scheme 11.25) [44]. C—O bond formation was proposed to occur by coupling of the acceptor-derived borinic ester with a π-allyl-Pd electrophile. This *de novo* regioselective

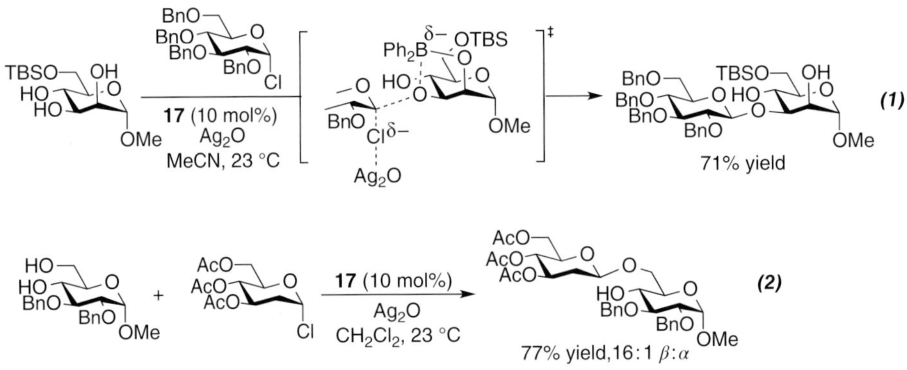

Scheme 11.24 Proposed mechanism of glycosylation catalyzed by **17**; regio- and stereochemical coupling of a 2-deoxyglycosyl chloride.

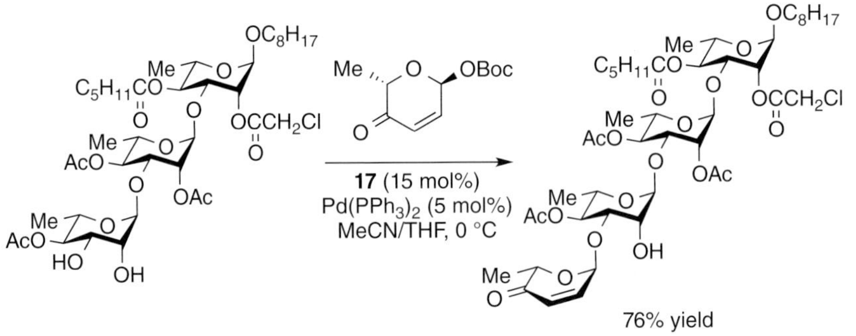

Scheme 11.25 Dual organoboron / palladium-catalyzed glycosylation in the synthesis of a tetrarhamnopyranoside.

glycosylation facilitated the synthesis of several members of the mezzettiaside family of oligorhamnopyranoside natural products.

Diorganotin-catalyzed regioselective glycosylations have also been developed (Scheme 11.26) [45]. One of the keys to achieving this result was the use of a diaryltin compound (Ph_2SnCl_2), rather than the dialkyltin species that had been employed in the stoichiometric variants (see Section 11.5). The optimal reaction conditions involved the use of glycosyl bromide or chloride donors, along with Ag_2O and 5,5′-dimethyl-2,2′-bipyridyl (DMBPY). Selective deprotonation of the more accessible tin-bound OH group of a five-membered chelate by DMBPY was proposed to account for the observed regioselectivity. An S_N2-type glycosylation mechanism was suggested, based on the formation of a β-linkage from a perbenzylated α-glycosyl chloride donor. A noteworthy feature of this catalyst system is the ability to carry out selective 3-*O*-glycosylations in the presence of a free 6-OH group.

11.8 Summary and Outlook

The first instinct of a chemist undertaking the synthesis of an oligosaccharide target might be to devise a protective group strategy ensuring that each glycosylation step employs an acceptor having only a single free OH group. The preceding discussion has

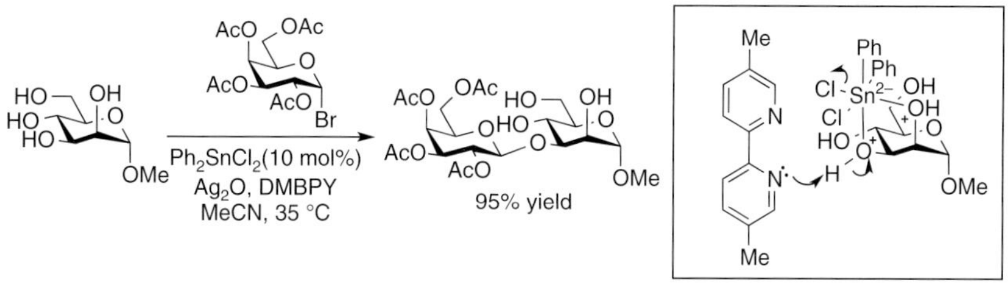

Scheme 11.26 Organotin-catalyzed regioselective glycosylation.

emphasized that an alternative approach based on regioselective glycosylation is worthy of consideration and that numerous options exist for practitioners interested in exploring this possibility.

For some targets, differences in the reactivity of acceptor OH groups can be relied upon in synthetic planning so as to minimize protective group installation and removal steps. The ability to fine-tune or even alter the relative reactivity of OH groups by variation of protective groups – either at nearby positions on the glycosyl acceptor or on the glycosyl donor – further broadens the scope of this approach. These types of substrate-controlled glycosylations present interesting and largely unexplored opportunities for detailed studies of reaction mechanism. Some general principles have been identified (e.g., the effects of intra- or intermolecular hydrogen bonding interactions on regioselectivity and differences in regiochemical outcomes of 2-*O*-acylated versus 2-*O*-alkylated glycosyl donors), but definitive experimental and computational support for the mechanistic bases of these observations is needed. In comparison to the level of mechanistic insight that has been attained for stereoselective glycosylations [46], the current understanding of regioselective variants is less advanced. Knowledge gained from mechanistic studies will facilitate refinements in methodology and novel applications.

Reagent- and catalyst-controlled glycosylations present opportunities to achieve regiochemical outcomes that would be difficult or impossible under substrate control. Although a gradual shift in emphasis toward catalytic methods appears to be underway, further development of efficient, reagent-controlled protocols is certainly warranted, especially since the high value of the products can often justify the cost of using a stoichiometric quantity of promoter. Refinements in chemoenzymatic and biocatalytic glycosylations, along with increased access to the requisite enzymes and glycosyl donors, have fueled impressive applications in recent years. It is now clear that enzyme-based methods can enable practical access to complex glycans for use as research tools or therapeutic agents. The emergence of synthetic catalysts for regioselective glycosylation has been another exciting development. Although these do not, at present, rival enzymes in terms of their functional group tolerance and high levels of regio- and stereoselectivity, they may offer complementary features in terms of the types of glycosyl donors and reaction conditions used, as well as the ability to systematically alter the catalyst structures. It can be anticipated that the discovery of new catalyst architectures and substrate activation modes will be instrumental in expanding the scope and applications of this approach.

References

1 (a) Ferrier, R.J. and Furneaux, R.H. (2009) *Aust. J. Chem.*, 62, 585–589; (b) Böttcher, S. and Thiem, J. (2014) *Curr. Org. Chem.*, 18, 1804–1817.
2 Tanaka, H., Tateno, Y., Nishiura, Y., and Takahashi, T. (2008) *Org. Lett.*, 10, 5597–5600.
3 Zhang, Y., Wang, K., Zhan, Z., Yang, Y., and Zhao, Y. (2011) *Tetrahedron Lett.*, 52, 3154–3157.
4 (a) Bilodeau, M.T., Park, T.K., Hu, S., Randolph, J.T., Danishefsky, S.J., Livingston, P.O., and Zhang, S. (1995) *J. Am. Chem. Soc.*, 117, 7840–7841; (b) Toepler, A. and Schmidt, R.R. (1992) *Tetrahedron Lett.*, 33, 5161–5164; (c) Boons, G.-J. and Zhu, T. (1997) *Synlett*, 1997, 809–811.
5 Plante, O.J., Palmacci, E.R., Andrade, R.B., and Seeberger, P.H. (2001) *J. Am. Chem. Soc.*, 123, 9545–9554.
6 Handa, M., Smith, W.J. III, and Roush, W.R. (2008) *J. Org. Chem.*, 73, 1036–1039.
7 Zhang, Y., Dausse, B., Sinaÿ, P., Afsahi, M., Berthault, P., and Desvaux, H. (2000) *Carbohydr. Res.*, 324, 231–241.
8 For another example, see: Deshpande, P.P. and Danishefsky, S.J. (1997) *Nature*, 387, 164–166.
9 Cheng, L., Chen, Q., and Du, Y. (2007) *Carbohydr. Res.*, 342, 1496–1501.
10 Ando, T., Ishida, H., and Kiso, M. (2001) *J. Carbohydr. Chem.*, 20, 425–430.
11 Yu, H.N., Furukawa, J.-I., Ikeda, T., and Wong, C.-H. (2004) *Org. Lett.*, 6, 723–726.
12 Vasella, A. (1993) *Pure Appl. Chem.*, 65, 731–752.
13 Kalikanda, J. and Li, Z. (2010) *Tetrahedron Lett.*, 51, 1550–1553.
14 Crich, D. and Dudkin, V. (2001) *J. Am. Chem. Soc.*, 123, 6819–6825.
15 Di Benedetto, R., Zanetti, L., Varese, M., Rajabi, M., Di Brisco, R., and Panza, L. (2014) *Org. Lett.*, 16, 952–955.
16 Numomura, S., Iida, M., Numata, M., Sugimoto, M., and Ogawa, T. (1994) *Carbohydr. Res.*, 263, C1–C6.
17 Lay, L., Manzoni, L., and Schmidt, R.R. (1998) *Carbohydr. Res.*, 310, 157–171.
18 Ellervik, U. and Magnusson, G. (1998) *J. Org. Chem.*, 63, 9314–9322.
19 Lawandi, J., Rocheleau, S., and Moitessier, N. (2011) *Tetrahedron*, 67, 8411–8420.
20 Mootoo, D.R., Konradsson, P., Udodong, U., and Fraser-Reid, B. (1988) *J. Am. Chem. Soc.*, 110, 5583–5584.
21 For a review: Bohé, L. and Crich, D. (2010) *Trends Glycosci. Glycotechnol.*, 22, 1–15.
22 Fraser-Reid, B., Lu, J., Jayaprakash, K.N., and López, J.C. (2006) *Tetrahedron: Asymmetry*, 17, 2449–2463.
23 (a) Anilkumar, G., Nair, L.G., and Fraser-Reid, B. (2000) *Org. Lett.*, 2, 2587–2589; (b) Uriel, C., Agocs, A., Gómez, A.M., López, J.C., and Fraser-Reid, B. (2005) *Org. Lett.*, 7, 4899–4902; (c) Jayaprakash, K.N. and Fraser-Reid, B. (2007) *Carbohydr. Res.*, 342, 490–498.
24 Jayaprakash, K.N., Lu, J., and Fraser-Reid, B. (2005) *Angew. Chem. Int. Ed.*, 44, 5894–5898; *Angew. Chem.*, 117, 6044–6048.
25 (a) Cid, M.B., Alonso, I., Alfonso, F., Bonilla, J.B., López-Prados, J., and Martín-Lomas, M. (2006) *Eur. J. Org. Chem.*, 2006, 3947–3959; (b) Uriel, C., Gómez, A.M., López, J.C., and Fraser-Reid, B. (2012) *Org. Biomol. Chem.*, 10, 8361–8370.
26 Wagner, D., Verheyden, J.P.H., and Moffatt, J. (1974) *J. Org. Chem.*, 39, 24–30.

27 Ogawa, T. and Matsui, M. (1976) *Carbohydr. Res.*, 51, C13–C18.

28 Augé, C. and Veyrières, A. (1979) *J. Chem. Soc., Perkin Trans. 1*, 1825–1832.

29 (a) Cruzado, C., Bernabe, M., and Martin-Lomas, M. (1990) *Carbohydr. Res.*, 203, 296–301; (b) Garegg, P.J., Maloisel, J.-L., and Oscarson, S. (1995) *Synthesis*, 1995, 409–414; (c) Kaji, E. and Harita, N. (2000) *Tetrahedron Lett.*, 41, 53–56; (d) Kaji, E., Shibayama, K., and In, K. (2003) *Tetrahedron Lett.*, 44, 4881–4885; (e) Maggi, A. and Madsen, R. (2013) *Eur. J. Org. Chem.*, 2013, 2683–2691.

30 Wu, B., Li, M., and O'Doherty, G.A. (2010) *Org. Lett.*, 12, 5466–5469.

31 Paulsen, H., Hadamczyk, D., Kutschker, W., and Bünsch, A. (1985) *Liebigs Ann. Chem.*, 1985, 129–141.

32 Tsvetkov, Y.E., Kitov, P.I., Backinowsky, L.V., and Kochetkov, N.K. (1996) *J. Carbohydr. Chem.*, 15, 1027–1050.

33 Kotov, P.I., Tsvetkov, Y.E., Backinowsky, L.V., and Kochetkov, K.N. (1993) *Russ. Chem. Bull.*, 42, 1909–1915.

34 Kaji, E., Nishino, T., Ishige, K., Ohya, Y., and Shirai, Y. (2010) *Tetrahedron Lett.*, 51, 1570–1573.

35 Fenger, T.H. and Madsen, R. (2013) *Eur. J. Org. Chem.*, 2013, 5923–5933.

36 Oshima, K. and Aoyama, Y. (1999) *J. Am. Chem. Soc.*, 121, 2315–2316.

37 (a) Breton, C., Snajdrová, L., Jeanneau, C., Koca, J., and Imberty, A. (2006) *Glycobiology*, 16, 29R–37R; (b) Hancock, S.M., Vaughan, M.D., and Withers, S.G. (2006) *Curr. Opin. Chem. Biol.*, 10, 509–519; (c) Schmaltz, R.M., Hanson, S.R., and Wong, C.-H. (2011) *Chem. Rev.*, 111, 4259–4307.

38 Wang, Z., Chinoy, Z.S., Ambre, S.G., Peng, W., McBride, R., de Vries, R.P., Glushka, J., Paulson, J.C., and Boons, G.-J. (2013) *Science*, 341, 379–383.

39 Vaughan, M.D., Johnson, K., DeFrees, S., Tang, X., Warren, A.J., and Withers, S.J. (2006) *J. Am. Chem. Soc.*, 128, 6300–6301.

40 (a) Gouliaras, C., Lee, D., Chan, L., and Taylor, M.S. (2011) *J. Am. Chem. Soc.*, 133, 13926–13929; (b) Taylor, M.S. (2015) *Acc. Chem. Res.*, 48, 295–305.

41 Lee, D., Williamson, C.L., and Taylor, M.S. (2012) *J. Am. Chem. Soc.*, 134, 8260–8267.

42 Beale, T.M. and Taylor, M.S. (2013) *Org. Lett.*, 15, 1358–1361.

43 Beale, T.M., Moon, P.J., and Taylor, M.S. (2014) *Org. Lett.*, 16, 3604–3607.

44 Bajaj, S.O., Sharif, E.U., Akhmedov, N.G., and O'Doherty, G.A. (2014) *Chem. Sci.*, 5, 2230–2234.

45 Muramatsu, W. and Yoshimatsu, H. (2013) *Adv. Synth. Catal.*, 355, 2518–2524.

46 (a) Krumper, J.R., Salamant, W.A., and Woerpel, K.A. (2008) *Org. Lett.*, 10, 4907–4910; (b) Crich, D. (2010) *Acc. Chem. Res.*, 43, 1144–1153; (c) Huang, M., Garrett, G.E., Birlirakis, N., Bohé, L., Pratt, D.E., and Crich, D. (2012) *Nat. Chem.*, 4, 663–667; (d) Chan, J., Sannikova, N., Tang, A., and Bennet, A.J. (2014) *J. Am. Chem. Soc.*, 136, 12225–12228.

12

Regioselective, One-Pot Functionalization of Carbohydrates

Suvarn S. Kulkarni

12.1 Introduction

Carbohydrates in the form of glycoconjugates are ubiquitously distributed on the cell surfaces and are thus involved in a plethora of vital life processes [1]. Structurally, they are complex and diverse. In order to study their biological roles, access to chemically pure and well-characterized materials is essential. However, the glycoconjugates exist in microheterogeneous forms in nature and therefore cannot be procured in acceptable purity and quantity from natural sources. Chemical synthesis of carbohydrates offers the opportunity to obtain pure glycoconjugates.

Chemical synthesis of a target glycan essentially involves two major challenges: regioselective protection of individual hydroxyl groups [2, 3] and stereoselective glycosylations [4]. Sometimes, the target molecule comprises rare deoxyamino sugars, which necessitates *a priori* synthesis of such uncommon monosaccharides [5]. Preparation of orthogonally protected monosaccharide building blocks, to be employed as glycosyl donors and acceptors in glycosylation, is a challenge, as all the secondary hydroxyls display comparable reactivity under various conditions leading to a mixture of regioisomers. Separation of the mixtures at each stage makes the overall synthesis inefficient and laborious. Moreover, a great deal of planning is involved in designing the overall protecting group strategy for the synthesis of a target glycan as it is crucial not only for obtaining the desired stereoselectivity in glycosylation but also for tuning the overall electronic properties of the donors and acceptors so as to match the donor–acceptor pair [6]. The orthogonal nature of protecting groups is also essential for further selective deprotections and glycosylations as well as functional group modification. A chemical synthesis of glycan carried out in a traditional way thus involves a number of steps and requires months to complete. Often, it is the regioselective functionalization of the monosaccharides that takes most of the effort and time. At the beginning of this century, it was realized that the glycan synthesis can be greatly simplified by carrying out certain consecutive, compatible transformations in a sequential, one-pot manner. Such procedures would obviate the need to carry out intermittent purification and thus reduce time, efforts, precious materials, and chemical wastes. Over the past 15 years, intense efforts have been devoted toward this goal, and as

Selective Glycosylations: Synthetic Methods and Catalysts, First Edition. Edited by Clay S. Bennett.

a result, several "one-pot" protocols have been established for regioselective functionalization of monosaccharides [7, 8].

For the success of a one-pot reaction, it is essential that each step should rapidly go to completion in the presence of stoichiometric reagents and generate a single regioisomer of the product at each stage. This requires absolute fine-tuning of the reaction conditions. Moreover, the reagents used must be compatible with the side products and the intermediates generated after each step. Further, the consecutive steps should be carried out using a single catalyst under either acidic or basic conditions; otherwise, a neutralization step is necessary for switching between the two. Alternatively, one can have a sequence of reactions with brief workups in between the steps and a single chromatographic purification at the end of the sequence. Such highly regioselective sequential reactions, although not really "one-pot," are also important for gaining expedient access to differentially protected building blocks. Over the past few years, efficient protocols for rapid discrimination of several hydroxyl groups have been developed, which give access to partially or fully protected monosaccharides either in a few steps or via one-pot protocols. This chapter describes several notable one-pot and semi-one-pot procedures for regioselective protection and/or functionalization of monosaccharides developed over the past 15 years. Efficient protocols to synthesize orthogonally protected rare deoxy amino sugars will be discussed as well.

12.2 Regioselective, Sequential Protection/Functionalization of Carbohydrate Polyols

Monosaccharides have several hydroxyl groups, which need to be differentiated according to the structure of the target glycan under consideration. Importantly, the hydroxyl groups of different sugars have different relative orientation in space. The stereo relationship among various hydroxyl groups and the electronic factors can be tuned under appropriate (reagent, solvent, and temperature) conditions to achieve regioselectivity in the protection of various hydroxyls. The regioselectivity may, however, differ from one sugar to another, owing to the aforementioned differences. The orientation of anomeric functionality also plays a key role. The dependence of regioselective protection on the relative stereochemistry of hydroxyl groups and the anomeric functionality in various sugars was noted by Kong and coworkers in 2000 during their initial explorations on one-pot protection of monosaccharides (Table 12.1) [9]. Their methodology involved sequential tritylation–silylation–diacylation of hexopyranoside tetraols under basic conditions. Accordingly, protection of the most reactive primary hydroxyl (O6) by using a bulky protecting group (Trityl or TBDPS) under DMAP catalysis, followed by selective mono TBDMS protection of the 2,3,4-triol using imidazole as a base (at O2 or O3), and subsequent acylation of the remaining diol under the prevalent basic conditions furnished fully protected diacylated monosaccharides in a one-pot manner (71–86% overall yields). An interesting feature of this method is that the regioselectivity of the intermediate silylation (second step) was found to be dependent on the monosaccharide type and the orientation of the substituent at the anomeric position. Thus, while the α-mannosides **1** and **2**, β-glucosides **3** and **4**, and β-galactoside **6** were silylated at O3, the α-glucoside **5** was silylated exclusively at O2, and poor O2/O3 selectivity was observed in the case of α-galactoside 7. Thioglycosides and *O*-alkyl glycosides showed similar results. Using this

Table 12.1 Kong's approach for one-pot protection of free sugars.

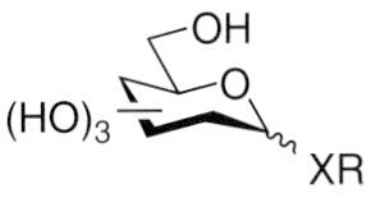

1. TrCl (1.25 eq), pyridine, DMAP, 80 °C, 16 h
2. Imidazole (2 eq), TBDMSCl (1.1 eq), DMF, 0 °C-RT, overnight
3. Ac_2O (3 eq) or BzCl (2.5 eq), pyridine, 50 °C, overnight

→ **Product**

Substrate	Product (% yield)	
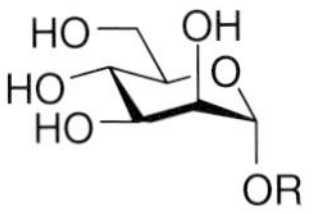 **1**. R = allyl **2**. R = Me	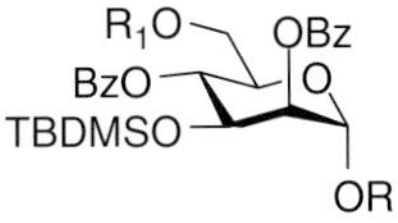 **8**. R = allyl, R_1 = Trityl (79%) **9**. R = Me, R_1 = TBDPS (71%)	
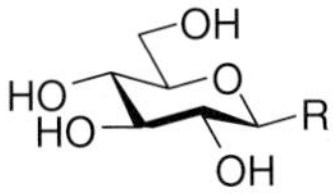 **3**. R = SPh **4**. R = O-*n*-C_8H_{17}	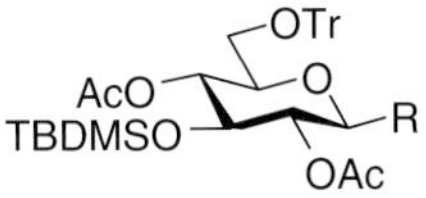 **10**. R = SPh (75%) **11**. R = O-*n*-C_8H_{17} (65%)	
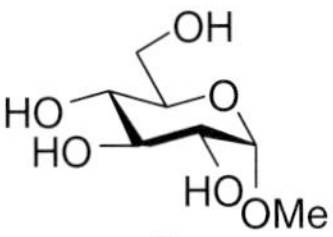 **5**	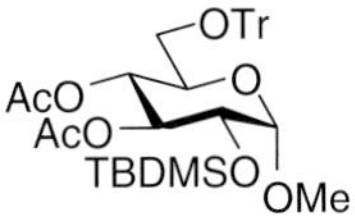 **12** (86%)	
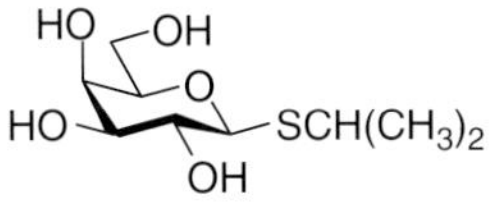 **6**	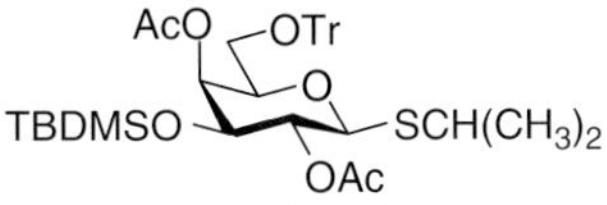 **13** (83%)	
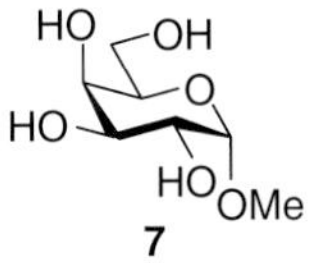 **7**	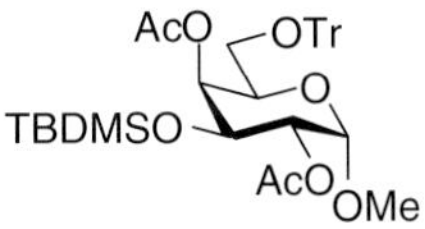 **14** (43%)	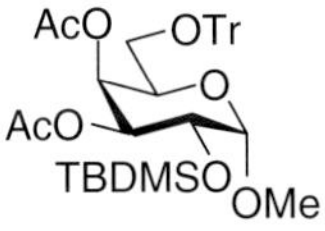 **15** (48%)

method, each of the fully protected building blocks **8–15** could be obtained from the respective tetraols in 3–4 days time, after a single-column chromatographic purification. This methodology was successfully applied in the synthesis of 3,6-branched oligosaccharides including mannopyranosyl pentasaccharide and hexasaccharide of β-1,3-glucans. It should be noted that TBDMS group at O3 position of glucose is prone to migration under strong basic conditions [10]. The complete switch in the O2 versus O3 regioselectivity observed in silylation of α- and β-glucosides **3, 4,** and **5** could be perhaps attributed to the steric effects, particularly, the ease of accessibility of the hydroxyl group for the approach of the reagent.

The effect of the orientation of the anomeric functionality on the outcome of regioselectivity was also observed by Onomura and coworkers [11], during their studies on the regioselective protection of monosaccharide polyols catalyzed by dimethyltin dichloride (Me_2SnCl_2). Regioselective monobenzoylation of various glycosides (tetraols and triols) employing 0.05 equiv. of Me_2SnCl_2 in THF or aqueous THF at RT cleanly afforded a single product in high yields (Scheme 12.1). The regioselectivity of benzoylation was perceived as an intrinsic character of the sugar based on the stereo relationship among the hydroxyl groups. The monobenzoylation predominantly took place at the 1,2-diol moiety except for the β-methyl glucoside. While 3-OBz products were obtained with α or β mannosides and galactosides, as anticipated based on the preferential coordination of Me_2SnCl_2 catalyst with the *cis*-1,2-diols, the α-methylglucoside **5** afforded 2-OBz product **16**, exclusively, whereas the corresponding β-methyl glucoside **17** furnished 6-OBz isomer **18** (Scheme 12.1a). This change in regioselectivity was explained based on the preferential approach of the reagent. First, Me_2SnCl_2 loosely interacts with the hydroxyl groups and can freely move around. This interaction leads to enhancement of the acidity of OH protons, which can be easily removed by mild base such as DIPEA. Thus, the most accessible OH group of the 1,2-diol moieties gets attacked by the base and gets benzoylated. The equatorial group adjacent to the reacting OH group hinders the approach of DIPEA. In the case of α-methyl glucoside **5**, the 2-OH group is freely accessible and thus gets benzoylated, whereas in β-methyl glucoside **17**, the equatorial methoxy group blocks the approach of the base, which then picks up the proton from the primary hydroxyl 6-OH group. Onomura and coworkers nicely extended these results to the sequential protection of sugar polyols as exemplified in Scheme 12.1b. The 2-OBz derivative **16** was selectively monotosylated at the primary O6 position to give 6-OTs derivative **19** under identical conditions. Subsequent bocylation of the remaining 3,4-diol was achieved at O3 regioselectively using Me_2SnCl_2 catalysis to obtain **20** in excellent yield, and the remaining 4-OH was phosphorylated under the prevailing basic conditions to construct the orthogonally protected building block **21**. Although the sequential reactions have not been carried out under one-pot setup, since all the reactions are done using a single catalyst, the sequence is certainly amenable to one-pot transformations. Onomura and coworkers further demonstrated a highly regioselective dioctyltin-dichloride-catalyzed thiocarbonylation of unprotected monosaccharides, followed by Barton-type deoxygenation to access various deoxy sugars [12]. Very recently, Iadonisi and coworkers [13] developed a first catalytic version of stannylene-mediated regioselective benzylation and allylation of sugar polyols using a catalytic amount of Bu_2SnO under solvent-free conditions. The Dong and Pei groups [14] further showed that the regioselective benzylation can be carried out using catalytic amount of any of the organotin reagents – Bu_2SnO, Bu_2SnCl_2, or Me_2SnCl_2 – in toluene and also proposed a catalytic cycle. Such novel methodologies could potentially be integrated and brought under the realm of one-pot reactions.

Of the various hydroxyl groups present on the sugar backbone, the *cis* 1,2-diols can be selectively protected in the form of cyclic orthoesters, which can be further rearranged with predictable regioselectivity. Field and coworkers [15] used the orthoester

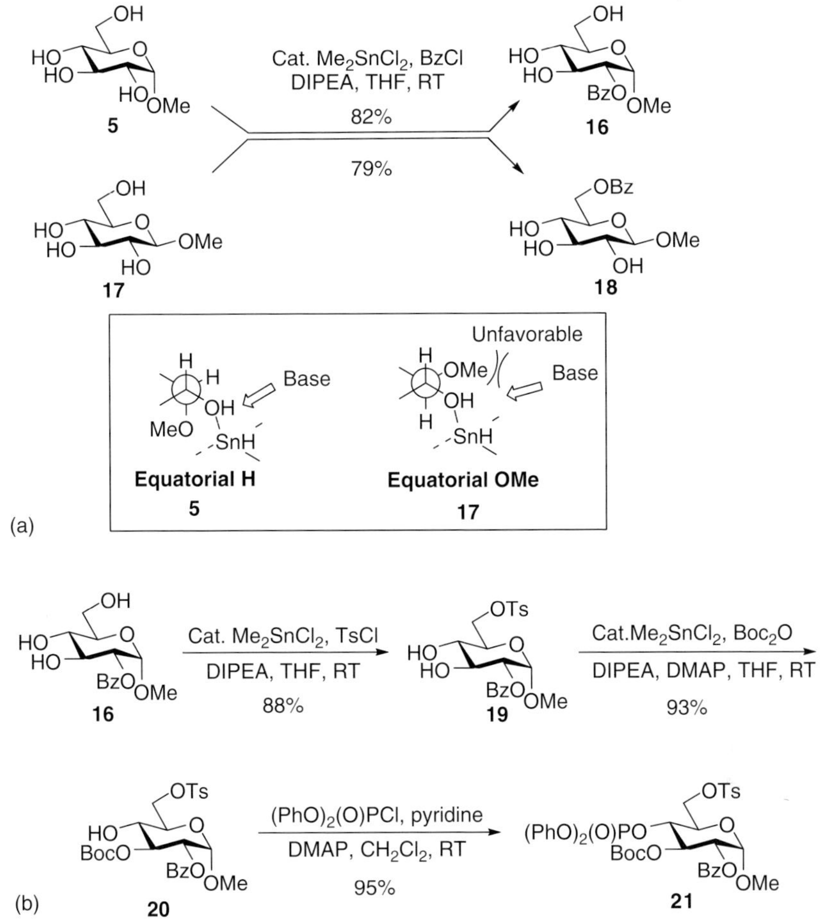

Scheme 12.1 Dimethyltin-dichloride-catalyzed regioslective esterifications.

rearrangement to establish a one-pot protocol for differentiation of sugar polyols. Their method includes sequential selective orthoesterification of *cis* 1,2-diols, followed by benzylation of the remaining hydroxyls and subsequent orthoester rearrangement (Scheme 12.2). For example, the orthoesterification of methyl rhamnoside **22** tied up the two adjacent *cis*-hydroxy groups (C2, C3-diol), and the remaining 4-OH in **23** was benzylated under basic conditions in the same vessel. Upon treatment of **24** with 1 M aq. HCl, the orthoester was selectively opened to reveal an axial acetate (at C2) and an equatorial free hydroxyl (at C3) to obtain **25**. The reaction worked well on various sugar triols and tetraols; alcohols **26–30** were obtained in very good yields from the respective monosaccharide polyols. Acetonitrile worked the best for reactions of methyl glycosides, whereas DMF gave the best results for thioglycosides.

One of the most common ways to protect sugar diols is the use of benzylidene acetals and acetonides, which can be installed with predictable regioselectivity. The benzylidene acetals are routinely used to protect the 4,6-diol moiety of hexopyranosides. Upon protection/functionalization of the remaining 2,3-diol, the benzylidene acetal can be readily

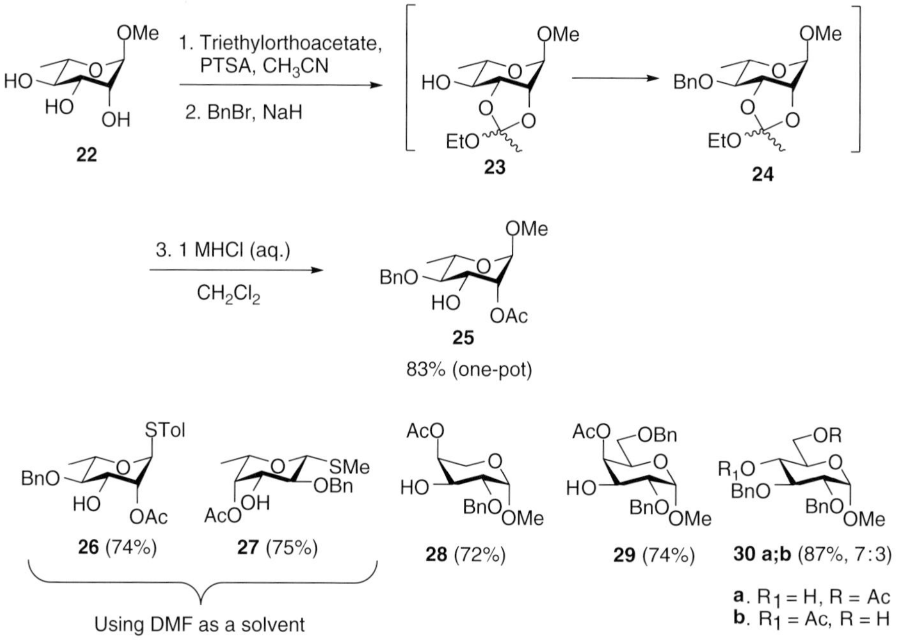

Scheme 12.2 One-pot protection via orthoester rearrangement.

opened at O6 or O4, with high regioselectivity and predictability, using an appropriate reducing agent, solvent, and Lewis acid combination [16]. One can also hydrolyze the acetal and subsequently protect the primary hydroxyl group with an orthogonal protecting group. Alternatively, the Hanessian–Hullar reaction of benzylidene cleavage using NBS results in the corresponding 4-OBz, 6-bromo derivatives in high regioselectivity [17]. A number of one-pot methods have been reported in the literature for sequential acetalation–acetylation using a single catalyst via the introduction of these acetals or ketals using the requisite dimethyl acetal/ketal reagent, followed by acetylation of the remaining diols using Ac_2O. These catalysts include immobilized $HClO_4$ [18] and H_2SO_4 [19] on silica, PTSA [20], I_2 [21], $Cu(OTf)_2$ [22], and cyanuric chloride [23]. Moreover, the acetonide and benzylidene acetals offer complementary regioselectivity. For example, using PTSA as a catalyst, the D-galactose-derived tetraol **31** was fully protected as the 4,6-acetal **32** and the 3,4-acetonide **33** in a one-pot manner under similar conditions (Scheme 12.3) [20]. A methodology for tandem acetonide formation–acetylation using isopropenyl acetate and catalytic iodine under solvent-free conditions has also been reported [24]. Furthermore, the regioselective Et_3SiH- or $NaCNBH_4$-reductive O4 opening of benzylidene acetals was incorporated in the acetalation–acetylation sequence using $Cu(OTf)_2$ and cyanuric chloride as catalysts, respectively. Thus, the D-gluco-, D-galacto-, and D-glucosamine-derived polyols **3**, **34**, and **35** were efficiently transformed into 4-hydroxy compounds **36–38**, respectively, over three steps in a sequential one-pot manner using cat. $Cu(OTf)_2$ (Scheme 12.4) [22].

An indirect route to access orthogonally protected D-glucosamine derivatives involves C2 inversion of D-mannose via combination of tin-mediated acylation and benzylidene formation. The Me_2SnCl_2-catalyzed regioselective acylation reaction was extended to

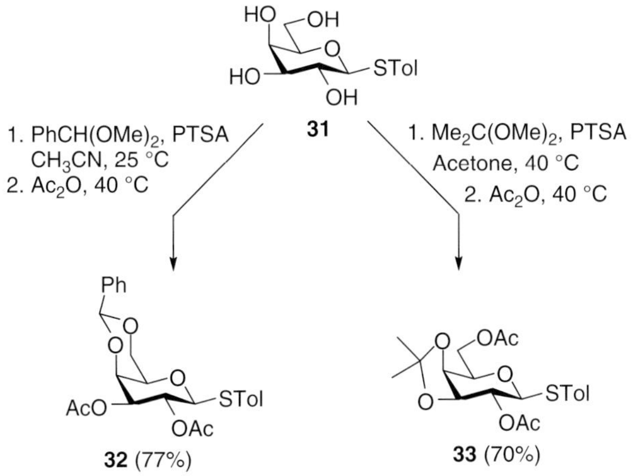

Scheme 12.3 One-pot acetalation–acetylation of glycosides.

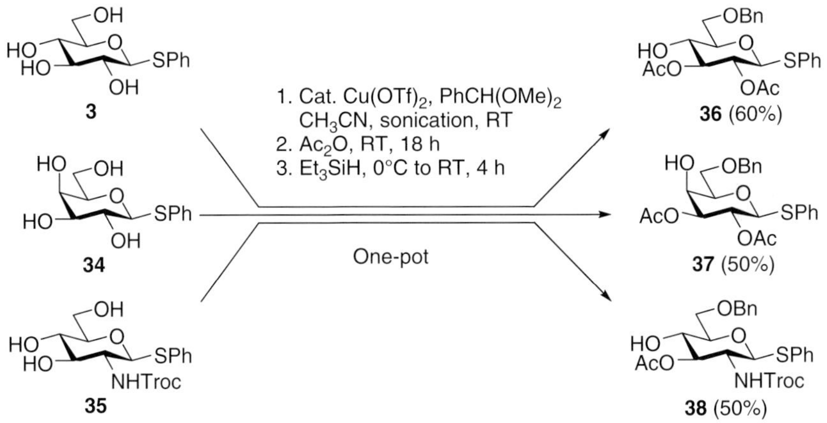

Scheme 12.4 One-pot protection of monosaccharides to access 4-alcohols.

β-thiomannoside **39**, which was efficiently transformed into glucosamine derivatives **42a** and **42b** via sequential 3-O-acylation (Ac, Bz), 4,6-O-benzyldination, O2 triflation, and concomitant azide displacement of 2-OTf in very good yields (Scheme 12.5) [25]. Although this is not a one-pot transformation, the route is efficient and simple. The entire sequence could be carried out in a single day without any intermittent purification; this way, compound **42b** was obtained in 44% yield over four steps after a single chromatographic purification. The D-glucosamine derivatives were converted to orthogonally protected D-galactosamine derivative through benzylidene hydrolysis, regioselective silylation at O6, and concomitant C4 inversion. This route bypasses the cumbersome diazotransfer reaction, which is usually employed to convert amine to azide to access 2-azido glycosides.

Regioselective acylation of monosaccharides has been a topic of great interest as evidenced from a number of methodologies published in recent times based on reagents

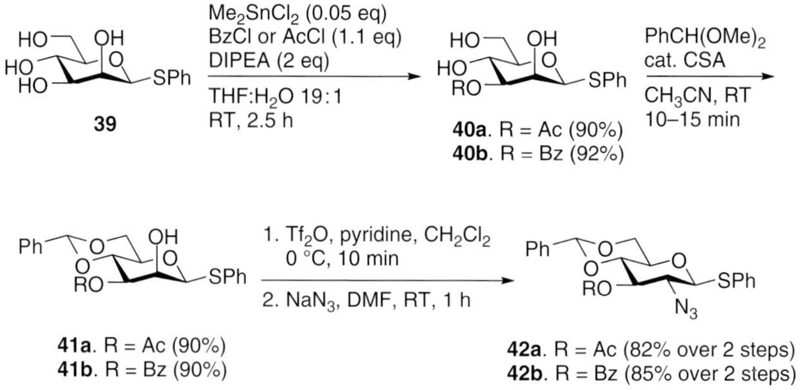

Scheme 12.5 From D-mannose to selectively protected D-glucosamine building blocks.

such as organotin [26], organoboron [27], organosilicon [28, 29], organobase [30, 31], and tetrabutylammonium salts [32, 33]. It is usually observed that the primary hydroxyl (6-OH) is the most reactive, followed by 2-OH/3-OH, whereas the 4-OH is the least reactive of all. Kawabata and coworkers, however, explored a C2-symmetric, chiral DMAP derivative **43** as a catalyst that led to O4 acylation of β-D-glucopyranosides, exclusively [34]. For instance (Scheme 12.6), using 1 mol% of **43**, the *n*-octyl β-D-glucopyranoside generated the O4 acylated derivative **45** in 99% yields. Subsequent regioselective TBDPS protection at O6 generated **46** (98%), which upon highly regioselective Boc protection of 2-OH, furnished **47** (98%). Finally, benzyloxymethyl (BOM) protection of the remaining 3-OH designed the orthogonally protected glucoside **48** (99%) [35]. A model involving multiple noncovalent catalyst–substrate interactions including hydrogen bonds between the most reactive 6-OH and 3-OH with the amide carbonyl bond and indole NH, respectively, has been proposed to explain this unusual selectivity. These interactions presumably fix the conformation of the substrate at the transition state for acylation in such a way that the 4-OH group comes in a close proximity to the reactive carbonyl group of the acylpyridinium ion, resulting in the selective acylation at 4-OH. The O4 regioselective acylation works well for β-glycosides with an equatorially oriented 4-OH group, for example, octyl β-glucopyranoside, octyl β-thioglucopyranoside, and octyl β-mannopyranoside, but the regioselectivity dropped for octyl α-glucopyranoside, and complete O6 regioselectivity was observed for octyl β-galactopyranoside. It also worked well on several disaccharides including one with seven hydroxyl groups [36] and has been also applied to achieve the selective monoacylation of cardiac glycoside digitoxin [37]. This off-the-track methodology gives a rapid access to a variety of differentially functionalized monosaccharides.

12.3 Regioselective, One-Pot Protection of Sugars via TMS Protection of Polyols

The methods discussed so far, use carbohydrate polyols as starting materials, which are not easily soluble in common organic solvents. To address this problem, Hung and

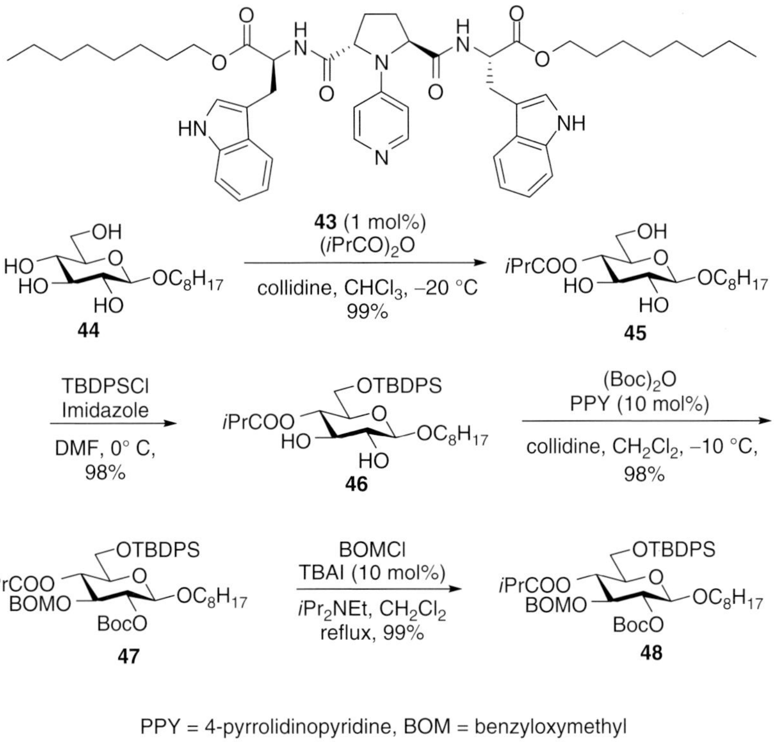

Scheme 12.6 Chiral DMAP catalysts for regioslective O4 acylation.

coworkers introduced trimethylsilyl (TMS) protection of the hydroxy groups to assist the sequential one-pot regioselective protection of sugars. The TMS protection of alcohols can be readily achieved using Et_3N and TMSCl, and they can be removed easily by using a fluoride source or acidic resin such as Dowex. The TMS-protected sugars are completely soluble in common organic solvents such as dichloromethane and offer thermodynamic advantage by the formation of TMSOTMS or TESOTMS in subsequent steps. In their pioneering approach, Hung and coworkers established a highly regioselective one-pot protection of per-*O*-TMS glucosides (α-OMe or β-STol) to access hundreds of 2-, 3-, 4-, or 6-alcohol as well as fully protected derivatives in excellent overall yields [38, 39]. The elegant method nicely integrates acetalation [40] and regioselective reductive etherification [41, 42] using TMS ethers as starting materials and TMSOTf as a common catalyst. The systematic approach entails a sequential regioselective installation of benzylidene-type acetal at O4 and O6 positions of the per-OTMS glucoside **49**, followed by introduction of a benzyl-type ether at O3 in the same vessel for *in situ* generation of the 2-OTMS derivative **50** (Scheme 12.7). This key transformation can be carried out using a combination of a variety of substituted aryl aldehydes to generate orthogonal sets of protecting groups. On the other hand, if the same ether (e.g., benzyl) is desired at the C3 position as that of the 4,6-*O*-arylidene acetal (e.g., benzylidene acetal), the process can be achieved in one step using 2 equiv. of the aldehyde (e.g., PhCHO). The acetal formation only requires the corresponding aldehyde (instead of the corresponding dimethyl acetal), and the equilibrium shift toward the acetal is largely driven by the formation of

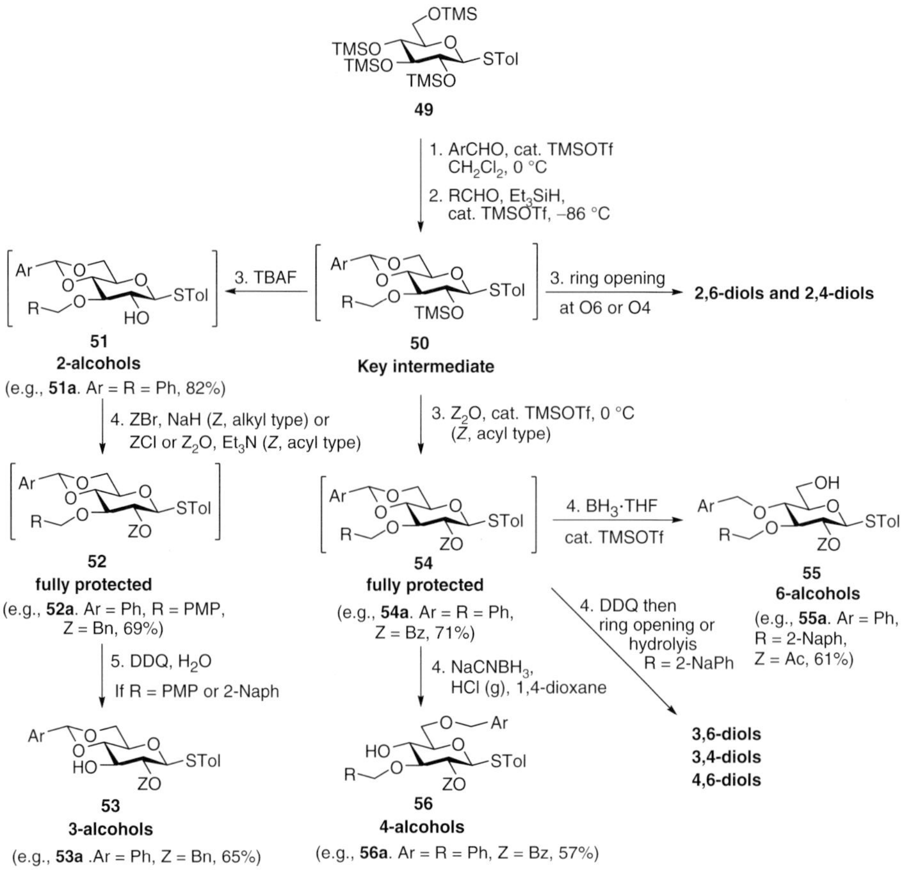

Scheme 12.7 Hung's one-pot protection strategy starting from per-OTMS glucosides.

the stable TMSOTMS molecule [40]. The high regioselectivity observed in the reductive 3-*O*-benzyl-type ether formation has been attributed to the aforementioned thermodynamic effect [43] and also to the higher nucleophilicity of O3 owing to the strong inductive effect experienced by O2 [42]. The key intermediate **50** can be further manipulated to access diverse building blocks. Subsequent treatment of **50** with TBAF affords 2-alcohols **51**, which can be alkylated or acylated under basic conditions to furnish fully protected building blocks **52**. Concomitant oxidative removal of the substituted benzyl-type group (e.g., PMB or 2-NAP) affords 3-alcohols **53** in excellent overall yields. Alternatively, acylation of the 2-OTMS ether **50** can be carried out under the prevailing acidic conditions using TMSOTf to obtain fully protected **54** bearing a more acid stable 2-naphthylmethyl (NAP) group at O3. The reaction can be extended further by regioselectively cleaving the 4,6-*O*-arylidiene acetal in **54**, under tandem catalysis of TMSOTf, using BH_3·THF as a reducing agent to form 6-alcohols **55**. Alternatively, treatment of **54** with $NaCNBH_3$ and HCl gas cleaves the acetal at O4 to afford 4-alcohols **56** exclusively.

The one-pot protection has also been extended to access a variety of diols [44]. The reaction course can be diverted from **54** by first removing the NAP group at the

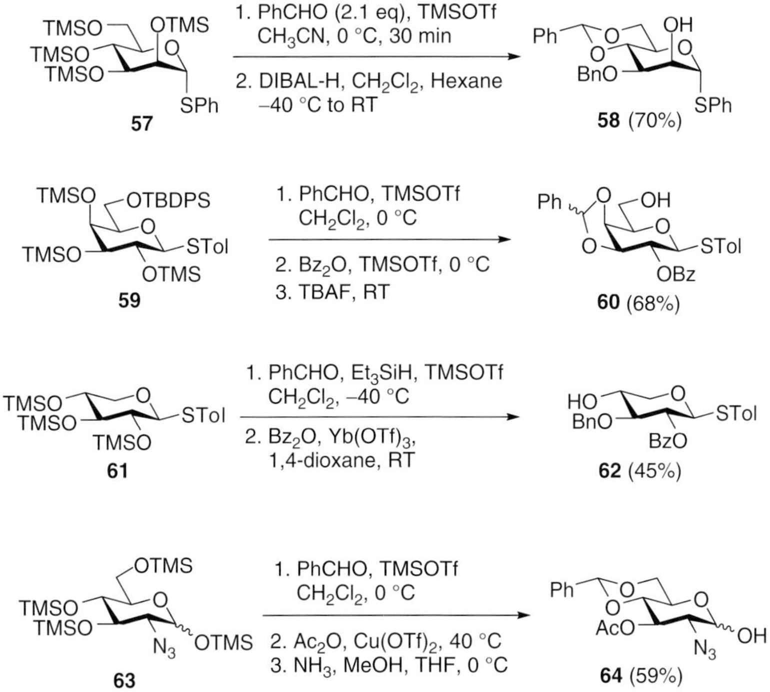

Scheme 12.8 One-pot protection of various per-*O*-TMS glycosides.

3-position by DDQ treatment followed by regioselective ring opening at O4 or O6 using appropriate reagents to generate 3,4 and 3,6 diols; alternatively, **54** can be hydrolyzed using trifluoroacetic acid to gain access to 4,6-diols. Similarly, ring opening of 2-OTMS derivative **50** furnishes 2,4 and 2,6 diols in excellent yields. Thus, a panoply of orthogonally protected building blocks can be created in a one-pot manner starting from TMS thioglucoside **49**. Moreover, the same chemistry can be carried out using the corresponding tetra-TMS derivative of α-methylglucoside. The one-pot protocol can accommodate three to five steps and generates differentially protected building blocks in excellent overall yields after a single-column chromatographic purification. Some of the building blocks have been successfully employed in the assembly of heparin and heparan sulfate [45, 46], alginates [47] as well as in the one-pot synthesis of a small library [48] of influenza trisaccharides.

The one-pot protocol works well with other sugars, with slight modification of the reagents and techniques depending on the arrangement of the hydroxyl groups, as exemplified in Scheme 12.8. For instance, in the case of D-mannoside, the *cis* orientation of O2 and O3 readily allows dibenzylidenation along with O4 and O6, but the endo/exo stereochemistry of the phenyl group in the generated dioxolane function, which decides the sense of ring opening in the succeeding step, is difficult to control. In the reaction of α-thiomannoside **57** under TMSOTf catalysis at low temperature, the *exo* isomer precipitated out from the CH_3CN solvent, and the subsequent reductive ring opening of the dioxolane ring using DIBAL-H at −40 °C furnished the 2-alcohol **58**, exclusively, keeping the 4,6-*O*-benzylidene acetal intact [49]. In the case of D-galactose, the O6 TBDPS-protected derivative **59** was employed to circumvent the competitive formation of 4,6-*O*-benzylidene acetal, which allowed the acetal formation at the alternate *cis*-oriented O3 and O4 positions [50]. Subsequent 2-O-benzoylation followed by treatment with TBAF in the same vessel furnished 6-OH derivative **60**. The D-xylose derivative **61** upon regioselective reductive benzylation at O3 generated the corresponding 2,4-diol, which upon chelating with $Yb(OTf)_3$ allowed selective benzoylation at O2 to furnish 4-OH derivative **62** [50]. The D-glucosamine derivative **63** was readily transformed into hemiacetal **64** via benzylidenation at O4 and O6 followed by $Cu(OTf)_2$-catalyzed 1,3-di-O-acetylation and subsequent selective removal of the anomeric acetate using ammonia gas [51]. The reaction works well on a large scale and has been utilized in the total synthesis of MECA-79 [52].

Beau and coworkers subsequently disclosed that $Cu(OTf)_2$ [53] as well as non-air-sensitive $FeCl_3{\cdot}6H_2O$ [54] can be used for such a tandem catalysis process on D-glucose. This is particularly advantageous as, while the effective use of TMSOTf as a catalyst requires subzero temperatures [38], $Cu(OTf)_2$ [53, 55] and $FeCl_3$ [54] give satisfactory results at room temperature. A different mechanism involving formation of 4,6-*O*- and 2,3-*O*-dibenzylidene acetals followed by regioselective Et_3SiH-reductive ring opening of the dioxolane ring at O2 to generate O3 benzyl ethers at RT is proposed. Such a pathway seems unlikely at −86 °C using cat. TMSOTf, and it has been proposed that the TMSOTf reaction goes through a regioselective reductive etherification of the more nucleophilic O3 [38]. By using $Cu(OTf)_2$ as a catalyst (Scheme 12.9), the tetra-*O*-TMS thioglucoside **65** was similarly transformed into fully protected derivative **66** through 4,6-O-benzylidenation followed by selective O3 benzylation and subsequent O2 acetylation in 79% overall yield in one-pot manner [53]. Similarly, the 2,4-diol derivative **68** and 2,6-diol derivative **69** were obtained via the intermediacy of the 2-OH glucosides **67** in a one-pot manner through tandem catalysis of $Cu(OTf)_2$ or $FeCl_3{\cdot}6H_2O$ at RT. The O6 opening of benzylidene acetal of **66** did not proceed using $BH_3{\cdot}THF$ as a reducing agent in the presence of $FeCl_3{\cdot}6H_2O$ probably due to complexation with the catalyst. The

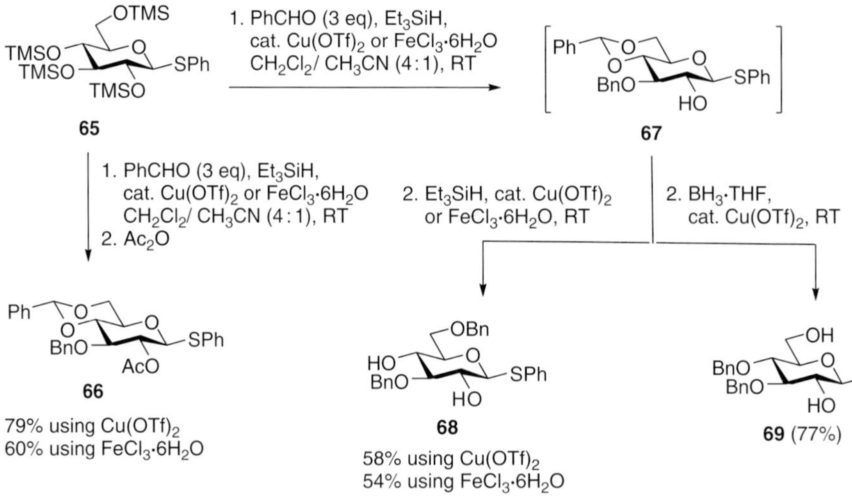

Scheme 12.9 One-pot protection using $Cu(OTf)_2$ and $FeCl_3{\cdot}6H_2O$ as catalysts.

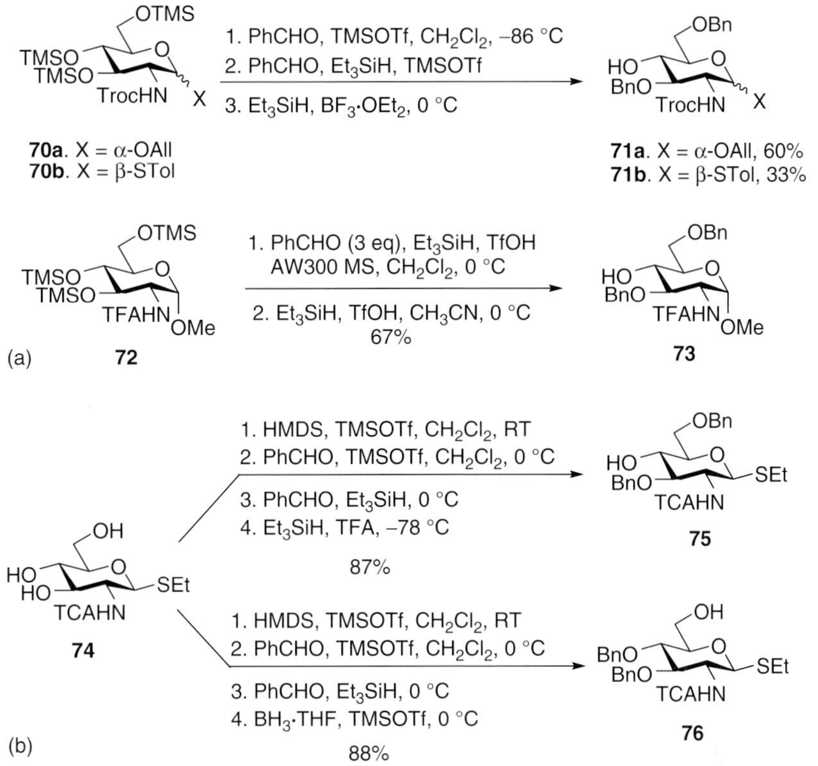

Scheme 12.10 One-pot protection of D-glucosamine carrying a variety of *N*-protecting groups.

one-pot reactions work similarly with the corresponding tetra-*O*-TMS α-methylglucoside. The $FeCl_3{\cdot}6H_2O$ catalyst has also been used in the regioselective one-pot protection of maltose as well as the C2 symmetric disaccharide trehalose [54]. However, $Cu(OTf)_2$ and $FeCl_3{\cdot}6H_2O$ do not work well for glucosamine derivatives, whereas TMSOTf works well for 2-azido derivatives as described previously.

Strategically, although the 2-azido derivatives are excellent for 1,2-cis glycosylation [56], they offer limited choice of conditions such as solvent participation when it comes to 1,2-trans glycosylation [57]. The nitrogen-protecting groups capable of providing anchimeric assistance (NHTroc, NPhth, NHTFA, NHTCA) are better suited for this purpose. Marques and coworkers [58] employed TMSOTf catalysis to transform NHTroc-protected α-*O*-allyl and β-SPh glucosamine derivatives **70a,b** into 4-OH derivatives **71a,b** via similar 4,6-O-benzylidenation, 3-O-benzylation, and regioselective Et_3SiH reductive ring opening of benzylidene acetal using $BF_3{\cdot}OEt_2$ in a one-pot manner (Scheme 12.10a). Using TfOH on molecular sieves as a catalyst [59], Beau's group obtained the 4-alcohol **73** from the glycoside **72** via a similar one-step tandem benzylidene formation and 3-O-benzylation sequence. Subsequent addition of Et_3SiH, TfOH, and CH_3CN in the same flask led to the 4-O-ring opening.

Wang and coworkers further streamlined the one-pot protection methodology by integrating the initial TMS protection step in the one-pot process. Preparation of per-OTMS starting materials requires the use of excess TMSCl and bases with long reaction times and produces large quantities of salts as by-products. Wang and coworkers addressed this issue by using near stoichiometric or slight excess of hexamethyldisilazane (HMDS) and catalytic TMSOTf to generate TMS ethers in a very short time [60]. Subsequent protection steps could be carried out in the same vessel obviating the need to do intermittent workup or purification of the per-*O*-TMS intermediate. The conditions work very well with all the sugars and are especially useful for D-glucosamine derivatives. As an example, the glucosamine triol derivative **74** was transformed into orthogonally protected 4-OH **75** and 6-OH **76** under TMSOTf catalysis in a one-pot manner employing HMDS silylation as the first step in very high overall yields (Scheme 12.10b). These advances have been applied in the synthesis of the heparin-based anticoagulant drug Fondaparinux [61].

A clever extension of the one-pot method is the use of Microwave (MW) irradiation to convert per-*O*-TMS derivatives into thioglycosides [62]. Accordingly, the free sugars were first converted to the per-*O*-TMS derivatives using HMDS, and their subsequent treatment with TMSSTol and ZnI_2 under microwave irradiation at 150 °C produced the corresponding per-*O*-TMS thioglycosides (major α-anomers), which can be similarly transformed into various differentially protected building blocks in a one-pot manner (Scheme 12.11). Under these conditions, D-glucose **77** was transformed into thioglycoside **79** (76%, $\alpha/\beta = 5:1$) via the intermediacy of per-*O*-TMS glucose **78**. Alternatively, the MW irradiation of the TMSOTf-treated intermediate **78** at 100 °C followed by HMDS treatment afforded 1,6-anhydroglucose **80** in 72% yield.

12.4 Orthogonally Protected D-Glycosamine and Bacterial Rare Sugar Building Blocks via Sequential, One-Pot Nucleophilic Displacements of O-Triflates

Regioselective inversion of stereocenters via nucleophilic displacements of sulfonates allows one to acquire sugar building blocks with a different configuration [63]. This is particularly helpful to access rare sugar building blocks and also the expensive ones such as D-galactosamine derivatives, which are often obtained through stereoinversion of C4 position of the cheaply available D-glucosamine. Some notable examples are shown in Scheme 12.12. For instance, Kulkarni and coworkers [25, 64] employed Lattrell–Dax reaction to displace C4-OTf using $TBANO_2$ (**81a**,**b** to **82a**,**b** and **83–84**), whereas Bundle and coworkers [65] employed intramolecular displacement of C4-OTf of glucosamine by Piv ester to form an orthoester, which undergoes in situ rearrangement to arrive at a galactosamine derivative (**85–86**). Toth and coworkers [66] observed that, by adjusting the N-protecting group, selective C4 or C3 mono displacements of D-glucosamine-derived 3,4,6-trimesylates **87** and **89** allow facile access to D-galactosamine and D-allosamine derivatives **88** and **90**, respectively. Ramström and coworkers performed some elegant studies to delineate mechanistic details of the Lattrell–Dax reaction, which involves nitrite displacement of *O*-triflates. They established that such displacements are favored by the presence of a neighboring equatorial acyl group [67], which apparently modulates the ambident nucleophilicity of the nitrite ion [68]. Moreover, they identified that the reaction rate is controlled by supramolecular anion recognition (host–guest) system involving complexation of the axial H1, H3, and H5

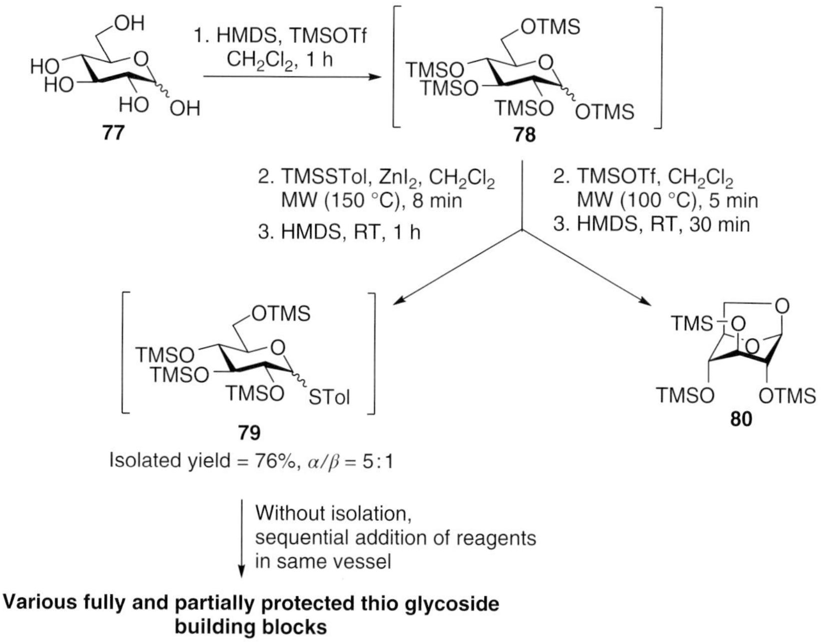

Scheme 12.11 Microwave-assisted synthesis of thioglycosides and 1,6-anhydrosugars.

protons of carbohydrate with the NO_2^- ion [69, 70]. These studies were extended to regioselective stannylene-mediated 3,6-diacylation of β-methyl D-glucoside and methyl D-galactoside followed by simultaneous or stepwise inversion of two stereocenters (C2 and C4) via displacement of *O*-triflates using nitrite or acetate nucleophiles to obtain β-talo (e.g., **92** from **91**) and β-manno (e.g., **94** from **93**) configured derivatives, respectively [71]. An important finding of this work is that the C4-OTf of D-gluco and D-galacto configured sugars is more reactive compared to C2-OTf. This reactivity difference allows one to carry out sequential nucleophilic displacements.

Recently, Kulkarni and coworkers disclosed an expedient protocol to access orthogonally protected D-galactosamine thioglycoside building blocks, from cheap and abundant D-mannose, which can as such be used as versatile donors in glycosylation [72]. Their methodology involves a sequential double serial inversion of β-thiomannoside at C2 and C4 by azide and nitrite nucleophiles, respectively (Scheme 12.13). The selective C2-*O*-triflate displacement required fine-tuning of the reaction conditions. First, the β-thiomannoside **39** is converted to a 2,4-diols **95a,b** via regioselective silylation at O6 followed by Me_2SnCl_2-catalyzed acylation at O3. Upon treatment with Tf_2O, the 2,4-diols **95a,b** are individually converted to the corresponding 2,4-bis triflates **96**, which can be made to undergo highly regioselective double displacement reactions under controlled conditions to generate a variety of building blocks **97**, depending on the nucleophiles used and their order of addition. As delineated in Scheme 12.14, triflation of diols **95a,b** and their concomitant displacement with excess NaN_3 in DMF give 2,4-diazido derivatives **98a,b**. For the double serial inversion, first the bis-triflate **96** is reacted with a stoichiometric amount of $TBAN_3$ at −30 °C in acetonitrile as a solvent to displace only

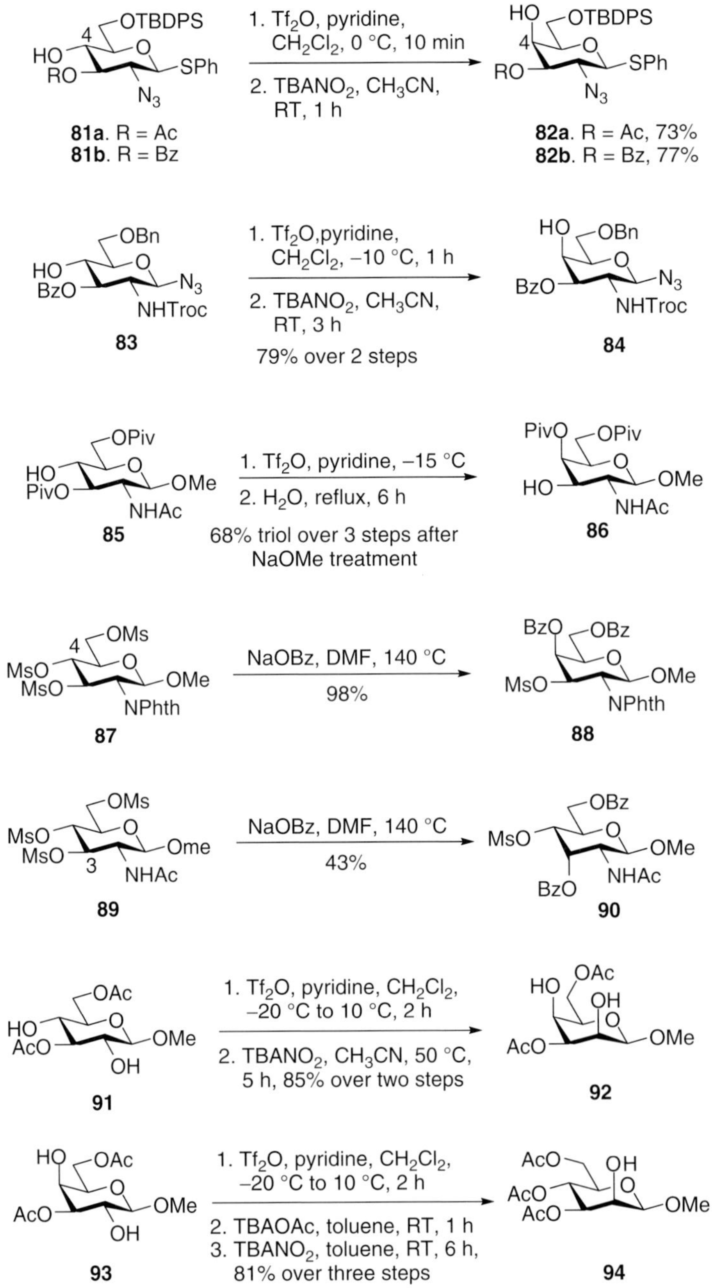

Scheme 12.12 Nucleophilic displacement of sulfonates to access GalNAc and other sugars.

the C2 triflate, exclusively, which allows the sequential addition of $TBANO_2$ in the same vessel to afford directly the orthogonally protected D-galactosamine derivatives **99a,b** in ~60% yields over three steps, after a single-column chromatographic purification. The regioselectivity of C2-OTf displacement is attributed to its trans axial effect [63]. The axially oriented C2-*O*-triflate group effectively shields the top face of D-mannoside allowing the reaction to take place at C2 from the easily accessible bottom face, preferentially. The regioselectivity is further tuned by using bulky nucleophiles and low temperature addition. The corresponding 4-azido, 2-hydroxy galactose derivatives **100a,b** can be obtained by switching the course of addition of reagents (first $TBANO_2$, then $TBAN_3$ at 0 °C). The 3-hydroxy $GalN_3$ derivative **101** could be accessed by employing the acetate migration protocol. Accordingly, triflation of **95a**, followed by azide displacement of the C2-OTf, addition of H_2O, and subsequent heating at 65 °C, affords **101** (56%), in a one-pot manner. In this case, the 3-*O*-acetyl group migrates from C-3 to C-4 along the top face of sugar to displace the C4-triflyloxy group to form an orthoester, which upon in situ rearrangement generates axial acetate **101**.

The efficient methodology also gives a rapid access to a variety of bacterial rare sugars such as 2,4-diacetamido-2,4,6-trideoxyhexose (DATDH), 2-acetamido-4-amino-2,4,6-trideoxy-D-galactose (AAT), D-fucosamine, and D-bacillosamine, which are not present on the host cells (Figure 12.1) [73, 74]. For the development of carbohydrate-based vaccines, procurement of rare sugar building blocks is essential [5]. For this purpose, the silylation step is replaced by selective O6 deoxygenation of β-D-thiomannoside (tosylation followed by LAH reduction), and regioselective O3 acylation of the resulting D-rhamnosyl triol affords the diol **102**, which can be used as a suitable starting material to design a variety of rare sugar building blocks **103–107**. As shown in Scheme 12.15, diol **102** can be similarly transformed via triflation followed by selective azide displacement of C2-OTf and concomitant nucleophilic displacement of C4-OTf in the same pot using N_3^-, $NPhth^-$, NO_2^-, and OAc (intramolecular) to afford the orthogonally protected rare sugar building blocks **103a,b** (DATDH), **104** (AAT), **105a,b,** and **106** (fucosamine), respectively, in excellent overall yields. An iteration of C-4 triflation and tandem azide displacement affords **107a,b** (bacillosamine). Thus,

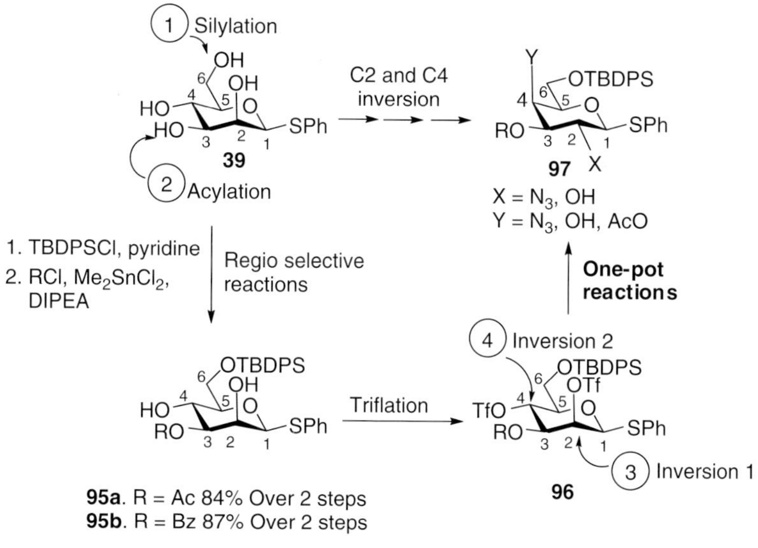

Scheme 12.13 A strategy to transform D-mannose into orthogonally protected $GalN_3$ thioglycosides.

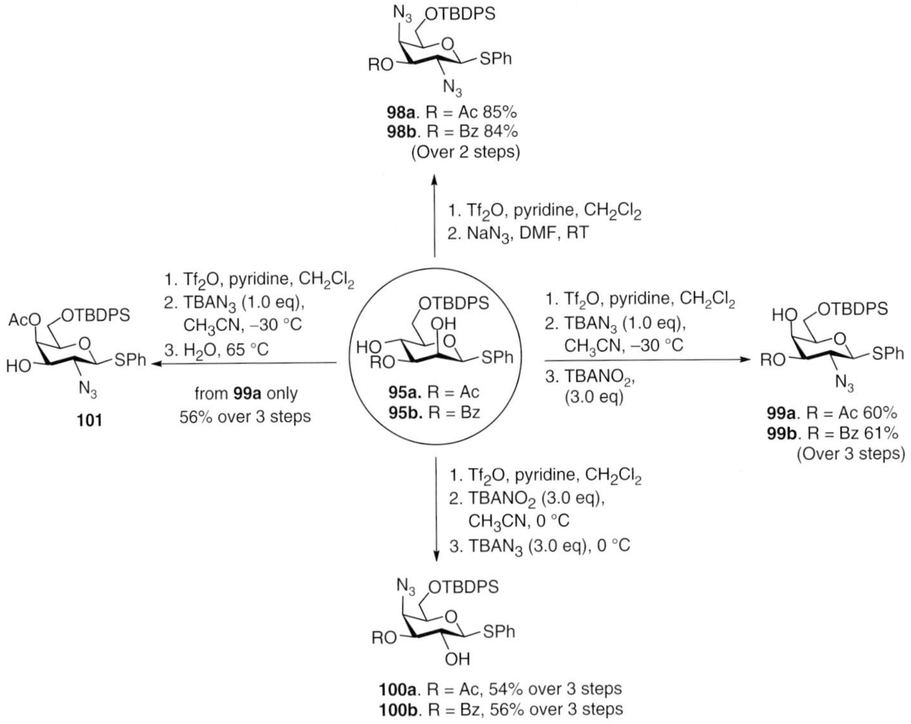

Scheme 12.14 One-pot nucleophilic double displacements of triflates to access glycosamine blocks.

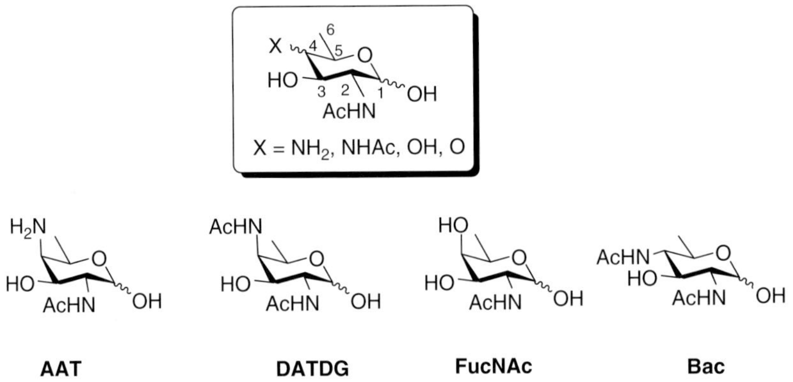

Figure 12.1 Bacterial rare sugars.

D-mannosyl- and D-rhamnosyl 2,4-diols can be efficiently transformed into various D-galactosamines as well as bacterial rare sugars, respectively, as orthogonally protected thioglycoside building blocks on a gram scale in 1–2 day time, in 54–85% overall yields, after a single chromatographic purification. The methodology has been applied to the synthesis of various bacterial *O*-glycans [57, 74, 75].

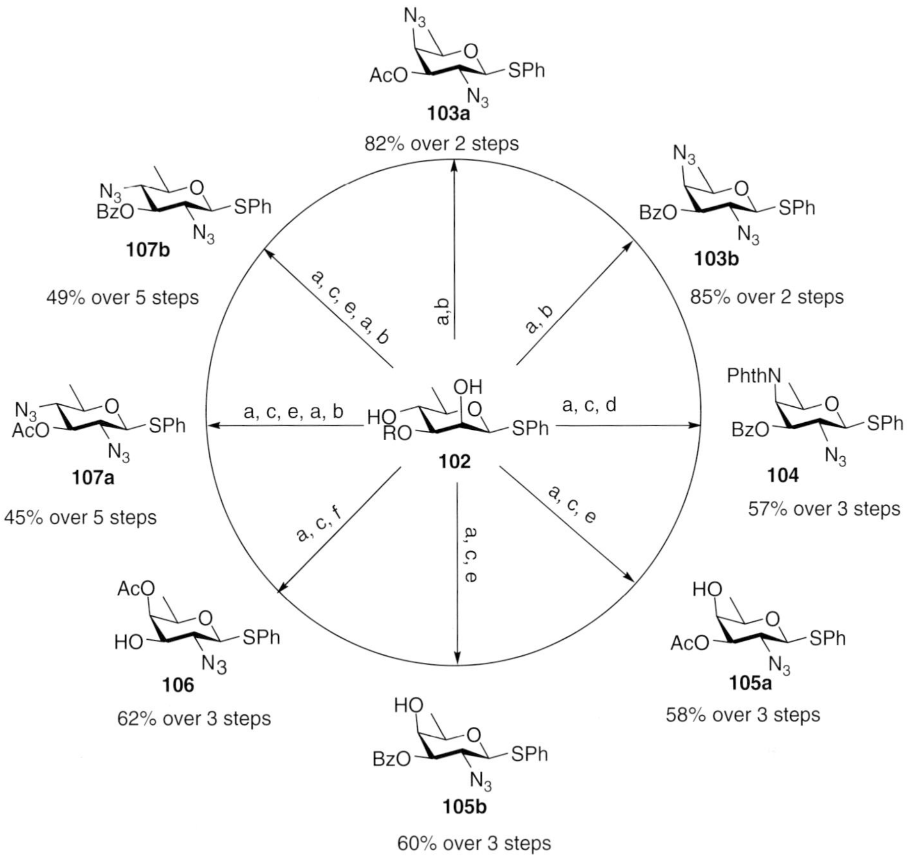

(a) Tf_2O, pyridine, CH_2Cl_2, 2 h; (b) NaN_3, DMF, 8 h; (c) $TBAN_3$ (1.0 eq), CH_3CN, –30 °C, 20 h; (d) PhthNK, DMF, 10 h; (e) $TBANO_2$, 1.5 h; (f) H_2O, 65 °C, 1.5 h

Scheme 12.15 From D-mannose to bacterial rare sugar building blocks.

12.5 Summary and Outlook

In this chapter, we discussed the recently developed one-pot and partial one-pot protocols to rapidly access regioselectively protected hexopyranoside building blocks as well as bacterial rare sugars. Ready availability of such thioglycoside building blocks in conjunction with the automated oligosaccharide synthesis [76, 77] is expected to expedite the glycan assembly and in turn speed up carbohydrate-based drug discovery in the years to come.

References

1 Varki, A., Cummings, R.D., Esko, J.D., Freeze, H.H., Stanley, P., Bertozzi, C.R., Hart, G.W., and Etzler, M.E. (eds) (2009) Essentials of Glycobiology, Cold Spring Harbor Press, New York, pp. 229–248.

2 Filice, M., Guisan, J.M., and Palomo, J.M. (2010) *Curr. Org. Chem.*, 14, 516–532.
3 Lee, D. and Taylor, M.S. (2012) *Synthesis*, 44, 3421–3431.
4 Demchenko, A.V. (2008) Handbook of Chemical Glycosylation: Advances in Stereoselectivity and Therapeutic Relevance, Chapter 1, Wiley-VCH Verlag GmbH, pp. 1–27.
5 Emmadi, M. and Kulkarni, S.S. (2014) *Nat. Prod. Rep.*, 31, 870–879.
6 Mydock, L.K. and Demchenko, A.V. (2010) *Org. Biomol. Chem.*, 8, 497–510.
7 Wang, C.-C., Zulueta, M.M.L., and Hung, S.-C. (2011) *Chimia*, 65, 54–58.
8 Zulueta, M.M.L., Janreddy, D., and Hung, S.-C. (2015) *Isr. J. Chem.*, 55, 347–359.
9 Du, Y., Zhang, M., and Kong, F. (2000) *Org. Lett.*, 2, 3797–3800.
10 Adak, S., Emmadi, M., and Kulkarni, S.S. (2014) *RSC Adv.*, 4, 7611–7616.
11 Demizu, Y., Kubo, Y., Miyoshi, H., Maki, T., Matsumura, Y., Moriyama, N., and Onomura, O. (2008) *Org. Lett.*, 10, 5075–5077.
12 Muramatsu, W., Tanigawa, S., Takemoto, Y., Yoshimatsu, H., and Onomura, O. (2012) *Chem. Eur. J.*, 18, 4850–4853.
13 Giordano, M. and Iadonisi, A. (2014) *J. Org. Chem.*, 79, 213–222.
14 Xu, H., Lu, Y., Zhou, Y., Ren, B., Pei, Y., Dong, H., and Pei, Z. (2014) *Adv. Synth. Catal.*, 356, 1735–1740.
15 Mukhopadhyay, B. and Field, R.A. (2003) *Carbohydr. Res.*, 338, 2149–2152.
16 Ohlin, M., Johnsson, R., and Ellervik, U. (2011) *Carbohydr. Res.*, 346, 1358–1370.
17 Crich, D., Yao, Q., and Bowers, A.A. (2006) *Carbohydr. Res.*, 341, 1748–1752.
18 Mukhopadhyay, B., Russell, D.A., and Field, R.A. (2005) *Carbohydr. Res.*, 340, 1075–1080.
19 Mukhopadhyay, B. (2006) *Tetrahedron Lett.*, 47, 4337–4341.
20 Mong, K.-K.T., Chao, C.-S., Chen, M.-C., and Lin, C.-W. (2009) *Synlett*, 2009, 603–606.
21 Jones, R.A., Davidson, R., Tran, A.T., Smith, N., and Galan, M.C. (2010) *Carbohydr. Res.*, 345, 1842–1845.
22 Tran, A.-T., Jones, R.A., Pastor, J., Boisson, J., Smith, N., and Galan, M.C. (2011) *Adv. Synth. Catal.*, 353, 2593–2598.
23 Tatina, M., Yousuf, S.K., and Mukherjee, D. (2012) *Org. Biomol. Chem.*, 10, 5357–5360.
24 Mukherjee, D., Ali Shah, B., Gupta, P., and Taneja, S.C. (2007) *J. Org. Chem.*, 72, 8965–8968.
25 Emmadi, M. and Kulkarni, S.S. (2011) *J. Org. Chem.*, 76, 4703–4709.
26 Dong, H., Pei, Z., Byström, S., and Ramström, O. (2007) *J. Org. Chem.*, 72, 1499–1502.
27 Lee, D. and Taylor, M.S. (2011) *J. Am. Chem. Soc.*, 133, 3724–3727.
28 Witschi, M.A. and Gervay-Hague, J. (2010) *Org. Lett.*, 12, 4312–4315.
29 Zhou, Y., Ramström, O., and Dong, H. (2012) *Chem. Commun.*, 48, 5370–5372.
30 Kurahashi, T., Mizutani, T., and Yoshida, J.I. (2002) *Tetrahedron*, 58, 8669–8677.
31 Sun, X., Lee, H., Lee, S., and Tan, K.L. (2013) *Nat. Chem.*, 5, 790–795.
32 Ren, B., Rahm, M., Zhang, X., Zhou, Y., and Dong, H. (2014) *J. Org. Chem.*, 79, 8134–8142.
33 Lu, Y., Wei, P., Pei, Y., Xu, H., Xina, X., and Pei, Z. (2014) *Green Chem.*, 16, 4510–4514.
34 Kawabata, T., Muramatsu, W., Nishio, T., Shibata, T., and Schedel, H. (2007) *J. Am. Chem. Soc.*, 129, 12890–12895.
35 Muramatsu, W., Mishiro, K., Ueda, Y., Furuta, T., and Kawabata, T. (2010) *Eur. J. Org. Chem.*, 2010, 827–831.

36 Ueda, Y., Muramatsu, W., Mishiro, K., Furuta, T., and Kawabata, T. (2009) *J. Org. Chem.*, 74, 8802–8805.
37 Yoshida, K., Furuta, T., and Kawabata, T. (2010) *Tetrahedron Lett.*, 51, 4830–4832.
38 Wang, C.-C., Lee, J.-C., Luo, S.-Y., Kulkarni, S.S., Huang, Y.-W., Lee, C.-C., Chang, K.-L., and Hung, S.-C. (2007) *Nature*, 446, 896–899.
39 Wang, C.-C., Kulkarni, S.S., Lee, J.-C., Luo, S.-Y., and Hung, S.-C. (2008) *Nat. Protoc.*, 3, 97–113.
40 Tsunoda, T., Suzuki, M., and Noyori, R. (1980) *Tetrahedron Lett.*, 21, 1357–1358.
41 Hatekeyama, S., Mori, H., Kitano, K., Yamada, H., and Nishizawa, M. (1994) *Tetrahedron Lett.*, 35, 4367–4370.
42 Wang, C.-C., Lee, J.-C., Luo, S.-Y., Fan, H.-F., Pai, C.-L., Yang, W.-C., Lu, L.-D., and Hung, S.-C. (2002) *Angew. Chem.*, 114, 2466–2468; *Angew. Chem. Int. Ed.* (2002) 41, 2360–2362.
43 Sassaman, M.B., Kotian, K.D., Prakash, G.K.S., and Olah, G.A. (1987) *J. Org. Chem.*, 52, 4314–4319.
44 Huang, T.-Y., Zulueta, M.M.L., and Hung, S.-C. (2014) *Org. Biomol. Chem.*, 12, 376–382.
45 Hu, Y.-P., Lin, S.-Y., Huang, C.-Y., Zulueta, M.M.L., Liu, J.-Y., Chang, W., and Hung, S.-C. (2011) *Nat. Chem.*, 3, 557–563.
46 Hu, Y.-P., Zhong, Y.-Q., Chen, Z.-G., Chen, C.-Y., Shi, Z., Zulueta, M.M.L., Ku, C.-C., Lee, P.-Y., Wang, C.-C., and Hung, S.-C. (2012) *J. Am. Chem. Soc.*, 134, 20722–20727.
47 Chi, F.-C., Kulkarni, S.S., Zulueta, M.M.L., and Hung, S.-C. (2009) *Chem. Asian J.*, 4, 386–390.
48 Chang, C.-H., Lico, L.S., Huang, T.-Y., Lin, S.-Y., Chang, C.-L., Arco, S.D., and Hung, S.-C. (2014) *Angew. Chem.*, 126, 10034–10037; *Angew. Chem. Int. Ed.* (2014) 53, 9876–9879.
49 Patil, P.S., Lee, C.-C., Huang, Y.-W., Zulueta, M.M.L., and Hung, S.-C. (2013) *Org. Biomol. Chem.*, 11, 2605–2612.
50 Huang, T.-Y., Zulueta, M.M.L., and Hung, S.-C. (2011) *Org. Lett.*, 13, 1506–1509.
51 Chang, K.-L., Zulueta, M.M.L., Lu, X.-A., Zhong, Y.-Q., and Hung, S.-C. (2010) *J. Org. Chem.*, 75, 7424–7427.
52 Behera, A., Emmadi, M., and Kulkarni, S.S. (2014) *RSC Adv.*, 4, 58573–58580.
53 Français, A., Urban, D., and Beau, J.-M. (2007) *Angew. Chem.*, 119, 8816–8819; *Angew. Chem. Int. Ed.* (2007) 46, 8662–8665.
54 Bourdreux, Y., Lemétais, A., Urban, D., and Beau, J.-M. (2011) *Chem. Commun.*, 47, 2146–2148.
55 Yang, W.-C., Lu, X.-A., Kulkarni, S.S., and Hung, S.-C. (2003) *Tetrahedron Lett.*, 44, 7837–7840.
56 Zulueta, M.M.L., Lin, S.-Y., Lin, Y.-T., Huang, C.-J., Wang, C.-C., Ku, C.-C., Shi, Z., Chyan, C.-L., Irene, D., Lim, L.-H., Tsai, T.-I., Hu, Y.-P., Liu, J.-Y., Chang, W., Arco, S.D., Wong, C.-H., and Hung, S.-C. (2012) *J. Am. Chem. Soc.*, 134, 8988–8995.
57 Podilapu, A.R. and Kulkarni, S.S. (2014) *Org. Lett.*, 16, 4336–4339.
58 Enugala, R., Carvalho, L.C.R., and Marques, M.M.B. (2010) *Synlett*, 2010, 2711–2716.
59 Despras, G., Urban, D., Vauzeilles, B., and Beau, J.-M. (2014) *Chem. Commun.*, 50, 1067–1069.
60 Joseph, A.A., Verma, V.P., Liu, X.-Y., Wu, C.-H., Dhurandhare, V.M., and Wang, C.-C. (2012) *Eur. J. Org. Chem.*, 2012, 744–753.

61 Hsu, Y., Ma, H.-H., Lico, L.S., Jan, J.-T., Fukase, K., Uchinashi, Y., Zulueta, M.M.L., and Hung, S.-C. (2014) *Angew. Chem.*, 126, 2445–2448; *Angew. Chem. Int. Ed.* (2014) 53, 2413–2416.
62 Ko, Y.-C., Tsai, C.-F., Wang, C.-C., Dhurandhare, V.M., Hu, P.-L., Su, T.-Y., Lico, L.S., Zulueta, M.M.L., and Hung, S.-C. (2014) *J. Am. Chem. Soc.*, 136, 14425–14431.
63 Hale, K.J., Hough, L., Manaviazar, S., and Calabrese, A. (2015) *Org. Lett.*, 16, 4838–4841.
64 Sanapala, S.R. and Kulkarni, S.S. (2014) *Chem. Eur. J.*, 20, 3578–3583.
65 Cai, Y., Ling, C.-C., and Bundle, D.R. (2009) *J. Org. Chem.*, 74, 580–589.
66 McGeary, R.P., Wright, K., and Toth, I. (2001) *J. Org. Chem.*, 66, 5102–5105.
67 Dong, H., Pei, Z., and Ramström, O. (2006) *J. Org. Chem.*, 71, 3306–3309.
68 Dong, H., Rahm, M., Thota, N., Deng, L., Brinck, T., and Ramström, O. (2013) *Org. Biomol. Chem.*, 11, 648–653.
69 Dong, H., Rahm, M., Brinck, T., and Ramström, O. (2008) *J. Am. Chem. Soc.*, 130, 15270–15271.
70 Ren, B., Dong, H., and Ramström, O. (2014) *Chem. Asian J.*, 9, 1298–1304.
71 Dong, H., Pei, Z., Angelin, M., Byström, S., and Ramström, O. (2007) *J. Org. Chem.*, 72, 3694–3701.
72 Emmadi, M. and Kulkarni, S.S. (2013) *Org. Biomol. Chem.*, 11, 4825–4830.
73 Emmadi, M. and Kulkarni, S.S. (2013) *Org. Biomol. Chem.*, 11, 3098–3102.
74 Emmadi, M. and Kulkarni, S.S. (2013) *Nat. Protoc.*, 8, 1870–1889.
75 Emmadi, M. and Kulkarni, S.S. (2014) *Carbohydr. Res.*, 399, 57–63.
76 Hsu, C.-H., Hung, S.-C., Wu, C.-Y., and Wong, C.-H. (2011) *Angew. Chem.*, 123, 12076–12129; *Angew. Chem. Int. Ed.* (2011) 50, 11872–11923.
77 Seeberger, P.H. (2008) *Chem. Soc. Rev.*, 37, 19–28.

Part V

Stereoselective Synthesis of Deoxy Sugars, Furanosides, and Glycoconjugate Sugars

13

Selective Glycosylations with Deoxy Sugars

Clay S. Bennett

13.1 Introduction

Many natural products possess oligosaccharide chains that are at least partially composed of deoxy sugars (Figure 13.1) [1]. These sugars are often critical for a molecule's biological activity, which can be profoundly altered by changing the composition and/or length of the attached oligosaccharides. For example, in the angucycline anticancer–antibiotic, landomycins the full-length deoxy-sugar hexasaccharide chain of landomycin A is necessary for maximal cytotoxicity [2]. Conversely, in the cardiac glycoside digitoxin, it has been shown by both the Thorson group and the O'Doherty group that it is possible to improve selectivity for anticancer activity over cardiotoxicity by replacing the trisaccharide side chain with different monosaccharides [3, 4]. Observations such as these not only point to the importance of deoxy sugars in natural products but also suggest that alteration of these oligosaccharides may result in new leads for the development of therapeutics. This process, termed glycorandomization,[1] has the potential to serve as a new platform for drug discovery; however, it has yet to be widely adopted. This is due in large part to the difficulties associated with oligosaccharide synthesis. Chemical synthesis remains the main avenue of producing oligosaccharides, especially in the case of unusual oligosaccharides from secondary metabolites or unnatural glycorandomized materials. As noted earlier, many of these oligosaccharides possess 2-deoxy-sugar linkages. Controlling the stereochemical outcome of glycosylations using 2-deoxy-sugar donors has been an active area of research for three decades, and early approaches have been reviewed extensively [5]. This chapter focuses on recent developments (mostly within the past 10 years) in the development of direct methods to construct 2-deoxy-sugar linkages.

1 Strictly speaking, glycorandomization refers to enzymatic modification of the sugars on a natural product; however, chemical synthesis still affords access to a greater diversity of possible oligosaccharide structures. We therefore apply the term to both enzymatic and chemical alterations of a natural product's oligosaccharides.

Selective Glycosylations: Synthetic Methods and Catalysts, First Edition. Edited by Clay S. Bennett.

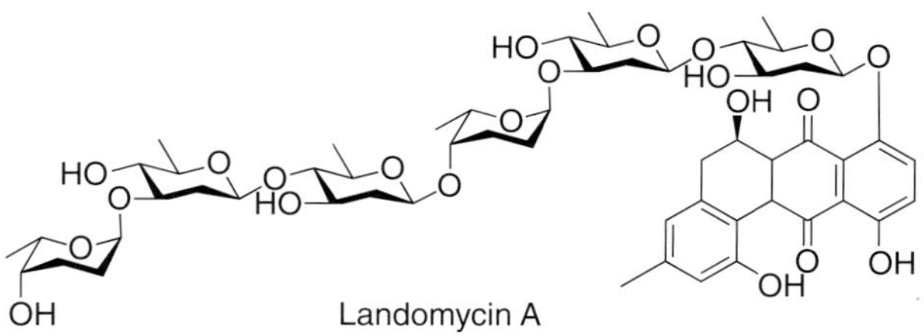

Landomycin A

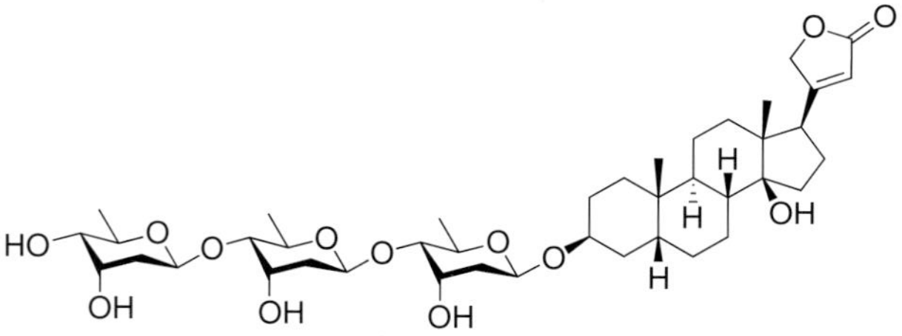

Digitoxin

Figure 13.1 Examples of deoxy sugar containing bioactive natural products.

13.2 Challenges in 2-Deoxy-Sugar Synthesis

2-Deoxy-sugars possess several characteristics that make their synthesis unique from and, at times more challenging than, classical oligosaccharide synthesis. The lack of substitution at C-2 (and often C-6) makes deoxy sugars more labile than their fully substituted counterparts. As a consequence, many classes of deoxy-sugar donors are so unstable that they must be generated and used immediately. Furthermore, many classes of deoxy-sugar donors decompose upon exposure to standard silica gel purification and must be isolated using vacuum line techniques [6] or using extremely expensive and specialized purification media [7]. Even in cases where the doxy-sugar donor is stable (e.g., hemiacetals of glycals), the glycosylation products are extremely sensitive and can be prone to decomposition [8].

In addition to the issues associated with deoxy-sugar stability, deoxy sugars tend to undergo nonselective glycosylations with alcohol nucleophiles in the absence of directing groups. For example, Woerpel has demonstrated that, while deoxy-sugar thioglycosides react with weak nucleophiles, such as 2,2,2-trifluroethanol, in the presence of NIS/TfOH to afford products with modest selectivity, more reactive alcohols provide products as a mixture of isomers [9]. This is attributed to the fact that alcohol nucleophiles, which react near the diffusion limit, are able to react with either conformer of the intermediate oxocarbenium cation. The reaction can be made to proceed through an S_N2-like manifold (which leads to more selective reactions) through judicious choice of solvent or leaving group. For example, in 1995, Hashimoto and coworkers demonstrated that activation of 2-deoxy-sugar phosphates at low temperatures (−94 °C) afforded products with high levels of β-selectivity, presumable through the intermediacy of a glycosyl triflate [10]. Later, during their studies on the landomycin trisaccharide repeat, Sulikowski and coworkers found that other 2-deoxy-sugar phosphates reacted with only modest selectivity (about 3 : 1 $\beta:\alpha$), highlighting the sensitivity of many methods to the structures of the coupling partners [11]. Selectivity in this case could be improved through the use of alternative solvents. For example, Woerpel and coworkers later demonstrated that conducting this chemistry in 2,2,2-trichloroethylene affords β-linked products resulted in good (9 : 1 $\beta:\alpha$) selectivity [12]. A general approach to controlling selectivity in glycosylation reactions with deoxy-sugar donors has yet to emerge, however.

One of the challenges in developing selective glycosylations with deoxy sugars is directly a result of their deoxygenation. Specifically, they often cannot take advantage of the tools for controlling selectivity developed for glycosylations with other classes of monosaccharides. For example, the lack of oxygenation at C-2 and usually C-6 precludes the use of strategies such as anchimeric assistance [13] or conformational locking [14]. This has led to a number of strategies for controlling the selectivity in glycosylation reactions using 2-deoxy sugars. These approaches can be broken into three broad categories: indirect synthesis, *de novo* synthesis, and direct synthesis (Scheme 13.1). While the focus of this chapter is on the methods for direct synthesis, a brief introduction to the other two approaches will follow.

In indirect synthesis, a temporary group, such as a halide or thioether, is attached to C-2 of the glycosyl donor in order to both control selectivity and increase donor stability [15, 16]. Alternatively, 2,3-anhydro sugar thioglycosides can be activated using either 4 Å

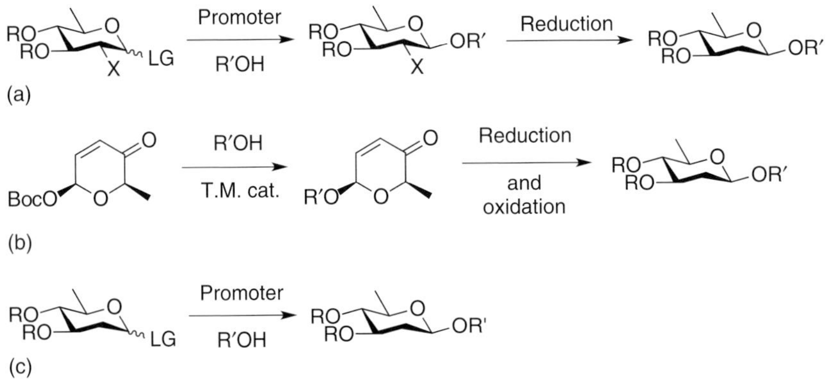

Scheme 13.1 Synthesis of deoxy sugars using indirect glycosylation (a), *de novo* synthesis (b), and direct glycosylation (c).

molecular sieves in refluxing CH_2Cl_2 or $Cu(OTf)_2$ to react with nucleophiles to afford β-linked 2-deoxy-2-thiotoylyl sugars in excellent yield as single isomers. Reductive removal of the thioether or iodide using Bu_3SnH, hydrogenolysis, or visible light then furnished the deoxy sugar [17–19]. This approach suffers from the drawback of the need to stereoselectively install and later remove the prosthetic group, adding additional steps to the synthesis. Still, it has found extensive use in deoxy-sugar oligosaccharide synthesis. Indeed, a number of impressive oligosaccharide synthesizes have relied exclusively on it, including approaches to the landomycin hexasaccharide by the Roush and Yu groups [20, 21].

On the other hand, *de novo* synthesis relies on forming the glycosidic linkages on substrates that do not possess the oxidation states found in the desired product, usually through a transition-metal-mediated process [22]. Manipulation of the oxidation states of the products of these reactions then affords the desired glycosides. Because the reactions in question proceed through a different mechanism than the classical glycosylation manifold, this approach can permit the construction of anomeric linkages with extremely high levels of selectivity. As with indirect synthesis, this is a powerful approach, especially for systems possessing unusual sugars, such as digitoxin and vineomycin B2 [23, 24]. It does suffer from a similar limitation as indirect synthesis, however, in that extensive manipulation of the oxidation states of the substrate must be carried out following the glycosylation in order to obtain the desired glycan product.

Efforts to streamline deoxy-sugar synthesis have led to increased interest in developing direct methods for the stereoselective glycosylation reactions with deoxy sugars. In this approach, all of the functionality in the product (and only functional groups in the product) are already present in both of the coupling partners. Thus, it has the potential to permit synthetic schemes that consist solely of glycosylation and deprotection steps (similar to peptide synthesis). As noted earlier, however, a direct approach to the direct stereoselective construction of deoxy-sugar oligosaccharides remains to be established. That being said, a number of promising approaches that hold enormous potential for addressing this issue have emerged in the past decade or so. The remainder of this chapter will focus on recent developments in this latter approach.

13.3 Protecting Group Strategies

Classical protecting group strategies have met with limited success in controlling the selectivity in glycosylation reactions with 2-deoxy-sugar donors. This has spurred many groups to adopt the use of unconventional protecting groups in 2-deoxy-sugar synthesis, an approach that has led to some successes. For example, Tanaka, Yoshizawa, and Takahashi found that olivose trichloroacetimidates possessing a benzenesulfonate at C4 undergo highly selective glycosylations (>95 : 5 β : α) promoted by triethylsilane and iodide at low temperatures (Scheme 13.2a) [7]. Although no explanation for the origin of the selectivity is provided, it appears that the C4 sulfonate group is necessary for optimal selectivity. Using this approach, Tanaka, Takahashi and coworkers prepared a library of landomycin A analogs through a [3+3] strategy (Scheme 13.2b). Again, trisaccharide donors that possessed a C4 sulfonate on the reducing sugar provided the highest selectivity in the reaction [25]. Interestingly, it appears that both the protecting group and the nature of the promoter system are necessary for selectivity. When olivose donors with this protecting group at C4 are activated with bromine and tetrabutylammonium bromide at 0 °C, the reactions proceed with high levels of α-selectivity [26].

More recently, Mong and coworkers have shown that it is possible to obtain highly selective glycosylation reactions using 2-deoxy-thioglycoside donors possessing a C6 picolyl (Pico) protecting group (Scheme 13.3) [27]. This protecting group was initially developed by Demchenko and coworkers to promote selective glycosylations through long-range interactions [28]. Rather than interact directly with the activated

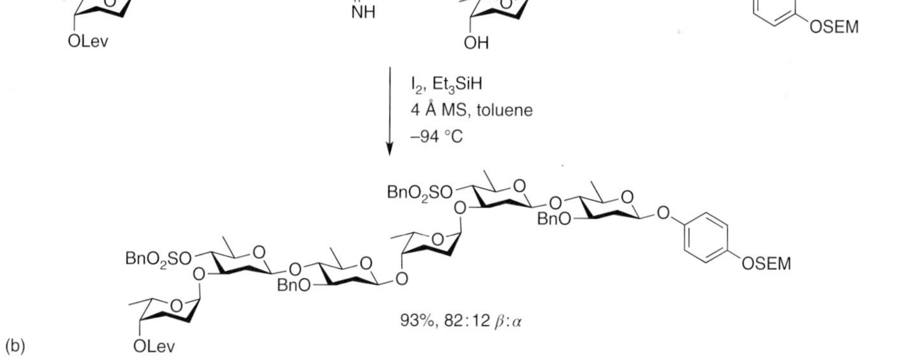

Scheme 13.2 (a) Tanaka and Takahashi's use of the C-4 benzenesulfonate protecting groups to affect β-selectivity in deoxy-sugar synthesis. (b) Use of this protecting group in the synthesis of the landomycin A hexasaccharide.

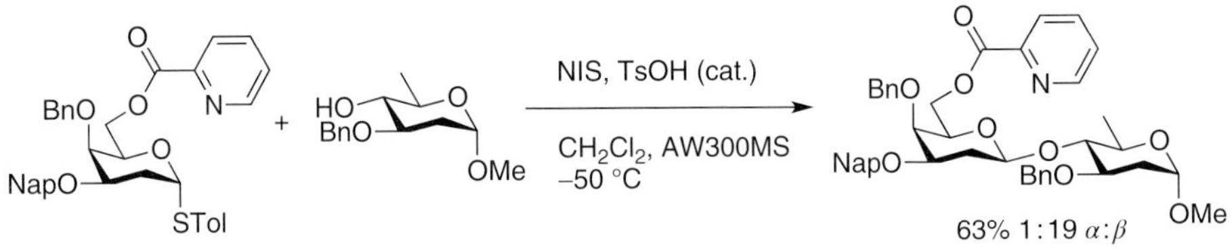

Scheme 13.3 Mong's use of the picolyl-protecting group in β-selective deoxy-sugar synthesis.

oxocarbenium cation, this picolyl nitrogen is thought to control selectivity by hydrogen bonding to the glycosyl acceptor. This results in the nucleophile attacking the same face of the molecule that the Pico directing group is located on. Mong demonstrated that this hydrogen-bond-mediated aglycone delivery approach can be successfully used to synthesize a number of β-linked deoxy sugars, including 2-deoxy-thioglucosides, 2-deoxy-thiogalactosides, and 2-deoxy-thioallosides. Furthermore, this group also showed that using the Pico group at the C4 position leads to α-selective glycosylations, thereby demonstrating that the stereochemical outcome of the reaction is one in which the newly formed glycosidic linkage is *syn* to the Pico group. Following glycosylation, this group can be selectively removed using $Cu(OAc)_2$ in methanol to expose the C6 alcohol for Barton–McCombie deoxygenation [29].

13.4 Addition to Glycals

In 1990, Mioskowski and Falck reported that treating a protected glycal with 5 mol% $PPh_3{\cdot}HBr$ in the presence of an alcohol led to the formation of α-linked-2-deoxy glycosides in moderate-to-good yields and with excellent selectivity (Scheme 13.4a) [30]. This reaction offered a key advantage over the previously reported additions to glycals in that no Ferrier rearrangement products were observed in the course of the reaction.

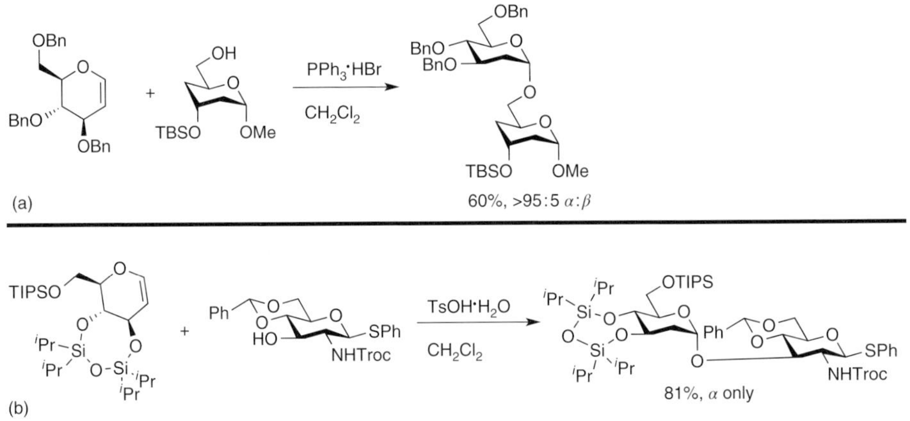

Scheme 13.4 (a) Acid-catalyzed activation of glycals for glycosylation. (b) The use of 3,4-*O*-tetraisopropyldisiloxane-protecting group for α-selective glycosylations.

Recent years have witnessed the development of several modifications to this chemistry in order to improve both its generality and selectivity. In 2004, Kirschning and coworkers increased the utility of the reaction by demonstrating that polymer-immobilized diphenylphosphine hydrobromide could not only efficiently promote the reaction but also be used to carry out multiple glycosylations in a single pot [31]. The conditions for the reaction are particularly mild; however, selectivity in the reaction does appear to depend on the nature of the coupling partners and the protecting groups on the glycal. For example, Wandzik and coworkers have shown that an allylic-TBS-protecting group on the glycal can permit superior levels of selectivity in the rhamnal series [32].

A number of other Lewis and Brønsted acids have been used to activate glycals for glycosylation; recent examples include the use of Re(V) catalysis [33], $AlCl_3$ with microwave irradiation [34], PPh_3/TMSI [35], and thiourea acids [36]. In general, the reactions tend to be α-selective, with galacto-configured glycals providing higher levels of selectivity compared to the corresponding gluco-configured donors. Selectivity with the latter group of donors can be low [30]; however, this issue can be overcome with specialized protecting groups. For example, Galan and McGarrigle demonstrated that constraining glucal donors with a 3,4-*O*-siloxane group permitted extremely high levels of selectivity in reactions promoted by a catalytic amount of *p*-toluenesulfonic acid (Scheme 13.4b) [37]. In this case, oxygenation at the C6 position of the donor is necessary for optimal selectivity, as reactions with similarly protected rhamnal donors proceeded with attenuated, although usually synthetically useful, selectivity.

The configuration of the donor plays an important role in the stereochemical outcome of reaction. For example, McDonald and coworkers demonstrated that 6-deoxy-allal donors in which a bulky *tert*-butyldimethylsilyl protecting group was present on the axial C-3 alcohol could be activated with PPh_3·HBr for highly β-selective glycosylation reactions (Scheme 13.5) [38]. Using this chemistry, along with their tungsten-mediated alkynol cyclization chemistry for *de novo* oligosaccharide synthesis [15], this group was able to rapidly synthesize the trisaccharide of digitoxin. More recently, the Zhu group demonstrated that the Re(V) promoters developed by Toste for α-selective glycal activation could also be used for β-selective reactions with allal donors. This approach was also applied to the synthesis of the digitoxin trisaccharide [39].

13.5 Additions to Glycosyl Halides

While glycosyl halides are commonly used as donors for a variety of glycosidic linkages, the use of 2-deoxy-glycosyl halides is far more limited, owing in part to their lability. 2-Deoxy-glycosyl chlorides are the most stable species and have been used as intermediates

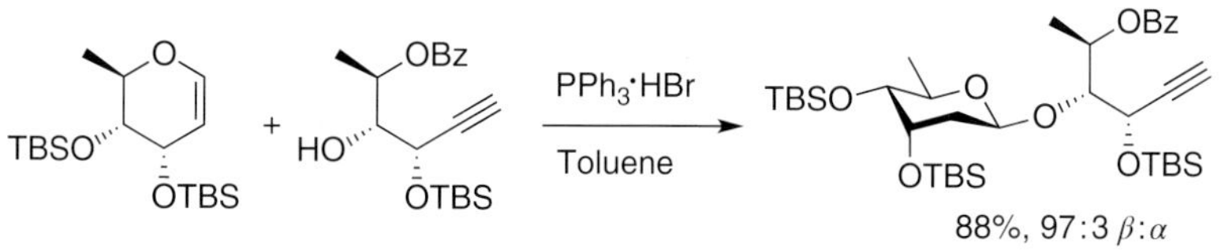

Scheme 13.5 The use of an axial TBS ether for β-selective acid-catalyzed addition to glycals.

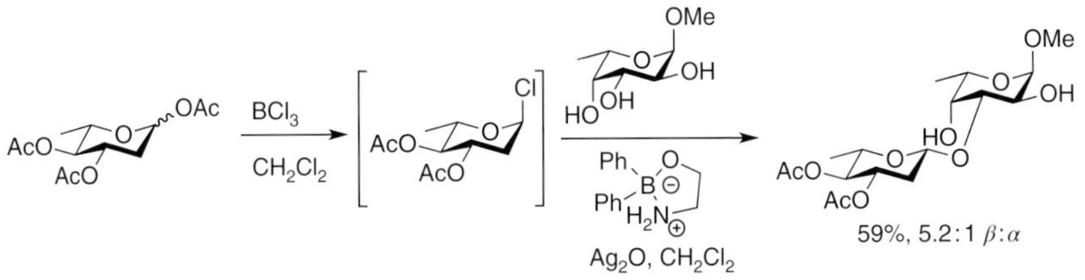

Scheme 13.6 Aminoethyl-diphenylboronate-catalyzed regioselective glycosylation of 2,6-dideoxy-sugar chlorides.

in the synthesis of other glycosyl donors, such as glycosyl phosphonates [40]. The species can be activated by a variety of silver salts, and the use of bulky insoluble silver salts, such as Ag_2O, can permit activation through an S_N2-like manifold to afford β-enriched products. Taylor and coworkers were able to take advantage of this property to activate glycosyl chlorides for regioselective glycosylation of diols using aminoethyl diphenylborinate catalysis to control selectivity (Scheme 13.6) [41]. As with most β-selective reactions with deoxy sugars, 2-deoxy sugars consistently provided higher levels of selectivity compared to the corresponding 2,6-dideoxy sugars. The reaction also requires the use of disarming protecting groups, presumably to inhibit complete ionization of the donor to the corresponding oxocarbenium cation. Still, the approach represents an extremely attractive avenue to β-linked 2-deoxy sugars.

Other classes of glycosyl halides (bromides and iodides) are less stable, and specialized techniques are often required for their synthesis and isolation. Thiem and Meyer first reported the synthesis of 2-deoxy-glycosyl bromides and iodides in 1980 [42]. The authors found that treating 2-deoxy-glycosyl acetates with trimethylsilyl bromide or iodide resulted in the formation of the corresponding glycosyl bromide or iodide, respectively. This group also demonstrated that the 2,6-dideoxy-1-methyl glycosides could also be converted to the corresponding glycosyl bromide using TMSBr [43]. The highly reactive glycosyl bromides underwent glycosylation with secondary acceptors in the presence of *sym*-collidine and 4 Å MS to afford the corresponding α-linked disaccharides in moderate yield (38–41%, Scheme 13.7). The highly reactive nature of these molecules meant that they had to be generated immediately prior to their use [44]. Despite this, there has been a resurgence of interest in these molecules over the past decade.

In 2003, Lam and Gervay-Hague demonstrated that 2-deoxy-glycosyl iodides could be generated in high yield from the corresponding glycals through a two-step sequence [45]. This process first involved conversion of the glycal into the corresponding glycosyl acetate using substoichiometric amounts of 30% HBr in acetic acid. These acetates

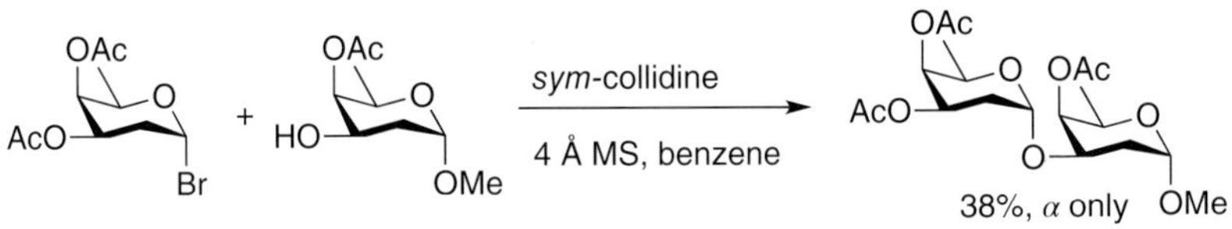

Scheme 13.7 The use of deoxy-sugar bromides for α-selective glycosylation.

could then be converted into the corresponding glycosyl iodides under Thiem's conditions (TMSI). Importantly, they found that treating 2-deoxy- and 2,6-dideoxy-glycosyl iodides with the potassium salts of aromatic alcohols afforded β-linked phenolic glycosides in moderate-to-excellent yield as single isomers (Scheme 13.8). The approach has yet to be extended to using glycoside acceptors. Nevertheless, it represents an example of an S_N2 glycosylation to afford aryl glycoside linkages commonly found in many natural products.

More recently, Kaneko and Hezron reported that α-linked glycosyl bromides can be activated using silver silicate for highly β-selective glycosylations with carbohydrate acceptors (Scheme 13.9) [6]. Although the bromides are not stable, the Hezron group was able to isolate these species using vacuum and inert gas techniques and fully characterize them by NMR. The authors found that it was not necessary to purify the bromide before glycosylation, which further simplified the experimental setup. The inherent instability of these donors required the use of a nitrogen dry box to set up the glycosylation, however, highlighting the challenges inherent in deoxy-sugar synthesis.

13.6 Latent Glycosyl Halides

While the highly reactive nature of 2-deoxy-sugar halides makes them attractive as glycosyl donors, this is somewhat attenuated by their instability. This has prompted searches for methods that permit the *in situ* generation of glycosyl halides from stable donors. In 2011, Nogueira and Bennett reported that preactivation of 2-deoxy-sugar hemiacetals with 3,3-dichloro-1,2-diphenylcyclopropene and tetrabutylammoniumiodide (TBAI) resulted in the generation of a species that underwent glycosylation with a number of donors in moderate-to-good α-selectivity (Scheme 13.10a) [46]. ^{1}H NMR studies on the reaction revealed that the dihalocyclopropene rapidly converted the hemiacetal into the

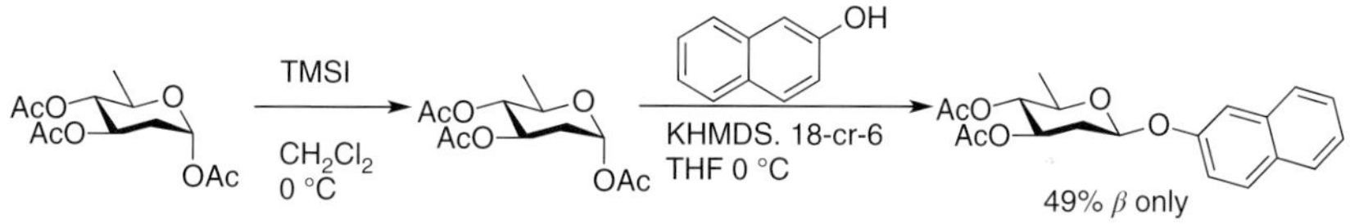

Scheme 13.8 The use of 2-deoxy-glycosyl iodides in the β-specific construction of aryl glycosides.

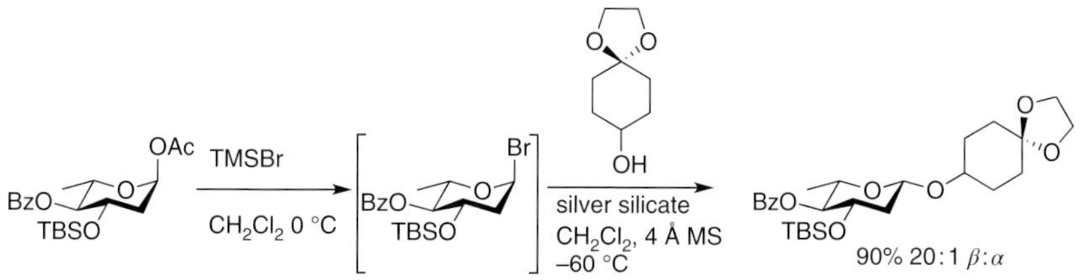

Scheme 13.9 Silver-silicate-promoted activation of *in situ* generated 2-deoxy-glycosyl bromides for β-selective glycosylation.

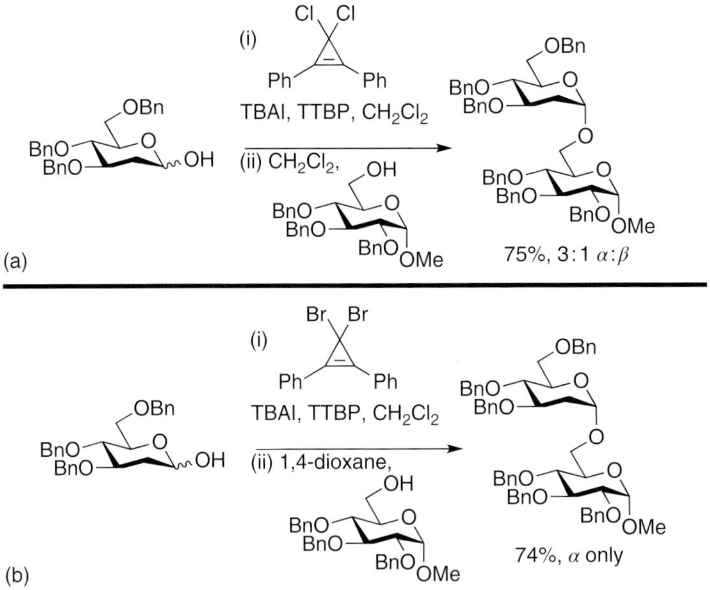

Scheme 13.10 First- (a) and second-generation (b) dihalocyclopropene/TBAI-promoted dehydrative glycosylations with 2-deoxy sugars.

corresponding glycosyl chloride, in line with other reports describing the reaction of this reagent with alcohols [47]. Under these conditions, it was assumed that the chloride reacted with the TBAI to generate a reactive glycosyl iodide *in situ* [48]. This latter species reacted with acceptors under halide ion conditions [49], to afford α-linked products. Such a mechanism was supported by the fact that both D- and L-donors reacted with D-acceptors in moderate-to-good α-selectivity. Reaction times were long (48 h), however, which could be accounted for by the fact that NMR studies showed that the glycosyl chloride reacted only slowly with the TBAI. In an effort to improve this, the Bennett lab subsequently investigated the use of the corresponding dibromocyclopropene in the reaction. Using this promoter in the presence of 1,4-dioxane led to both a substantial reduction in reaction times and an increase in α-selectivity (Scheme 13.10b) [50]. Again the stereochemical outcome of the reaction was independent of the nature of the coupling partners, providing evidence for an S_N2-like reaction.

Subsequent to these studies, Verma and Wang reported that 2-deoxy- and 2,6-dideoxy-thioglycosides can be preactivated with a combination of Ag(OTf) and *p*-TolSCl for α-selective glycosylation reactions (Scheme 13.11a) [51]. Although the authors initially suspected that the reaction was proceeding through the intermediacy of a glycosyl triflate, low-temperature NMR studies revealed that the thioglycoside was instead being converted into the corresponding glycosyl chloride. This latter species is activated by the silver salt for glycosylation. Interestingly, the authors noted that the reaction was somewhat sensitive to the protecting group patterns on the donor. While benzyl-ether-protected donors underwent highly selective reactions, the use of the 4,6-benzylidene protecting group led to an erosion of selectivity. Since the reaction relies on preactivation of the donor, the authors investigated if it would be possible to use this chemistry

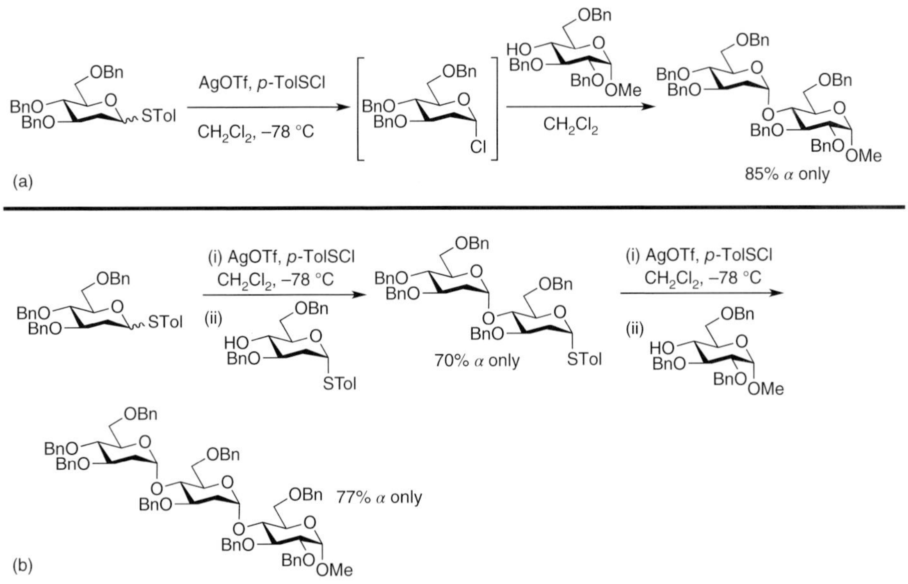

Scheme 13.11 The use of 2-deoxy-thioglycosides as latent glycosyl chlorides for α-selective glycoside synthesis.

in iterative oligosaccharide synthesis. This indeed proved to be the case, and they were able to demonstrate the use of this chemistry in the synthesis of a trisaccharide in excellent yield as a single α,α-isomer (Scheme 13.11b).

13.7 Reagent-Controlled Approaches

In the latent halide activation methods described earlier, the selectivity in the reaction is completely independent of the stereochemical information in the coupling partners. Such processes are highly desirable, because they have the potential to eliminate the need for specialized protecting group patterns or problems with stereochemical mismatch in chemical glycosylation reactions [52]. This has in turn led to a search for other promoter systems that are able to dictate the stereochemical outcome of the glycosylation reaction. In 2013, the Bennett lab reported that metalating a hemiacetal at low temperature with KHMDS, followed by treating the resulting anion with *N*-tosylimidazole, resulted in the *in situ* formation of a species that underwent β-specific glycosylation reactions with thiolate and phenolate nucleophiles (Scheme 13.12a) [53]. It was hypothesized that the reactive species in solution was an α-glycosyl tosylate that underwent displacement by the nucleophile through an S_N2-like manifold. This observation was confirmed in 2014, when the same group used low-temperature NMR to study the reaction between a metalated hemiacetal and a *p*-toluenesulfonic anhydride [54]. The initially formed species under these conditions possesses an anomeric proton at δ 6.11 ppm in THD-d_8, which correlated in the heteronuclear single-quantum correlation (HSQC) spectrum to an anomeric carbon at δ 102.3 ppm, both of which are

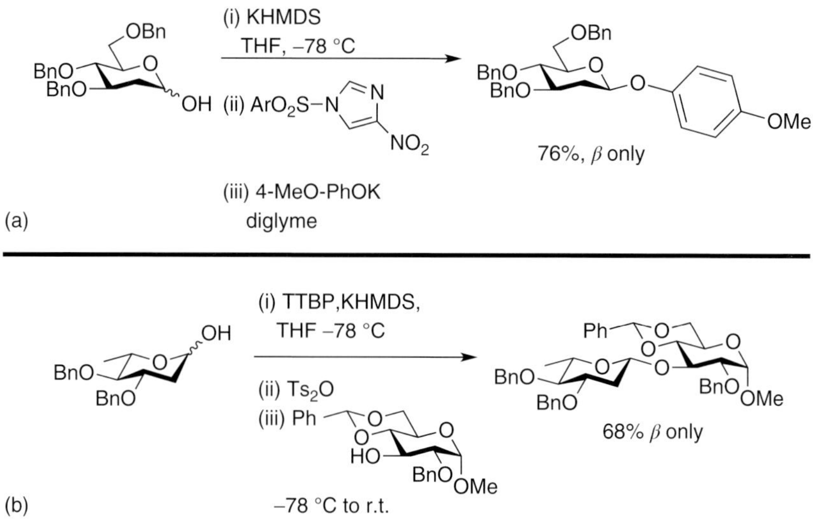

Scheme 13.12 β-Specific deoxy-sugar synthesis through the intermediacy of a glycosyl tosylate promoted by either *p*-toluenesulfonyl-4-nitroimidazole (a) or *p*-toluenesulfonic anhydride (b).

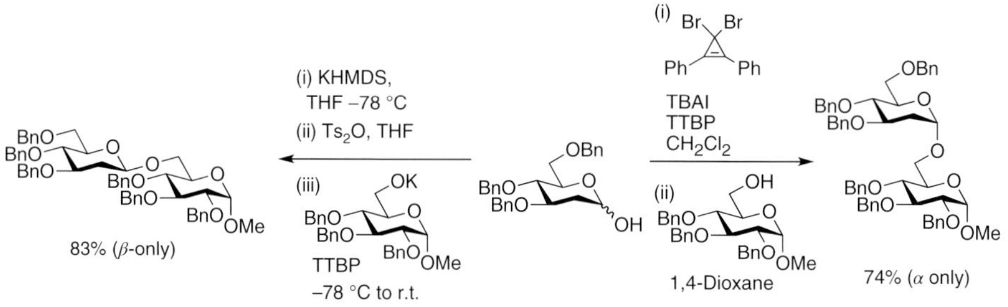

Scheme 13.13 "Reagent-controlled" synthesis of 2-deoxy sugars, where the stereochemical outcome of the reaction is controlled entirely by the glycosylation promoter.

diagnostic for a glycosyl sulfonate [55]. This species was also shown to undergo β-specific reactions with carbohydrate nucleophiles (Scheme 13.12b). Importantly, under these conditions, the same coupling partners that underwent α-selective coupling reactions in the dihalocyclopropene promoted the reactions described earlier (Scheme 13.10, Section 13.6), reacted in a β-specific manner. Taken together, these studies demonstrate that it is possible to place the stereochemical outcome of the reaction entirely under control of the promoter (Scheme 13.13), a process that the Bennett lab has described as "reagent control."

Reagent-controlled approaches have also been applied to the construction of α-linked deoxy sugars. In 2014, the Mong lab demonstrated that preactivation of thioglycosides with *N*-iodosuccinimide and TfOH in the presence of *N*,*N*-dimethylformamide (DMF) results in the formation of species that consistently undergo α-selective glycosylation reactions (Scheme 13.14a) [56]. NMR studies demonstrated

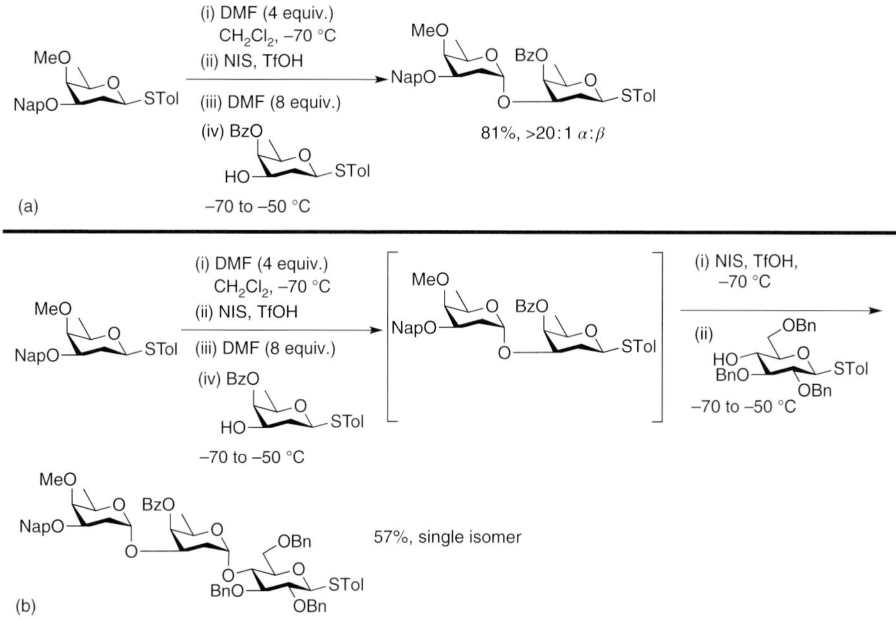

Scheme 13.14 (a) DMF-modulated α-selective glycosylation with deoxy sugars. (b) Application to "one-pot" oligosaccharide synthesis.

that the DMF functions by capturing the oxocarbenium cation generated by thioglycoside activation as an α-glycosyl imidate. This mechanism is in line with this lab's previous work on DMF-modulated glycosylation with fully substituted sugars [57]. This species presumably exists in equilibrium with the corresponding β-imidate, which reacts through an S_N2-like manifold to afford α-linked products. The conditions provide an approach to both 2-deoxy- and 2,6-dideoxy sugars with excellent levels of α-selectivity. A larger amount of DMF (up to 32 equiv.) and lower temperatures are required for selective reactions with the 2,6-dideoxy sugars; however, the excess reagent can be removed by evaporation. Importantly, since the approach relies on preactivation, it can be used with thioglycoside acceptors, permitting one-pot oligosaccharide synthesis (Scheme 13.14b).

13.8 Umpolung Reactivity

A conceptually different approach to deoxy-sugar synthesis is the one in which the "donor" is the nucleophilic coupling partner. This type of umpolung approach was first described by Schmidt and coworkers in the early 1980s, who demonstrated that metalated hemiacetals can react with triflates to generate glycosidic linkages [58]. The reaction worked well for primary triflates; however, secondary triflates required special conditions (HMPA, low temperatures), and the reactions were not selective [59].

In 2009, Morris and Shair demonstrated that this approach could be applied to the synthesis of 2-deoxy sugars (Scheme 13.15) [60]. They found that metalating hemiacetals with NaH at room temperature followed by the addition of the triflates resulted in

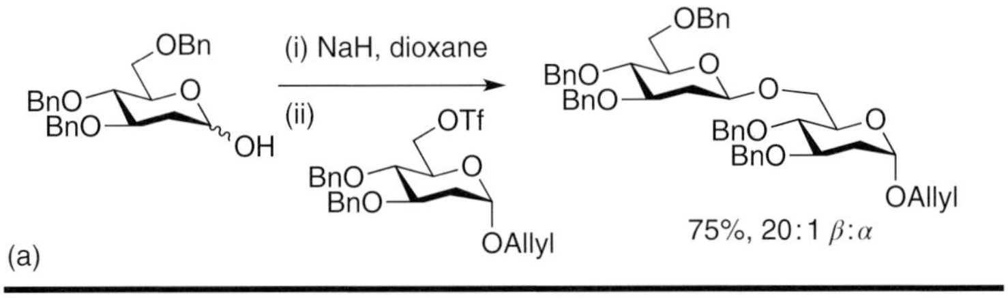

Scheme 13.15 Umpolung activation of hemiacetals for deoxy-sugar disaccharide synthesis (a) and S_NAr construction of phenolic glycosides (b).

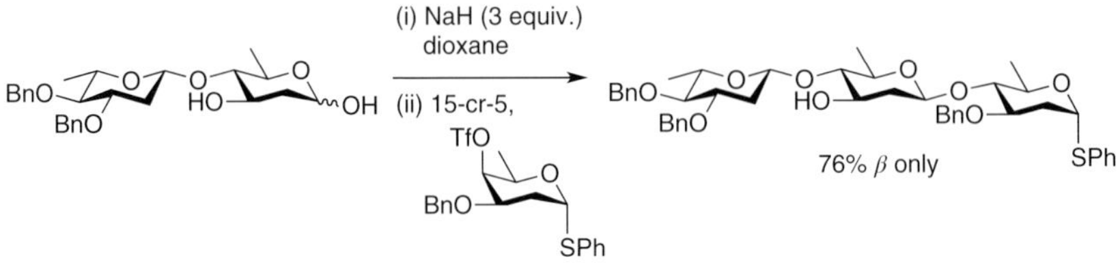

Scheme 13.16 Umpolung approach using dianion formation for the construction of secondary glycosidic linkages.

highly β-selective reactions. Yields were generally good for primary triflates; however, secondary triflates failed to undergo productive glycosylations, reacting instead at the sulfur atom on the triflate leaving group. Interestingly, however, the approach could be applied to S_NAr reactions for aryl glycoside construction, provided a suitably reactive electrophile was used in the reaction.

In 2014, Zhu and coworkers revisited this approach and found that coordinating the sodium counter ion with 15-crown-5 permitted reactions with secondary triflates. Under these conditions, yields were moderate to low, owing to opening of the pyranose ring and elimination of the protected C-3 alcohol. In an effort to circumvent this issue, the authors examined an approach where the C-3 hydroxyl was unprotected (Scheme 13.16) [61]. By treating the diol with excess sodium hydride, the authors were able to generate a dianion, which reacted selectively at the anomeric position in high yield and with near-perfect β-selectivity. Using this approach, the authors were able to synthesize several deoxy-sugar tri- and tetrasaccharides. As a consequence, this approach represents an attractive alternative to conventional glycosylation reactions. Interestingly, the Zhu group found that when digitoxose or boivinose donors were used in these conditions, the reaction was α-specific (Scheme 13.17) [62]. The authors attribute this to a situation where the sodium cation is coordinated between the axial C3 hydroxyl and the anomeric hydroxyl, forcing the latter to adopt an axial configuration. Treatment with the electrophile then leads to the observed selectivity.

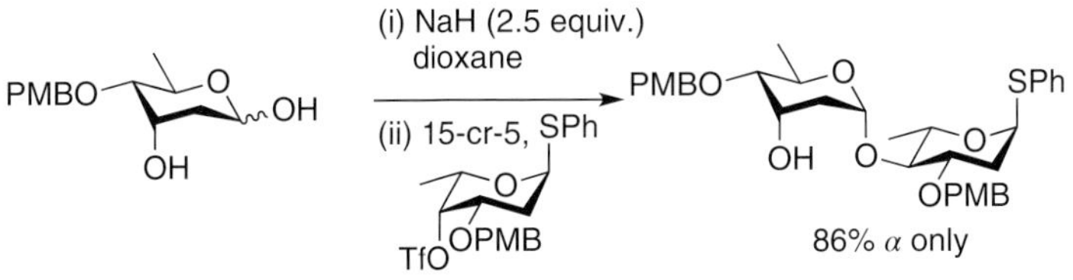

Scheme 13.17 Effect of axial C3 hydroxyl on the stereochemical outcome of dianion umpolung glycosylation.

13.9 Conclusion

Despite numerous advances in the course of the past few decades, carbohydrate synthesis remains an extremely challenging endeavor. This is especially true in the case of the construction of stereodefined deoxy-sugar linkages, which has generally been considered to be one of the most difficult challenges in carbohydrate chemistry [63]. As such, it has attracted increased attention over the past several years. This has led to the development of several new methodologies, which represent the cutting edge of oligosaccharide synthesis, including catalytic methods, reagent-controlled glycosylations, and umpolung approaches. Importantly, many of these approaches may have the potential to function with other classes of glycans, as has been demonstrated in certain cases. Despite this considerable progress, however, a general approach to the stereoselective construction of deoxy-sugar linkages has yet to emerge. Thus, much work remains in the field, which continues to be a fruitful and exciting area for methodology development.

References

1 (a) McCranie, E.K. and Bachmann, B.O. (2014) *Nat. Prod. Rep.*, 31, 1026–1042; (b) Elshahawi, S.I., Shaaban, K.A., Kharel, M.K., and Thorson, J.S. (2015) *Chem. Soc. Rev.*, 44, 7591–7697.

2 (a) Zhu, L., Luzhetskyy, A., Luzhetska, M., Mattingly, C., Adams, V., Bechthold, A., and Rohr, J. (2007) *ChemBioChem*, 8, 83–88; (b) Crow, R.T., Rosenbaum, B., Smith, R. III, Guo, Y., Ramos, K.S., and Sulikowski, G.A. (1999) *Bioorg. Med. Chem. Lett.*, 9, 1663–1666.

3 Langenhan, J.M., Peters, N.R., Guzei, I.A., Hoffman, F.M., and Thorson, J.S. (2005) *Proc. Natl. Acad. Sci. U.S.A.*, 32, 12305–12310.

4 Iyer, A.K.V., Zhou, M., Azad, N., Elbaz, H., Wang, L., Rogalsky, D.K., Rojanasakul, Y., O'Doherty, G.A., Langenhan, J.M., and Med, J.M.A.C.S. (2010) *Chem. Lett.*, 1, 326–330.

5 (a) Marzabadi, C.H. and Franck, R.W. (2000) *Tetrahedron*, 56, 8385–8417; (b) Hou, D. and Lowary, T.L. (2009) *Carbohydr. Res.*, 344, 1911–1940; (c) Borovika, A. and Nagorny, P. (2012) *J. Carbohydr. Chem.*, 31, 255–283.

6 Kaneko, M. and Herzon, S.B. (2014) *Org. Lett.*, 16, 2776–2779.

7 Tanaka, H., Yoshizawa, Y., and Takahashi, T. (2007) *Angew. Chem. Int. Ed.*, 46, 2505–2507.

8 Shan, M., Sharif, E.U., and O'Doherty, G.A. (2010) *Angew. Chem. Int. Ed.*, 49, 9492–9495.

9 Beaver, M.G. and Woerpel, K.A. (2010) *J. Org. Chem.*, 75, 1107–1118.
10 Hashimoto, S.-i., Sano, A., Sakamoto, H., Nakajima, M., Yanagiya, Y., and Ikegami, S. (1995) *Synlett*, 1995, 1271–1273.
11 Guo, Y. and Sulikowski, G.A. (1998) *J. Am. Chem. Soc.*, 120, 1392–1397.
12 Kendale, J.C., Valentín, E.M., and Woerpel, K.A. (2014) *Org. Lett.*, 16, 3684–3687.
13 Nukada, T., Berces, A., Zgierski, M.Z., and Whitfield, D.M. (1998) *J. Am. Chem. Soc.*, 120, 13291–13295.
14 Crich, D. (2010) *Acc. Chem. Res.*, 43, 1144–1153.
15 Roush, W.R. and Bennett, C.E. (1999) *J. Am. Chem. Soc.*, 121, 3541–3542.
16 Yu, B. and Yang, Z. (2001) *Org. Lett.*, 3, 377–379.
17 Hou, D. and Lowary, T.L. (2009) *J. Org. Chem.*, 74, 2278–2289.
18 Wang, H., Tao, J., Cai, X., Chen, W., Zhao, Y., Xu, Y., Yao, W., Zeng, J., and Wan, Q. (2014) *Chem. Eur. J.*, 20, 17319–17323.
19 Yu, B. and Wang, P. (2002) *Org. Lett.*, 4, 1919–1922.
20 Roush, W.R. and Bennett, C.E. (2000) *J. Am. Chem. Soc.*, 122, 6124–6125.
21 Yang, X., Fu, B., and Yu, B. (2011) *J. Am. Chem. Soc.*, 133, 12433–12435.
22 (a) McDonald, F.E. and Zhu, H.Y.H. (1998) *J. Am. Chem. Soc.*, 120, 4246–4247; (b) Haukaas, M.H. and O'Doherty, G.A. (2002) *Org. Lett.*, 4, 1771–1774.
23 Zhou, M. and O'Doherty, G.A. (2008) *Org. Lett.*, 10, 2283–2286.
24 Yu, X. and O'Doherty, G.A. (2008) *Org. Lett.*, 10, 4529–4532.
25 Tanaka, H., Yamaguchi, S., Yoshizawa, A., Takagi, M., Shin-ya, K., and Takahashi, T. (2010) *Chem. Asian J.*, 5, 1407–1424.
26 Tanaka, H., Yoshizawa, A., Chijiwa, S., Ueda, J.-y., Takagi, M., Shun-ya, K., and Takahashi, T. (2009) *Chem. Asian J.*, 4, 1114–1125.
27 Ruei, J.-H., Venukumar, P., Ingle, A.B., and Mong, K.-K.T. (2015) *Chem. Commun.*, 51, 5394–5397.
28 Yasomanee, J.P. and Demchenko, A.V. (2012) *J. Am. Chem. Soc.*, 134, 20097–20102.
29 Barton, D.H.R. and McCombie, S.W. (1975) *J. Chem. Soc., Perkin Trans. 1*, 1574–1585.
30 Bolitt, V., Misokowski, C., Lee, S.-G., and Falck, J.R. (1990) *J. Org. Chem.*, 55, 5812–5813.
31 Jaunzems, J., Kashin, D., Schönberger, A., and Kirschning, A. (2004) *Eur. J. Org. Chem.*, 2004, 3435–3446.
32 Wandzik, I. and Bieg, T. (2006) *Carbohydr. Res.*, 341, 2702–2707.
33 Sherry, B.D., Loy, R.N., and Toste, F.D. (2004) *J. Am. Chem. Soc.*, 126, 4510–4511.
34 Lin, H.-C., Pan, J.-F., Chen, Y.-B., Lin, Z.-P., and Lin, C.-H. (2011) *Tetrahedron*, 67, 6362–6368.
35 Cui, X.-K., Zhong, M., Meng, X.-B., and Li, Z.-J. (2012) *Carbohydr. Res.*, 358, 19–22.
36 Balmond, E.I., Coe, D.M., Galan, M.C., and McGarrigle, E.M. (2012) *Angew. Chem. Int. Ed.*, 51, 9152–9155.
37 Balmond, E.I., Benito-Alifonso, D., Coe, D.M., Alder, R.W., McGarrigle, E.M., and Galan, M.C. (2014) *Angew. Chem. Int. Ed.*, 53, 8190–8194.
38 McDonald, F.E. and Reddy, K.S. (2001) *Angew. Chem. Int. Ed.*, 40, 3653–3655.
39 Baryal, K.N., Adhikari, S., and Zhu, J. (2013) *J. Org. Chem.*, 78, 12469–12476.
40 Niggemann, J., Lindhorst, T.K., Walfort, M., Laupichler, L., Sajus, H., and Thiem, J. (1993) *Carbohydr. Res.*, 246, 173–183.
41 Beale, T.M., Moon, P.J., and Taylor, M.S. (2014) *Org. Lett.*, 16, 3604–3607.
42 Thiem, J. and Meyer, B. (1980) *Chem. Ber.*, 113, 3075–3085.

43 Thiem, J. and Meyer, B. (1980) *Chem. Ber.*, 113, 3058–3066.
44 Streicher, H., Latxague, L., Wiemann, T., Rolllin, P., and Thiem, J. (1995) *Carbohydr. Res.*, 278, 257–270.
45 Lam, S.N. and Gervay-Hague, J. (2003) *Org. Lett.*, 5, 4219–4222.
46 Nogueira, J.M., Nguyen, S.H., and Bennett, C.S. (2011) *Org. Lett.*, 13, 2814–2817.
47 Kelly, B.D. and Lambert, T.H. (2009) *J. Am. Chem. Soc.*, 131, 13930–13931.
48 Kronzer, F.J. and Schuerch, C. (1974) *Carbohydr. Res.*, 34, 71–78.
49 Lemieux, R.U., Hendriks, K.B., Stick, R.V., and James, K. (1975) *J. Am. Chem. Soc.*, 97, 4056–4062.
50 Nogueira, J.M., Issa, J.P., Chu, A.-H.A., Sisel, J.A., Schum, R.S., and Bennett, C.S. (2012) *Eur. J. Org. Chem.*, 2012, 4927–4930.
51 Verma, V.P. and Wang, C.-C. (2013) *Chem. Eur. J.*, 19, 846–851.
52 Spijker, N.M. and von Boeckel, C.A.A. (1991) *Angew. Chem. Int. Ed. Engl.*, 30, 180–183.
53 Issa, J.P., Lloyd, D., Steliotes, E., and Bennett, C.S. (2013) *Org. Lett.*, 15, 4170–4173.
54 Issa, J.P. and Bennett, C.S. (2014) *J. Am. Chem. Soc.*, 136, 5740–5744.
55 Eby, R. and Schuerch, C. (1974) *Carbohydr. Res.*, 34, 79–90.
56 Chen, J.-H., Ruei, J.-H., and Mong, K.-K.T. (2014) *Eur. J. Org. Chem.*, 2014, 1827–1831.
57 Lu, S.-R., Lai, Y.-H., Chen, J.-H., Liu, C.-Y., and Mong, K.-K.T. (2011) *Angew. Chem. Int. Ed.*, 50, 7315–7320.
58 Schmidt, R.R. and Reichrath, M. (1979) *Angew. Chem. Int. Ed. Engl.*, 18, 466–467.
59 Tsvetkov, Y.E., Klotz, W., and Schmidt, R.R. (1992) *Liebigs Ann. Chem.*, 1992, 371–375.
60 Morris, W.J. and Shair, M.D. (2009) *Org. Lett.*, 11, 9–12.
61 Zhu, D., Baryal, K.N., Adhikari, S., and Zhu, J. (2014) *J. Am. Chem. Soc.*, 136, 3172–3175.
62 Zhu, D., Adhikari, S., Baryal, K.N., Abdullag, B.N., and Zhu, J. (2014) *J. Carbohydr. Chem.*, 33, 438–451.
63 Crich, D. (2011) *J. Org. Chem.*, 76, 9193–9209.

14

Selective Glycosylations with Furanosides

Carola Gallo-Rodriguez and Gustavo A. Kashiwagi

14.1 Introduction

Carbohydrates in the furanose form are ubiquitous in nature. In mammals, furanoses are only restricted to ribose as components of nucleic acids. Nevertheless, furanoses can be found in bacteria, fungi, protozoa, and plants [1–7]. The interest in the synthesis of furanose-containing oligosaccharides has been growing rapidly in the past few years because furanoses are often constituents of pathogenic microorganisms. *Mycobacterium tuberculosis* is a very interesting example, which produces a cell wall containing an arabinogalactan (AG) with D-arabinofuranose (D-Ara*f*) and D-galactofuranose (D-Gal*f*), and a lipoarabinomannan (LAM) with D-Ara*f* [8, 9]. Other pathogenic bacteria such as *Klebsiella pneumoniae, Escherichia coli,* or *Streptococcus pneumoniae* also contain D-Gal*f* [5]. This unit is also present in the pathogenic protozoan *Trypanosoma cruzi,* as part of glycoinositolphospholipids (GIPLs) and mucins and in *Leishmania* as part of lipophosphoglycan (LPG) and GIPLs [2]. D-Gal*f* has also been found in pathogenic fungi such as *Aspergillus fumigatus, Paracoccidioides brasiliensis,* and *Cryptococcus neoformans.* L-Arabinose is commonly found in plants [6].

In many cases, D-Gal*f*- or D-Ara*f*-containing glycoconjugates from pathogenic microorganism are essential for its viability or survival, which prompts one to consider its metabolism as potential target of chemotherapy [4, 5]. The enzymes involved in the biosynthesis of these glycoconjugates cannot be present in humans due to the xenobiotic nature of these units. Synthetic oligosaccharides would provide tools for studying their metabolic pathway whether pursuing a chemotherapeutic agent, or a precise antigen epitope for immunological studies, or a vaccine candidate [10–12]. Recently, immobilized Gal*f*-containing glycans were recognized by Human interlectin-1 [13]. The use of multivalent gold nanoparticles carrying Gal*f* demonstrated that this unit is able to modulate the innate immune response via dendritic cells [14].

The glycosylation reaction has a primordial role in the synthesis of oligosaccharides; however, mechanistic studies have been mainly focused on pyranoses [15–20]. The use of a benzylidene-protecting group in β-mannopyranosylation and subsequent rationalization of how it provides selectivity provided important evidence of the mechanism of the glycosylation reaction [18, 21]. Although several uncertainties still remain, it is now accepted that glycosylation reactions could be considered in terms of a continuum of

Selective Glycosylations: Synthetic Methods and Catalysts, First Edition. Edited by Clay S. Bennett.

mechanism from pure concerted S_N2 reaction to a perfect S_N1 reaction with a discrete glycosyl oxocarbenium ion intermediate if a separated solvent ion pair is considered [15, 16].

The findings in the furanose glycosylation field usually follow the findings on pyranose compounds. The reactivities of pyranoses and furanoses are quite different. By first sight, this is logical to expect if the difference on conformation is taken into consideration. The first aspect to observe in furanose is the flexibility of the ring. Five-membered rings possess much higher ring strains than six-membered rings, and as a consequence, furanoses tend to adopt several conformational states, which are close in energy [22] rather than a single low-energy chair conformation [23]. This fact affects the stereoselectivity of a glycosylation reaction as will be discussed later. In spite of the flexibility of furanose, experimental evidence of the anomeric effect was found on conformationally restricted furanosides ("norbornane-furanosides") in which the anomeric *O*- or *S*-substituent preferred a pseudoaxial disposition [24].

The furanose form is thermodynamically less stable than the pyranose form. For that reason, the synthesis of the furanosyl-containing oligosaccharides requires not only a stereoselective construction of the furanosyl linkage but also the proper choice of furanosyl precursor, which often has to be synthesized *de novo.* Furthermore, the selection of the furanosyl template will be in close relationship to the glycosylation method to be used. A simplified furanosylation mechanism is shown in Scheme 14.1.

As in pyranosylation, the construction of a new furanosyl linkage involves the activation of the anomeric center of a furanose donor **1** to give an oxocarbenium ion intermediate **2**. Complete stereocontrol of nucleophilic attack would provide either the 1,2-*trans* **3** or the 1,2-*cis* **4** as the sole product. While the 1,2-*trans* glycosidic linkage is easily constructed by anchimeric assistance, the stereoselective synthesis of 1,2-*cis* furanosyl units constitutes one of the most challenging issues in oligosaccharide synthesis. As a matter of fact, the difficulty in achieving 1,2-*cis* arabino or galactofuranosyl linkages has been compared to the difficulty in accessing β-mannopyranosyl linkages [17].

This chapter aims to present the strategies and mechanistic aspects for the stereoselective construction of furanosyl linkages. Although furanoses are found in many natural products, nucleosides have been excluded. In furanoses, there are many aspects to study from the mechanistic point of view, and it is easy to find gray zones. Several aspects of the furanose field have been reviewed in the past [1, 3–5, 22, 25–28], including a seminal report by Imamura and Lowary [28]. Accordingly, this chapter focuses on recent advances in the furanose glycosylation field, including its scope and limitations.

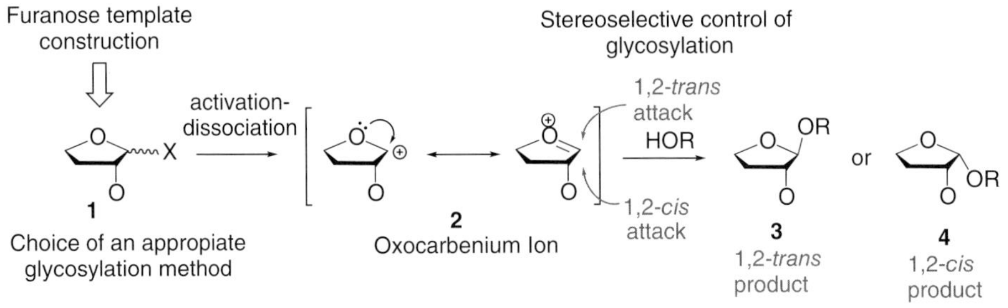

Scheme 14.1 General glycosylation reaction with a furanoside donor.

14.2 Construction of the Furanose Template

The approaches developed to access to furanosyl templates include the following: kinetically controlled Fisher glycosylation, dithioacetal cyclization, high-temperature acylation, glycal ozonolysis, aldonolactone reduction, formation of borate complexes, and formation of 1,4-anhydrosugars [28]. Other methods have been recently employed to obtain the galactofuranose template such as one-step anomeric O-alkylation of galactose [29–32] (**5**, Scheme 14.2), sodium borohydride–iodine reduction of furanoside derivatives of D-galacturonic acid [33], and the convenient one-step global t-butyldimethylsilylation of galactose in N,N-dimethylformamide (DMF) developed by Baldoni and Marino (**6**, Scheme 14.2) [34]. Other bulky silylating agents such as *tert*-butyldiphenylsilyl (BPS)-protecting groups have been recently used for D-arabinose, giving rise to the D-Ara*f* template in high yield (**7**, Scheme 14.2) [35]. In view of the success of global silylation in DMF for obtaining the D-Ara*f* and D-Gal*f* templates, a general procedure was explored for several pentoses and hexoses. The persilylated furanose derivatives were obtained only in sugars structurally related to D-Gal, such as D/L-Ara*f*, D/L-Fuc*f*, D-GalA, and D-Gal*f*NAc, in the sense that they share a common spatial 1,2-, 2,3-, and 3,4-*trans* substituent dispositions. Moreover, the tert-butyldimethylsilyl (TBS) groups can subsequently be converted to acetyl groups by treatment with catalytic *p*-toluensulfonic acid without any conversion to the pyranose form. As an example, persilylated D-fucofuranosyl derivative **8** was transformed to acetylated analog **9** (Scheme 14.2) [36].

More recently, D-galactofuranosyl- and L-fucofuranosyl-protected precursors **12**, **13**, and **15** were obtained by a pyranoside-into-furanoside (PIF) rearrangement (Scheme 14.3) from 3-O-substituted allyl glycopyranosides **10**, **11**, and **14**, respectively [37–39]. This procedure, inspired by the finding of a furanose by-product in acid-promoted *O*-sulfation of oligofucosides, includes an acid-promoted sulfation of **10** to give intermediate **16**. After protonation, a rearrangement occurs from the pyranoside ring **17** through an open-form intermediate **19** via a proposed transition state **18**. Recyclization of **19** followed by a sulfate transfer ends the rearrangement to give sulfated furanoside ring **20**, which is further *O*-desulfated by acid hydrolysis to give **12** (Scheme 14.3) [37]. This process could be compared to the enzymatic isomerization of nucleotide sugar mediated by the pyranose mutases [4, 40]. PIF-rearranged furanosyl precursors were employed for the synthesis of a galactomannan

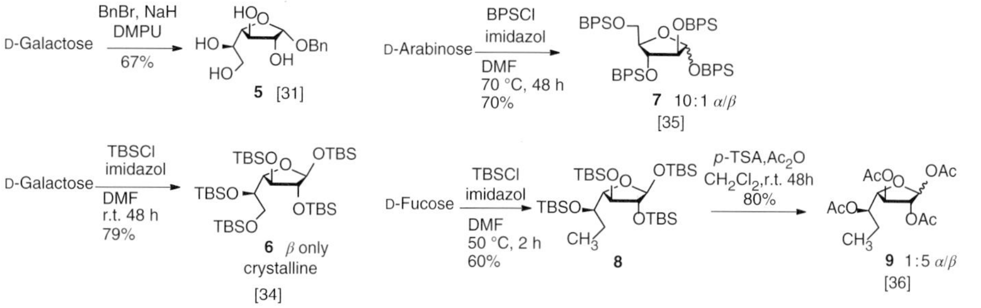

Scheme 14.2 Some recent examples of the synthesis of novel furanosyl templates. For the synthesis of other furanosides precursors, see [28].

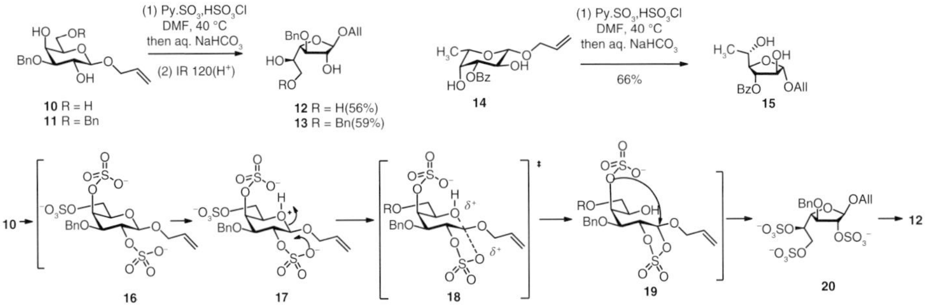

Scheme 14.3 Synthesis of D-galactofuranoside and L-fucofuranoside precursors by the novel pyranose-into-furanoside (PIF) rearrangement.

oligosaccharide fragment from the pathogen fungus *A. fumigatus* [38] and for the synthesis of rare L-fucofuranosyl containing oligosaccharides from brown seaweed *Chordaria flagelliformis* [39].

14.3 Stereoselective Glycosylation with Furanoside Donors

As mentioned earlier, the selection of the furanosyl template is in close relationship with the glycosylation method chosen. The next step involves the stereoselective control of the furanosylation reaction. As with pyranoses, the 1,2-*trans* glycosidic linkage is easily constructed by anchimeric assistance; however, the stereoselective synthesis of 1,2-*cis* furanosyl units constitutes one of the most challenging issues in oligosaccharide synthesis and will be described in Section ***14.3.2***. Nevertheless, side reactions in 1,2-*trans*-furanoside synthesis could be an important issue and should be considered. This issue of side reactions has been reviewed recently; however, the major focus was on pyranose derivatives [41].

14.3.1 1,2-*trans* Furanosides

As in pyranosides, neighboring group participation by an acyl group attached to *O*-2 is the common procedure to achieve 1,2-*trans* glycofuranosides, and many methods of glycosylation have been reported [1, 3–5, 27, 42, 43]. As far as we know, there is no exception for this rule in contrast to what has been observed with 2-*O*-acyl-substituted gluco- and galactopyranoside donors [44–48]. Some recent examples are described here. As part of the synthesis of oligosaccharide mucin family from *T. cruzi*, Kashiwagi *et al.* have synthesized an internal Gal*f*-containing hexasaccharide **21** [31] (Scheme 14.4) recently employed as substrate of trans-sialylation reaction [49]. Benzyl α-D-Gal*f* (**5**) was used as novel galactofuranose precursor to permit the construction of the trisaccharide **27** from the reducing end to the nonreducing end with complete diastereoselection of the glycosidic linkages. The levulinoyl group at *O*-2 of furanose trichloroacetimidate **22** provided the 1,2-*trans* β-linkage D-Gal*f*(1-4)-D-GlcNAc diastereoselectively. Removal of the levuniloyl group of **24** and glycosylation with galactofuranosyl imidate **26** gave the trisaccharide **27** [31]. Interestingly, trisaccharide **27** had been synthesized previously using D-galactone-1,4-lactone as D-Gal*f*

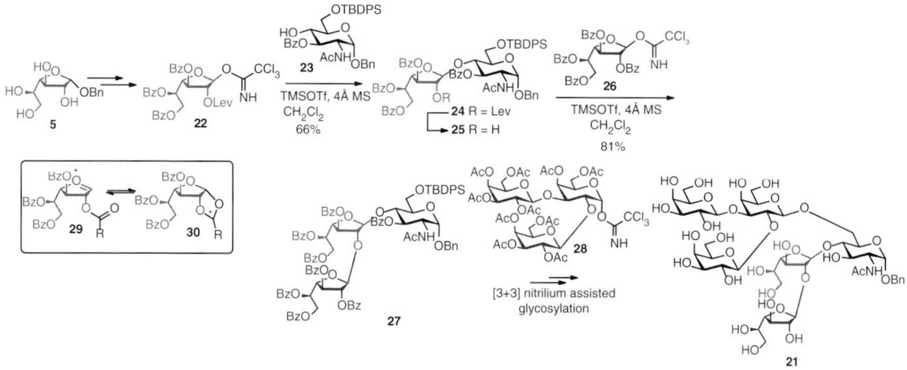

Scheme 14.4 Synthesis of 1,2-*trans* Gal*f*-containing hexasaccharide constituent of *T. cruzi* mucin.

precursor. In this case, synthesis proceeded in the opposite direction (nonreducing end to reducing end) because of the straightforwardness of the synthesis (aldonolactone approach for furanose). In the latter case, low diastereoselectivity was achieved for the 1,2-*trans*-linkages when it was not possible to take advantage of anchimeric assistance [50].

Furanosyl orthoesters have also been employed as donors for 1,2-*trans* glycosylation [4, 5, 51, 52]. Recently, gold(III)-catalyzed pentafuranosylation has been developed starting from the propargyl 1,2-orthoester derived from D-arabinose, D-lyxose, D-ribose, or D-xylose, yielding the corresponding 1,2-*trans* glycoside stereoselectively [53–55].

1,2-*trans* Furanosides have also been prepared using donors lacking participating group at *O*-2, expanding the possibilities on orthogonal protection of donors required for complex oligosaccharide synthesis. For instance, glycosylation using per-*O*-*tert*-butyldimethyl-β-D-galactofuranose (**6**) by anomeric iodination with trimethylsilyl iodide (TMSI) followed by treatment with acceptors in the presence of *N,N*-diisopropylethylamine (DIPEA) as a base gave solely 1,2-*trans* β-products **33** with primary or secondary acceptors (Scheme 14.5). The reaction involves initial activation of the donor as β-iodide **31**, which is the reactive intermediate [34]. Oxocarbenium cation **32** is probably involved in the reaction (in this case, a solvent separated ion pair (SSIP) [16, 18]) as indicated by the formation of the 1,2-*trans* β-glycosylation product. The attack of the carbohydrate nucleophile to C-1 from the β face could be due the steric effect of the bulky substituent at C-2 and the lateral chain at C-4. However, in contrast to carbohydrate acceptors,

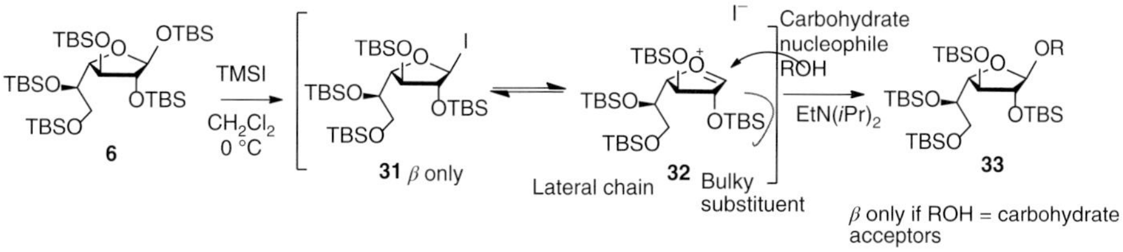

Scheme 14.5 1,2-*trans* Galactofuranosides from donor **6** lacking 2-*O*-acyl group.

moderate diastereoselectivities were obtained with primary nonsugar nucleophiles. It is worth mentioning that, in contrast to the pyranose counterparts where a conformational change is enforced by bulky silyl groups giving rise to super-armed glycosyl donor [56, 57], persilylated furanosyl **6** presents almost the similar vicinal coupling constants to the peracylated analog [34].

In glycopyranosylation, the influence of the solvent of the *gluco* or *galacto* series is well studied when the pyranosyl donor lacks 2-O participating group. For example, acetonitrile can be a participating solvent to help afford 1,2-*trans* linkages [16, 58, 59]. The β-selectivity is attributed to the formation of α-D-glycopyranosyl nitrilium ion. However, in the furanose case, the nitrile effect is not that pronounced. For instance, the nature of the solvent determined the stereochemical course of galactofuranosylation when using *O*-(2,3,5,6-tetra-*O*-benzyl-galactofuranosyl) trichloroacetimidate (**34**). 1,2-*trans* β-Products were obtained as the major product when acetonitrile was used, but the selectivity was highly influenced by the nature of the acceptor [60]. This observation was previously observed with thioglycosides [61]. More recently, Mong *et al.* proposed a revised mechanism for glycosylation in nitrile solvents with the participation of ether functions in *O*-2 [62]. This participation is based on the interaction between the lone electron pair of the oxygen atom of the 2-O substituent and α-position of the nitrilium intermediate forming a 1,2-*cis* oxazolinium intermediate **35** (Scheme 14.6). After optimizing conditions for highly diastereoselective 1,2-*trans* glycosylation of **37** with donor **36** ($CH_2Cl_2/CH_3CN/EtCN$ 1 : 2 : 1, −70 °C, Scheme 14.6), the efficient one-pot synthesis of α-(1-5)-arabinan oligomers was achieved.

Acetonitrile also favors the β-1,2-*trans* glycosylation when using pyranosyl donors carrying a nonparticipating 2-*O*-glycosyl substituent [31, 49, 63, 64]. However, differences were observed when using galactofuranosyl donors substituted either by 2-*O*-Gal*p* or by 2-*O*-Gal*f* and poorly nucleophilic 4-OH GlcNAc acceptors. For instance, β-stereochemistry is favored in dichloromethane when employing a Gal*f*(1-2)Gal*f* trichloroacetimidate donor, whereas the β-linkage is favored in acetonitrile with the Gal*p*(1-2) Gal*f* analog donor [50].

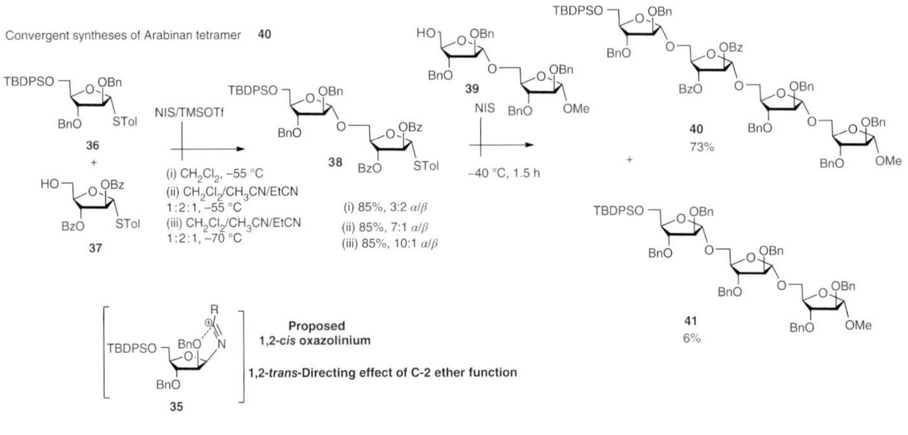

Scheme 14.6 Proposed neighboring group participation by C-2 ether functions in glycosylation directed by nitrile solvents. Synthesis of arabinan **40**.

14.3.2 1,2-*cis* Furanosides

The synthesis of stereodefined 1,2-*cis* furanosyl linkages is considerably more difficult than their 1,2-*trans* counterparts. There is currently no general method to achieve this. The first requirement is the presence of a nonparticipating group at *O*-2. Due to the flexibility of the furanose ring, the kinetic anomeric effect is not as pronounced as in the pyranoses. Mixtures of anomers are usually obtained due to transition states of similar energies. Flexible donors carrying 2-O-nonparticipating groups were first examined based on empirical results. In an attempt to obtain diastereoselectivity, other methods have been proposed including the use of conformationally locked donors and indirect methods such as intramolecular aglycon delivery (IAD) and oxidation/reduction. On the other hand, the work of Woerpel on C-glycosylation of oxocarbenium intermediates has provided new insights into O-glycosylation. Furthermore, analytical techniques such as variable-temperature nuclear magnetic resonance (VT-NMR) have helped increase our understanding of the mechanism glycosylation [65]. An example is the identification of glycosyl triflate intermediates, such as described by Crich in his 1,2-*cis*-β-mannosylation work. In this regard, a few furanosyl triflates (conformationally restricted arabinofuranosyl triflates) have been detected as intermediates by VT-NMR and will be discussed later [66, 67].

14.3.2.1 Flexible Donors

The preparation of 1,2-*cis* β-arabinofuranosides using 2-O-benzylated glycosyl halides and simple acceptors in the absence of promoters was reported in the 1960s by Glaudeman and Fletcher [68–70]. 1,2-*cis* Arabinofuranosides were obtained as major products irrespective of the anomeric configuration of the halide suggesting an ion-pair S_N1 mechanism of glycosylation [68]. Nevertheless, the drastic conditions for the generation of glycosyl halides were a major drawback. It was not until the beginning of the century that new methodologies for the stereocontrolled synthesis of β-D-Ara*f* using 2-O-benzylated donors were explored. Yin and Lowary reported the use of stable thioglycoside donors activated with NIS/AgOTf in CH_2Cl_2 at −78 °C to achieve 1,2-*cis* selectivity with good yields. The preliminary study also concluded that best yields and stereoselectivity could be obtained with fully alkylated donors [71]. Under these conditions, Lowary *et al.* synthesized more than 10 oligosaccharides containing a terminal β-D-Ara*f* related to *M. tuberculosis* AG (Scheme 14.7) [72]. The β-D-Ara-(1→2)-D-Ara*f* disaccharide unit is present in *M. tuberculosis* AG and lipoarabinomannan [8].

Subsequent investigations of conformationally flexible thioglycoside donor **43** showed the influence on the selectivity by the protecting group on O-5 and the acceptors used [73, 74]. The best results were obtained with electron-rich acceptors and “armed” 5-*O*-*p*-methoxybenzyl (PMB) alkylated donors [74, 75]. The reason for this tendency remains unclear. Very recently, upon continuation of the synthesis of the motif A arabinan hexasaccharide cap from *M. tuberculosis*, the influence of the steric crowding in the arabinofuranosyl acceptors was studied. 1,2-*cis* Ara*f* products were favored with less reactive and more sterically demanding acceptors [76].

The construction of 1,2-*cis* α-D-Gal*f* linkages with flexible 2,3,5,6-tetra-*O*-benzyl-β-D-galactofuranosyl donors depends deeply on the glycosylation method. For instance, pentenyl glycosides gave mainly 1,2-*trans* β-products, and solvents did not affect the stereochemical course of the reaction [77]. However, trichloroacetimidates [32, 60, 78,

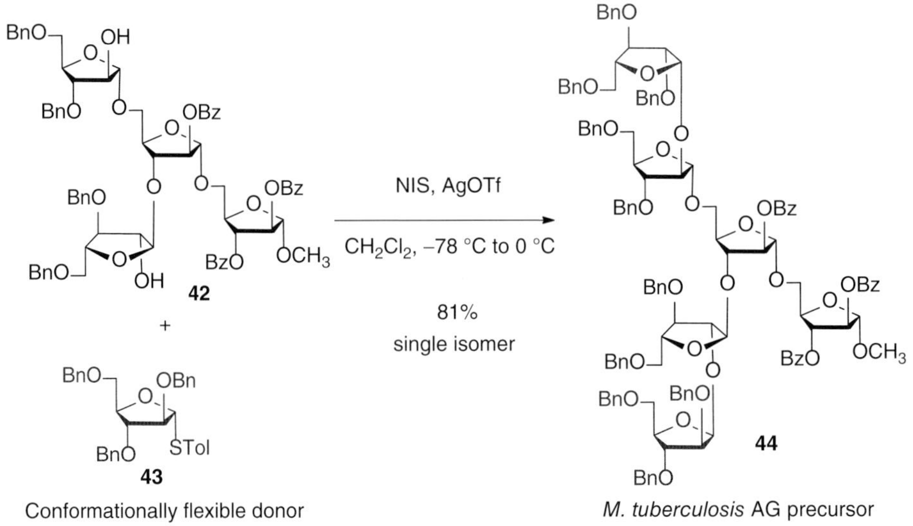

Scheme 14.7 Synthesis of the hexasaccharide **44** of *M. tuberculosis*.

79] and thioglycosides [61] gave moderate-to-good 1,2-*cis* α-selectivities, depending on the nature of the acceptor. The influence of the solvent has been evaluated in glycosylation reactions of *O*-(2,3,5,6-tetra-*O*-benzyl-β-D-galactofuranosyl) trichloroacetimidate (**34**) with several acceptors at −78 °C. A nonparticipating solvent such as CH_2Cl_2 at −78 °C favors 1,2-*cis* α-products. In contrast, acetonitrile strongly favors β-linked products, whereas no selectivities were observed with Et_2O [60]. Interestingly, high stereoselectivity (α:β 10 : 1) has been observed in the construction of the α-D-Gal*f*-(1-2)-L-Rha linkage using allyl 3,4-di-*O*-benzyl-α-L-rhamnopyranoside (**47**) as acceptor at −78 °C in CH_2Cl_2 with TMSOTf as catalyst, whereas the same conditions in acetonitrile gave the β-product as a single diastereomer. This finding allowed the synthesis of the trisaccharide α-D-Glc*p*-(1-3)-α-D-Gal*f*-(1-2)-L-Rha*p* (**54**) constituent of *S. pneumoniae* 22 F with complete diastereoselectivity using allyl α-D-galactofuranoside (**45**) as novel precursor of the internal Gal*f* (Scheme 14.8) [32]. The influence of the 3-O-substituent (PMB, Bz, PFBz, PMBz, TIPS) was evaluated, and complete stereoselectivity was observed with all protecting groups used in Gal*f* donor, but the 3-*O*-benzoyl substitution gave the best yield. Remote participation [80] by the benzoate could be involved in controlling selectivity.

Very recently, remote stereocontrol has been demonstrated in the construction of 1,2-*cis* α-L-fucofuranoside (6-deoxy-L-galactofuranoside) linkages present in a fucoidan from *C. flagelliformis*. In this case, the 3-OBz donor **55** gave the 1,2-*cis* product **58** as the major isomer via the proposed intermediate **60**, whereas the 3-OBn donor **56** afforded **59** as the major product with the opposite configuration at the anomeric center (Scheme 14.9) [39].

Excellent selectivity for α-ribofuranosylation was achieved under acid- and heavy-metal-free conditions with several acceptors by activation of iodide **63**, using DIPEA as a mild base and triphenylphosphine oxide (TPPO) as a promoter (Scheme 14.10). [81]

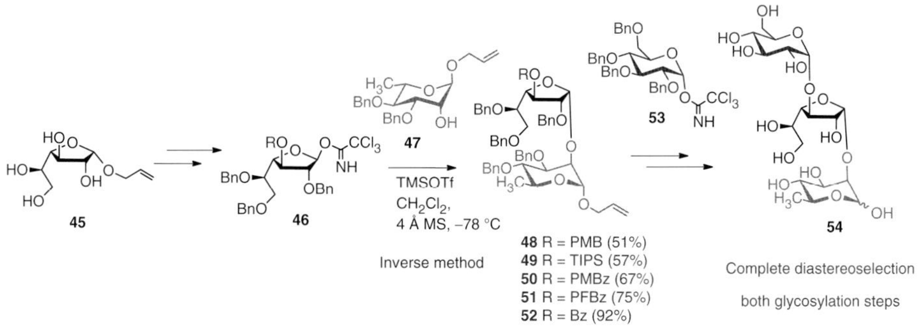

Scheme 14.8 Diastereoselective synthesis of an internal 1,2-*cis* α-D-Gal*f*-containing trisaccharide from *S. pneumoniae* 22 F. Effect of 3-O-substitution on Gal*f* donor.

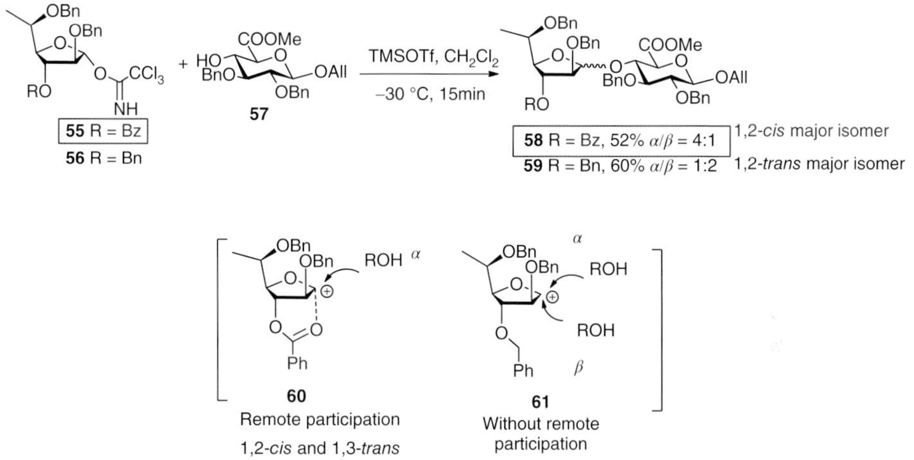

Scheme 14.9 Remote participation in L-fucofuranosylation. The 3-OBz substitution reverses the selectivity obtained by the 3-OBn substitution.

The selectivity of the reaction was explained [82] on the basis of the "inside attack" [83] of the nucleophile alcohol at the "exploded" transition state [81].

In 2001, Kim *et al.* introduced the 2-carboxybenzyl (CB) glycoside donors to construct the challenging β-D-mannopyranosyl linkages with high selectivity after activation with triflic anhydride at −78 °C [84]. Later, this method was applied for β-D-arabinofuranosylation in the 1,2-*cis*-selective synthesis of an octaarabinofuranoside found in mycobacterial cell wall [85]. However, the selectivity depended on the protecting group pattern of the arabinofuranosyl donor and acceptor. The mechanism involves a triflate-mediated lactonization of the aglycone in the presence of diterbutylmethylpyridine (DTBMP) as base (−78 °C), to give oxocarbenium ion **68**, which might be in equilibrium with the α-triflate **69** (Scheme 14.11). This method was also used for α-D-galactofuranosylation [86, 87] in the synthesis of the antineoplastic glycosphingolipid agelagalastatin, which presents a terminal α-D-Gal*f*-(1 → 2)-D-Gal*f* unit and constituted the first example of a completely diastereoselective α-galactofuranosylation [86]. More recently, di- and tetrasaccharide subunits of the cell wall polysaccharide of *Talaromyces flavus* were

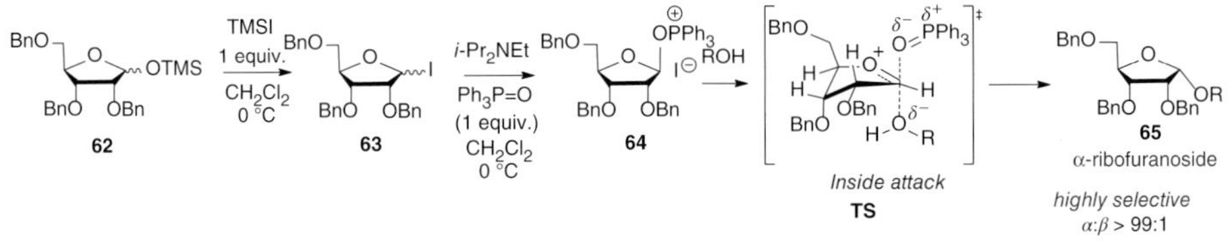

Scheme 14.10 α-Selective Ribofuranosylation. Proposed mechanism under the addition of TPPO.

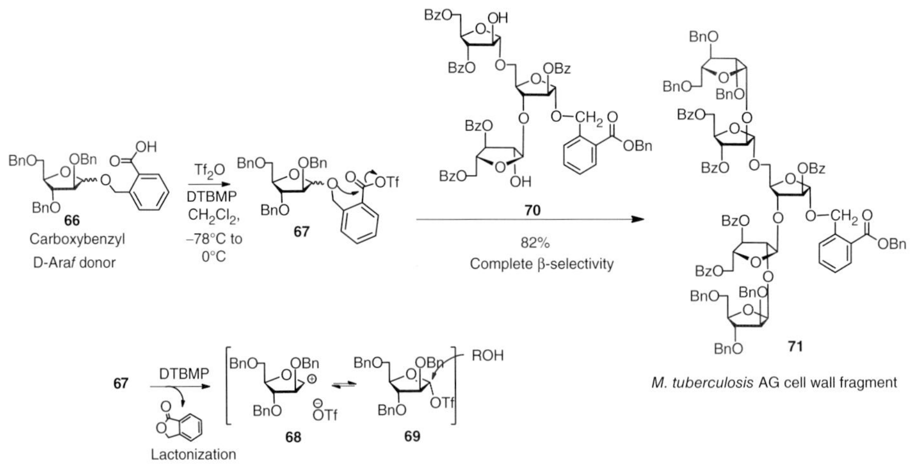

Scheme 14.11 Kim's 2-O-Bn carboxybenzyl furanoside approach for selective β-D-furanosylation.

obtained by employing the same method. In this case, the α-D-Gal*f*-(1 → 2)-Man linkage was obtained as a single diastereomer after optimizing the protecting group pattern in the acceptor and the donor [87]. The reaction conditions (triflate ion, nonparticipating solvent, and strict low temperature) share many similarities with the conditions used to activate Lowary's 2-*O*-benzyl thioglycoside donor **43** (Scheme 14.7) [71]. Hence, both glycosylation mechanisms could involve a displacement of an α-triflate in an S_N2-like manner.

Yang *et al.* applied the H-bond-mediated aglycone delivery (HAD) methodology developed by Demchenko [88] to 1,2-*cis* β-stereoselective D- and L-arabinofuranosylation [89]. Several 5-*O*-directing groups and glycosylation methods were examined. 5-*O*-(2-Quinolinecarbonyl) (5-O-Quin)-substituted arabinosyl ethylthioglycoside **72** gave high-to excellent selectivities (7 : 1 to 1 : 0 α/β ratio) with a range of acceptors including arabinofuranose, gluco-, galacto-, and mannopyranose derivatives. Selectivity is thought to arise through the proposed intermediate **73**. This strategy was validated through the synthesis of the mannose-capped octasaccharide fragment of *M. tuberculosis* LAM. Glycosylation of tetrasaccharide acceptor **75** with donor **72** gave **76** carrying an orthogonal Quin group for further modification (Scheme 14.12). The β-directing effect of the 5-*O*-Quin group is somewhat related to the 5-NAP (5-*O*-(2-naphthyl) methyl)-mediated IAD strategy [90] *vida infra.* Very recently, the HAD strategy was applied for the introduction of two terminal 1,2-*cis* α-D-Gal*f* units in the challenging synthesis of the glycosphingolipid Vesparioside B [91].

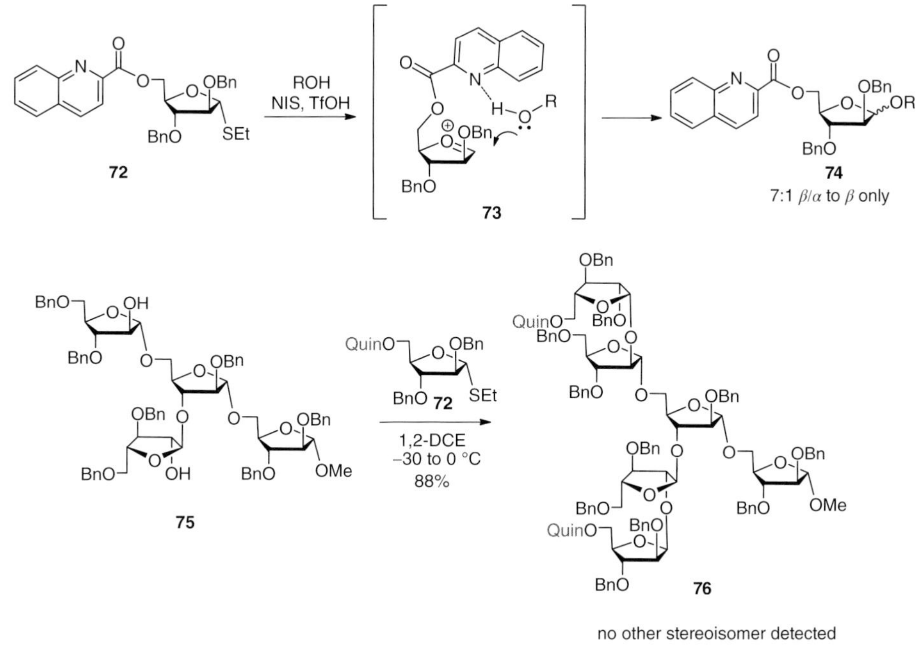

Scheme 14.12 Proposed HAD mechanism by 5-*O*-Quin donor **72**.

14.3.2.2 Conformationally Restricted Donors

Due to the small energy differences between furanose conformers, several approaches focusing on conformational restriction of the ring were developed in order to increase the energy barrier of the transition states to favor the formation of either 1,2-*cis* or 1,2-*trans* linkages. The first approach using conformationally restricted furanosyl derivatives was developed by Lowary *et al.* for the stereoselective construction of 1,2-*cis* β-arabinofuranosides. In this indirect approach, 2,3-anhydro thioglycoside **77** or glycosyl sulfoxide **78** were used as donors to afford the 1,2-*cis* O-glycosylation product **79** with the new glycosidic bond *cis* to the 2,3-epoxy group (Scheme 14.13). A regioselective opening of the epoxide with lithium benzyl alkoxide in the presence of (−)-sparteine subsequently afforded the corresponding 1,2-*cis* furanoside **80** with excellent-to-perfect selectivity. Interestingly, the chirality of the base was not important for the regiochemical outcome of the nucleophilic opening of 2,3-anhydro-D-lyxofuranosides (Scheme 14.13) [92, 93].

In the case of thiosulfoxide derivative **78**, the selectivity was due to the formation of a single glycosyl triflate intermediate **85** (determined by VTNMR) that would react with the nucleophile alcohol via S_N2-like displacement [66]. While the 2,3-anhydro-β-D-*lyxo* thioglycoside **77** or its thiosulfoxide **78** is used to synthesize β-D-arabinofuranosides **80**, the 2,3-anhydro-β-D-*gulo* thioglycoside **81** or thiosulfoxide **82** is suitable to obtain α-D-galactofuranosides **84**. This methodology was applied to the synthesis of Gal*f*(α1-2)Gal*f*(α1-4)Gal*p* **86**, a trisaccharide structurally related to the antigenic polysaccharide from *Eubacterium saburreum* strain T19 (Scheme 14.14) [94]. Whereas the each glycosylation reaction afforded only one stereoisomer (**88** and **90**), the further ring-opening reactions were not completely regioselective. The challenging synthesis of the

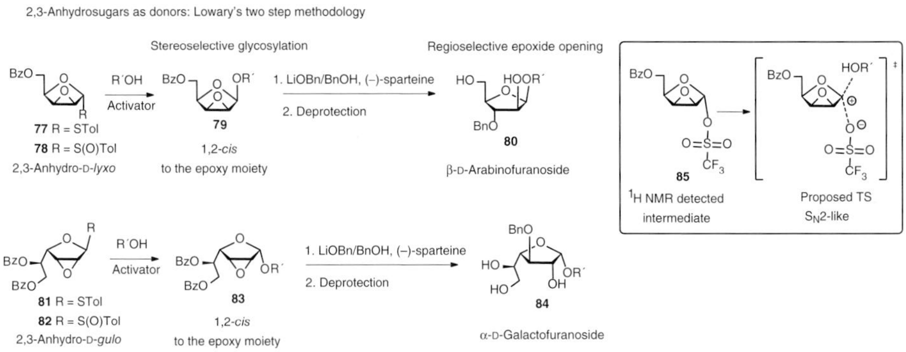

Scheme 14.13 2,3-Anhydro-furanose derivatives in the synthesis of 1,2-*cis* β-D-arabinofuranosides and α-D-galactofuranosides. ^{1}H NMR detected intermediate **85**.

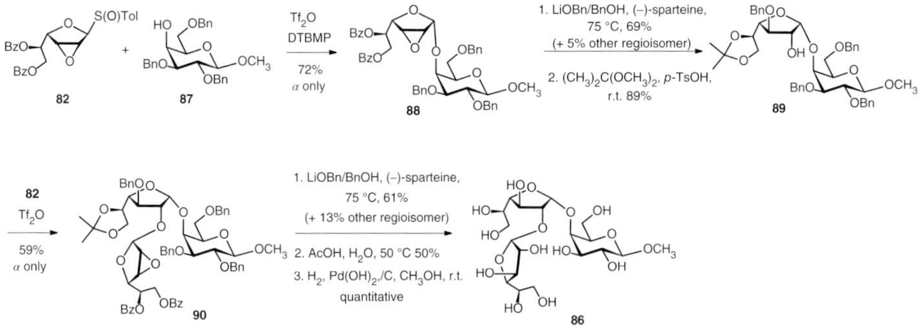

Scheme 14.14 Synthesis of the trisaccharide **86** from *Eubacterium saburreum* strain T19 by the anhydro-sugar methodology.

pentasaccharide repeating unit of varianose, which also possesses an internal Gal*f* 1,2-*cis* unit, was also accomplished using this approach [95].

A new perspective on the mechanism of O-furanosylation arose from the earlier studies of Woerpel *et al.* on C-furanosylation, which examined the stereoselective addition of a nucleophile (allyl silane) to oxocarbenium ions [83, 96–98]. *C*-glycosylations of ribose derivatives are highly 1,2-*cis* selective regardless of anomeric configuration, suggesting a common oxocarbenium ion intermediate. The sp^2 character of the O-atom and adjacent C-1 situate C-2 and C-4 in a coplanar disposition. The oxocarbenium ion could adopt two possible envelope conformations, which are in equilibrium: ^3E conformation with C-3 above the plane defined by C-4, endocyclic O, C-2, and C-1, or E_3 with the C-3 below that plane (Scheme 14.15). The attack of the nucleophile occurs from "inside" the envelope (*inside attack*) in order to avoid eclipsing interactions with H-2 (derived from the outside attack), to give a staggered product rather than the eclipsed product (Scheme 14.16) [97, 98].

The factors that stabilize the conformation of the oxocarbenium ion were determined for several ribose derivatives [83]. For example, C-glycosylation of ribose derivative **91** is highly 1,2-*cis* α-selective to give **92** through inside attack on the E_3 conformer. In this case, the intermediate adopts the lowest energy conformer E_3 with the C-3 substituent

Scheme 14.15 Conformations of furanosyl oxocarbenium.

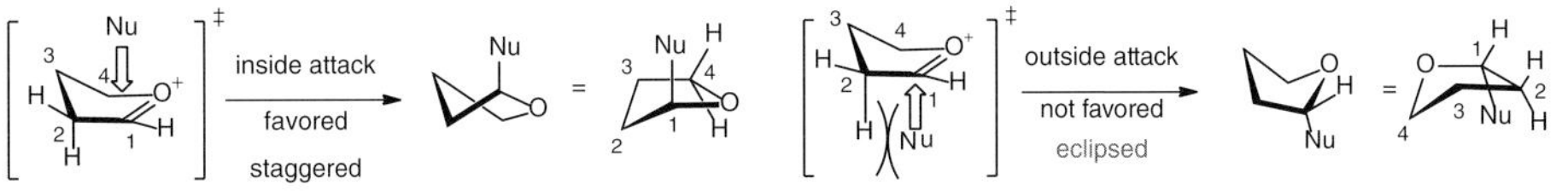

Scheme 14.16 Nucleophile trajectory of inside or outside attacks of the oxocarbenium ^{3}E conformer.

in pseudoaxial disposition minimizing the resulting dipole, whereas the pseudoequatorial C-2 substituent allows the efficient superposition of the electron donating σ_{C-H} orbital with the unoccupied oxocarbenium C-1 orbital. A slight influence is exerted by the C-4 substituent (Scheme 14.17a). In the case of arabinofuranose derivative **93**, mixtures of anomers of **94** (α:β 1 : 1) were obtained due to the opposing stereoelectronic influence of C-2 and C-3 substituents in *trans*-relationship with both conformers ^{3}E and E_3 possessing similar energies (Scheme 14.17b).

Codeé and coworkers examined Lewis-acid-mediated substitution reactions using [D]triethylsilane as a nucleophile (rather than allyltrimethylsilane). The reaction with acetyl 2,3,5-tri-*O*-benzyl-D-arabinose (**92**) provided the 1,2-*cis* product stereoselectively. Similar results were found for ribose, xylose, and lyxose derivatives. These results

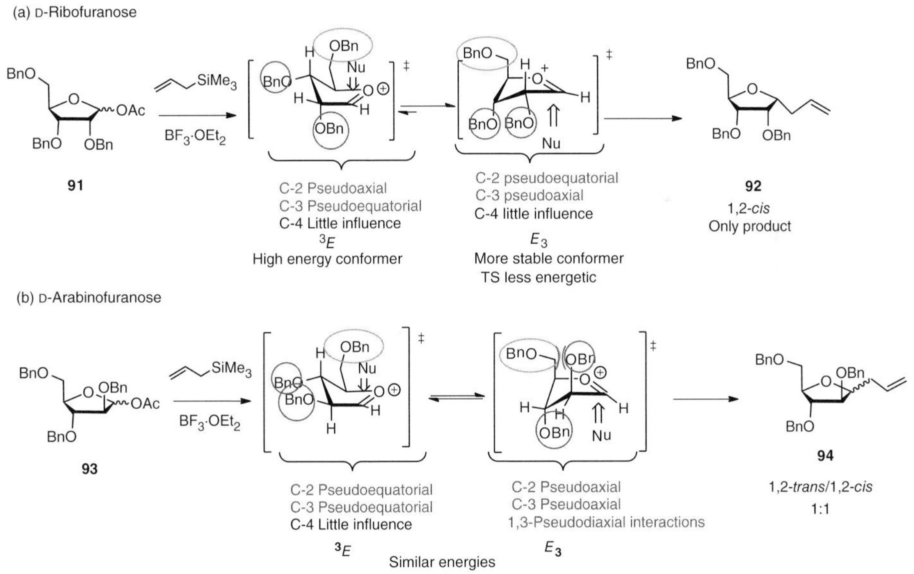

Scheme 14.7 C-glycosylation of D-Rib*f* and D-Ara*f* derivatives. Stabilizing factors of oxocarbenium.

were explained by the formation of a furanosyl oxocarbenium ion, and detailed furanosyl oxocarbenium ion energy maps have been calculated in agreement with the experiment. The authors concluded that these results would be of great impact in stereoselective *cis*-O-glycosylation if the reaction is forced to proceed via the furanosyl oxocarbenium ion as intermediate [99].

In 2006, Boons *et al.* designed the conformationally constrained donor L-**95** in order to provide highly selective 1,2-*cis* β-L-arabinofuranosylation (Scheme 14.18). The introduction of the 3,5-*O*-di-*tert*-butylsilylene (3,5-O-DTBS)-protecting group in L-arabinofuranosyl thioglycoside derivative L-**95** locked the oxocarbenium intermediate ring **96** in a E_3 conformation, which directed the attack of the acceptor to the β face (L-series) giving rise to 1,2-*cis* linkage in **97** [100]. Density functional theory (DFT) calculations on the methyl substituted oxocarbenium ion analog of **96** were in agreement with this model. High-to-complete 1,2-*cis* stereoselectivity was achieved with several manno-, galacto-, and glucopyranosyl acceptors (**98–102**) by activation with NIS/AgOTf at −30 °C. Glycosylation with the flexible analog L-**103** with the same experimental conditions gave lower selectivities proving that selectivity was due to the conformational control exerted by the 3,5-*O*-DTBS-protecting group. L-**95** was used for the synthesis of an AG fragment constituent of a plant cell wall by the introduction of two arabinofuranosyl units with complete stereoselectivity [100].

Simultaneously, Crich *et al.* followed a similar approach for 1,2-*cis* D-arabinofuranosylation starting with 3,5-*O*-benzylidene derivative D-**104** (Scheme 14.19). The crystal structure of both the conformationally restricted 3,5-*O*-DTBS D-**105** and the 3,5-benzylidene derivative D-**104** shows the furanose ring in a nearly perfect E_4 envelope conformation [67, 101]. Despite this, the donor gave low selectivities and low yields, in contrast to what was observed in 4,6-*O*-benzylidene-directed β-mannopyranosylation. Further glycosylation studies employing D-**105** with several acceptors showed that both the activation method of glycosylation and the promoter were crucial for stereoselectivity [67].

Variable-temperature NMR studies on the reaction of sulfoxide L-**106** with Tf_2O showed a complex mixture at −70 °C that coalesced into anomeric trifate **111**

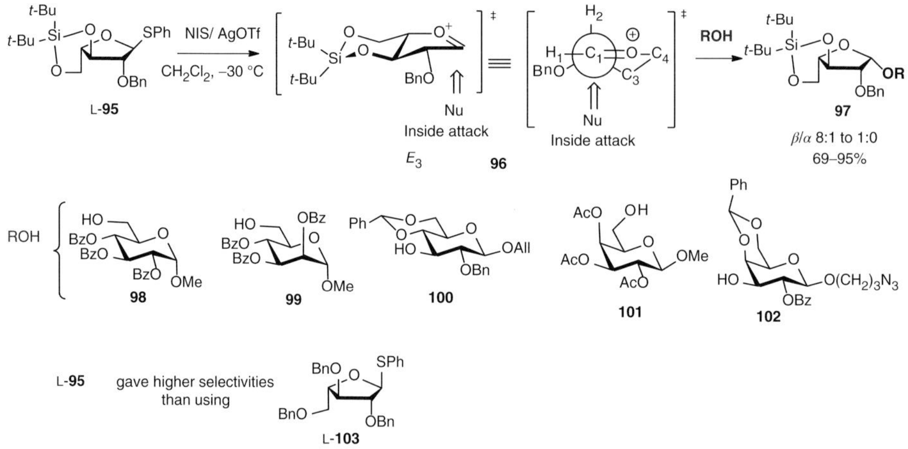

Scheme 14.18 Conformationally locked 3,5-*O*-DTBS-L-Ara*f* donor L-**95** developed by Boons *et al.*

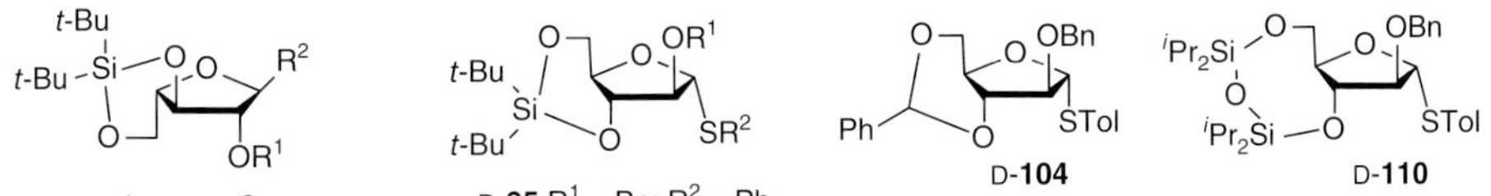

L-**95** R^1 = Bn; R^2 = SPh
L-**105** R^1 = Bn; R^2 = STol
L-**106** R^1 = Bn; R^2 = S(O)Ph
L-**107** R^1 = Bn; R^2 = OC=NHCCl$_3$
L-**108** R^1 = Nap; R^2 = SPh

D-**95** R^1 = Bn; R^2 = Ph
D-**105** R^1 = Bn; R^2 =Tol
D-**109** R^1 = TIPS; R^2 = Ph

Scheme 14.19 Examples of conformationally constrained donors used for 1,2-*cis* β-Ara*f* linkage construction.

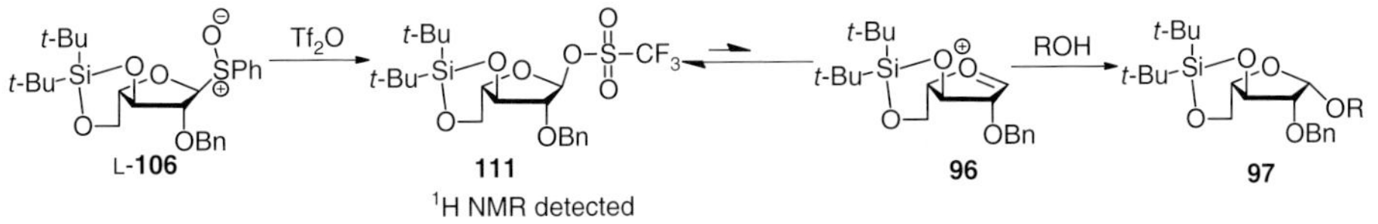

Scheme 14.20 Glycofuranosylation via the anomeric triflate. Compound **111** detected by Crich *et al.*

(Scheme 14.20) at −50 °C. This latter species reacted with alcohols to afford products with high β-selectivity. As in the case of β-mannopyranosylation, the anomeric triflate **111** could be in equilibrium with the oxocarbenium species **96**. NIS/AgOTf produced the same selectivity irrespective of the configuration of the donor; thus, an oxocarbenium ion intermediate may be involved, as described by Boons.

The NIS/AgOTf-promoted D-arabinofuranosylation was less selective compared to L-arabinofuranosylation [102], which can be attributed to a mismatched double diastereoselection [103]. The development of donor D-**105** [67, 101] allowed the synthesis of the 22-residue arabinan oligosaccharide **114** from mycobacterial AG and related compounds (Scheme 14.21) [104, 105].

Constrained trichloroacetimidate donor L-**107** [106, 107] (Scheme 14.19) was used in the synthesis of a hexasaccharide and related fragments of rhamnogalacturonan II (RGII), a highly complex pectic oligosaccharide of the primary cell wall of higher plants. In this case, the β-anomer was obtained as a single diastereomer, whereas the L-thioarabinofuranoside analog L-**95** gave a slightly reduced anomeric selectivity. D-**109** has been employed for the synthesis of the 2-azidoethyl glycoside of the Ara*f* pentasaccharide fragment of mycobacterial arabinan [108]. L-**108** was recently employed for the synthesis of a tetrameric β(1–2)-linked arabinofuranoside that is capped by an α(1–3)-arabinofuranoside. In this case, β-1,2-*cis* selectivity was achieved provided that there was not a 3,5-*O*-DTBS-protecting group on the acceptor as well [109].

Simultaneously, Ito *et al.* performed a study with a wide range of conformationally constrained donors containing either a 3,5-*O*-DTBS or a 3,5-*O*-tetra-*i*-propyldisiloxanyllidene (3,5-O-TIPDS) protecting group (D-**110**, Scheme 14.19) using mainly arabinofuranosyl acceptors. The 3,5-*O*-TIPDS group was optimum for β-selectivity, in agreement with the previous work by Woerpel *et al.* on C-glycosylation of eight-five bicyclic acetate detivatives. D-**110** [110, 111] was employed in a synthesis of docosasaccharide arabinan motif of mycobacterium cell wall [90].

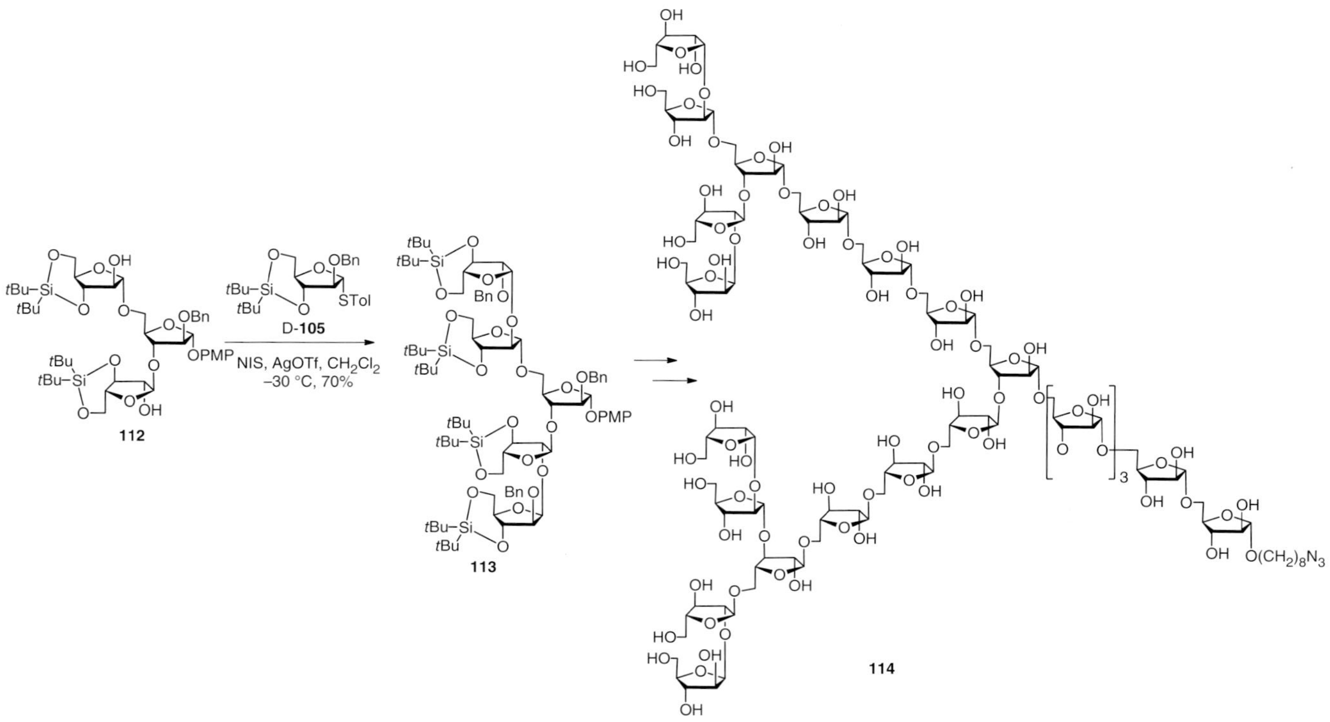

Scheme 14.21 Synthesis of the 22-fragment *M. tuberculosis* arabinan using 3,5-*O*-DTBS donor D-**105**.

In view of the stereochemical relationship between L-arabinose and D-galactose, conformationally restricted 3,5-*O*-(di-*tert*-butylsilylene)-D-galactofuranosyl trichloroacetimidate donors **115** and **116** (Scheme 14.22) have been evaluated for 1,2-*cis* α-galactofuranosylation with a wide variety of acceptors [112]. The trichloroacetimidate method was chosen due to the high stereoselectivity previously achieved with 3,5-*O*-DTBS-L-Ara*f* donor L-**107** (Scheme 14.19) [106, 107]. Unexpectedly, almost no 1,2-*cis* α-selectivity was observed in the nonparticipating solvent CH_2Cl_2 commonly used for 1,2-*cis* β-L-arabinofuranosylation. On the other hand, modest-to-high selectivity was observed using diethylether as the solvent at −78 °C, suggesting a participating effect of the solvent on the intermediate. Interestingly, the opposite situation had been observed with the flexible analog, trichloroacetimidate 2,3,5,6-tetra-*O*-benzyl-β-D-galactofuranoside (**34**, Scheme 14.22) in which ethereal solvents gave no selectivity or slightly favored the β-product, whereas CH_2Cl_2 favored the 1,2-*cis* α-product [60]. This contrasting difference in the solvent effects could be due to the change in the conformation of the oxocarbenium intermediate **118**. Higher 1,2-*cis* α-selectivity was achieved with the 6-*O*-acetyl-substituted constrained trichloroacetimidate donor **115** compared to 6-*O*-benzyl analog **116** excluding remote participation and suggesting a stereoelectronic effect.

Considering the influence of the glycosylation method previously observed in 1,2-*cis* β-arabinofuranosylation [67], conformationally restricted thioglycoside **119** (Scheme 14.22) was evaluated. The NIS/AgOTf system [67, 100] was not satisfactory for 1,2-*cis* α-D-galactofuranosylation because of the high temperature required for the reaction to proceed. However, moderate-to-excellent α-selectivity was achieved with all the acceptors employed with *p*-NO_2PhSCl/AgOTf as a promoting system [113], in CH_2Cl_2 at −78 °C. In contrast to the trichloroacetimidate donor **115**, the best

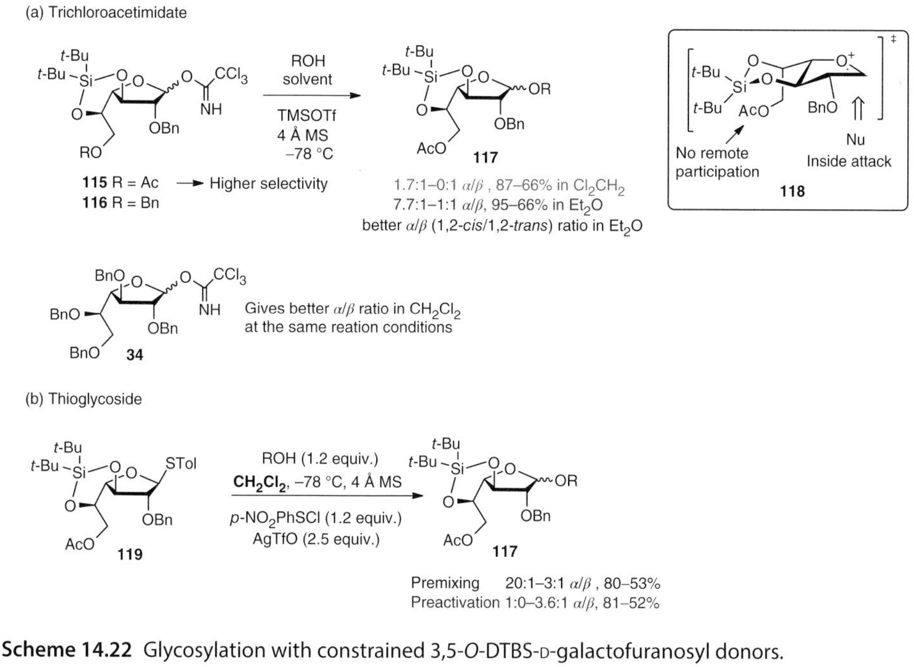

Scheme 14.22 Glycosylation with constrained 3,5-*O*-DTBS-D-galactofuranosyl donors.

selectivity was achieved by using a nonparticipating solvent (CH_2Cl_2) rather than using ethyl ether, regardless of the reaction temperature. In toluene, the 1,2-*trans* β-glycoside was obtained exclusively. In contrast to the 4,6-di-*O*-benzylidene-mannopyranosyl case [15, 16], donor preactivation did not substantially affect the stereochemical course of the glycosylation. The α-D-Gal*f*-(1 → 6)-D-Man linkage was synthesized with complete diastereoselectivity by preactivation of the conformationally constrained thioglycoside donor **119**.

As described earlier, the development of methods for stereoselective 1,2-*cis* β-D-arabinofuranosylation was a requirement for the studies on the biosynthesis of the *M. tuberculosis* AG. The use of 3,5-*O*-di-*tert*-butylsilylene- or 3,5-*O*-tetra-*i*-propyldisiloxane-protected thioglycosides (Scheme 14.19) would complicate further elaboration on 5-O position of the 1,2-*cis* arabinofuranosyl residue. Thus, a conformationally locked 2,3-*O*-xylylene arabinofuranosyl thioglycoside **120** was proposed as donor (Scheme 14.23), which would allow orthogonal 5-O-protection [114]. Moderate-to-high 1,2-*cis* selectivity was achieved using **120** as donor with NIS-AgOTf [115] as the promoter, depending on the acceptor used in the reaction. Particularly high selectivity was achieved with 2-OH-arabinofuranosyl acceptor **123** (Scheme 14.23). Moreover, 5-O-electron-donating protecting groups such as PMB in **122** increased the diastereoselectivity to 12 : 1 β/α in **126**. This effect has also been observed with other flexible arabinofuranosyl donors [73–75]. An inversion of the stereoselectivity (β/α 1 : 2) was observed in the absence of triflate ion when the acid catalyst AgOTf was replaced with $BF_3{\cdot}OEt_2$. The presence of a triflate ion is crucial for the β-selectivity suggesting that a triflate intermediate **127** (or a contact ion pair (CIP)) could be involved in the mechanism. The development of conformationally constrained donor **122** allowed the synthesis of the mannose-capped lipoarabinomannan (LAM) heptasaccharide **131** present in *M. tuberculosis* through a double-site β-selective glycosylation of acceptor **128** (Scheme 14.23).

14.3.2.3 Indirect Methods

Besides the conformationally restricted anhydro-sugar methodology, other indirect methods of glycosylation have been used for selective glycosylation with furanoses. One example is IAD, in which the nucleophile acceptor is temporally tethered to the donor prior its activation, providing the exclusive formation of the 1,2-*cis* glycosidic bond under kinetic control [116, 117]. Various types of tethers have been used. The first report on the use of IAD on a furanose derivative was a long-range type performed by a 5-*O*-silylene-tethered ribofuranosyl thioglycoside [118]. The 2-iodopropylidene-tethered IAD methodology was also evaluated in the synthesis of 1,2-*cis* α-D-glucofuranoside. Modest yields were obtained with secondary acceptors but with complete stereocontrol [119].

In 1994, Ito and Ogawa introduced the use of 2-*O*-PMB-substituted donors as precursor for IAD through the formation of *p*-methoxybenzylidene acetal tether for the challenging 1,2-*cis* β-mannopyranosylation [120]. This methodology was employed for the first time in the furanose field for the synthesis of 1,2-*cis* β-fructofuranosides [121, 122]. The synthesis of the disaccharide **135**, a precursor of the disaccharide constituent of a capsular polysaccharide of *Haemophilus influenzae* type e, is shown in Scheme 14.24. 2-*O*-PMB donor **133**α and 3-OH mannosamine acceptor **132** were

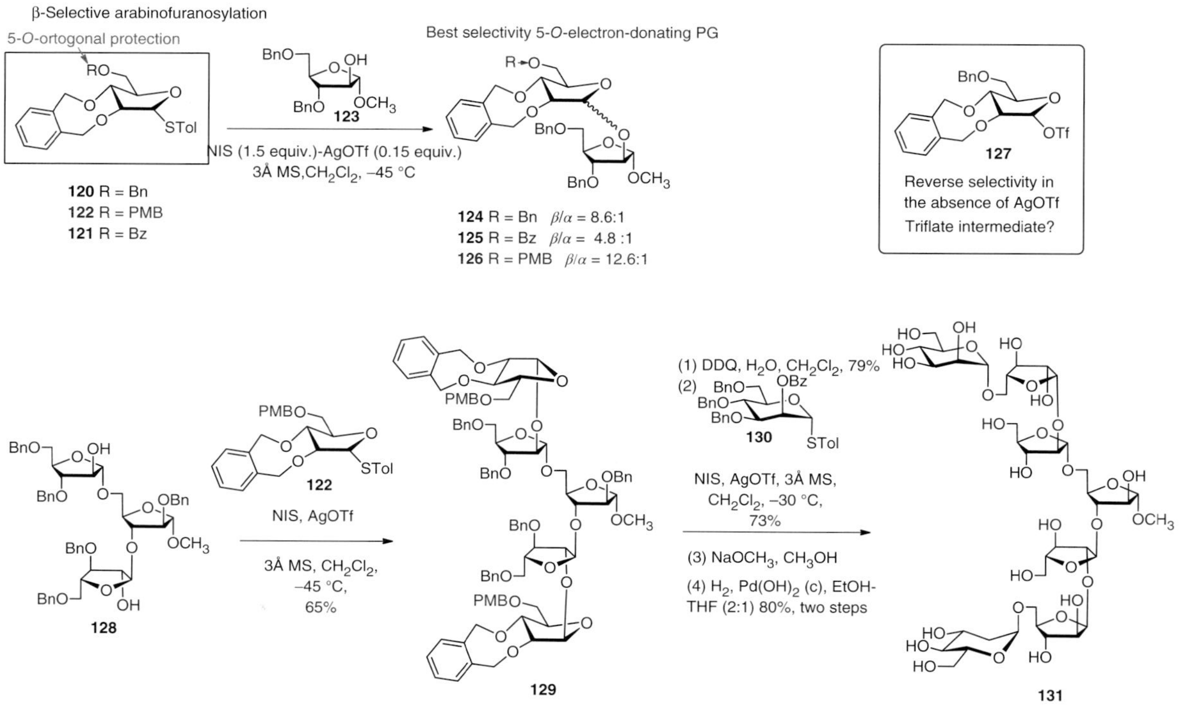

Scheme 14.23 Constrained 2,3-*O*-xylylene-D-arabinofuranosyl thioglycoside donor. Synthesis of the heptasaccharide fragment **131** in LAM.

linked by oxidative 2,3-dichloro-5,6-dicyanobenzoquinone (DDQ) treatment to give the mixed acetal **134**. Activation with the thiophilic promoter dimethyl(methylthio) sulfonium trifluoromethanesulfonate (DMTST) allowed the delivery of the acceptor from the 3-*O*-acetal on the β-side of the anomeric center giving the 1,2-*cis* β-anomer as the sole product. Interestingly, the same stereochemistry was obtained using a participating solvent such as acetonitrile or by using the β-anomer **133β** as the starting material rather than using **133α** [122].

In view of the importance of the lipoarabinomannan (LAM) and AG of *M. tuberculosis*, Prandi *et al.* employed the PMB-mediated IAD for the synthesis of 1,2-*cis* β-D-Ara*f* linkages [123–126]. 2-*O*-PMB-substituted thioarabinofuranosyl donor **136** was used for the synthesis of the tetrasaccharide cap of the lipoarabinomannan **140** (Scheme 14.25) as well as for the synthesis of the terminal hexasaccharide of the arabinan core of AG [126]. Recently, the PMB-mediated IAD strategy was also employed for the stereoselective synthesis of CLV3 glycopeptide, a secreted peptide hormone found in *Arabidopsis thaiana* plants, which contains three 1,2-*cis* β-L-Ara*f*(1-2)linked units [127].

The PMB-mediated IAD strategy was also employed for the synthesis of a 1,2-*cis* α-D-fucofuranosyl (a 6-deoxy analog of α-D-galactofuranoside) disaccharide constituent of a polysaccharide antigen of *E. saburreum*, a bacteria from the human oral cavity. In this case, *n*-pentenyl glycosides were used as donors, but yields in the reaction were low [128]. The synthesis of α-D-galactofuranosides by the IAD strategy has not yet been described.

The 2-naphthylmethyl (NAP) group was also proposed as tether by Ishiwata *et al.* as an extension of the PMB-based strategy [129]. Not only the 2-*O*-NAP donor **141** but also the 5-*O*-NAP-substituted donor **142** gave high stereoselectivities proving that the 5-O-tethering also directs the nucleophile approach from the β-face. Both **141** and **142** were used on the challenging synthesis of a 22-residue arabinan motif of *M. tuberculosis* cell wall [90]. The stereochemical homogeneity of the NAP-mediated IAD glycosylation product was confirmed by the use of perdeuterated acceptors [90]. This strategy was pivotal for the synthesis of 1,2-*cis* β-L-Ara*f*-containing peptides found in plants such as CVL3 [130] as well as the impressive synthesis of the hydrophilic repeating motif typical of extensions Ser(Galp_1)-Hyp-(Araf_4)-Hyp(Araf_4)-Hyp(Araf_3)-Hyp(Araf_1) (Scheme 14.26) [131].

The IAD-based methodology was also employed for the preparation of *p*-nitrophenyl β-L arabinofuranoside as substrates of β-L-arabinofuranosidase [132]. For instance, the use of 2-*O*-NAP-L-Ara*f* donors L-**144** and L-**149** gave the corresponding 1,2-*cis* products L-**148** and L-**151**, respectively, with complete diastereoselectivity (Scheme 14.27). Unexpectedly, the use of conformationally locked L-arabinofuranosyl donors L-**144** and L-**143** as donors (NIS/AgOTf) led to a reversal of selectivity in the reaction to give L-**146** and L-**145**, respectively, with the undesired 1,2-*trans* product as major product. This example demonstrates the importance of exploring complementary glycosylation methods and strategies.

14.3.2.4 Oxidation–Reduction

Stereoselective reduction of 2-uloses has been successfully applied to the construction of 1,2-*cis* β-mannopyranosyl linkages [133]. This procedure has inspired Field *et al.* for the construction of a 1,2-*cis* furanosyl linkage [134] in their synthesis of a rare

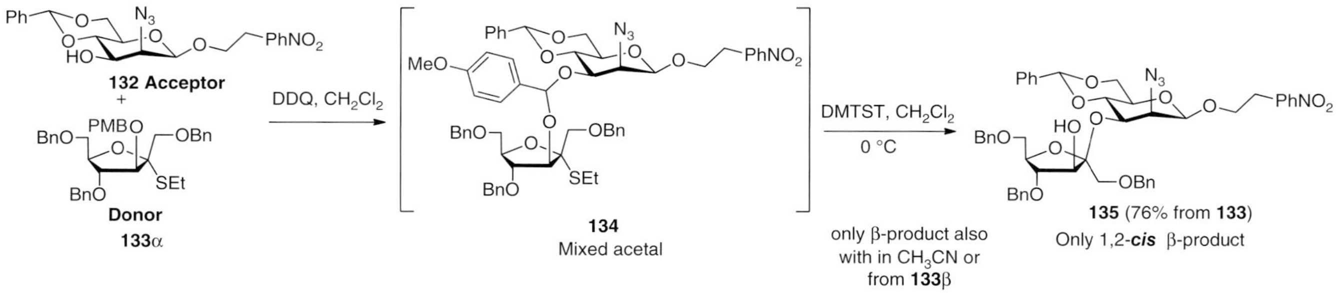

Scheme 14.24 PMB-mediated IAD for 1,2-*cis* β-fructofuranoside.

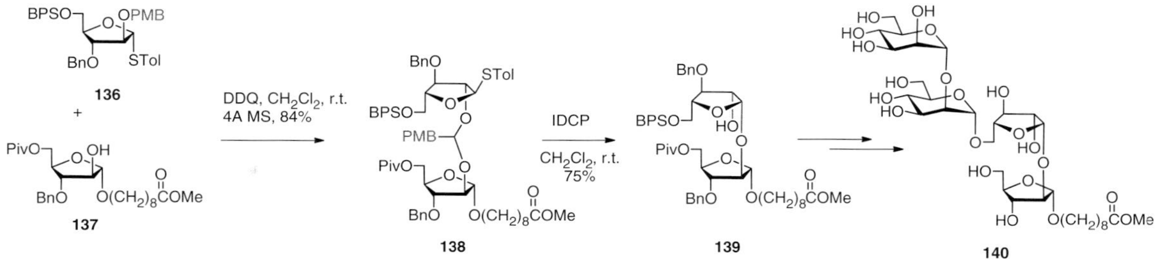

Scheme 14.25 PMB-mediated IAD for 1,2-*cis* β-D-arabinofuranosyl linkage in LAM tetrasaccharide **140**.

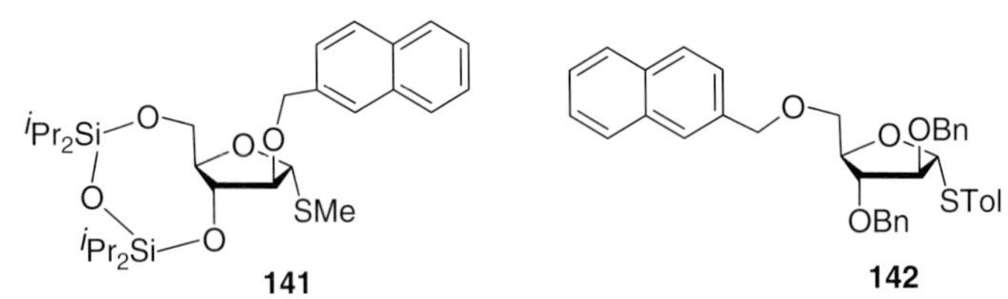

Scheme 14.26 2-*O*-NAP-D-Ara*f* and 5-*O*-NAP-D-Ara*f* donors in IAD strategy for β-D-arabinofuranosylation.

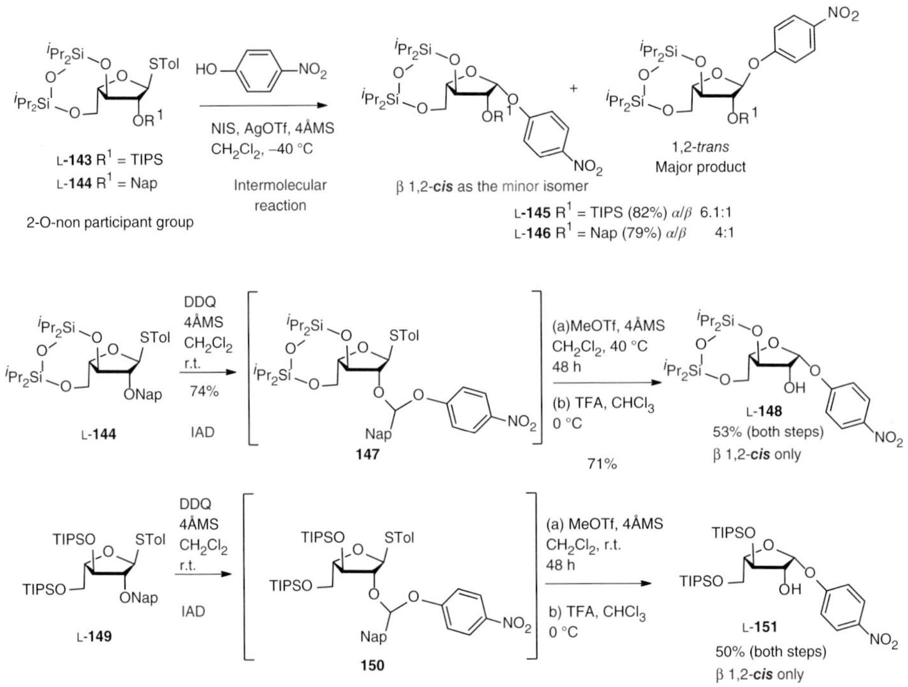

Scheme 14.27 Comparison between the IAD and intermolecular conformationally restricted glycosylation.

acetic-acid-containing disaccharide **157**, which is component of the RGII. By this strategy, the 1,2-*cis* linkage is installed after epimerization of C-2 by an oxidation–reduction sequence (Scheme 14.28). The 1,2-*trans* glycofuranoside **154** was obtained stereoselectively by anchimeric assistance. Deprotection of 2-*O*-acetyl group followed by Dess–Martin oxidation gave 2-oxo-glycoside **156**, which was reduced from the less hindered β-face by L-selectride to give the 1,2-*cis* disaccharide **157**. In this case, the formation of the undesired epimer was not observed [134].

The oxidation–reduction procedure was also used for the synthesis of (+)-sucrose (β-D-fructofuranosyl α-D-glucopyranoside), via β-selective coupling of a D-psicosyl phthalate donor with a glucopyranosyl acceptor. The 3-OH of the furanosyl ring was

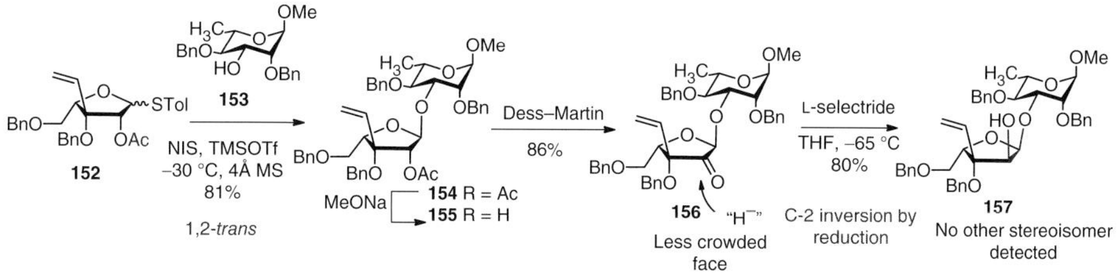

Scheme 14.28 1,2-*cis* Linkage by the 2-ulose reduction.

epimerized by a Swern oxidation–$NaBH_4$ reduction sequence [135]. In another example, the synthesis of the terminal 1,2-*cis* arabinofuranosyl linkage of the hexasaccharide motif of the *M. tuberculosis* AG **161** employed ribofuranosyl propargyl 1,2-orthoester donor **159** (Scheme 14.29). Following coupling, the two benzoates at C2 of the terminal riboses were removed to reveal **160**. An oxidation/reduction sequence then afforded **161**. In this case, the less crowded $NaBH_4$ was used for the stereoselective reduction [53]. In addition, the 1,2-*cis* glycosidic linkages α-D-Rib*f* and β-D-Lyx*f* were also synthesized by the oxidation–reduction sequence from the corresponding 1,2-*trans* 2-OH glycoside α-D-Ara*f* and β-D-Xyl*f*, respectively [55].

14.4 Reactivity Tuning of Furanosides for Oligosaccharide Synthesis

Protecting groups play important roles in glycosylation reactions [16–18, 20, 80], and their stereoelectronic effects can be modified not only to increase the selectivity of the glycosylation reaction but also to facilitate the oligosaccharide synthesis by limiting protecting group manipulation, thus shortening the reaction steps [136]. In this sense, the influence of silyl-protecting groups on the reactivity of D- and L-arabinofuranosyl thioglycosides was determined by three-component competition glycosylation reactions [137]. Whereas the silyl ether substituents in donor **162** (Scheme 14.30) enhanced the glycosylation reactivity by increasing the electron density on the anomeric center, the 3,5-*O*-di-*tert*-butylsilylene-protecting group in **165** disarmed the donor, which was rationalized as a torsional disarming effect. Scheme 14.30 shows a possible explanation for this phenomenon in the case of L-Ara*f* **166**. The participating benzoyl group would stabilize the oxocarbenium ion, and the nucleophile acceptor could attack from the top face to give a 1,2-*trans* α-glycoside. A less reactive donor could be the consequence of a high-energy transition state produced by the eclipsing interaction of the incoming nucleophile with the bicyclic cation. These findings allowed the one-pot synthesis of trisaccharides. The synthesis of **169** is shown in Scheme 14.30. Interestingly, the three glycosylating agents and the activating reagents were mixed simultaneously and not added sequentially [137].

More recently, a similar study was performed on stereochemically related D-galactofuranosyl thioglycoside donors carrying several silyl, benzoyl, and 3,5-*O*-di-*tert*-butylsilylene-protecting groups. Once again, whereas the silyl ethers of the D-Gal*f* ring

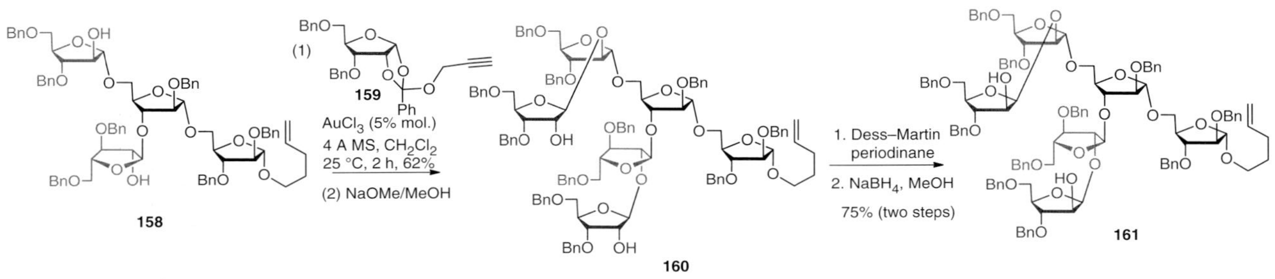

Scheme 14.29 1,2-*cis* β-Ara*f* construction by C-2 epimerization of 1,2-*trans* ribofuranosyl moiety.

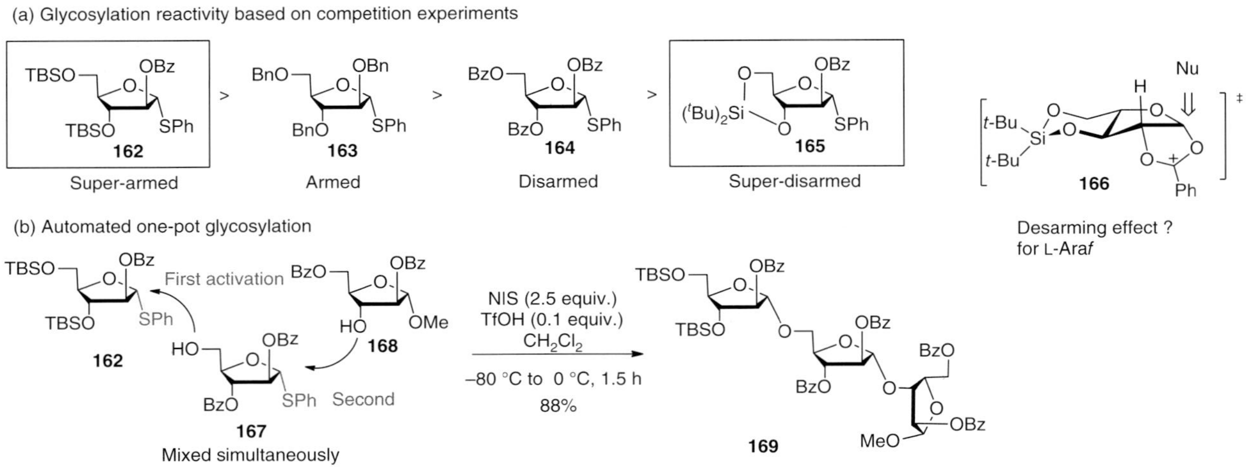

Scheme 14.30 Reactivity tuning of silyl-protecting groups on glycosylation reactivity of D-Ara*f* thioglycosides.

were found to arm the donor, the 3,5-*O*-DTBS-protecting group produced a disarming effect. In this sense, the 2,6-di-*O*-benzoyl-3,5-di-*O*-*tert*-butyldimethylsilyl-substituted Gal*f* derivative was a useful donor and enabled the rapid and high-yielding assembly of several β-D-Gal*f* containing tri- and tetrasaccharides of biological relevance [138]. Partially protected thiogalactofuranoside and thioarabinofuranoside donors were used as key glycosylating building blocks in one-pot syntheses of several oligosaccharides constituents of pathogenic microorganisms [43].

14.5 Conclusion

Furanosyl-containing carbohydrates are found in a wide range of pathogenic microorganisms and play crucial roles in biology. As a consequence, the synthesis of precise oligosaccharides can provide tools for understanding important biological processes. The glycosylation reaction is pivotal for this purpose. In this chapter, the challenges and difficulties in stereoselective furanoside synthesis were covered. The differences between the pyranose and furanose forms have been described. The flexibility of the furanoside ring emerges as the main reason of such differences. In this context, the development of new methodology to construct the furanose template in parallel with an appropriate glycosylation method is essential. 1,2-*trans* Furanoside linkages are easily built by anchimeric assistance to provide a single anomer; however, the construction of 1,2-*cis* furanosyl linkage remains a challenge. While diverse methodologies have been developed to tackle this problem, including conformationally restricted donors, remote participation, IAD, HAD, and oxidation–reduction sequences, there is no general method to accomplish the glycosylation reaction with complete control of the stereoselectivity. In almost all cases, tuning of the substitution patterns of the donor and acceptor was required.

In order to direct the glycosylation outcome to a complete stereoselective manner, new methodology is required to fully understand the mechanism of glycosylation in furanoses. Side reactions, the influences of solvent, temperature, and protecting groups have to be studied more extensively to provide insights into the reaction. In addition, the development of new methodology to reduce the protecting group manipulation in oligosaccharide synthesis is still needed. The recent advances on the reactivity tuning of furanoses highlight the importance of this. Such studies are important because the synthesis of oligosaccharides has a direct impact on glycobiology, and this knowledge could bring alternatives in glycomedicine.

References

1 Houseknecht, J.B. and Lowary, T.L. (2001) *Curr. Opin. Chem. Biol.*, 5, 677–682.
2 de Lederkremer, R.M. and Colli, W. (1995) *Glycobiology*, 5, 547–552.
3 Peltier, P., Euzen, R., Daniellou, R., Nugier-Chauvin, C., and Ferrieres, V. (2008) *Carbohydr. Res.*, 343, 1897–1923.
4 Richards, M.R. and Lowary, T.L. (2009) *ChemBioChem*, 10, 1920–1938.

5 Marino, C., Gallo-Rodriguez, C., and de Lederkremer, R.M. (2012) in Glycans: Biochemistry, Characterization and Applications (ed. H.M. Mora Montes), Nova Science, pp. 207–268.
6 Estévez, J.M., Kieliszewski, M.J., Khitrov, N., and Somerville, C. (2006) *Plant Physiol.*, 142, 458–470.
7 Tefsen, B., Ram, A.F., van Die, I., and Routier, F.H. (2012) *Glycobiology*, 22, 456–469.
8 Kaur, D., Guerin, M.E., Škovierová, H., Brennan, P.J., and Jackson, M. (2009) in Advances in Applied Microbiology, vol. 69 (eds A.I. Laskin, S. Sariaslani, and G.M. Gadd), Academic Press, pp. 23–78.
9 Tam, P.-H. and Lowary, T.L. (2009) *Curr. Opin. Chem. Biol.*, 13, 618–625.
10 Boltje, T.J., Buskas, T., and Boons, G.-J. (2009) *Nat. Chem.*, 1, 611–622.
11 Wang, L.-X. and Davis, B.G. (2013) *Chem. Sci.*, 4, 3381–3394.
12 Stallforth, P., Lepenies, B., Adibekian, A., and Seeberger, P.H. (2009) *J. Med. Chem.*, 52, 5561–5577.
13 Wesener, D.A., Wangkanont, K., McBride, R., Song, X., Kraft, M.B., Hodges, H.L., Zarling, L.C., Splain, R.A., Smith, D.F., Cummings, R.D., Paulson, J.C., Forest, K.T., and Kiessling, L.L. (2015) *Nat. Struct. Mol. Biol.*, 22, 603–610.
14 Chiodo, F., Marradi, M., Park, J., Ram, A.F.J., Penadés, S., van Die, I., and Tefsen, B. (2014) *ACS Chem. Biol.*, 9, 383–389.
15 Bohé, L. and Crich, D. (2015) *Carbohydr. Res.*, 403, 48–59.
16 Bohé, L. and Crich, D. (2011) *C.R. Chim.*, 14, 3–16.
17 Crich, D. (2011) *J. Org. Chem.*, 76, 9193–9209.
18 Crich, D. (2010) *Acc. Chem. Res.*, 43, 1144–1153.
19 Mydock, L.K. and Demchenko, A.V. (2010) *Org. Biomol. Chem.*, 8, 497–510.
20 Nigudkar, S.S. and Demchenko, A.V. (2015) *Chem. Sci.*, 6, 2687–2704.
21 Aubry, S., Sasaki, K., Sharma, I., and Crich, D. (2011) in Topics in Current Chemistry, Reactivity Tuning in Oligosaccharide Assembly, vol. 301 (eds B. Fraser-Reid and J.C. López), Springer, Berlin, Heidelberg, pp. 141–188.
22 Taha, H.A., Richards, M.R., and Lowary, T.L. (2013) *Chem. Rev.*, 113, 1851–1876.
23 Angyal, S.J. (1984) in Advances in Carbohydrate Chemistry and Biochemistry, vol. 42 (eds R.S. Tipson and D. Horton), Academic Press, pp. 15–68.
24 Ellervik, U. and Magnusson, G. (1994) *J. Am. Chem. Soc.*, 116, 2340–2347.
25 Lowary, T.L. (2001) in Glycochemistry (eds P.G. Wang and C.R. Bertozzi), Marcel Dekker, pp. 131–162.
26 Lowary, T.L. (2003) *Curr. Opin. Chem. Biol.*, 7, 749–756.
27 Marino, C. and Baldoni, L. (2014) *ChemBioChem*, 15, 188–204.
28 Imamura, A. and Lowary, T. (2011) *Trends Glycosci. Glycotechnol.*, 23, 134–152.
29 Klotz, W. and Schmidt, R.R. (1993) *Liebigs Ann. Chem.*, 683–690.
30 Gola, G., Libenson, P., Gandolfi-Donadio, L., and Gallo-Rodriguez, C. (2005) *Arkivoc*, 12, 234–242.
31 Kashiwagi, G.A., Mendoza, V.M., de Lederkremer, R.M., and Gallo-Rodriguez, C. (2012) *Org. Biomol. Chem.*, 10, 6322–6332.
32 Gola, G. and Gallo-Rodriguez, C. (2014) *RSC Adv.*, 4, 3368–3382.
33 Bordoni, A., de Lederkremer, R.M., and Marino, C. (2010) *Bioorg. Med. Chem.*, 18, 5339–5345.
34 Baldoni, L. and Marino, C. (2009) *J. Org. Chem.*, 74, 1994–2003.

35 Kraft, M.B., Martinez Farias, M.A., and Kiessling, L.L. (2013) *J. Org. Chem.*, 78, 2128–2133.
36 Dureau, R., Legentil, L., Daniellou, R., and Ferrières, V. (2012) *J. Org. Chem.*, 77, 1301–1307.
37 Krylov, V.B., Argunov, D.A., Vinnitskiy, D.Z., Verkhnyatskaya, S.A., Gerbst, A.G., Ustyuzhanina, N.E., Dmitrenok, A.S., Huebner, J., Holst, O., Siebert, H.-C., and Nifantiev, N.E. (2014) *Chem. Eur. J.*, 20, 16516–16522.
38 Argunov, D.A., Krylov, V.B., and Nifantiev, N.E. (2015) *Org. Biomol. Chem.*, 13, 3255–3267.
39 Vinnitskiy, D.Z., Krylov, V.B., Ustyuzhanina, N.E., Dmitrenok, A.S., and Nifantiev, N.E. (2016) *Org. Biomol. Chem.*, 14, 598–611.
40 Chlubnova, I., Legentil, L., Dureau, R., Pennec, A., Almendros, M., Daniellou, R., Nugier-Chauvin, C., and Ferrières, V. (2012) *Carbohydr. Res.*, 356, 44–61.
41 Christensen, H.M., Oscarson, S., and Jensen, H.H. (2015) *Carbohydr. Res.*, 408, 51–95.
42 Zhu, S.-Y. and Yang, J.-S. (2012) *Tetrahedron*, 68, 3795–3802.
43 Deng, L.-M., Liu, X., Liang, X.-Y., and Yang, J.-S. (2012) *J. Org. Chem.*, 77, 3025–3037.
44 Zeng, Y., Ning, J., and Kong, F. (2002) *Tetrahedron Lett.*, 43, 3729–3733.
45 Chen, L. and Kong, F. (2003) *Tetrahedron Lett.*, 44, 3691–3695.
46 Imamura, A., Kimura, A., Ando, H., Ishida, H., and Kiso, M. (2006) *Chem. Eur. J.*, 12, 8862–8870.
47 Imamura, A., Ando, H., Ishida, H., and Kiso, M. (2005) *Org. Lett.*, 7, 4415–4418.
48 Imamura, A., Ando, H., Korogi, S., Tanabe, G., Muraoka, O., Ishida, H., and Kiso, M. (2003) *Tetrahedron Lett.*, 44, 6725–6728.
49 Agustí, R., Giorgi, M.E., Mendoza, V.M., Kashiwagi, G.A., de Lederkremer, R.M., and Gallo-Rodriguez, C. (2015) *Bioorg. Med. Chem.*, 23, 1213–1222.
50 Mendoza, V.M., Kashiwagi, G.A., de Lederkremer, R.M., and Gallo-Rodriguez, C. (2010) *Carbohydr. Res.*, 345, 385–396.
51 Lu, J. and Fraser-Reid, B. (2005) *Chem. Commun.*, 862–864.
52 Fraser-Reid, B., Lu, J., Jayaprakash, K.N., and López, J.C. (2006) *Tetrahedron: Asymmetry*, 17, 2449–2463.
53 Thadke, S.A., Mishra, B., and Hotha, S. (2013) *Org. Lett.*, 15, 2466–2469.
54 Thadke, S.A. and Hotha, S. (2014) *Org. Biomol. Chem.*, 12, 9914–9920.
55 Thadke, S.A., Mishra, B., and Hotha, S. (2014) *J. Org. Chem.*, 79, 7358–7371.
56 Pedersen, C.M., Nordstrøm, L.U., and Bols, M. (2007) *J. Am. Chem. Soc.*, 129, 9222–9235.
57 Pedersen, C.M., Marinescu, L.G., and Bols, M. (2011) *C.R. Chim.*, 14, 17–43.
58 Pougny, J.-R. and Sinaÿ, P. (1976) *Tetrahedron Lett.*, 17, 4073–4076.
59 Ratcliffe, A.J. and Fraser-Reid, B. (1990) *J. Chem. Soc., Perkin Trans. 1*, 747–750.
60 Gola, G., Tilve, M.J., and Gallo-Rodriguez, C. (2011) *Carbohydr. Res.*, 346, 1495–1502.
61 Gelin, M., Ferrières, V., and Plusquellec, D. (2000) *Eur. J. Org. Chem.*, 2001, 1423–1431.
62 Chao, C.-S., Lin, C.-Y., Mulani, S., Hung, W.-C., and Mong, K.-K.T. (2011) *Chem. Eur. J.*, 17, 12193–12202.
63 Mendoza, V.M., Agusti, R., Gallo-Rodriguez, C., and de Lederkremer, R.M. (2006) *Carbohydr. Res.*, 341, 1488–1497.
64 van Well, R.M., Collet, B.Y.M., and Field, R.A. (2008) *Synlett*, 2008, 2175–2177.
65 Frihed, T.G., Bols, M., and Pedersen, C.M. (2015) *Chem. Rev.*, 115, 4963–5013.

66 Callam, C.S., Gadikota, R.R., Krein, D.M., and Lowary, T.L. (2003) *J. Am. Chem. Soc.*, 125, 13112–13119.
67 Crich, D., Pedersen, C.M., Bowers, A.A., and Wink, D.J. (2007) *J. Org. Chem.*, 72, 1553–1565.
68 Glaudemans, C.P.J. and Fletcher, H.G. (1965) *J. Am. Chem. Soc.*, 87, 2456–2461.
69 Glaudemans, C.P.J. and Fletcher, H.G. (1965) *J. Am. Chem. Soc.*, 87, 4636–4641.
70 Subramaniam, V. and Lowary, T.L. (1999) *Tetrahedron*, 55, 5965–5976.
71 Yin, H. and Lowary, T.L. (2001) *Tetrahedron Lett.*, 42, 5829–5832.
72 Yin, H., D'Souza, F.W., and Lowary, T.L. (2002) *J. Org. Chem.*, 67, 892–903.
73 Ishiwata, A., Akao, H., Ito, Y., Sunagawa, M., Kusunose, N., and Kashiwazaki, Y. (2006) *Bioorg. Med. Chem.*, 14, 3049–3061.
74 Liu, C., Richards, M.R., and Lowary, T.L. (2010) *J. Org. Chem.*, 75, 4992–5007.
75 Liu, C., Richards, M.R., and Lowary, T.L. (2011) *Org. Biomol. Chem.*, 9, 165–176.
76 Islam, M., Gayatri, G., and Hotha, S. (2015) *J. Org. Chem.*, 80, 7937–7945.
77 Arasappan, A. and Fraser-Reid, B. (1995) *Tetrahedron Lett.*, 36, 7967–7970.
78 Gelin, M., Ferrieres, V., and Plusquellec, D. (1997) *Carbohydr. Lett.*, 2, 381–388.
79 Gandolfi-Donadio, L., Gola, G., de Lederkremer, R.M., and Gallo-Rodriguez, C. (2006) *Carbohydr. Res.*, 341, 2487–2497.
80 Kim, K.S. and Suk, D.H. (2011) in Topics in Current Chemistry, Reactivity Tunung in Oligosaccharide Assembly, vol. 301 (eds B. Fraser-Reid and J.C. Lopez), Springer, Berlin, Heidelberg, 109–140.
81 Oka, N., Kajino, R., Takeuchi, K., Nagakawa, H., and Ando, K. (2014) *J. Org. Chem.*, 79, 7656–7664.
82 Prévost, M., St-Jean, O., and Guindon, Y. (2010) *J. Am. Chem. Soc.*, 132, 12433–12439.
83 Larsen, C.H., Ridgway, B.H., Shaw, J.T., Smith, D.M., and Woerpel, K.A. (2005) *J. Am. Chem. Soc.*, 127, 10879–10884.
84 Kim, K.S., Kim, J.H., Lee, Y.J., Lee, Y.J., and Park, J. (2001) *J. Am. Chem. Soc.*, 123, 8477–8481.
85 Lee, Y.J., Lee, K., Jung, E.H., Jeon, H.B., and Kim, K.S. (2005) *Org. Lett.*, 7, 3263–3266.
86 Lee, Y.J., Lee, B.-Y., Jeon, H.B., and Kim, K.S. (2006) *Org. Lett.*, 8, 3971–3974.
87 Baek, J.Y., Joo, Y.J., and Kim, K.S. (2008) *Tetrahedron Lett.*, 49, 4734–4737.
88 Yasomanee, J.P. and Demchenko, A.V. (2012) *J. Am. Chem. Soc.*, 134, 20097–20102.
89 Liu, Q.-W., Bin, H.-C., and Yang, J.-S. (2013) *Org. Lett.*, 15, 3974–3977.
90 Ishiwata, A. and Ito, Y. (2011) *J. Am. Chem. Soc.*, 133, 2275–2291.
91 Gao, P.-C., Zhu, S.-Y., Cao, H., and Yang, J.-S. (2016) *J. Am. Chem. Soc.*, 138, 1684–1688.
92 Gadikota, R.R., Callam, C.S., and Lowary, T.L. (2001) *Org. Lett.*, 3, 607–610.
93 Gadikota, R.R., Callam, C.S., Wagner, T., Del Fraino, B., and Lowary, T.L. (2003) *J. Am. Chem. Soc.*, 125, 4155–4165.
94 Bai, Y. and Lowary, T.L. (2006) *J. Org. Chem.*, 71, 9658–9671.
95 Bai, Y. and Lowary, T.L. (2006) *J. Org. Chem.*, 71, 9672–9680.
96 Smith, D.M. and Woerpel, K.A. (2004) *Org. Lett.*, 6, 2063–2066.
97 Smith, D.M., Tran, M.B., and Woerpel, K.A. (2003) *J. Am. Chem. Soc.*, 125, 14149–14152.
98 Larsen, C.H., Ridgway, B.H., Shaw, J.T., and Woerpel, K.A. (1999) *J. Am. Chem. Soc.*, 121, 12208–12209.
99 van Rijssel, E.R., van Delft, P., Lodder, G., Overkleeft, H.S., van der Marel, G.A., Filippov, D.V., and Codée, J.D.C. (2014) *Angew. Chem. Int. Ed.*, 53, 10381–10385.

100 Zhu, X., Kawatkar, S., Rao, Y., and Boons, G.-J. (2006) *J. Am. Chem. Soc.*, 128, 11948–11957.
101 Nacario, R.C., Lowary, T.L., and McDonald, R. (2007) *Acta Crystallogr.*, E63, o498–o500.
102 Wang, Y., Maguire-Boyle, S., Dere, R.T., and Zhu, X. (2008) *Carbohydr. Res.*, 343, 3100–3106.
103 Bohé, L. and Crich, D. (2010) *Trends Glycosci. Glycotechnol.*, 22, 1–15.
104 Joe, M., Bai, Y., Nacario, R.C., and Lowary, T.L. (2007) *J. Am. Chem. Soc.*, 129, 9885–9901.
105 Rademacher, C., Shoemaker, G.K., Kim, H.-S., Zheng, R.B., Taha, H., Liu, C., Nacario, R.C., Schriemer, D.C., Klassen, J.S., Peters, T., and Lowary, T.L. (2007) *J. Am. Chem. Soc.*, 129, 10489–10502.
106 Rao, Y. and Boons, G.-J. (2007) *Angew. Chem. Int. Ed.*, 46, 6148–6151.
107 Rao, Y., Buskas, T., Albert, A., O'Neill, M.A., Hahn, M.G., and Boons, G.-J. (2008) *ChemBioChem*, 9, 381–388.
108 Fedina, K.G., Abronina, P.I., Podvalnyy, N.M., Kondakov, N.N., Chizhov, A.O., Torgov, V.I., and Kononov, L.O. (2012) *Carbohydr. Res.*, 357, 62–67.
109 Kaeothip, S. and Boons, G.-J. (2013) *Org. Biomol. Chem.*, 11, 5136–5146.
110 Ishiwata, A., Akao, H., and Ito, Y. (2006) *Org. Lett.*, 8, 5525–5528.
111 Ishiwata, A. and Ito, Y. (2009) *Trends Glycosci. Glycotechnol.*, 21, 266–289.
112 Tilve, M.J. and Gallo-Rodriguez, C. (2011) *Carbohydr. Res.*, 346, 2838–2848.
113 Crich, D., Cai, F., and Yang, F. (2008) *Carbohydr. Res.*, 343, 1858–1862.
114 Imamura, A. and Lowary, T.L. (2010) *Org. Lett.*, 12, 3686–3689.
115 Konradsson, P., Udodong, U.E., and Fraser-Reid, B. (1990) *Tetrahedron Lett.*, 31, 4313–4316.
116 Cumpstey, I. (2008) *Carbohydr. Res.*, 343, 1553–1573.
117 Ishiwata, A., Lee, Y.J., and Ito, Y. (2010) *Org. Biomol. Chem.*, 8, 3596–3608.
118 Bols, M. and Hansen, H.C. (1994) *Chem. Lett.*, 23, 1049–1052.
119 Cumpstey, I., Fairbanks, A.J., and Redgrave, A.J. (2004) *Tetrahedron*, 60, 9061–9074.
120 Ito, Y. and Ogawa, T. (1994) *Angew. Chem. Int. Ed. Engl.*, 33, 1765–1767.
121 Krog-Jensen, C. and Oscarson, S. (1996) *J. Org. Chem.*, 61, 4512–4513.
122 Krog-Jensen, C. and Oscarson, S. (1998) *J. Org. Chem.*, 63, 1780–1784.
123 Désiré, J. and Prandi, J. (1999) *Carbohydr. Res.*, 317, 110–118.
124 Sanchez, S., Bamhaoud, T., and Prandi, J. (2000) *Tetrahedron Lett.*, 41, 7447–7452.
125 Bamhaoud, T., Sanchez, S., and Prandi, J. (2000) *Chem. Commun.*, 659–660.
126 Marotte, K., Sanchez, S., Bamhaoud, T., and Prandi, J. (2003) *Eur. J. Org. Chem.*, 2003, 3587–3598.
127 Shinohara, H. and Matsubayashi, Y. (2013) *Plant Cell Physiol.*, 54, 369–374.
128 Gelin, M., Ferrières, V., Lefeuvre, M., and Plusquellec, D. (2003) *Eur. J. Org. Chem.*, 2003, 1285–1293.
129 Ishiwata, A., Munemura, Y., and Ito, Y. (2008) *Eur. J. Org. Chem.*, 2008, 4250–4263.
130 Kaeothip, S., Ishiwata, A., and Ito, Y. (2013) *Org. Biomol. Chem.*, 11, 5892–5907.
131 Ishiwata, A., Kaeothip, S., Takeda, Y., and Ito, Y. (2014) *Angew. Chem. Int. Ed.*, 53, 9812–9816.
132 Kaeothip, S., Ishiwata, A., Ito, T., Fushinobu, S., Fujita, K., and Ito, Y. (2013) *Carbohydr. Res.*, 382, 95–100.

133 Lichtenthaler, F.W. (2011) *Chem. Rev.*, 111, 5569–5609.
134 de Oliveira, M.T., Hughes, D.L., Nepogodiev, S.A., and Field, R.A. (2008) *Carbohydr. Res.*, 343, 211–220.
135 Uenishi, J.I. and Ueda, A. (2008) *Tetrahedron: Asymmetry*, 19, 2210–2217.
136 Premathilake, H. and Demchenko, A. (2011) in Topics in Current Chemistry, Reactivity Tuning in Oligosaccharide Assembly, vol. 301 (eds B. Fraser-Reid and J.C. López), Springer, Berlin, Heidelberg, pp. 189–221.
137 Liang, X.-Y., Bin, H.-C., and Yang, J.-S. (2013) *Org. Lett.*, 15, 2834–2837.
138 Wang, S., Meng, X., Huang, W., and Yang, J.-S. (2014) *J. Org. Chem.*, 79, 10203–10217.

15

De novo Asymmetric Synthesis of Carbohydrate Natural Products

Pei Shi and George A. O'Doherty

15.1 Introduction

The synthesis of carbohydrate-based natural products has been an integral part of organic synthesis for many decades. For the purposes of this review, the definition of "*de novo*" or "*de novo* asymmetric" routes is one that starts from achiral starting materials where asymmetric catalysis is used to install the stereocenters [1]. Thus, approaches that start with chiral starting materials (sugars, amino acids, and chiral auxiliaries) will not be reviewed herein. In addition, this chapter tries to steer clear of the material reviewed in the other book chapters. As earlier reviews cover the *de novo* asymmetric approaches to carbohydrates, this review is limited to carbohydrate-based natural products [2]. Of course, this designation incorrectly implies that carbohydrates are not natural products, but for the purposes of this review, we limit the discussion to the *de novo* asymmetric synthesis of carbohydrate-based secondary metabolites. Thus, in this review, we do not include syntheses of natural products from carbohydrates or other chiral starting materials.

Similarly to the first *de novo* syntheses of the hexoses by Sharpless and Masamune [3], the initial *de novo* asymmetric routes to carbohydrate natural products serve as test to the power of the asymmetric method. While they tend not to be addressed to carbohydrate-containing natural product syntheses, other groups have developed asymmetric strategies for the *de novo* synthesis of carbohydrates. These include the Sharpless and Masamune iterative epoxidation approach, the Sharpless Wong dihydroxylation/enzymatic aldol approach [4], the Johnson enzymatic resolution approach [5, 6], the Hudlicky [5b, 7] and Banwell enzymatic oxidation approach [8], and the Vogel furan Diels–Alder approach [9]. As the synthetic utility of these *de novo* asymmetric methods advances, the aim of these efforts evolves to address medicinal chemistry needs. Thus, one can see in the more recent examples (*vide infra*) that the inspiration for the synthesis comes from the medicinal chemistry community as these new routes often provide access to unnatural sugars, which could be of use in structure–activity relationship (SAR) studies.

Selective Glycosylations: Synthetic Methods and Catalysts, First Edition. Edited by Clay S. Bennett.

15.2 Danishefsky Hetero-Diels–Alder Approach

At the same time when Sharpless was developing his asymmetric epoxidation chemistry [3] and with Masamune its application to the hexoses, Danishefsky was developing his Diels–Alder chemistry for the synthesis of six-membered ring containing natural products [10]. A particularly powerful application of the Danishefsky Diels–Alder reaction was its use for the stereoselective synthesis of pyran natural products via the hetero-Diels–Alder reaction of 1,3-dialkoxydienes (aka, Danishefsky dienes) and aldehydes (Scheme 15.1). This in turn led to the development of a flexible asymmetric approach to various hexoses (**3a/b**) [11]. The success of this approach eventually inspired further studies by Danishefsky toward oligosaccharide synthesis, although the glycals used in these oligosaccharide assemblies are mostly derived from chiral carbohydrate starting materials [11].

The Danishefsky approach began with the synthesis of either a 3-siloxydiene with an alkoxy group(s) at the 1- or 1,3-position (i.e., Danishefsky dienes **1** and **4**) [10]. When these Danishefsky dienes are reacted with aldehydes in a Lewis-acid-catalyzed hetero-Diels–Alder cycloaddition/elimination, they afford *cis-* and *trans*-dihydropyrans (i.e., **2a/b**) [11, 12]. Subsequent post-Diels–Alder transformation was used to install the carbohydrate functionality from the initial Diels–Alder adduct. It is important to note that the approaches by Danishefsky used both a chiral auxiliary and a chiral Lewis acid. Thus, according to our definition, they are not *de novo* asymmetric approaches. However, it should be noted that Danishefsky was aware of the potential for a truly catalytic asymmetric variant with the use of more selective asymmetric catalysis. In 2002, his dream was realized with the discovery, by Jacobsen, of the (salen)-manganese complexes that could effectively catalyze the Danishefsky hetero-Diels–Alder/elimination reaction (e.g., **4** to **2a/b**) in excellent yields and enantioexcesses (Scheme 15.1) [13]. As a result, this work by Jacobsen and others transforms the early synthetic approach by Danishefsky (*vide infra*) into formal *de novo* asymmetric approach.

In addition to the numerous spiroketal-based natural product applications, Danishefsky also recognized the power of this asymmetric hetero-Diels–Alder for the synthesis of rare sugars. Excellent examples of this approach can be seen in Danishefsky's synthesis of the KDO [12] and papulacandin sugars [14] (Schemes 15.2 and 15.3). Danishefsky's

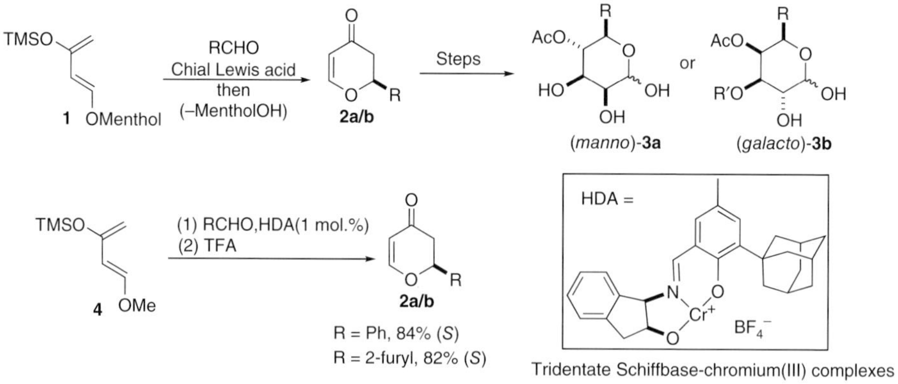

Scheme 15.1 Jacobsen-catalyzed asymmetric Danishefsky hetero-Diels–Alder.

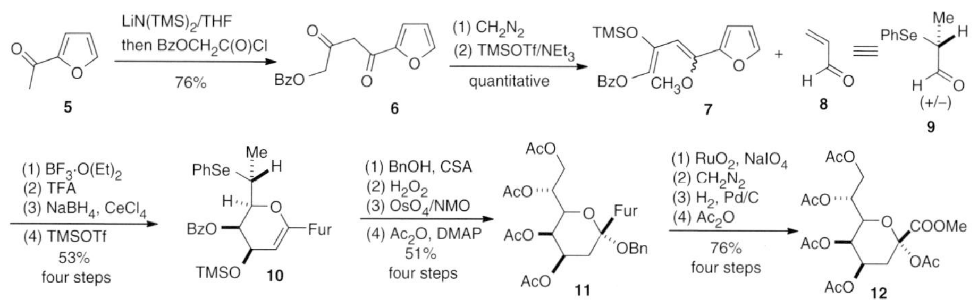

Scheme 15.2 Synthesis of a KDO-sugar derivative.

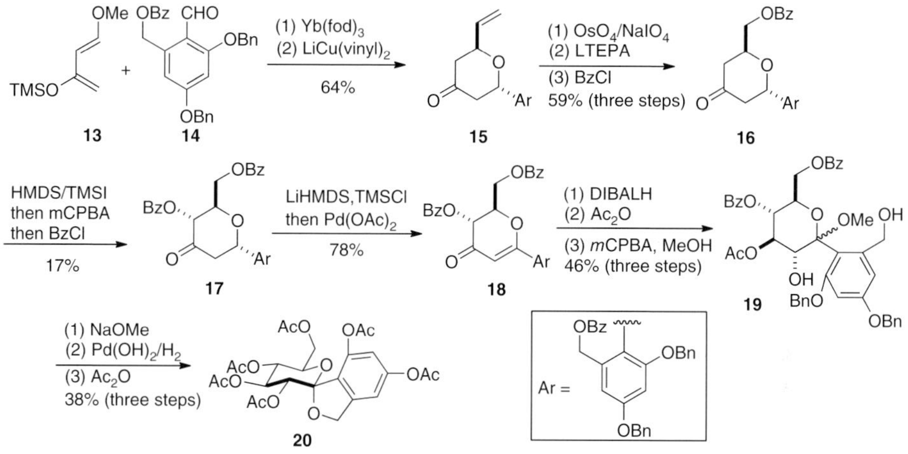

Scheme 15.3 Synthesis of the ring system of papulacandin D.

approach to KDO sugar **12** began with the synthesis of tri-alkoxy diene **7** (as a mixture of *E,Z*-isomers) from acetylfuran **1**. In this application, the hetero-Diels–Alder reaction occurred via a BF_3-catalyzed cycloaddition between **7** and aldehyde **9** (as a chiral acrolein equivalent). Following a TFA-promoted elimination of methanol, Luche reduction and trimethylsilyl (TMS) ether formation, the Diels–Alder adduct was converted into glycal **10**. Addition of benzyl alcohol across the glycal double bond followed by selenide oxidation/elimination, subsequent diastereoselective dihydroxylation and acylation were used to install the C-7/8 acetoxy groups in **11**. An oxidative cleavage of the electron-rich furan ring was used to form a carboxylic acid, which was then converted to a methyl ester with diazomethane. Finally, hydrogenolysis and acylation reactions were used to give the peracylated KDO sugar **12**.

In a related approach, the Danishefsky group also applied the hetero-Diels–Alder reaction for a synthesis of the papulacandins (Scheme 15.3) [14]. The papulacandins are an important group of antifungal antibiotics extracted [15], which have been shown to inhibit 1,3-β-D-glucan synthase. The specific hetero-Diels–Alder reaction was between diene **13** and aldehyde **14**, which was catalyzed by $Yb(fod)_3$. After a Lewis-acid-catalyzed loss of methanol, a vinyl-cuprate addition was used to diastereoselectively install the C-6 carbon affording **15**. In three steps, the vinyl group in **15** was easily converted into the Bz-protected hydroxymethyl ether in **16** by a Johnson–Lemieux oxidative cleavage [16], aldehyde reduction with lithium tris(3-ethyl-3-pentyl)oxoaluminum hydride (LTEPA), and benzoylation sequence. Following an enol ether formation, Rubottom oxidation and benzoylation sequence on **16** were used to provide **17** with the C-4 stereocenter in place. Subsequent enol ether generation and Saegusa oxidation [17] provided enone **18**, which was reduced with DIBAL-H and acylated to give a glycal intermediate. Finally, the papulacandin ring system in **20** was assembled by epoxidation and ring opening in methanol to give the *gluco*-sugar intermediate **19**, which was completed by a benzoylation, spiroketal formation, debenzylation, and peracylation sequence.

15.3 MacMillan Proline Aldol Approach

At the turn of the century, an alternative method for the practical *de novo* asymmetric synthesis of several hexoses was developed by MacMillan (Scheme 15.4) [18]. This approach featured the use of chiral amines (e.g., proline) for the asymmetric aldol reaction, which had been previously studied by MacMillan and others [19]. The MacMillan solution to the hexose problem involved the iterative use of aldol reactions (i.e., a proline-catalyzed aldol followed by a subsequent diastereoselective aldol reaction) to produce various hexoses (**24a–c**). Specifically, they found conditions where proline can be used catalytically to induce an aldol dimerization reaction between two aldehydes to give aldol products with erythrose stereochemistry (e.g., **23** from 2 × **22**). A second Lewis-acid-catalyzed Mukaiyama aldol reaction (**23** + **26**) was then used to convert the C-4 sugar **23** into hexo-pyranose **24** with various C-2/3 hexose stereochemistries (e.g., allose **24a**, glucose **24b**, and mannose **24c**). Thus, three different hexose diastereomers were produced in seven total steps (four longest linear). The relatively high amount (~30%) of proline used in the aldol reactions can easily be rationalized by the ready availability of L-proline, which provides the L-enantiomer of the sugars [18–20].

MacMillan was the first to demonstrate the power of his variant of the iterative aldol approach to the hexoses with the synthesis of two sugar-containing natural products. The first was Littoralisone **27** (Scheme 15.5) [21], and the second was Callipeltoside **38**/**39** (Scheme 15.7) [22]. A unique feature of both of these routes is how both the

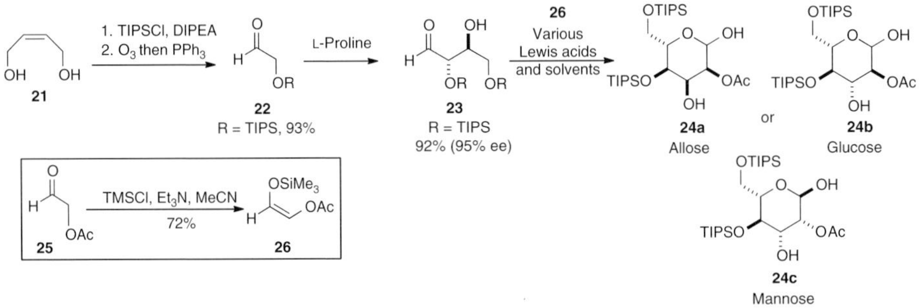

Scheme 15.4 Proline aldol approach to L-allose **24a**, L-glucose **24b**, and L-mannose **24c**.

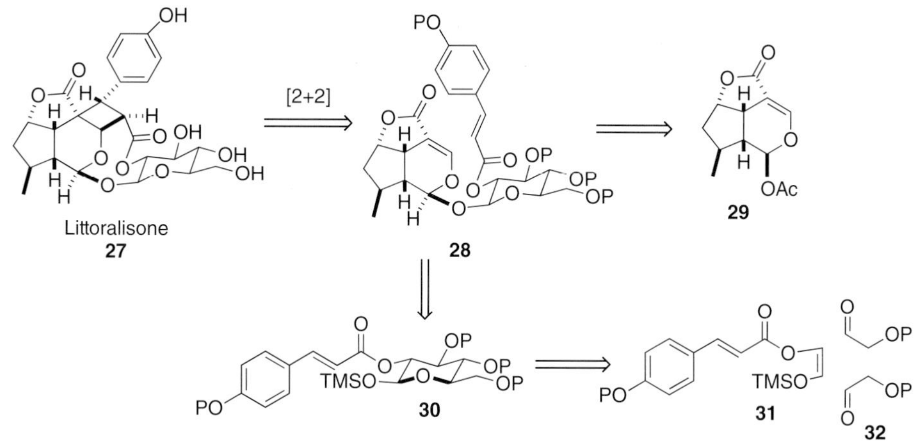

Scheme 15.5 Retrosynthetic of Littoralisone **27**.

aglycon and the carbohydrate portions of these two natural products derive their asymmetry by proline-catalyzed reactions.

Retrosynthetically, the MacMillan approach to Littoralisone **27** involves a glycosylation union between aglycon **29** and TMS-glycoside **30** to form **28**, which upon a photoinduced [2 + 2]-cycloaddition afforded Littoralisone **27**. Both the aglycon **29** and the D-sugar carbohydrate donor **30** were prepared via proline catalysis. The synthesis of the D-sugar component **30** began with the aldol dimerization of aldehyde **33** with L-proline to form **34** (Scheme 15.6). A Lewis acid $MgBr_2$-promoted Mukaiyama-type *anti*-aldol reaction [23] of **34** with the TMS enol ether **35** gave the *gluco*-sugar **36**, which after per-benzyl ether protection, hydrolysis of the anomeric position, and TMS protection gave **37**.

In contrast to the common D-glucose sugar chemistry in Littoralisone **27**, the carbohydrate portion of Callipeltoside C consists of the rare sugar **41** in unknown stereochemistry (Scheme 15.7). The unknown nature of the absolute stereochemistry for the sugar portion of the natural product made it an ideal candidate for *de novo* asymmetric synthesis. The unknown nature of the absolute stereochemistry for the sugar portion of the

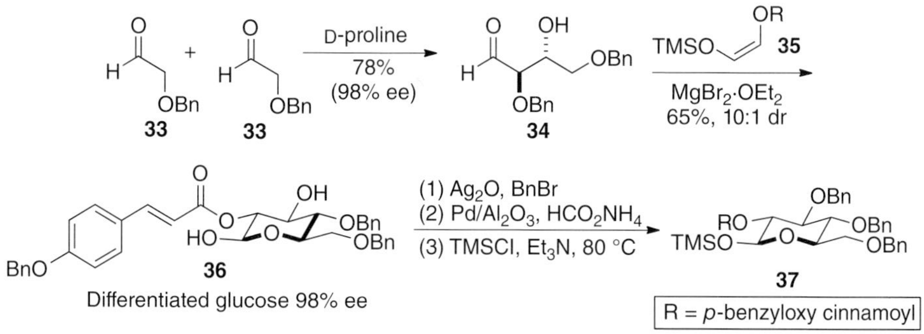

Scheme 15.6 Synthesis of glucose donor.

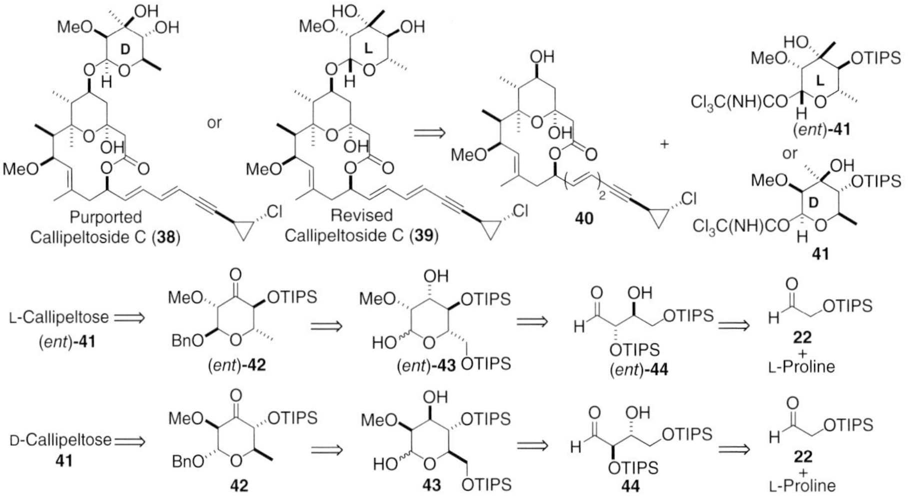

Scheme 15.7 Retrosynthetic analysis of Callipeltoside C **38/39**.

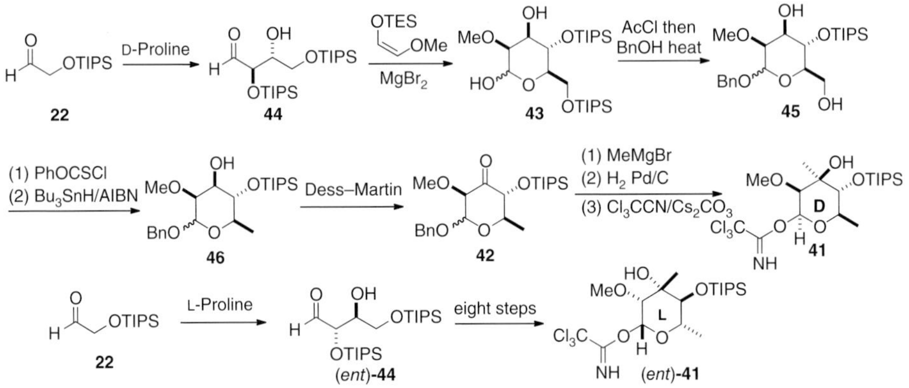

Scheme 15.8 Synthesis of D- or L-callipeltose donors **41** and (*ent*)-**41**.

natural product made it an ideal candidate for de novo asymmetric synthesis, as either enantiomer of the glycosyl-donor 41 could be ultimately obtained by the dimerization of aldehyde 22 with D- or L-proline to form 44 or (ent)-44. Finally, the MacMillan approach to Callipeltoside C **38/39** involves a glycosylation of aglycon **40** with either enantiomer of glycosyl donor **41**. Access to both diastereomers **38** and **39** enabled the MacMillan group to revise the relative stereochemistry of Callipeltoside C from **38** to structure **39**.

The MacMillan *de novo* synthesis of the D-glycosyl donor **41** proceeded as outlined in Scheme 15.8. Once again, the synthesis began with the dimerization of aldehyde **22** with D-proline to form **44**. A magnesium-enolate-promoted *anti*-aldol of **44** provided *manno*-sugar **43**. An acid-catalyzed deprotection of the C-6 TIPS group and concomitant mixed acetal formation gave **45**. A regioselective Barton-type deoxygenation [24] of the C-6 hydroxyl group in **45** gave 6-deoxysugar **46**. A Dess–Martin oxygenation of the C-3 hydroxyl group gave keto-sugar **42**. Finally, the D-glycosyl donor **41** was produced by a methyl anion addition, hydrogenolysis of the anomeric benzyl group, and trichloroacetamide formation. Of course, the L-glycosyl donor (*ent*)-**41** could similarly be prepared by simply switching the L-proline for D-proline [25].

Chandrasekhar has also used the MacMillan iterative aldol approach to the hexoses for the synthesis of the papulacandin ring system (Schemes 15.9 and 15.10) [26]. In addition to testing the MacMillan methodology, the Chandrasekhar synthesis serves as a good comparison of the MacMillan methodology with the Danishefsky approach to the same ring system. Unlike the Danishefsky approach, the Chandrasekhar route was designed to be stereodivergent, as it targets three different stereoisomers of the papulacandin ring system.

The Chandrasekhar route began with the D-proline dimerization of **47** to form D-erythrose sugar **48** (Scheme 15.9). Thus, exposing **48** to the titanium enolate of **49** induced a tandem aldol and spiroketal formation to form **50** as a mixture of sugar diastereomers. Deprotection of the silyl-protecting groups in **50** with excess tetrabutylaluminium fluoride (TBAF) gave a 15:4:1 mixture of *allo-*, *altro-*, and *gluco*-papulacandins (**51a–c**). While this

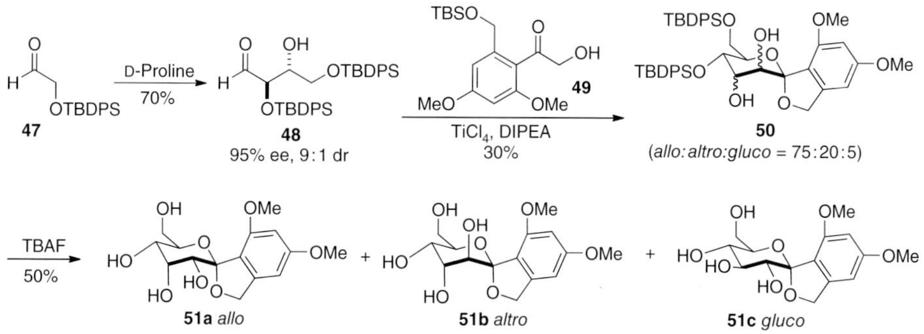

Scheme 15.9 Synthesis of three stereoisomers of the papulacandin ring system.

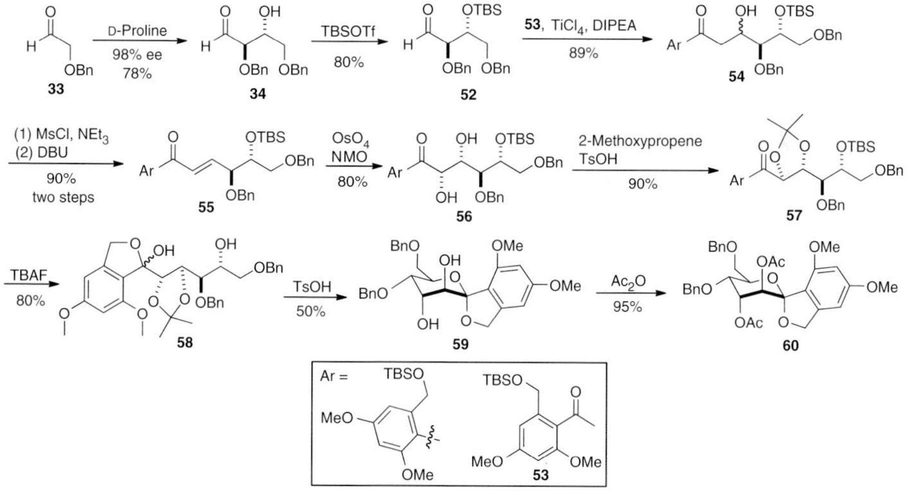

Scheme 15.10 Synthesis of the *altro*-papulacandin.

route produced the papulacandin ring system in a relatively short route, it suffered from poor stereocontrol.

To address the stereocontrol problem, Chandrasekhar developed a second-generation approach. In the second-generation synthesis, Chandrasekhar targeted a stereoselective synthesis of the *altro*-diastereomer of the papulacandin (Scheme 15.10). The revised route began with the proline-catalyzed dimerization of benzyloxyacetaldehyde **33** to form **34**. tert-butyldimethylsilyl (TBS) protection of the β-hydroxy group gave aldehyde **52**, which was reacted with the titanium enolate of **53** to afford **54** as a mixture of diastereomers. A dehydration of **54** gave enone **55**, which underwent a substrate-controlled diastereoselective (98:2 *anti*/*syn*) dihydroxylation to give **56**. Acetonide protection gave **57**, and TBS deprotection gave **58**. Acid-catalyzed acetonide deprotection and ketal formation gave *altro*-papulacandin **59**, which upon acetylation gave **60**.

15.4 The O'Doherty Approaches

15.4.1 O'Doherty Iterative Dihydroxylation Approach

Concurrent to the MacMillan iterative asymmetric aldol approach to the hexoses, the O'Doherty group developed two alternative *de novo* approaches to the hexoses. The first approach involved an iterative dihydroxylation of dienoates to produce various sugar lactones (Scheme 15.11) [27]. In its most efficient form, it required only one step to a racemic sugar, whereas three steps are required for sugars in optically active form. In its most simple variation, ethyl sorbate **61** was converted under the Upjohn conditions (OsO_4/NMO) [28] in one-pot reaction to afford the racemic γ-*galactono*-lactones **63**. In its asymmetric variant, the initial dihydroxylation with the Sharpless reagent [26] gave diol **64**, which when dihydroxylated with the pseudo-enantiomeric Sharpless reagent allows for a highly stereoselective three-step synthesis of the γ-*galactono*-lactone **66,** which is compatible with diverse C-6 substitution and near perfect enantio- and diastereocontrol.

The O'Doherty group also applied the iterative dihydroxylation approach to the *de novo* asymmetric synthesis of the papulacandins (Scheme 15.12) [29]. This synthesis provided two *galacto*-papulacandins in its pyranose **72** and furanose **73** forms. Their route began with the synthesis of dienone **69** via an Horner–Wadsworth–Emmons olefination between **67** and **68**. A regioselective asymmetric dihydroxylation of dienone **69** provided diol **70**, and a second substrate-controlled dihydroxylation provided tetraacetate **71** after

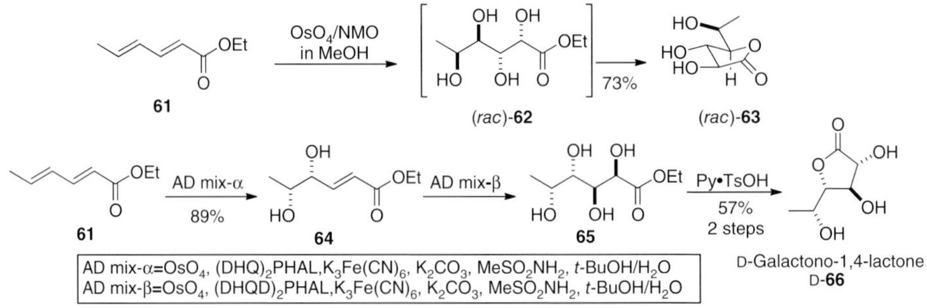

Scheme 15.11 Iterative dihydroxylation approach to carbohydrates.

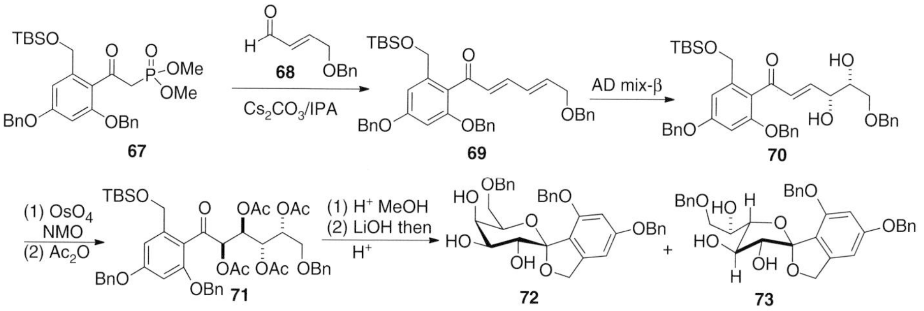

Scheme 15.12 Alternative enantioselective synthesis of the papulacandin ring system.

peracylation. A one-pot deprotection/spiroketalization provides a 2:1 mixture of the *galacto*-papulacandins in both its pyranose form and furanose form (**72** and **73**, respectively), which could be interconverted upon acid-catalyzed equilibration.

15.4.2 O'Doherty Achmatowicz Approach

The O'Doherty group has generated an alternative *de novo* approach to the hexoses and more generally to oligosaccharides. This alternative approach involves a *de novo* asymmetric synthesis of a furan alcohol coupled with an Achmatowicz reaction (Scheme 15.13) [30]. In addition to being synthetically orthogonal to the other approach mentioned earlier, this approach appears to be particularly well positioned to address medicinal chemistry issues [2, 31]. This Achmatowicz approach, similar to the iterative dihydroxylation approach, is ideally suited for the synthesis of hexo-pyranoses with variable C-6 substitutions. Of the two, the iterative asymmetric dihydroxylation of dienoates (Schemes 15.11 and 15.12) is the most efficient in terms of steps, whereas, the Achmatowicz approach is superior in terms of synthetic scope (*vide infra*). This scope is uniquely seen in the application for the efficient synthesis of oligosaccharides (e.g., mono-, di-, tri-, tetra-, and

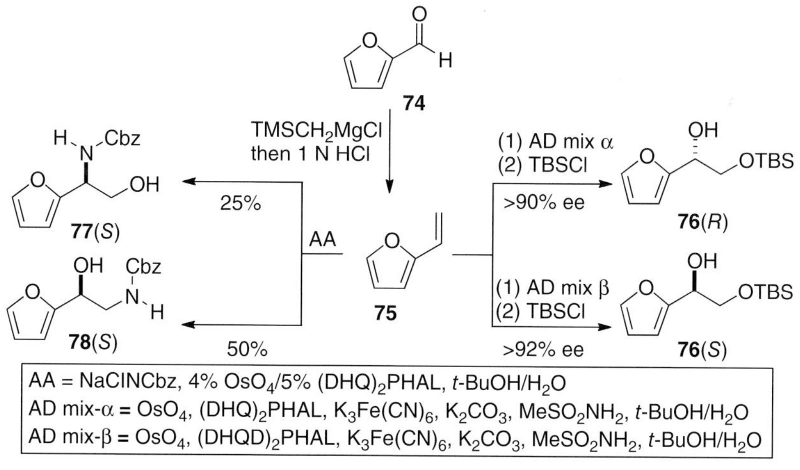

Scheme 15.13 *De novo* asymmetric oxidation approaches to chiral furan alcohols.

hepta-saccharide motifs) [32]. A consistent advantage to these approaches is the minimal use of protecting groups.

15.4.3 *De novo* Use of the Achmatowicz Approach to Pyranose

This *de novo* Achmatowicz approach began with the practical large-scale access to enantiomerically pure furan alcohols from achiral furans (e.g., **74** and **79**). O'Doherty's first approach to furan alcohols involves the asymmetric oxidation of vinylfuran **75** (Scheme 15.13) [33]. The synthesis began with an *in situ* Petersen olefination of furfural **74** to generate vinylfuran solutions, which was compatible with both the Sharpless asymmetric dihydroxylation (**75** to **76**) [29] and amino-hydroxylation (**75** to **77** and **78**) protocols [34].

An alternative furan alcohol synthesis involved the use of a Noyori hydrogen transfer reaction for the asymmetric reduction of acylfurans (**79** to **80**, Scheme 15.14) [35]. Of the two routes, the Sharpless route is most amenable to the synthesis of hexoses with a C-6 hydroxy group, whereas the Noyori route distinguishes itself in its flexibility to virtually any substitution at the C-6 position [36].

A unique feature of the Achmatowicz approach to the hexoses is the use of alkene and ketone π-bonds functionality as triol atomless protecting group (Scheme 15.15) [31]. For example, a four-step protocol can be used to diastereoselectively convert the optically pure alcohols, such as **76**, into *manno*-hexoses (e.g., **84**) [29]. The route was also amenable to *talo*- and *gulo*-diastereomers (**85** and **86**) [37]. In practice, an NBS promoted oxidative

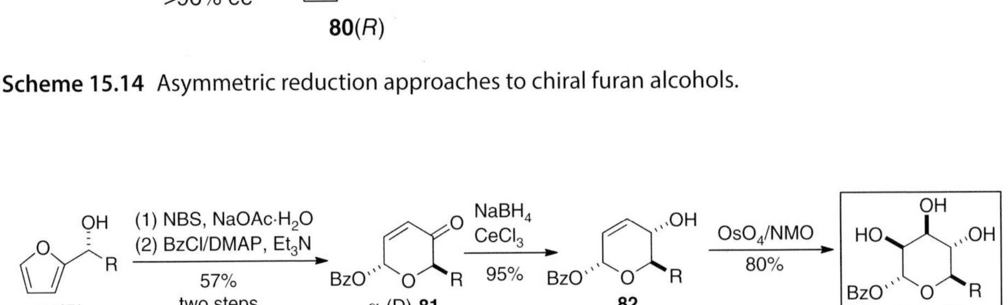

Scheme 15.14 Asymmetric reduction approaches to chiral furan alcohols.

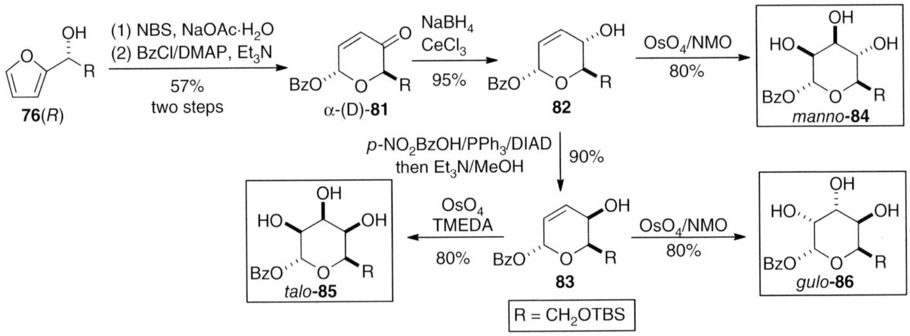

Scheme 15.15 Syntheses of the D-enantiomers of *manno*-, *gulo*-, and *talo*-pyranoses.

hydration of furan alcohol **76** to a 6-hydroxy-2*H*-pyran-3(6*H*)-ones (aka, Achmatowicz rearrangement) [38]. An α-selective benzoylation afforded benzoyl-protected **81,** which was followed by a highly diastereoselective Luche reduction ($NaBH_4/CeCl_3$) to convert **81** to allylic alcohol **82**. Finally, a highly diastereoselective Upjohn dihydroxylation ($OsO_{4(cat)}$/NMO) gave *manno*-pyranose **84**. The *talo*- and *gulo*-pyranoses were similarly prepared by incorporating a Mitsunobu/hydrolysis (**82** to **83**) sequence into the route. Exposure of the diastereomeric allylic alcohol **83** to the same Upjohn dihydroxylation ($OsO_{4(cat)}$/NMO) provides the *gulo*-pyranose **86**, whereas a hydroxy-directed dihydroxylation (OsO_4/TMEDA) affords the diastereomeric *talo*-pyranose **85** [29, 30]. Of course, deoxy- and deoxyamino-sugar variants were also prepared by this same approach [39].

15.4.4 *De novo* Access to Monosaccharide Natural Products

The O'Doherty group also applied the Achmatowicz approach to the *de novo* synthesis of the papulacandin ring system (Scheme 15.16) [40], which serves as an excellent comparison to the other *de novo* approaches (Schemes 15.3, 15.9, 15.10, and 15.12). The route began with the Pd-catalyzed coupling of **87** and **88** to form furan aldehyde **89**. A Wittig olefination on **89** provided vinylfuran **90**. A Sharpless dihydroxylation of **90** gave furan alcohol **91** after selective pivaloylation of the primary alcohol. An Achmatowicz oxidation of **91** was followed by spiro-ketalization, Luche reduction, and TBS protection to form spiroketal **92**. The use of an Upjohn dihydroxylation of **92** was followed by Piv- and TBS deprotection to provide a 4 : 1 ratio of *manno*- and *allo*-papulacandins, **94**

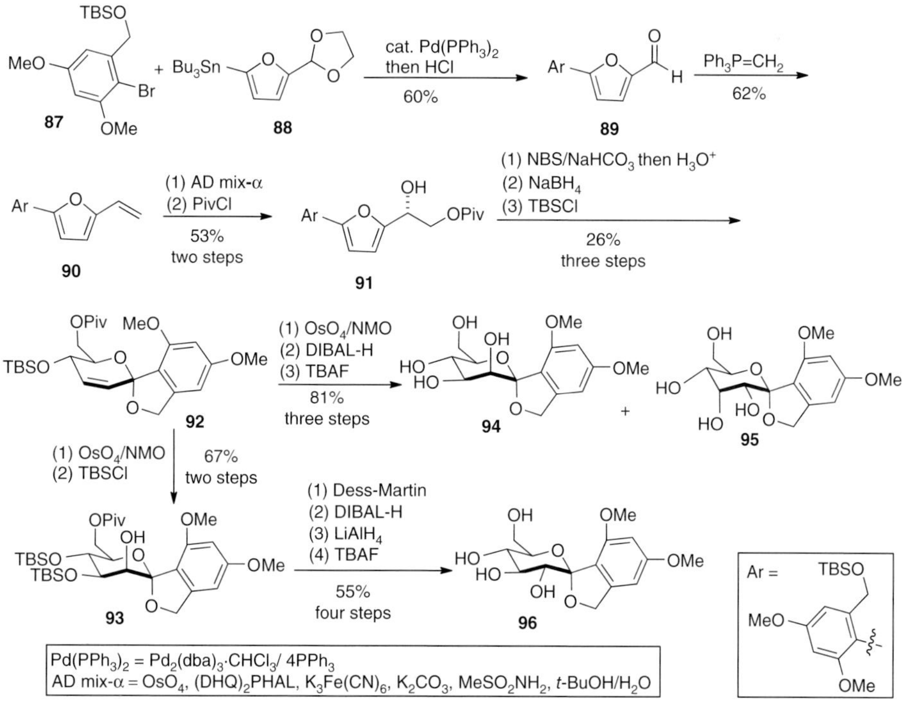

Scheme 15.16 *De novo* synthesis of the papulacandin ring system.

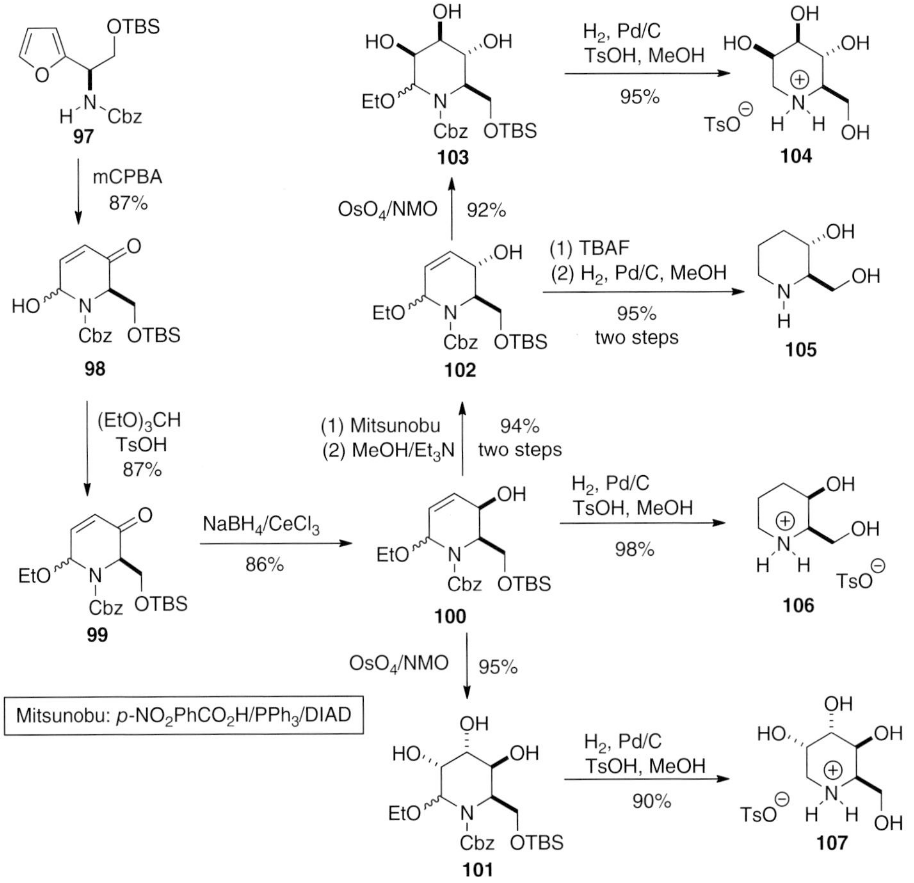

Scheme 15.17 Synthesis of D- and L-deoxymannojirimycin.

and **95**, respectively. The incorporation a protection step after the dihydroxylation provided the *manno*-papulacandin **93**, which was inverted by an oxidation (Dess-Martin periodinane), reduction (DIBAL-H), and deprotection (LAH/TBAF) sequence to give *gluco*-papulacandin **96**.

This *de novo* Achmatowicz approach was also applied to iminosugars (Scheme 15.17) [41], by using the Cbz-protected furan amine **97**. In this application, they found that the aza-Achmatowicz reaction was best accomplished using meta-chloroperbenzoic acid (mCPBA) to form **98** from **97**. After protection of the anomeric hydroxyl group as an ethoxy group, the resulting enone was reduced to form allylic alcohol **100**. A highly diastereoselective dihydroxylation of **100** gave **101,** which after hydrogenolysis gave deoxy-*gulo*-nojirimycin **107**. In addition, the approach could be expanded to the synthesis deoxymannojirimycin **104,** by incorporating Mitsunobu/hydrolysis (**100** to **102**) into the sequence.

The direct comparison of these various *de novo* routes is difficult in terms of number of steps, availability of starting materials, and/or atom economy. In our view, the best metric for this evaluation is in terms of synthetic utility and variability. In this regard, it is the compatibility of the *de novo* Achmatowicz approach with the Pd-π-allyl-catalyzed glycosylation that distinguishes it from the other *de novo* approach (Schemes 15.1–15.12) [42, 43]. To these ends, the O'Doherty group developed a general and stereospecific Pd(0)-catalyzed glycosylation [44, 45] (e.g., α-**108** to α-**109** and β-**108** to β-**109**, Scheme 15.18). This reaction, which at first glance does not resemble a typical glycosylation reaction, has the ability to use the enone functionality as precursor to mono-, di-, and triol products. These postglycosylation reactions transform the enone into the functional equivalent of variously substituted alcohol/polyol and hence function as atomless protecting groups.

One of the first applications of this Pd-catalyzed glycosylation to natural product synthesis was its use in the synthesis of daumone **117** (Scheme 15.19) [46]. Daumone is a rare ascarylose-containing pheromone natural product associated with the induction of a dauer state in *Caenorhabditis elegans* [47]. As both the aglycon and the sugar components were chiral, both halves required the asymmetric synthesis, which in this case involved the use of Noyori reductions of acylfuran (**79**, R = Me) and ynone **111**. This convergent route used the Pd-glycosylation at the point of convergence (**113** + **114** to **115**). The daumone ascarylose stereochemistry was introduced via two-step epoxidation, reduction, and ring-opening sequence (**115** to **116**). The synthesis was finished with TBS removal and selective oxidation of the terminal hydroxyl group to the carboxylic acid (**116** to **117**). The overall utility of this synthesis can be seen in its use mode of action studies [45].

The Achmatowicz approach was also used in the synthesis of the polyhydroxylated indolizidine alkaloids (i.e., D-Swainsonine **128,** Scheme 15.20) [48]. The interest in

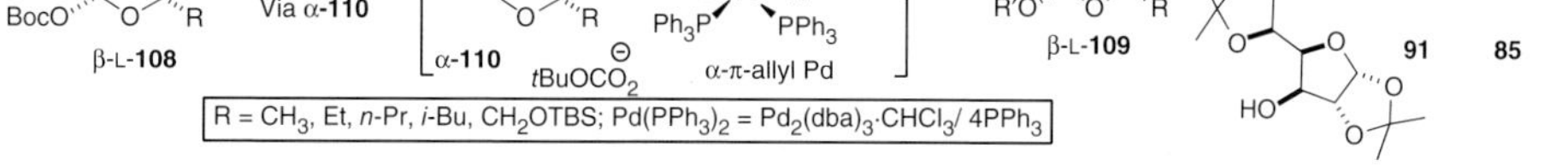

Scheme 15.18 Stereospecific palladium-catalyzed glycosylation.

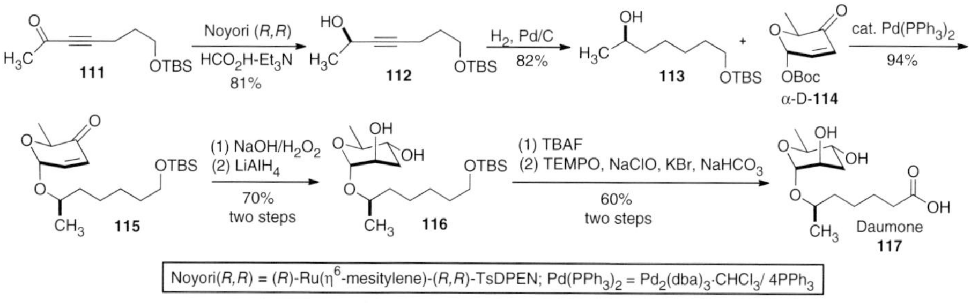

Scheme 15.19 Asymmetric synthesis of daumone.

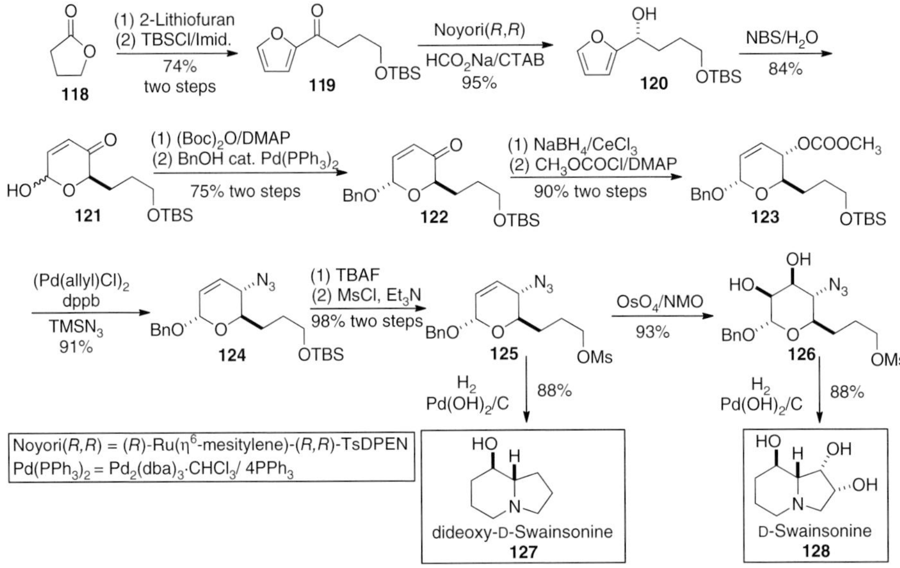

Scheme 15.20 Syntheses of dideoxy-D-Swainsonine and D-Swainsonine.

Swainsonine was derived from its potent inhibitory activity of both lysosomal α-mannosidase [49] and mannosidase II [50, 51]. In addition, as its enantiomer (L-Swainsonine) was also known to be a selective inhibitor of narginase (L-rhamnosidase, $K_i = 0.45\,\mu M$), it was an ideal candidate for *de novo* asymmetric synthesis [52]. Thus, both D- and L-Swainsonine and several epimers and analogs have become attractive targets for syntheses [48, 53]. The route began with the synthesis of achiral acylfuran **119** from 2-lithiofuran and buterolactone **118,** which has all the carbons required for Swainsonine. Noyori reduction (**119** to **120**) was used to install the asymmetry, whereas Achmatowicz reaction and Boc protection converted it into pyranone **121**. A Pd-catalyzed glycosylation of benzyl alcohol with **121** was used to install the required anomeric protecting group in **122**. A three-step Luche reduction, carbonate formation, and Pd-π-allyl allylic azide displacement sequence (**122** to **124**) was used to install the azido group in **124** at the C-4 position. The TBS ether was converted into a good leaving group (**124** to **125**), and the double bond in **125** was dihydroxylated to give the Swainsonine precursor **126**. Finally, an exhaustive hydrogenolysis (**126** to **128**) cleanly provided D-Swainsonine in optically pure form. As a result, this *de novo* route readily provided ample quantities of either D- or L-Swainsonine, as well as diastereomers and simpler dideoxy-D analogs (e.g., **127**) for future biological investigations.

An example of the application of the *de novo* Achmatowicz route to the synthesis of *O*-aryl-glycosides can be seen in the synthesis of SL0101 (Scheme 15.21) [54]. This synthesis has enabled significant medicinal chemistry studies, which include the SAR studies of absolute stereochemistry, C-6 alkyl substitution, as well as pyran ring oxygen substitution to a methylene (i.e., 5a-carbasugars) [55]. The synthesis of the acylated Kaempferol rhamnoside natural product began with a Pd-catalyzed glycosylation between the Bn-protected aglycon **129** and the pyranone donor L-**114**. The C-4 acetoxy-rhamnose functionality was introduced by a three-step $NaBH_4$ reduction, acylation, and

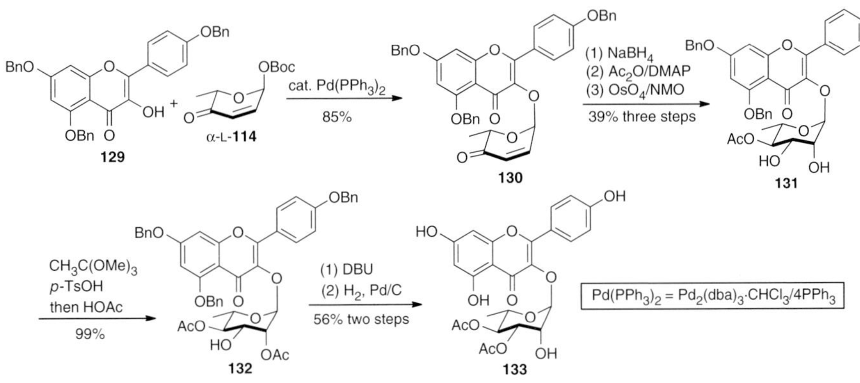

Scheme 15.21 *De novo* synthesis of the RSK inhibitor SL0101.

Upjohn dihydroxylation to form the C-2/3 diol **131**. Ortho-ester chemistry was used to regioselectively acylate at the C-2 position to give **132**. When the C-2 acetate was treated with DBU, it readily equilibrated to the equatorial C-3 position, after which the natural product **133** was revealed by means of hydrogenolysis.

The *de novo* Achmatowicz approach was also used to prepare other mono- and disaccharide natural products and related biologically important motifs. The monosaccharide targets include Methymycin [56], Jadomycin [57], Gilvocarcin [58], and Homo-adenosine [59]. The disaccharide targets include Trehalose [60] and Mannopeptimycin-ε [61].

15.4.5 *De novo* Access to Oligosaccharide Natural Products

This *de novo* Achmatowicz approach was also used for the synthesis and study of many oligosaccharide-containing natural products. For example, the synthesis of the trisaccharide portion of several rare sugar-containing natural products has been accomplished. This includes the tris-digitoxose portion of digitoxin [62], the olivose/rhodinose/acrolose portion of PI-080 [63], and the olivose/olivose/rhodinose portion of Landomycin E [64]. The establishment of a viable synthetic route to these three classes of trisaccharide natural products has enabled extensive medicinal chemistry study of these structural motifs [32e].

The approach to the Landomycin E trisaccharide is outlined in Scheme 15.22 [61]. The approach to the two olivose sugars began with the same postglycosylation strategy that installed the β-D-digitoxose sugar stereochemistry (**135** and **138**). Key to the success of applying this approach to Landomycin E was the discovery of a highly regioselective Mitsunobu-like inversion of the axial alcohol of a *cis*-1,2-diol (**135** to **136** and **138** to **139**) [60, 61]. This net transformation of **135** to **136** constitutes a good solution to the problem of a 1,2-*trans*-diequatorial addition across a cyclohexene. The Mitsunobu chemistry also differentially protects the diol for further glycosylation (e.g., **136** to **137**). In the latter case, the enone functionality undergoes a 1,2-reduction, Mitsunobu inversion, ester hydrolysis (**139** to **140**), and reduction installs the final rhodinose sugar in **141**.

This *de novo* Achmatowicz approach was also used for the synthesis of the even more complex oligosaccharide structural motifs. Examples of the utility of this methodology were nicely on display in the synthesis of three classes of oligo-rhamnose-containing

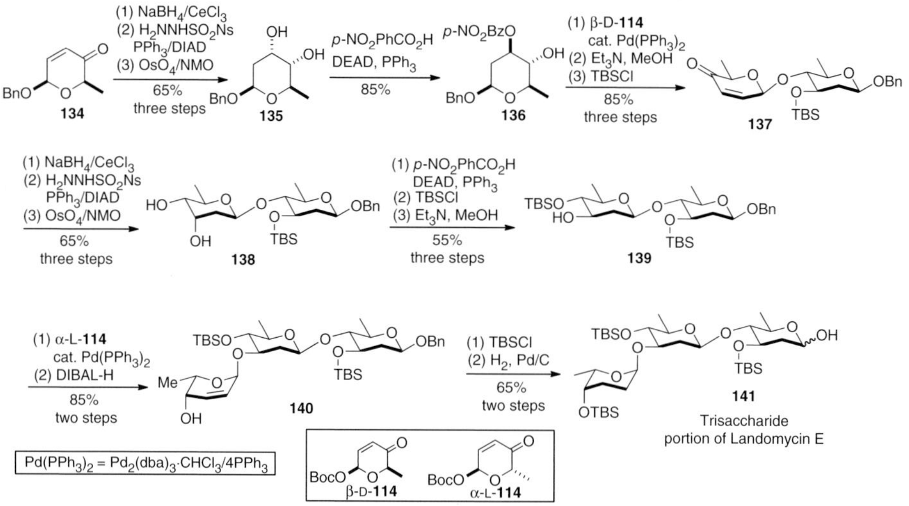

Scheme 15.22 *De novo* synthesis of the Landomycin E trisaccharide.

natural products (Scheme 15.23): the Cleistriosides (**143** and **144**), the Cleistetrosides (**142** and **145** to **149**) [65], and the Mezzettiasides (**150–159**) [66]. These partially acylated oligo-rhamnose natural products have been reported to possesses both interesting antibacterial and anticancer properties [67–69]. In addition to clarifying the structural issues, the *de novo* synthesis of these oligosaccharides was instrumental in supplying sufficient quantities of pure material for extensive SAR-type medicinal chemistry studies [70].

A unique feature of these three routes is how they enable access to all 18 of the compounds in Scheme 15.23, while only using two or fewer types of protecting groups. At most, only one acetonide and two chloroacetates were needed to address the various regiochemical issues. In addition, the routes featured the first use of cyclic tin acetals and boronate esters in the regioselective glycosylation reaction. To exemplify the flexibility of these approaches, the *de novo* Achmatowicz route to the Cleistrioside, the Cleistetroside classes of natural products were outlined in Schemes 15.24 and 15.25.

The route to Cleistetroside-2 (Scheme 15.24) began with pyranone α-L-**114**, which in four steps (glycosylation, reduction, dihydroxylation, and acetonide protection) was converted into protected rhamnose **160**. A four-step glycosylation (**160** to **161**), reduction, acylation, and dihydroxylation sequence was used to afford disaccharide **162**. A novel tin-directed regioselective (7 : 1) glycosylation of the diol in **162** and chloro-acylation was used to install a sugar at the C-3 position (**163**). A four-step postglycosylation transformation protocol was used to further elaborate **163** into the required rhamnose substitution in **164**, which in only six steps was converted into Cleistetroside-2 (**142**) in a 49% overall yield.

While the aforementioned *de novo* route to Cleistetroside-2 was comparable to the two previous routes to Cleistetroside-2, in terms of total number of steps, what distinguished it from these more traditional routes is its flexibility to diverge to any member of the Cleistrioside and Cleistetroside family of natural products. For instance, as outlined in Scheme 15.25, trisaccharide **163** was further elaborated into the Cleistriosides-5 and -6 as well as the five remaining Cleistetrosides-3, -4, -5, -6, and -7.

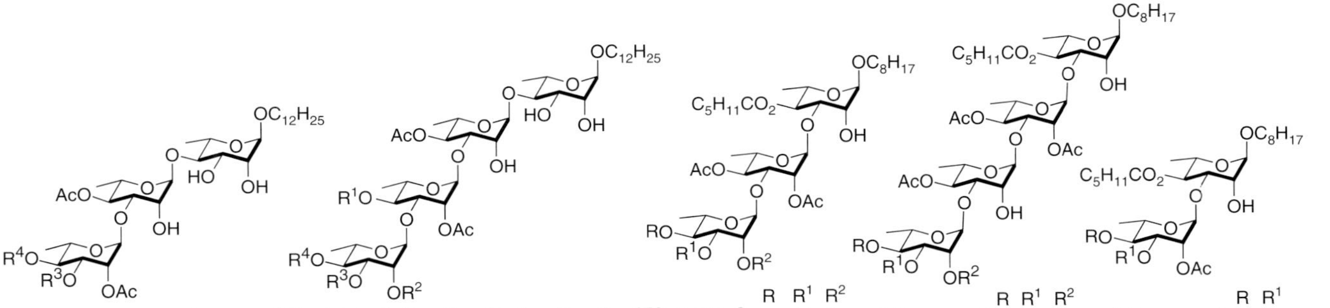

143: cleistri-5: R_3= H, R_4= Ac
144: cleistri-6: R_3= Ac, R_4 = H

celestri = cleistrioside
cleistet = cleistetroside
mezze = mezzettiaside

142: cleistet-2: $R_1 = R_4 = Ac$, $R_2 = R_3 = H$
145: cleistet-3: $R_1 = R_2 = Ac$, $R_3 = R_4 = H$
146: cleistet-4: $R_1 = Ac$, $R_2 = R_3 = R_4 = H$
147: cleistet-5: $R_1 = R_2 = R_3 = R_4 = H$
148: cleistet-6: $R_1 = R_2 = R_4 = Ac$, $R_3 = H$
149: cleistet-7: $R_1 = R_2 = R_3 = R_4 = Ac$

	R	R^1	R^2
150: mezze-2:	Ac	H	Ac
151: mezze-3:	Ac	Ac	H
152: mezze-4:	Ac	H	H
153: mezze-8:	H	H	Ac

	R	R^1	R^2
154: mezze-5:	H	Ac	H
155: mezze-6:	H	H	Ac
156: mezze-7:	H	H	H

	R	R^1
157: mezze-9:	H	Ac
158: mezze-10:	Ac	H
159: mezze-11:	Ac	Ac

Scheme 15.23 The Cleistriosides, Cleistetrosides, and Mezzettiasides.

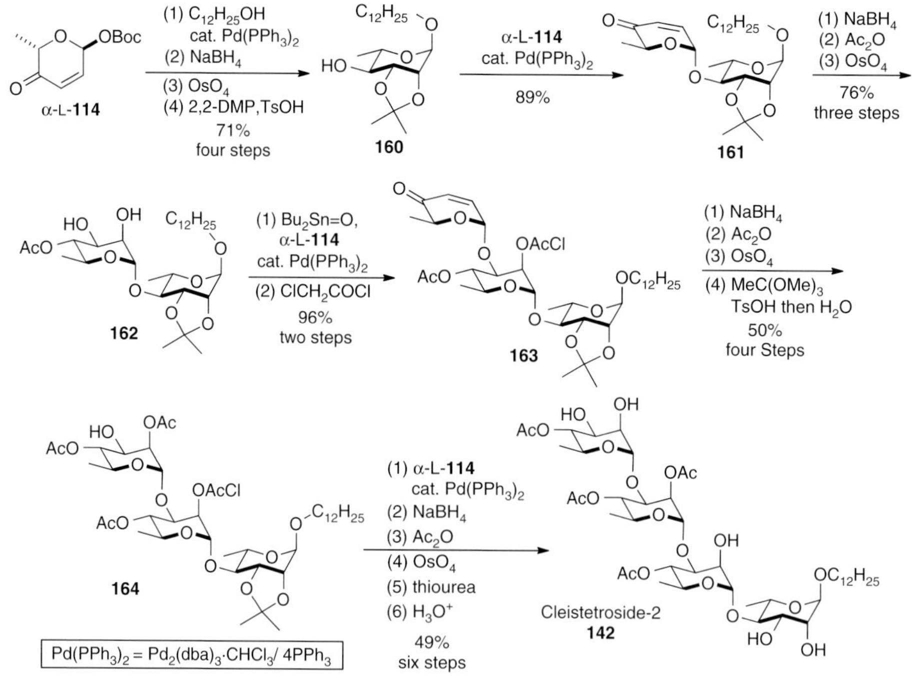

Scheme 15.24 Synthesis of Cleistetroside-2.

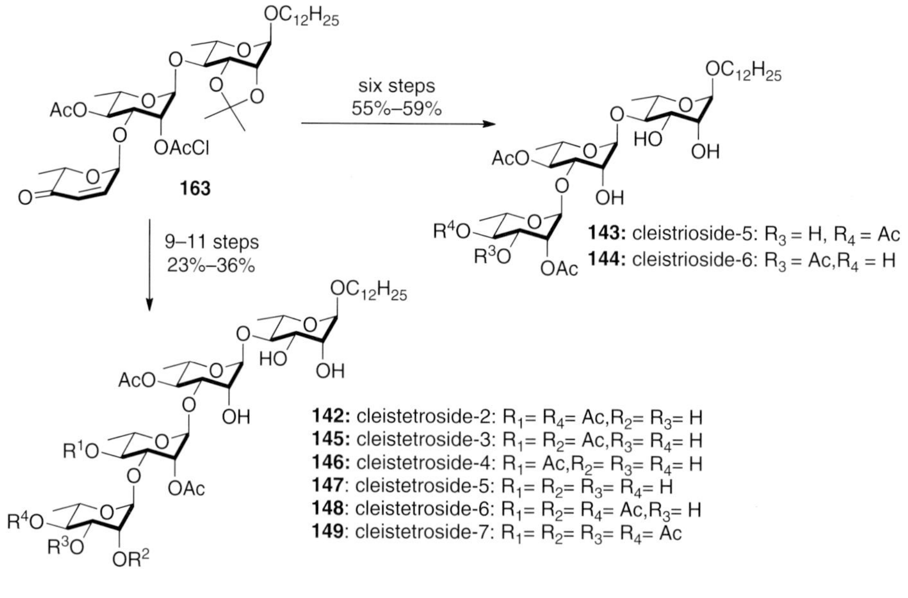

Scheme 15.25 Syntheses of Cleistrioside and Cleistetroside natural product families.

The *de novo* Achmatowicz approach was also successfully applied to the related oligorhamnan natural product, anthrax tetrasaccharide **165** (Scheme 15.26). Anthrax is the disease that is caused by *Bacillus anthracis* [71]. In an effort to find a unique carbohydrate motif that is uniquely associated with the *B. anthracis*, tetrasaccharide **165** was discovered. The anthrax tetrasaccharide **165** consists of three L-rhamnose sugars and a rare sugar, D-anthrose. Several carbohydrate approaches to the anthrax tetrasaccharide and one to a related trisaccharide have been reported [72]. The other routes drew all the products' stereochemistry from L-rhamnose and the rare D-fucose. In contrast, this *de novo* Achmatowicz approach to the tetrasaccharide **165** used only asymmetric catalysis to install the absolute stereochemistry. The approach merged the Pd-allylation chemistry to prepare C-4 amino mannose sugars, which was developed for the Swainsonine synthesis (Scheme 15.20) with the Pd-glycosylation chemistry for the synthesis of rhamnose-containing oligosaccharides (**174**) [73].

The fact that the anthrax tetrasaccharide was prepared by both *de novo* and traditional carbohydrate approaches provides an opportunity to compare the practicality of both approaches (Scheme 15.26). While the carbohydrate route required more total steps, they were virtually the same in terms of the longest linear synthesis, (carbohydrate route with 20 steps vs 19 steps for the *de novo* route). What is a little more surprising is that the *de novo* route emerges from significantly cheaper starting materials (acetylfuran, \$0.09/g) compared to the traditional carbohydrate approach (D-fucose, \$70/g, and D-rhamnose, \$15/g). Clearly, the price advantage for this *de novo* approach

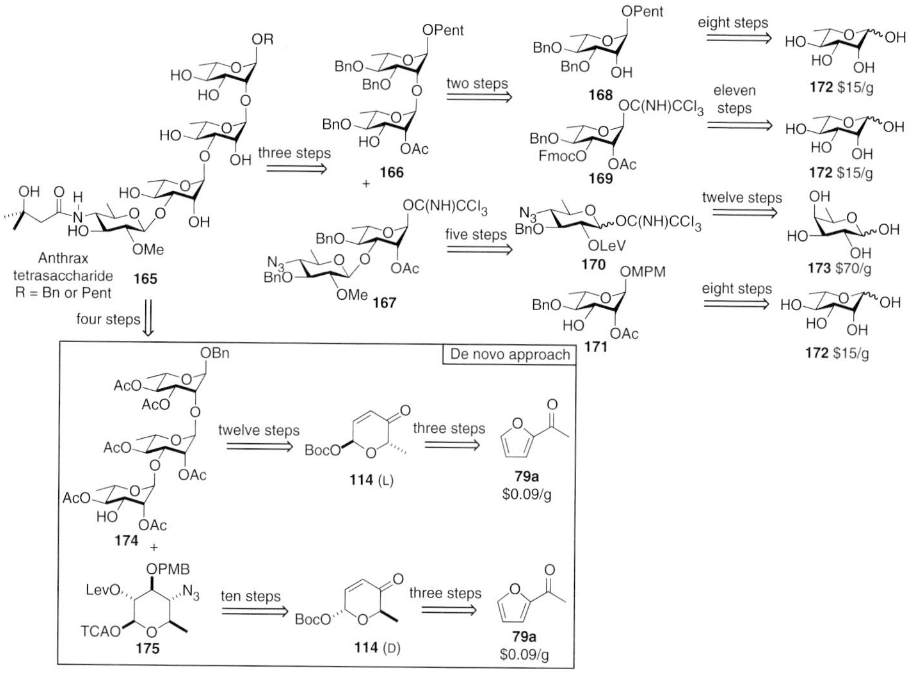

Scheme 15.26 Traditional carbohydrate versus *de novo* approach to anthrax tetrasaccharide.

was a result from the need in the traditional approach to prepare the rare anthrose sugar from the rare and expensive D-fucose.

Outlined in Scheme 15.27 is the *de novo* synthesis of the anthrose monosaccharide **184**, which involved the use of two Pd-π-allylation reactions. The route began with the synthesis of the para-methoxybenzyl (PMB)-protected pyranone **176** from α-pyranone **114**. Luche reduction and methyl carbonate formation were used to prepare **177** from **176**. The C-4 azide was introduced by using the Pd-catalyzed C-4 allylic azide chemistry (**177** to **178**). The allylic azide **178** was dihydroxylated to give *rhamno*-sugar **179**. Next, the 6-deoxy-*gluco*-stereochemistry was installed by a protection (**179** to **180**), C-2 inversion strategy (**180** to **182**, via **181**) to give anthrose sugar **182**. Finally a Lev-protection, PMB-deprotection strategy, and trichloroacetimidate formation (**183** to **184**) are used to convert **182** into the glycosyl-donor sugar **184**.

More recently, the *de novo* Achmatowicz approach was used for the synthesis of Merremoside D (Scheme 15.28) [74], which is a member of the resin glycoside family of natural products. These structurally complex tetrasaccharides possess a macrolactone (20 or 21-membered), which consists of a disaccharide bridged at the C-1 and C-2′ or C-3′ by a Jalapinolic acid. The amphiphilic nature of the resin glycosides has been suggested to be the source of its ionophoretic activity (i.e., membrane transporter) as observed in human erythrocyte membranes [75]. In addition to confirming its structure, the *de novo* synthesis Merremoside D also served as an opportunity to better understand its biological activities. Intrigued by the possibility that enantiomeric analogs of these target compounds would possess the ion transport properties, yet would lack the same target protein interactions, the O'Doherty group decided to develop a *de novo* asymmetric approach to the Merremosides.

The *de novo* Achmatowicz synthesis of Merremoside D (**185**) was accomplished from achiral starting materials. This was true for the 1,4-oligo-rhamnose tetrasaccharide portion of the natural product as well as the aglycon. The tetrasaccharide **186** was constructed by a convergent glycosylation between macrolactone disaccharide **190** and the imidate disaccharide **187**. This convergent route was desired to have both halves of the natural product for SAR study. Conveniently, the same Noyori catalyst that was used to install the sugar chemistry in the asymmetric reduction of acetylfuran **79a** was also used in the asymmetric reduction of ynone **192**.

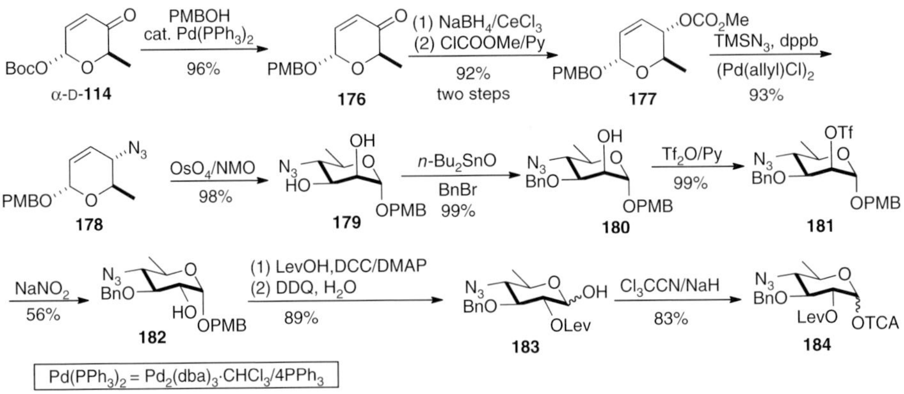

Scheme 15.27 Synthesis of anthrose glycosyl donor **184**.

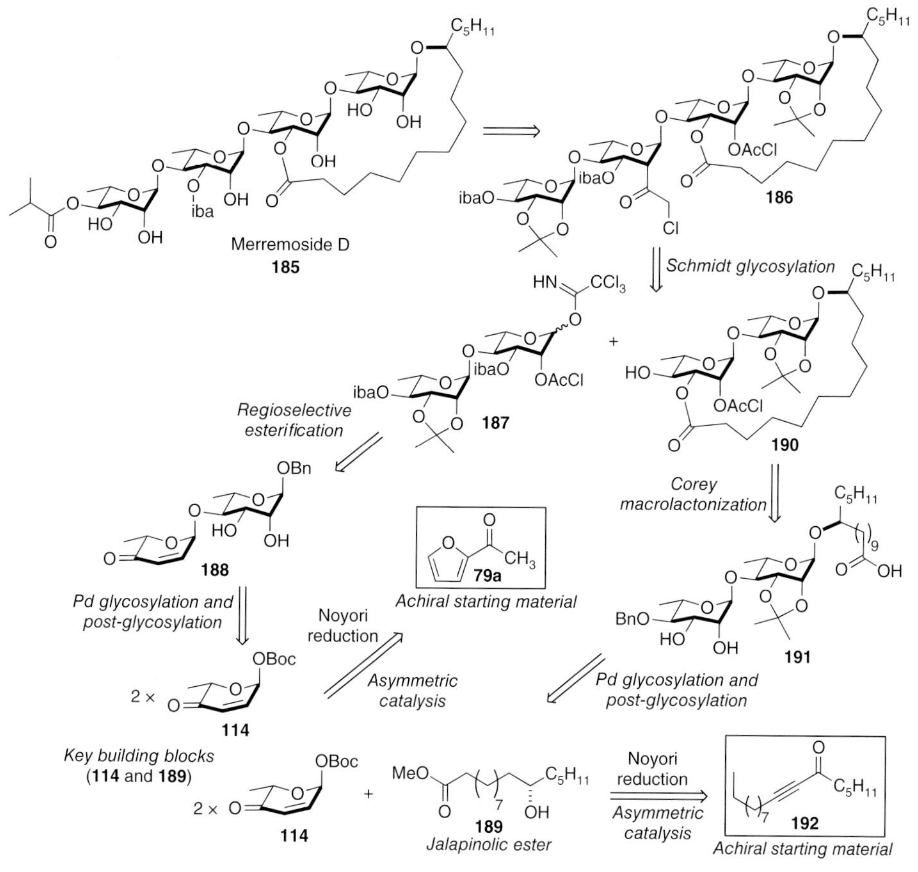

Scheme 15.28 *De novo* retrosynthesis for Merremoside D.

15.5 Conclusion

In conclusion, the once commonplace view that limited the role of the carbohydrate *de novo* synthesis to academic exercises must be discounted. This review demonstrated a wide range of structural motifs that are available by these approaches. In addition, the utility of these *de novo* asymmetric approaches has great potential for both synthesis and biological study of complex natural product structural motifs. In fact, in many cases, these approaches are more competitive than the routes via more traditional carbohydrate approaches. This superiority in utility is particularly evident when these approaches are applied in medicinal-chemistry-type SAR studies.

References

1 (a) Gijsen, H.J.M., Qiao, L., Fitz, W., and Wong, C.-H. (1996) *Chem. Rev.*, 96, 443–473; (b) Hudlicky, T., Entwistle, D.A., Pitzer, K.K., and Thorpe, A.J. (1996) *Chem. Rev.*, 96, 1195–1220; (c) Yu, X. and O'Doherty, G.A. (2008) *ACS Symp. Ser.*, 990, 3–28.

2 Aljahdali, A.Z., Shi, P., Zhong, Y., and O'Doherty, G.A. (2013) in De Novo Asymmetric Synthesis of the Pyranoses: From Monosaccharides to Oligosaccharides, Advances in Carbohydrate Chemistry and Biochemistry, Elsevier vol. 69 (ed. D. Horton), pp. 55–123.
3 (a) Ko, S.Y., Lee, A.W.M., Masamune, S., Reed, L.A. III, Sharpless, K.B., and Walker, F.J. (1983) *Science*, 220, 949–951; (b) Ko, S., Lee, A.W.M., and Masamune, S. (1990) *Tetrahedron*, 46, 245–264.
4 Henderson, I., Sharpless, K.B., and Wong, C.H. (1994) *J. Am. Chem. Soc.*, 116, 558–561.
5 (a) Johnson, C.R., Golebiowski, A., Steensma, D.H., and Scialdone, M.A. (1993) *J. Org. Chem.*, 58, 7185–7194; (b) Hudlicky, T., Pitzer, K.K., Stabile, M.R., Thorpe, A.J., and Whited, G.M. (1996) *J. Org. Chem.*, 61, 4151–4153.
6 Backvall, J.K., Bystrom, S.E., and Nordberg, R.E. (1984) *J. Org. Chem.*, 49, 4619–4631.
7 (a) Gibson, D.T., Koch, J.R., and Kallio, R.E. (1968) *Biochemistry*, 7, 2653–2662; (b) Gibson, D., Hemsley, M., Yoshioka, H., and Mabry, T. (1970) *Biochemistry*, 9, 1626–1630.
8 Banwell, M., Blakey, S., Harfoot, G., and Longmore, R. (1998) *J. Chem. Soc., Perkin Trans. 1*, 3141–3142.
9 Vogel, P. (2001) in Glycoscience, vol. 2 (eds B.O. Fraser-Reid, K. Tatsuta, and J. Thiem), Springer, Berlin, pp. 1023–1174.
10 Danishefsky, S.J. (1989) *Chemtracts*, 273–297.
11 Danishefsky, S.J. and DeNinno, M.P. (1987) *Angew. Chem. Int. Ed. Engl.*, 26, 15–23.
12 Danishefsky, S.J., Pearson, W.H., and Segmuller, B.E. (1985) *J. Am. Chem. Soc.*, 107, 1280–1285.
13 (a) Joly, G.D. and Jacobsen, E.N. (2002) *Org. Lett.*, 4, 1795–1798; (b) Schaus, S.E., Branalt, J., and Jacobsen, E.N. (1998) *J. Org. Chem.*, 63, 403–405.
14 Danishefsky, S.J., Phillips, G., and Ciufolini, M. (1987) *Carbohydr. Res.*, 171, 17–327.
15 (a) Traxler, P., Gruner, J., and Auden, J.A.L. (1977) *J. Antibiot.*, 30, 289–296.
16 Pappo, R., Allen, D.S. Jr., Lemieux, R.U., and Johnson, W.S. (1956) *J. Org. Chem.*, 21, 478–479.
17 Ito, Y., Hirao, T., and Saegusa, T. (1978) *J. Org. Chem.*, 43, 1011–1013.
18 Northrup, A.B. and MacMillan, D.W.C. (2004) *Science*, 305, 1752–1755.
19 Northhrup, A.B., Mangion, I.K., Hettche, F., and MacMillan, D.W.C. (2004) *Angew. Chem. Int. Ed.*, 43, 2152–2154.
20 Figueirido, R.M.F. and Christmann, M. (2007) *Eur. J. Org. Chem.*, 2007, 2575.
21 Mangion, I.K. and MacMillan, D.W.C. (2005) *J. Am. Chem. Soc.*, 127, 3696–3697.
22 Carpenter, J., Northrup, A.B., Chung, de.-M., Wiener, J.J.M., Kim, S.-G., and MacMillan, D.W.C. (2008) *Angew. Chem. Int. Ed.*, 47, 3568–3572.
23 Evans, D.A., Downey, C.W., Shaw, J.T., and Tedrow, J.S. (2002) *Org. Lett.*, 4, 1127–1130.
24 Barrett, A.G.M., Prokopiou, P.A., and Barton, D.H.R. (1979) *J. Chem. Soc., Chem. Commun.*, 1175.
25 Ibrahem, I., Zou, W., Xu, Y., and Cordova, A. (2006) *Adv. Synth. Catal.*, 348, 211–222.
26 Mainkar, P.S., Johny, K., Rao, T.P., and Chandrasekhar, S. (2012) *J. Org. Chem.*, 77, 2519–2525.
27 (a) Ahmed, M.M. and O'Doherty, G.A. (2005) *J. Org. Chem.*, 67, 10576–10578; (b) Ahmed, M.M. and O'Doherty, G.A. (2005) *Tetrahedron Lett.*, 46, 3015–3019; (c) Gao, D. and O'Doherty, G.A. (2005) *Org. Lett.*, 7, 1069–1072; (d) Zhang, Y. and O'Doherty, G.A. (2005) *Tetrahedron*, 61, 6337–6351; (e) Ahmed, M.M., Berry, B.P., Hunter, T.J., Tomcik, D.J., and O'Doherty, G.A. (2005) *Org. Lett.*, 7, 745–748.

28 VanRheenen, V., Kelly, R.C., and Cha, D.Y. (1976) *Tetrahedron Lett.*, 17, 1973–1976.
29 Ahmed, M.M. and O'Doherty, G.A. (2005) *Tetrahedron Lett.*, 46, 4151–4155.
30 Harris, J.M., Keranen, M.D., and O'Doherty, G.A. (1999) *J. Org. Chem.*, 64, 2982–2983.
31 Yu, X. and O'Doherty, G.A. (2008) in Chemical Glycobiology, ACS Symposium Series, vol. 990 (eds X. Chen, R. Halcomb, and P.G. Wang), ACS, Washington, DC, pp. 3–22.
32 (a) Cuccarese, M.F. and O'Doherty, G.A. (2012) in Asymmetric Synthesis II: More Methods and Applications (eds M. Christmann and S. Braese), Wiley-VCH Verlag GmbH & Co. KGaA, Weinheim, pp. 249–259; (b) Cuccarese, M.F., Li, J.J., and O'Doherty, G.A. (2014) in Modern Synthetic Methods in Carbohydrate Chemistry (eds S. Vidal and D.B. Werz), Wiley-VCH Verlag GmbH & Co. KGaA, Weinheim, pp. 1–28; (c) Babu, R.S., Chen, Q., Kang, S.-W., Zhou, M., and O'Doherty, G.A. (2012) *J. Am. Chem. Soc.*, 134, 11952–11955; (d) Babu, R.S., Zhou, M., and O'Doherty, G.A. (2004) *J. Am. Chem. Soc.*, 126, 3428–3429; (e) Zhou, M. and O'Doherty, G.A. (2008) *Curr. Top. Med. Chem.*, 8, 114–125.
33 Harris, J.M., Keranen, M.D., Nguyen, H., Young, V.G., and O'Doherty, G.A. (2000) *Carbohydr. Res.*, 328, 17–36.
34 (a) Bushey, M.L., Haukaas, M.H., and O'Doherty, G.A. (1999) *J. Org. Chem.*, 64, 2984–2985; (b) Haukaas, M.H., Li, M., Starosotnikov, A.M., and O'Doherty, G.A. (2008) *Heterocycles*, 76 (2), 1549–1559.
35 (a) Li, M. and O'Doherty, G.A. (2004) *Tetrahedron Lett.*, 45, 6407–6411; (b) Li, M., Scott, J.G., and O'Doherty, G.A. (2004) *Tetrahedron Lett.*, 45, 1005–1009.
36 (a) Mrozowski, R.M., Sandusky, Z.M., Vemula, R., Wu, B., Zhang, Q., Lannigan, D.A., and O'Doherty, G.A. (2014) *Org. Lett.*, 16, 5996–5999; (b) Mrozowski, R.M., Vemula, R., Wu, B., Zhang, Q., Schroederd, B.R., Hilinski, M.K., Clarke, D.E., Hecht, S.M., O'Doherty, G.A., and Lannigan, D.A. (2013) *ACS Med. Chem. Lett.*, 4, 175–179; (c) Yu, H., Wang, L., Wu, B., Zhang, Q., Kang, S.-W., Rojanasakul, Y., and O'Doherty, G.A. (2011) *ACS Med. Chem. Lett.*, 2, 259–263.
37 (a) Wang, H.-Y.L. and O'Doherty, G.A. (2011) *Chem. Commun.*, 47, 10251–10253; (b) Shan, M., Xing, Y., and O'Doherty, G.A. (2009) *J. Org. Chem.*, 74, 5961–5966.
38 Harris, J.M., Li, M., Scott, J.G., and O'Doherty, G.A. (2004) Achmatowicz approach to 5,6-dihydro-2*H*-pyran-2-one containing natural products, in Strategy and Tactics in Natural Product Synthesis (ed. M. Harmata), Elsevier, London.
39 (a) Haukaas, M.H. and O'Doherty, G.A. (2002) *Org. Lett.*, 4, 1771–1774; (b) Haukaas, M.H. and O'Doherty, G.A. (2001) *Org. Lett.*, 3, 3899–3992.
40 (a) Balachari, D. and O'Doherty, G.A. (2000) *Org. Lett.*, 2, 4033–4036; (b) Balachari, D. and O'Doherty, G.A. (2000) *Org. Lett.*, 2, 863–866; (c) Balachari, D., Quinn, L., and O'Doherty, G.A. (1999) *Tetrahedron Lett.*, 40, 4769–4773.
41 Haukaas, M.H. and O'Doherty, G.A. (2001) *Org. Lett.*, 3, 401–404.
42 Babu, R.S. and O'Doherty, G.A. (2003) *J. Am. Chem. Soc.*, 125, 12406–12407.
43 (a) Comely, A.C., Eelkema, R., Minnaard, A.J., and Feringa, B.L. (2003) *J. Am. Chem. Soc.*, 125, 8714–8715; (b) Kim, H., Men, H., and Lee, C. (2004) *J. Am. Chem. Soc.*, 126, 1336–1337.
44 Kim, H. and Lee, C. (2002) *Org. Lett.*, 4, 4369–4372.
45 Evans, P.A. and Kennedy, L.J. (2000) *Org. Lett.*, 2, 2213–2215.
46 Guo, H. and O'Doherty, G.A. (2005) *Org. Lett.*, 7, 3921–3924.
47 Baiga, T.J., Guo, H., Xing, Y., O'Doherty, G.A., Parrish, A., Dillin, A., Austin, M.B., Noel, J.P., and La Clair, J.J. (2008) *ACS Chem. Biol.*, 3, 294–304.

48 (a) Guo, H. and O'Doherty, G.A. (2006) *Org. Lett.*, 8, 1609–1612; (b) Coral, J.A., Guo, H., Shan, M., and O'Doherty, G.A. (2009) *Heterocycles*, 79, 521–529; (c) Abrams, J.N., Babu, R.S., Guo, H., Le, D., Le, J., Osbourn, J.M., and O'Doherty, G.A. (2008) *J. Org. Chem.*, 73, 1935–1940.
49 Liao, Y.F., Lal, A., and Moremen, K.W. (1996) *J. Biol. Chem.*, 271, 28348–28358.
50 (a) Elbein, A.D., Solf, R., Dorling, P.R., and Vosbeck, K. (1981) *Proc. Natl. Acad. Sci. U.S.A.*, 78, 7393–7397; (b) Kaushal, G.P., Szumilo, T., Pastuszak, I., and Elbein, A.D. (1990) *Biochemistry*, 29, 2168–2176; (c) Pastuszak, I., Kaushal, G.P., Wall, K.A., Pan, Y.T., Sturm, A., and Elbein, A.D. (1990) *Glycobiology*, 1, 71–82.
51 (a) Goss, P.E., Baker, M.A., Carver, J.P., and Dennis, J.W. (1995) *Clin. Cancer Res.*, 1, 935–944; (b) Das, P.C., Robert, J.D., White, S.L., and Olden, K. (1995) *Oncol. Res.*, 7, 425–433.
52 Davis, B., Bell, A.A., Nash, R.J., Watson, A.A., Griffiths, R.C., Jones, M.G., Smith, C., and Fleet, G.W.J. (1996) *Tetrahedron Lett.*, 37, 8565–8568.
53 (a) Mezher, H.A., Hough, L., and Richardson, A.C. (1984) *J. Chem. Soc., Chem. Commun.*, 447–448; (b) Fleet, G.W.J., Gough, M.J., and Smith, P.W. (1984) *Tetrahedron Lett.*, 25, 1853–1856.
54 Shan, M. and O'Doherty, G.A. (2006) *Org. Lett.*, 8, 5149–5152.
55 Li, M., Li, Y., Mrozowski, R.M., Sandusky, Z.M., Shan, M., Song, X., Wu, B., Zhang, Q., Lannigan, D.A., and O'Doherty, G.A. (2015) *ACS Med. Chem. Lett.*, 16, 95–99.
56 Borisova, S.A., Guppi, S.R., Kim, H.J., Wu, B., Liu, H.-w., and O'Doherty, G.A. (2010) *Org. Lett.*, 12, 5150–5153.
57 (a) Sharif, E.U. and O'Doherty, G.A. (2012) *Eur. J. Org. Chem.*, 2012 (11), 2095–2108; (b) Shan, M., Sharif, E.U., and O'Doherty, G.A. (2010) *Angew. Chem. Int. Ed.*, 49, 9492–9495.
58 (a) Tibrewal, N., Downey, T.E., Van Lanen, S.G., Sharif, E.U., O'Doherty, G.A., and Jurgen Rohr, J. (2012) *J. Am. Chem. Soc.*, 134, 12402–12405; (b) Sharif, E.U. and O'Doherty, G.A. (2014) *Heterocycles*, 88, 1275–1285.
59 Guppi, S.R., Zhou, M., and O'Doherty, G.A. (2006) *Org. Lett.*, 8, 293–296.
60 Babu, R.S. and O'Doherty, G.A. (2005) *J. Carbohydr. Chem.*, 24, 169–177.
61 (a) Babu, R.S., Guppi, S.R., and O'Doherty, G.A. (2006) *Org. Lett.*, 8, 1605–1608; (b) Guppi, S. and O'Doherty, G.A. (2007) *J. Org. Chem.*, 72, 4966–4969.
62 (a) Zhou, M. and O'Doherty, G.A. (2006) *Org. Lett.*, 8, 4339–4342; (b) Zhou, M. and O'Doherty, G.A. (2007) *J. Org. Chem.*, 72, 2485–2493; (c) Iyer, A., Zhou, M., Azad, N., Elbaz, H., Wang, L., Rogalsky, D.K., Rojanasakul, Y., O'Doherty, G.A., and Langenhan, J.M. (2010) *ACS Med. Chem. Lett.*, 1, 326–330.
63 Yu, X. and O'Doherty, G.A. (2008) *Org. Lett.*, 10, 4529–4532.
64 Zhou, M. and O'Doherty, G.A. (2008) *Org. Lett.*, 10, 2283–2286.
65 Wu, B., Li, M., and O'Doherty, G.A. (2010) *Org. Lett.*, 12, 5466–5469.
66 (a) Bajaj, S.O., Shi, P., Beuning, P.J., and O'Doherty, G.A. (2014) *MedChemComm*, 5, 1138–1142; (b) Bajaj, S.O., Sharif, E.U., Akhmedov, N.G., and O'Doherty, G.A. (2014) *Chem. Sci.*, 5, 2230–2234.
67 Tané, P., Ayafor, J.P., Sondengam, B.L., Lavaud, C., Massiot, G., Connolly, J.D., Rycroft, D.S., and Woods, N. (1988) *Tetrahedron Lett.*, 29, 1837–1840.
68 Seidel, V., Baileul, F., and Waterman, P.G. (1999) *Phytochemistry*, 52, 465–472.
69 Hu, J.-F., Garo, E., Hough, G.W., Goering, M.G., O'Neil-Johnson, M., and Eldridge, G.R. (2006) *J. Nat. Prod.*, 69, 585–590.

70 Shi, P., Silva, M., Wu, B., Wang, H.Y.L., Akhmedov, N.G., Li, M., Beuning, P., and O'Doherty, G.A. (2012) *ACS Med. Chem. Lett.*, 3, 1086–1090.

71 (a) Mock, M. and Fouet, A. (2001) *Annu. Rev. Microbiol.*, 55, 647–671; (b) Sylvestre, P., Couture-Tosi, E., and Mock, M. (2002) *Mol. Microbiol.*, 45, 169–178.

72 (a) Werz, D.B. and Seeberger, P.H. (2005) *Angew. Chem. Int. Ed.*, 44, 6315–6318; (b) Adamo, R., Saksena, R., and Kovac, P. (2005) *Carbohydr. Res.*, 340, 2579–2582; (c) Saksena, R., Adamo, R., and Kovac, P. (2006) *Bioorg. Med. Chem. Lett.*, 16, 615–617; (d) Adamo, R., Saksena, R., and Kovac, P. (2006) *Helv. Chim. Acta*, 89, 1075–1089; (e) Mehta, A.S., Saile, E., Zhong, W., Buskas, T., Carlson, R., Kannenberg, E., Reed, Y., Quinn, C.P., and Boons, G.J. (2006) *Chem. Eur. J.*, 12, 9136–9149; (f) Crich, D. and Vinogradova, O. (2007) *J. Org. Chem.*, 72, 6513–6520; (g) Werz, D.B., Adibekian, A., and Seeberger, P.H. (2007) *Eur. J. Org. Chem.*, 2007, 1976–1982.

73 (a) Guo, H. and O'Doherty, G.A. (2007) *Angew. Chem. Int. Ed.*, 46, 5206–5208; (b) Guo, H. and O'Doherty, G.A. (2008) *J. Org. Chem.*, 73, 5211–5220; (c) Wang, H.-Y.L., Guo, H., and O'Doherty, G.A. (2013) *Tetrahedron*, 69, 3432–3436.

74 Sharif, E.U., Wang, H.-Y.L., Akhmedov, N.G., and O'Doherty, G.A. (2014) *Org. Lett.*, 16, 492–495.

75 (a) Kitagawa, I., Baek, N.I., Kawashima, K., Yokokawa, Y., Yoshikawa, M., Ohashi, K., and Shibuya, H. (1996) *Chem. Pharm. Bull.*, 44, 1680; (b) Kitagawa, I., Baek, N.I., Yokokawa, Y., Yoshikawa, M., Ohashi, K., and Shibuya, H. (1996) *Chem. Pharm. Bull.*, 44, 1693; (c) Kitagawa, I., Ohashi, K., Kawanishi, H., Shibuya, H., Shinkai, H., and Akedo, H. (1989) *Chem. Pharm. Bull.*, 37, 1679.

16

Chemical Synthesis of Sialosides

Yu-Hsuan Lih and Chung-Yi Wu

16.1 Introduction

Sialic acids, also called *neuraminic acids*, are normally found in nature at the terminal positions of *N*-glycans, *O*-glycans, glycosphingolipids, and glycophosphoinositol anchors attached to protein or lipid moieties [1]. Sialic acids are a family of α-keto acids with a nine-carbon backbone. In nature, sialic acids usually do not exist in free sugar forms. They are commonly α-2,3- and α-2,6-linked to terminal galactose, *N*-acetylgalactosamine, or α-2,8- and α-2,9-linked to another sialic acid residue [2–4]. Sialic acids are usually the outermost residues and direct participants in many molecular recognitions and interactions. Moreover, the sialic acid-containing carbohydrates, or sialosides, play significant biological roles in cellular recognition, cell communication, pathogen infections, tumor metastasis, and disease states, which have led to extensive studies in sialobiology [5–7]. Because of their exposed position and diversity, sialosides represent a significant biological target in either masking recognition sites or facilitating cell recognition and adhesion [3]. Additionally, the type of sialyl linkage between the sialic acid and its adjacent carbohydrate moiety can decisively influence the function of sialosides. For example, avian influenza viruses favor the α 2,3-sialic acid receptor, whereas human-adapted influenza A viruses use the α 2,6-sialic acid receptor [8, 9]. Obviously, the biological significance of sialosides is enormous; however, because of their structural complexity and diversity, sialosides are extremely difficult to gather from natural sources in amounts abundant enough for biological studies. Recent advances in chemical synthesis and chemoenzymatic approaches successfully produced structurally defined sialosides in homogeneous forms [10, 11]; even so, unlike common monosaccharides such as galactose and glucose, the three additional carbons make sialic acids more complicated for structural modification than any other monosaccharides [11]. Because of the various sialyl linkages and different adjacent carbohydrate moieties, obtaining homogeneous sialosides remains a tremendous task.

Selective Glycosylations: Synthetic Methods and Catalysts, First Edition. Edited by Clay S. Bennett.

16.2 Chemical Synthesis of Sialosides

Preparation of structurally diverse sialosides can be achieved through chemical synthesis, enzymatic synthesis, or modification of the natural sialosides [11–13]. Advances in both chemical and enzymatic syntheses have provided reliable procedures to produce complex sialosides. Based on different modifications at C5 position, sialic acids are classified into three basic forms: Neu5Ac, Neu5Gc, and KDN. Besides the C5 position, single or multiple substitutions can occur on the hydroxyl groups at C4, C7, C8, and C9 positions (substituted by acetyl, sulfate, methyl, lactyl, phosphate, etc.). These modifications generate a diverse family of more than 50 structurally distinct forms of sialosides in nature [2], which are of key importance to cellular recognition, pathogen infections, and disease states [3]. The following pages concisely summarize up-to-date progress on the chemical synthesis of sialosides. The discussion also includes breakthrough developments in the study of conformational and stereoselective effects of sialosides synthesis.

Chemical sialylation has been considered to be one of the most challenging glycosylation reactions. The challenges come from the presence of the C1 electron-withdrawing carboxyl group at the tertiary anomeric center, which reduces the reactivity of sialic acid as a donor, and the lack of a participating group at C3 to direct the stereochemical outcome of glycosylation. As a result, low yield, undesired 2,3-elimination, and the formation of unnatural β glycosidic bond occur. Furthermore, the separation of α- and β-isomers is tedious and sometimes very challenging. Moreover, each sialic acid carries at least six hydroxyl groups, which must be protected and deprotected during synthesis. Glycosylation in each step generates a new stereocenter at the anomeric carbon, and there are no universal methods for the introduction of a desired glycosidic linkage in a stereocontrolled manner. The deciding factors in chemical synthesis of sialosides include protection schemes, functional substitutions, promoter choice, and acceptor architecture. In particular, significant efforts have been directed toward the development of sialyl donors for efficient α-sialylation [11, 14–20], including the use of anomeric leaving groups, such as halides [21–25], phosphites [26–29], phosphates [26, 30], sulfides [31–35], xanthates [36–39], and phenyltrifluoroacetimidates [40–42], the introduction of auxiliary groups at C1 [43–47] and C3 [23, 48–56], the modification of the *N*-acetyl functional group at C5 [57, 58], or the optimized combinations of the leaving group with positional modification [59–81]. Among them, glycosylation using various sialyl donors with different leaving groups and C5-modifications is the most powerful method to increase yield and improve α-selectivity, as summarized in Table 16.1. Acceptors listed in Table 16.1 are frequently used for direct comparison. Two glycosylation results, one with primary acceptors and the other with secondary acceptors, were included for every sialyl donor. Generally, the yields of sialylation by using donors with different leaving groups (entries 1–10) do not exceed 70%, and a higher α-selectivity was achieved when a less hindered glycosyl acceptor was used (entries 2, 5, 6, and 8). Notably, most of these reactions need to be conducted in acetonitrile. The second type of sialyl donor is with C5-modification, which often greatly enhances not only the α-selectivity but also the yield toward primary or secondary acceptors (entries 11–30). After comparing the glycosylation results of N-modified donors by coupling with primary acceptor **II** (entries 11–14 and 22–25),

Table 16.1 Review of sialyl donors with different leaving groups and C5-modifications.

Entry	Donor	Acceptor	Yield (%) α	Yield (%) β	Promoter	Solvent[a]	References
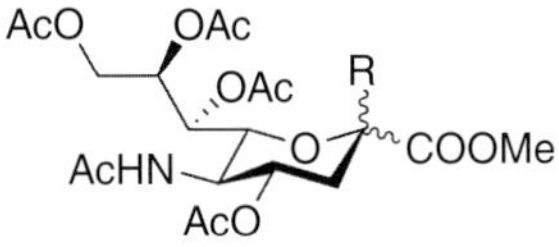							
1	R = Cl	**I** (1°)	67	0	Ag_2CO_3	e	[25]
		XIV (2°)	12	19	$Hg(CN)_2$/ $HgBr_2$	f	[24]
2	R = $OP(OBn)_2$	**II** (1°)	67	13	TMSOTf	a	[26]
		XXVI (2°)	67	11	TMSOTf	a	[26]
3	R = $OP(OEt)_2$	**II** (1°)	56	14	TMSOTf	a	[27]
		XXVII (2°)	53	0	TMSOTf	c	[29]
4	R = 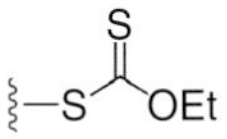	**III** (1°)	48	16	DMTST	a	[37]
		XX (2°)	75	0	AgOTf/ DTBP/ PhSCl	c	[39]
5	R = SMe	**IV** (1°)	70	0	DMTST	a	[33]
		XIX (2°)	52	0	DMTST	a	[33]
6	R = SPh	**II** (1°)	47	8	NIS/TfOH	a	[34]
		XIX (2°)	70	0	NIS/TfOH	a	[35]
7	R = 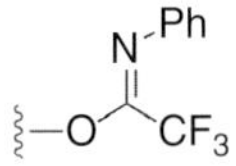	**II** (1°)	69	10	TMSOTf	c	[40]
		XV (2°)	61	20	TMSOTf	c	[40]
8	R = SBox	**I** (1°)	60	30	MeOTf	a	[31]
		XIX (2°)	71	4	MeOTf	a	[31]
9	R = $OPO(OBn)_2$	**II** (1°)	26	9	TMSOTf	a	[26]
10	R = $OPO(OEt)_2$	**XXV** (2°)	21	0	TMSOTf	a	[30]
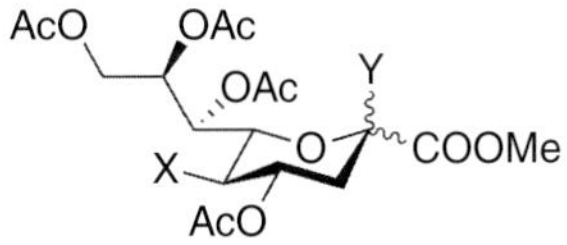							
11	X = NAc_2, Y = SPh	**II** (1°)	40	25	NIS/TfOH	a	[34]
	X = NAc_2, Y = SMe	**XIX** (2°)	72	0	NIS/TfOH	a	[78]
12	X = NBn_2, Y = STol	**II** (1°)	7	84	Ph_2SO/ Tf_2O/ TTBP	c	[74]

(Continued)

Table 16.1 (Continued)

Entry	Donor	Acceptor	Yield (%) α	Yield (%) β	Promoter	Solvent[a]	References
13	X = NHTroc, Y = SPh	**II** (1°)	81	10	NIS/TfOH	a	[34]
		XIX (2°)	35	8	NIS/TfOH	a	[79]
14	X = NHTFA, Y = SPh	**II** (1°)	85	7	NIS/TfOH	a	[34]
	X = NHTFA, Y = SMe	**XXI** (2°)	84	0	NIS/TfOH	a	[80]
15	X = NHTroc, Y = 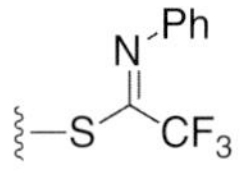	**VIII** (1°)	87	0	TMSOTf	c	[72]
16	X = NHTFA, Y = SBox	**VI** (1°)	19	77	AgOTf	b	[62]
		I (1°)	65	13	NIS/TfOH	a	[62]
17	X = N_3, Y = STol	**V** (1°)	65	0	NIS/TfOH	a	[70]
18	X = NPhth, Y = 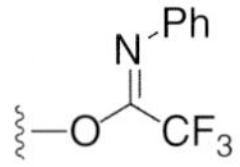	**VII** (1°)	92	0	TMSOTf	d	[68]
		XVI (2°)	75	2	TMSOTf	d	[68]
19	X = NHTCA, Y = 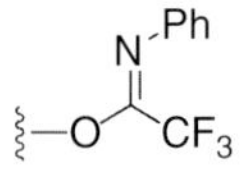	**V** (1°)	77	0	TMSOTf	c	[66]
		XXII (2°)	68	0	TMSOTf	c	[66]
20	X = NHTCA, Y = $OP(OBn)_2$	**VIII** (1°)	89	0	TMSOTf	c	[73]
		XXII (2°)	79	0	TMSOTf	c	[73]
21	X = N=C=S, Y = 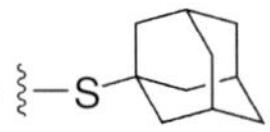	**XI** (1°)	80	0	NIS/TfOH	c	[71]
		XXI (2°)	87	0	NIS/TfOH	c	[71]
		XIII (1°)	58	0	NIS/TfOH	c	[71]
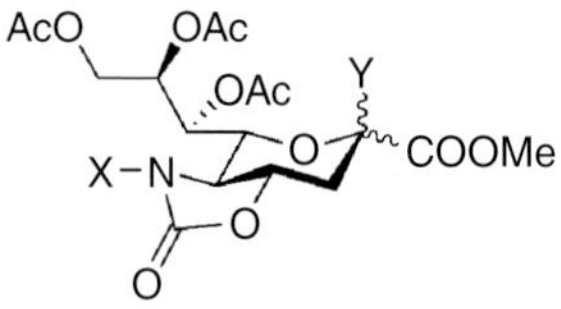							
22	X = H, Y = SPh	**II** (1°)	100	0	NIS/TfOH	b	[81]
		XIX (2°)	50	0	NIS/TfOH	a	[59]
23	X = Ac, Y = SPh	**II** (1°)	92	0	NIS/TfOH	b	[60]
		XXI (2°)	10	75	NIS/TfOH	b	[60]

Table 16.1 (Continued)

Entry	Donor	Acceptor	Yield (%) α	Yield (%) β	Promoter	Solvent[a]	References
24	X = Ac, Y = STol	**II** (1°)	81	12	NIS/TfOH	b	[69]
		XXI (2°)	44	37	NIS/TfOH	c	[64]
		XVII (2°)	20	66	Ph_2SO/ Tf_2O	b	[69]
25	X = Ac	**II** (1°)	91	0	NIS/TfOH	b	[59]
	Y = 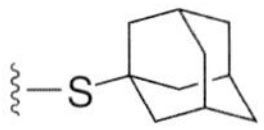	**XXI** (2°)	64	21	NIS/TfOH	c	[59]
26	X = Ac, Y = 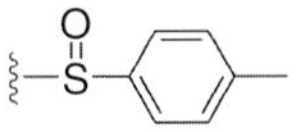	**IX** (1°)	80	0	Tf_2O/ Tol_2SO	b	[77]
		XVIII (2°)	70	0	Tf_2O/ Tol_2SO	b	[77]
27	X = H, Y = SBox	**VI** (1°)	63	3	$Bi(OTf)_3$	g	[62]
28	X = H, Y = $OPO(OBu)_2$	**X** (1°)	93	0	TMSOTf	c	[76]
		XXIV (2°)	53	16	TMSOTf	c	[76]
29	X = Ac, Y = $OPO(OBu)_2$	**V** (1°)	85	0	TMSOTf	b	[63]
		XXIII (2°)	83	0	TMSOTf	b	[63]
30	X = Boc, Y = $OPO(OBu)_2$	**XXIII** (2°)	59	3	TMSOTf	b	[75]

a) Solvents: a = MeCN; b = CH_2Cl_2; c = MeCN + CH_2Cl_2; d = EtCN; e = $CHCl_3$; f = $Cl(CH_2)_2Cl$; and g = CH_2Cl_2 + THF.

Acceptors

1° OH

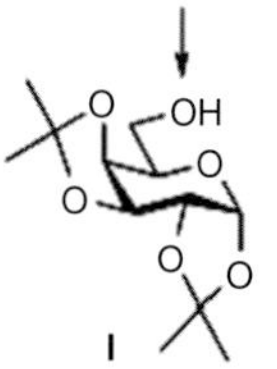

I

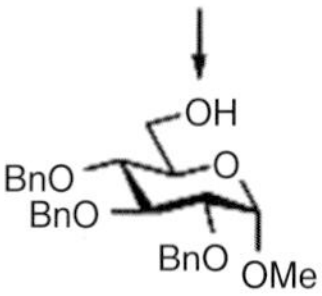

II

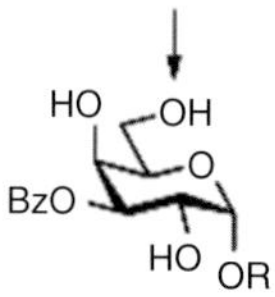

III, R = Me
IV, R = SE

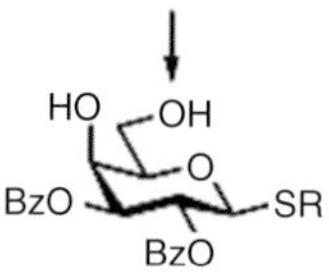

V, R = Tol
VI, R = Et

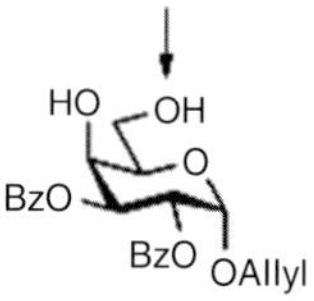

VII

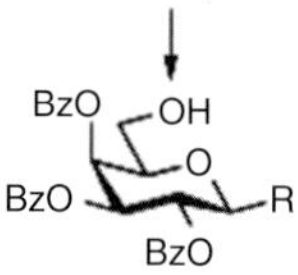

VIII, R = STol
IX, R = OMe

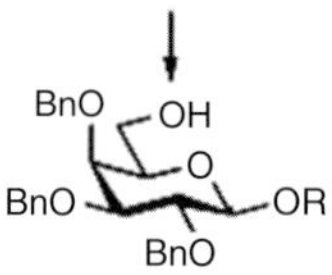

X, R = MP
XI, R = Me

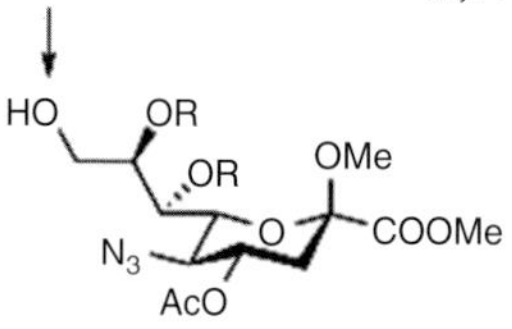

XII, R = *p*-ClBz

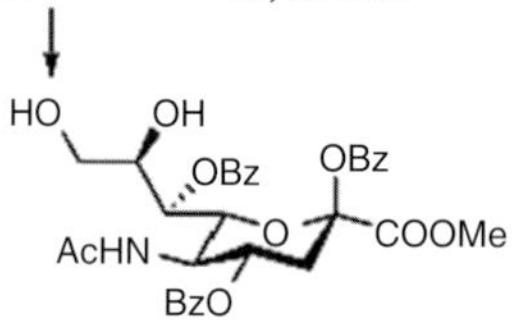

XIII

2° OH

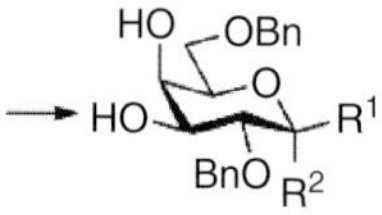

XIV, R^1 = OBn, R^2 = H
XV, R^1 = H, R^2 = OBn

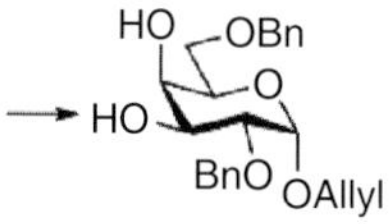

XVI

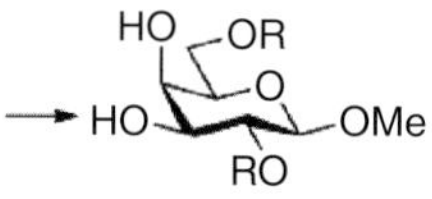

XVII, R = Bn
XVIII, R = Bz

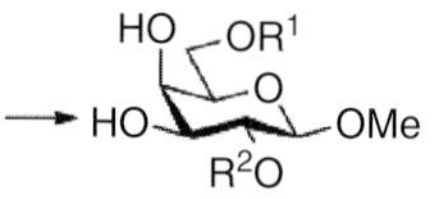

XIX, R^1 = Bz, R^2 = H
XX, R^1 = R^2 = Bn

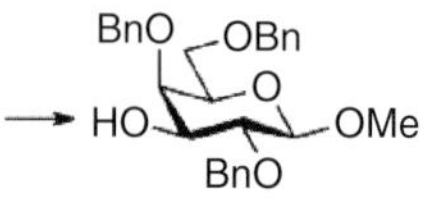

XXI

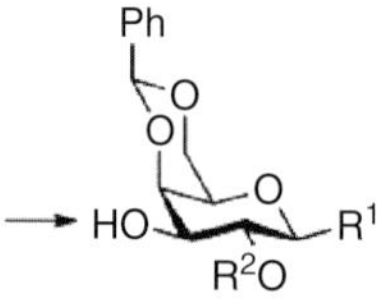

XXII, R^1 = STol, R^2 = H
XXIII, R^1 = STol, R^2 = Bz
XXIV, R^1 = OSE, R^2 = Bz

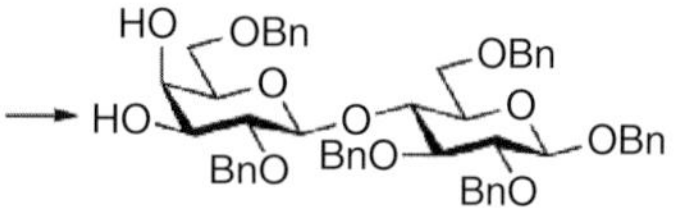

XXV

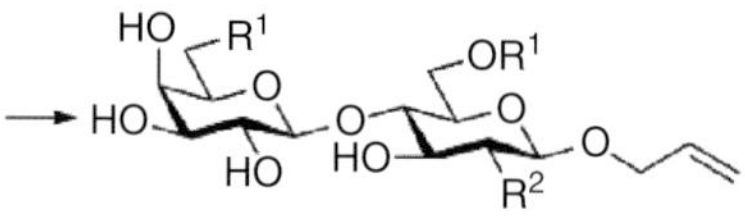

XXVI, R^1 = OTBDPS, R^2 = NHAc
XXVII, R^1 = R^2 = OPiv

5-*N*,4-*O*-oxazolidinone-protected donors show the best results (entries 22–25). On the other hand, for less hindered secondary acceptor **XIX** (entries 11, 13, and 22), the *N*,*N*-diacetyl sialyl donor gives the best outcome. In addition, for more hindered secondary acceptor **XXI** (entries 14, 21, 23, 24, and 25), *N*-TFA and isothiocyanate sialyl donors seem to be good choices, and conversion of the leaving group to the adamantanylthio group shows further improvement. Next, more improvement was achieved by using a combination of C5-modification and efficient leaving group (entries 18 and 19).

Among the sialylation donors listed in Table 16.1, the introduction of 5-*N*,4-*O*-oxazolidinone has drawn much attention, which has been applied to the synthesis of different sialosides by many groups during the past several years. The major advantages of such kind of modification include reducing the production of 2,3-elimination side product due to its high ring strain and increasing the α-selective sialylation. The 5-*N*,4-*O*-oxazolidinone thiophenyl sialoside donor was first reported by the Takahashi group [81]. The corresponding *S*-aryl thiosialoside [59–61, 64, 69, 81–86] and *O*-phosphoryl sialoside [63] donors were developed and successfully applied to the α-stereocontrol of sialylation. Crich and coworkers have shown that the *N*-acetyl-5-*N*,4-*O*-carbonyl thiosialoside can provide high α-selectivity and excellent yield in sialylation reactions; more importantly, the oxazolidinone group can be readily cleaved under mild conditions leaving the acetamide intact [60]. Moreover, the obtained peracetylated *N*-acetylneuraminic acid (NeuAc) sialosides can be treated with nitrosyl tetrafluoroborate (NOBF4) to form the *N*-nitroso sialosides **4**, and a library of NeuAc and/or KDN analogs modified at the 5-position can be quickly synthesized by the deamination followed by introduction of nucleophile. Mechanistically, the process involves a selective removal of the acetyl group from the *N*-nitrosoacetamide with sodium trifluoroethoxide to give a diazo derivative of NeuAc that is then substituted by the incoming nucleophile (Scheme 16.1) [87].

The *N*-acetyl-5-*N*,4-*O*-carbonyl thiosialoside was widely treated with *N*-iodosuccinimide/trifluoromethanesulfonic acid (NIS/TfOH) as the promoter for the glycosylations to afford high yields and α-stereoselectivities. Xing *et al.* thoroughly investigated the sialylation reaction of *N*-acetyl-5-*N*,4-*O*-oxazolidinone-protected *p*-tolyl 2-thio-sialoside donor using Ph_2SO/Tf_2O/2,4,6-tri-*tert*-butylpyridine (TTBPy) as promoter with various acceptors. They found that the stereoselectivity of sialylation was dependent on the various reaction conditions, such as preactivation time, reaction time, the amount of Ph_2SO, and TTBPy. For instance, Ph_2SO (2.0–3.0 equiv.)/TTBPy (0–1.0 equiv.) promotion achieved higher α-selective sialylation in dichloromethane, while Ph_2SO (4–5 equiv.)/TTBPy (0 equiv.) or Ph_2SO (2.0 equiv.)/TTBpy (2.0 equiv.) afforded lower stereoselectivity [69]. In addition to the standard sialylation conditions that involve activating the donor in the presence of the acceptor, Sun *et al.* reported that the *N*-acetyl-5-*N*,4-*O*-oxazolidinone-protected *p*-tolyl thiosialoside donor could be preactivated using AgOTf and *p*-toluenesulfenyl chloride, followed by coupling with other *p*-tolyl thioglycosides to prepare the thiosialosides without leaving group manipulation. This method will be quite useful for the one-pot sialylation reaction to prepare sialosides [88].

The *N*-acetyl-5-*N*,4-*O*-oxazolidinone-protected donor indeed shows excellent α-selectivity as well as high yield toward both primary and secondary acceptors, and the phosphate leaving group appears to be an excellent choice (Table 16.1, entries 28–30). Another advantage of the phosphate-based methodology includes the use of tolylthio glycosides as acceptors with defined relative reactivity values (RRVs) for sialylation to

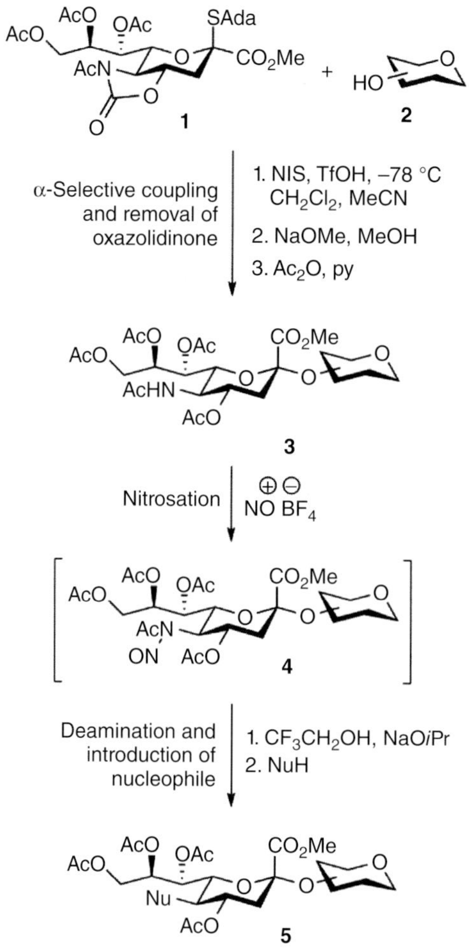

Scheme 16.1 Chemical synthesis of sialoside libraries with late-stage modification.

give sialyl disaccharides as building blocks for the subsequent reactivity-based programmable one-pot synthesis. In this way, the limitations of the relatively low reactivity of the thiosialoside donor and the difficulties in controlling the stereoselectivity of sialylation could be resolved. Moreover, the RRVs of sialylated disaccharides can be programmed by manipulating the protecting groups of the second sugar residue at the reducing end and can be applied to the synthesis of α-2,3-linked sialylated pentasaccharide **9** (Scheme 16.2) [63].

More importantly, the phosphate donor can also be applied to the convergent block synthetic strategy to synthesize the various lengths of α-2,9-linked oligosialic acids up to dodecamer in good yield and α-selectivity (Scheme 16.3) [89]. Using the same method, the synthesized α-2,9-di-, tri-, tetra-, and pentasialic acids have been conjugated to carrier protein for vaccine development. Access to these structures permitted studies in mice, which suggest that larger glycans are not necessarily better immunogens [90].

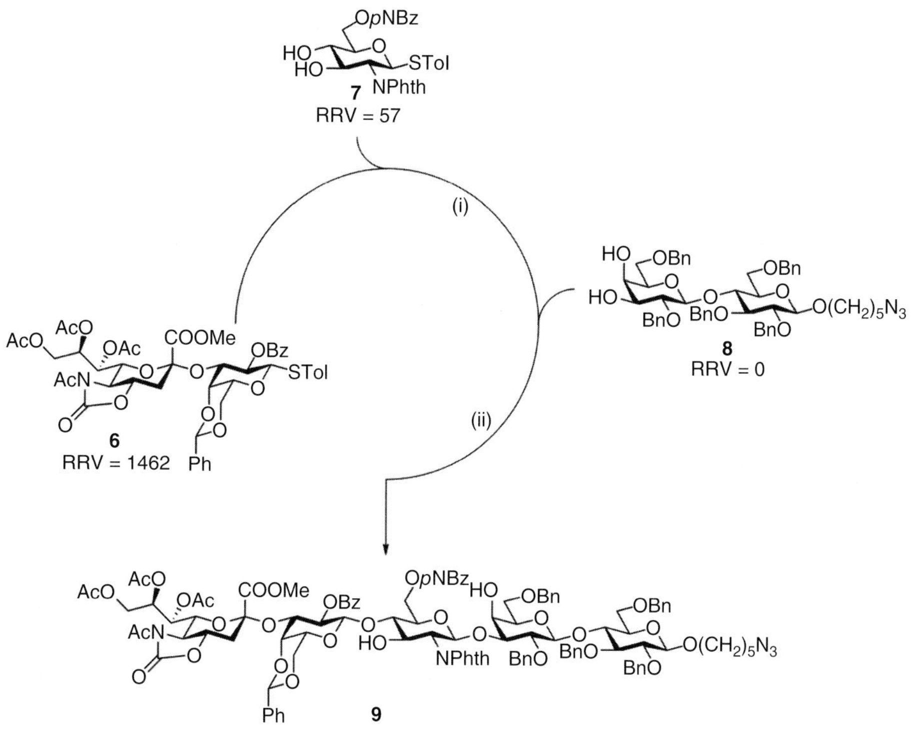

Scheme 16.2 Programmable one-pot synthesis of sialylated pentasaccharide **9**. (i) NIS, TfOH, 4 Å molecular sieves, CH_2Cl_2, −78 °C. (ii) NIS, TfOH, −20 °C to r.t.

Recently, Wong and Wu *et al.* developed a modular synthetic approach to construct both symmetric and asymmetric *N*-glycans. Using the sialyl phosphate donor, a set of modular building blocks with sialic acid, such as the examples shown in Figure 16.1, were prepared by total chemical synthesis. These modular sialylated building blocks can efficiently lead to diverse bi-, tri-, and tetra-antennary complex-type *N*-glycans, including asymmetric ones, which usually require multiple steps. With the specific linkage-fixed modules in hand, various *N*-glycans with mono-, di-, and tri-sialosides at distinct positions can be obtained. Moreover, further application to glycan array library construction will be more facile based on this approach [91].

Besides the synthesis of *O*-sialosides, Crich's group reported that the phosphate donor **21** can be activated under standard Lewis acidic conditions and treated with electron-rich olefins to afford C-sialosides with excellent α-selectivity. They found that when the more nucleophilic allyltributylstannane or trimethylsilyl enol ethers were used as nucleophiles, only the α-anomer formed (Scheme 16.4) [92]. Moreover, activating donor **21** to treat with primary, secondary, and tertiary thiols including galactosyl 3-, 4-, and 6-thiols, the α-*S*-sialosides were successfully synthesized at −78 °C in dichloromethane with good yields. The reactions proceeded under typical Lewis-acid-promoted glycosylation conditions and did not suffer from competing elimination of the phosphate. Deacetylation under standard conditions then afforded unprotected S-linked glycans (Scheme 16.5) [93].

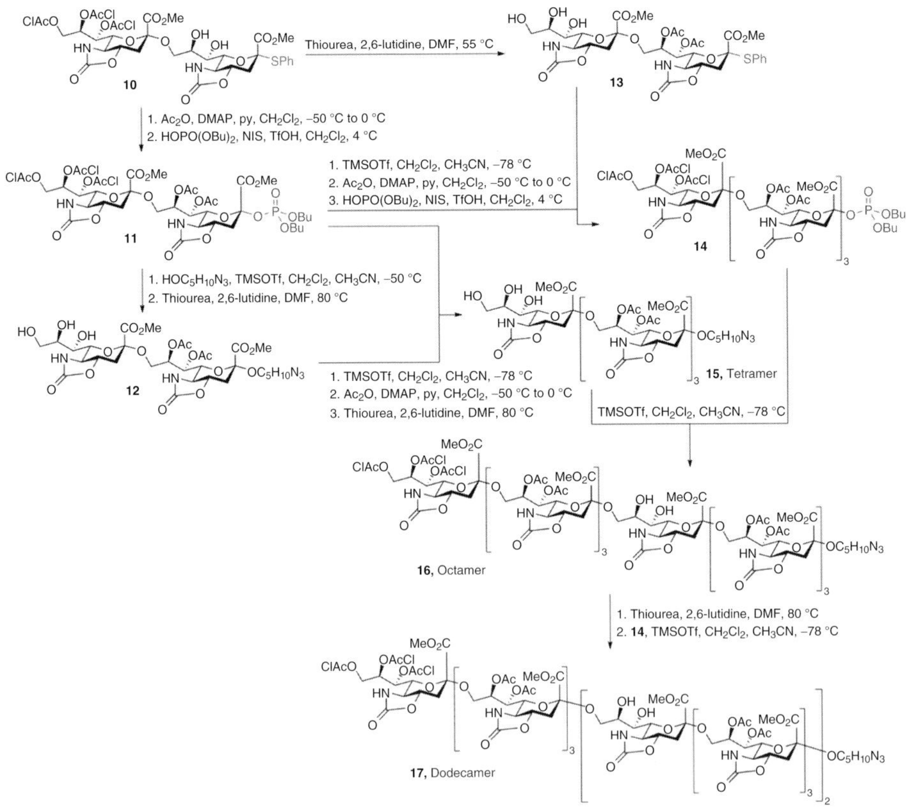

Scheme 16.3 Preparation of α 2,9-oligosialic acids up to dodecamer.

In addition to O-linked oligosialosides, S-linked α-2,8- and 2,9-oligosialic acids have been synthesized by Lin *et al.* The approach utilized *tert*-butyl disulfide protection to form anomeric thiol donor **28** with stable α-configuration *in situ.* This species reacted with iodosialyl acceptors in an S_N2 manner (Scheme 16.6) [94, 95]. Since polysialic acids are susceptible to degradation by sialidases *in vivo,* these S-linked oligosialosides, hydrolysis-resistant mimics, can be used for model studies or therapeutic interventions. However, due to the differences in length between the carbon–oxygen and carbon–sulfur bonds, O-sialosides and S-sialosides may possess different attributes.

The high α-selectivity in sialylation using 5-*N*,4-*O*-carbonyl sialoside as donor is believed as a result of its powerful electron-withdrawing nature, which is directed by their dipole moment. Recently, Crich *et al.* reported that the oxazolidinone groups are powerful electron-withdrawing protecting groups when *trans*-fused to pyranose ring systems. By virtue of their strong dipole moments in the plane of the ring system, these groups lead to increased α-selectivity in glycosylations. The estimation has been established by measuring the minimum cone voltage required for fragmentation in the mass spectrometer (Figure 16.2). Based on Crich's studies for the different oxazolidinone-based sialyl donors, the 5-*N*,4-*O*-oxazolidinone-protected phosphate donor possesses the strongest dipole moment effect and will react with the best α-selectivity [96].

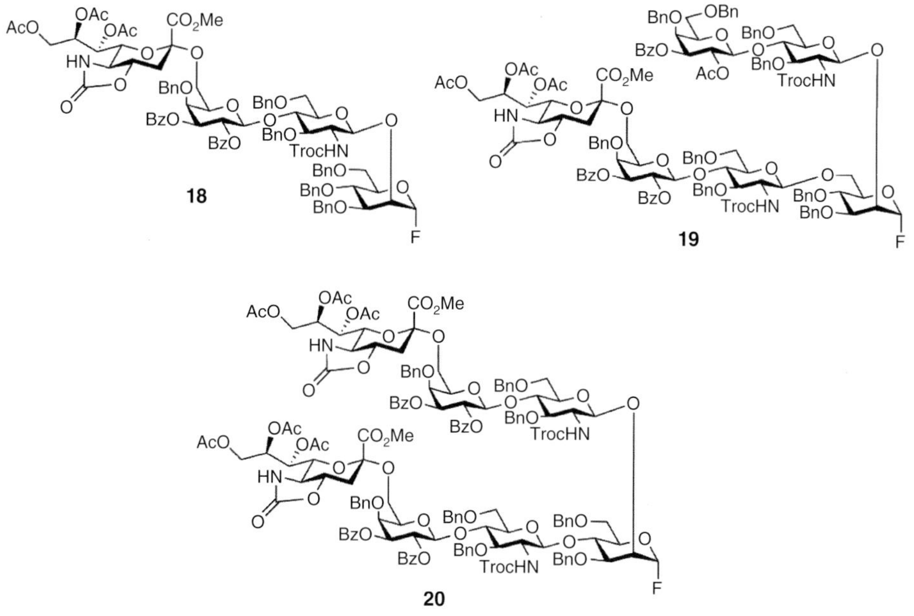

Figure 16.1 Structure of sialic-acid-containing modules.

Similarly to the dipole moment effect of the oxazolidinone on sialyl donor, the isothiocyanate group is considered to serve as a strong electron-withdrawing group, which can increase the α-selectivity of sialylation reaction. Recently, Crich *et al.* developed the C5-isothiocyanate-protected peracetyl adamantanyl thiosialoside **38** as a new sialyl donor (Figure 16.3). Unlike other sialyl donors, which only provide α-selectivity for either primary or secondary alcohols, this new isothiocyanate donor can undergo

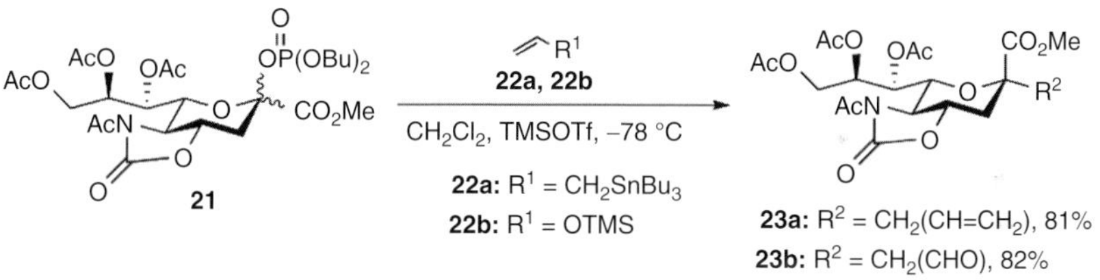

Scheme 16.4 Electrophilic C-sialylation.

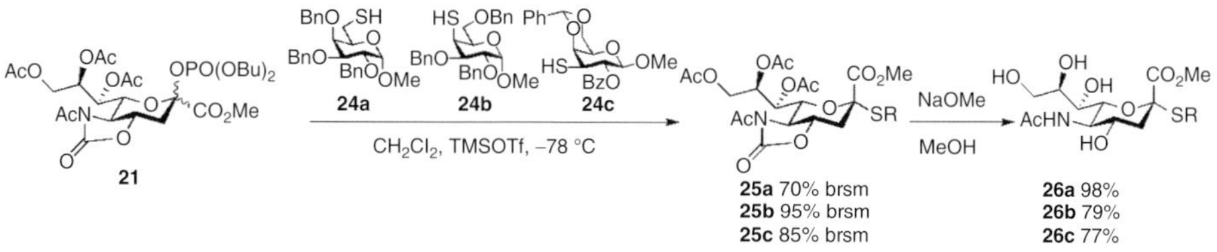

Scheme 16.5 Electrophilic S-sialylation.

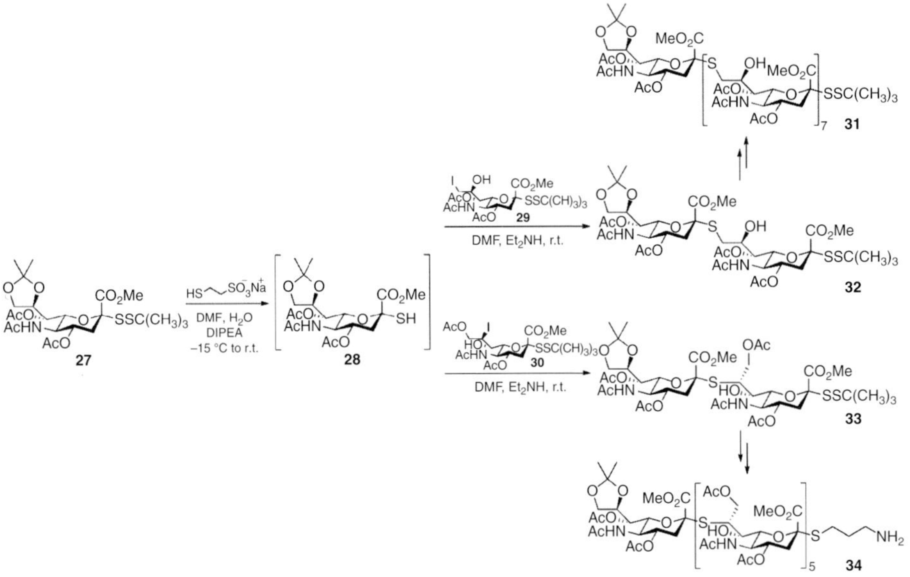

Scheme 16.6 Preparation of α S-linked oligosialosides.

α-specific sialosides with both primary and secondary alcohol acceptors [71]. It is considered the best sialyl donor for the synthesis of α-sialosides reported to date.

As mentioned previously, the electron-withdrawing nature reduces the reactivity of sialyl donors. In order to tackle the problem, Wong's group converted the carboxyl group of sialic acid into the hydroxymethyl group and prepared the acetyl derivative **39**. The yield of the reaction with this sialyl donor dramatically increased; however, the sialylation gave predominantly β-selectivity when promoted by dimethyl(methylthio) sulfonium triflate (DMTST) in acetonitrile, ether, toluene, or dichloromethane, probably due to significant anomeric effect (Scheme 16.7) [97].

Based on the high selectivity achieved with the use of 5-*N*,4-*O*-oxazolidinone-protected sialyl donors, benzylidene acetal mimic 5-*N*,7-*O*-oxazinanone-protected donors were also investigated. The cyclic protecting group is highly disarming and imparts conformational restriction through the fused ring. Much to our surprise, the glycosylation of the oxazinanone donor **42** with secondary alcohols resulted in β-sialosides (Scheme 16.8) [98].

Another approach to a more efficient donor relies on using electron-withdrawing protecting groups. Ye *et al.* prepared the OBz-protected sialyl donor **45** for testing.

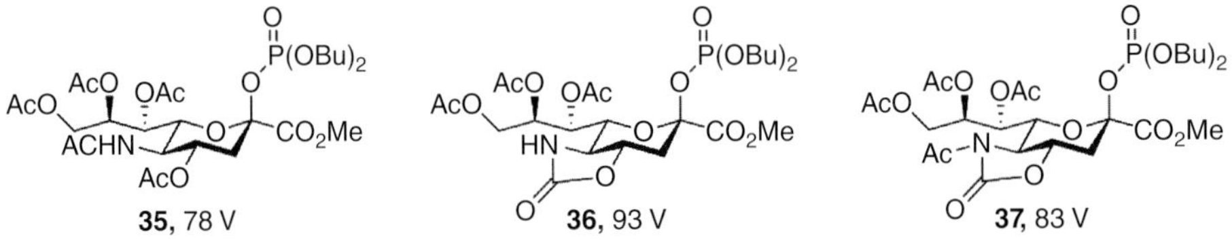

Figure 16.2 Onset cone voltages for fragmentation of sialyl phosphates.

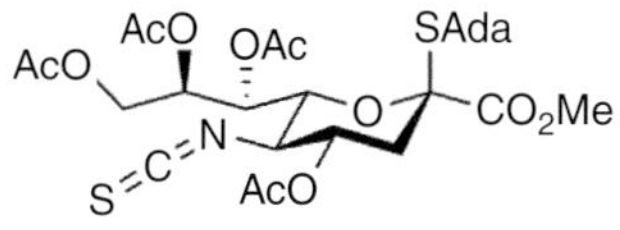

38

Figure 16.3 Structure of isothiocyanate-protected peracetyl adamantanyl thiosialoside **38** (Ada = 1-adamantanyl).

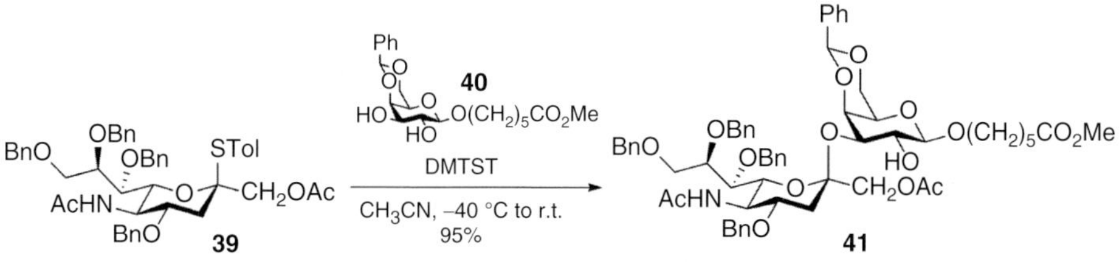

Scheme 16.7 Glycosylation of carboxyl group replaced sialyl donor **39**.

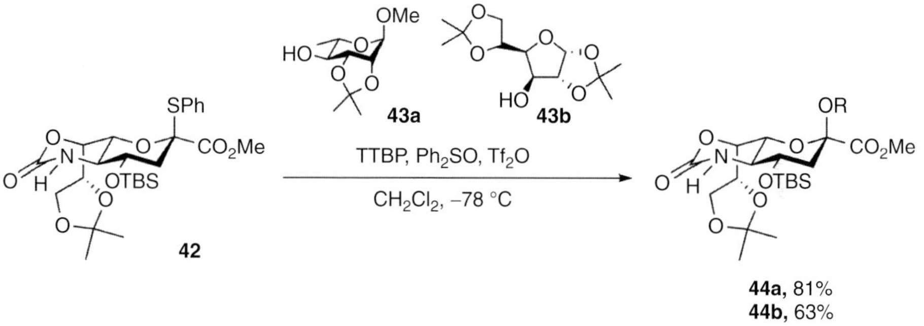

Scheme 16.8 Glycosylation of 5-*N*,7-*O*-oxazinanone-protected sialyl donor **42**.

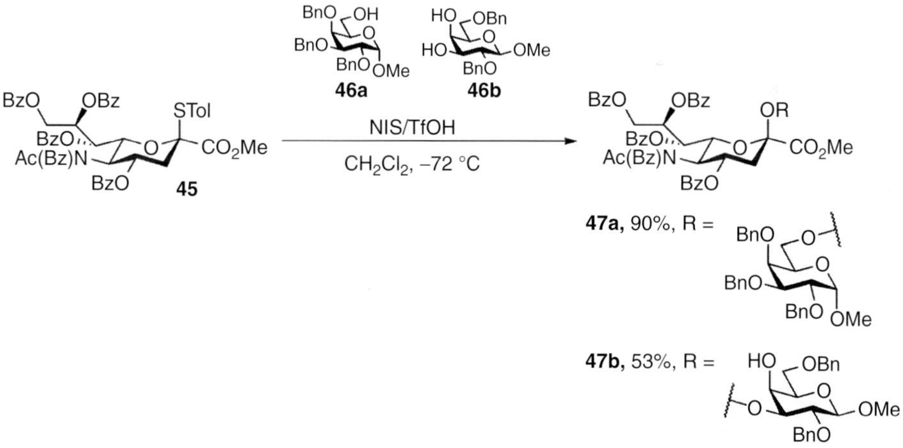

Scheme 16.9 Glycosylation of full benzoyl-protected sialyl donor **45**.

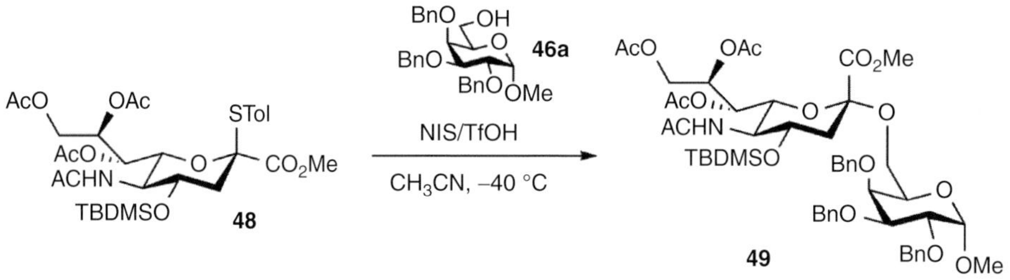

Scheme 16.10 Glycosylation of C4-bulky substituted sialyl donor.

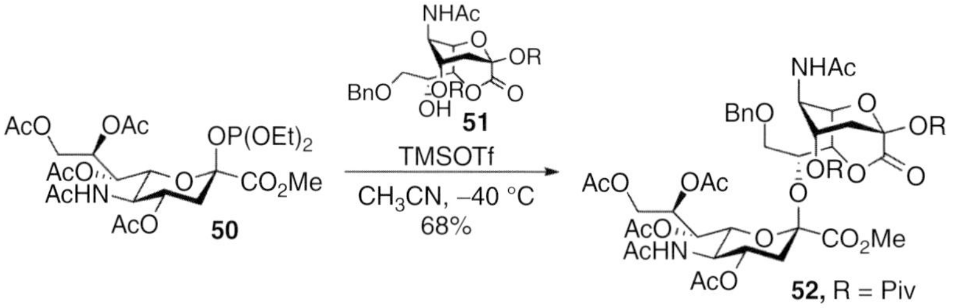

Scheme 16.11 β-Sialylation of Neu5Ac 1,7-lactone acceptor **51**.

When NIS/TfOH-promoted glycosylation of sialyl donor **45** was conducted in dichloromethane, unnatural β-sialosides were obtained, with both primary and secondary hydroxyl acceptors (Scheme 16.9) [99]. In additional studies, the effect of O-modification has been investigated, focusing on the O-4 and O-7 positions. It was demonstrated that a bulky group at O-7 alone cannot lead to preferential β-sialylation, while a bulky substituent at C4 can exert a strong influence on the stereoselectivity of sialylations and successfully lead to α-sialosides in acetonitrile (Scheme 16.10) [100]. In addition to the donor structure, β-sialylation can be achieved through the use of a specialized acceptor, such as Neu5Ac 1,7-lactone (Scheme 16.11) [101].

16.3 Conclusions

Modern methods in the chemical synthesis of sialosides have clearly made a major contribution toward improved production of many complex sialosides, especially with advancements in the formation of α-selective glycoside bond. Synthetic access to structurally homogeneous sialosides with specific lengths would provide a high-throughput method for analyzing the structure–activity relationships of this class of biomolecules. Although it is quite apparent that sialosides play diverse and crucial roles in a wide variety of biological systems, the advanced progress made thus far in this exciting field of sialoside chemistry and biology is only the beginning. With the help of powerful synthetic methods to access the bio-related sialic-acid-containing carbohydrates, we will be able to gradually unravel their roles in biological systems and human diseases.

References

1 Traving, C. and Schauer, R. (1998) *Cell. Mol. Life Sci.*, 54 (12), 1330–1349.
2 Angata, T. and Varki, A. (2002) *Chem. Rev.*, 102 (2), 439–469.
3 Varki, A. (1993) *Glycobiology*, 3 (2), 97–130.
4 Schauer, R. ed (1982) Sialic Acids: Chemistry, Metabolism and Function, Cell Biology Monographs, vol. 10, Springer-Verlag Wien, New York.
5 Schauer, R. (2009) *Curr. Opin. Struct. Biol.*, 19 (5), 507–514.
6 Varki, A. (2008) *Trends Mol. Med.*, 14 (8), 351–360.
7 Varki, A. (2007) *Nature*, 446 (7139), 1023–1029.
8 Connor, R.J., Kawaoka, Y., Webster, R.G., and Paulson, J.C. (1994) *Virology*, 205 (1), 17–23.
9 Rogers, G.N. and D'Souza, B.L. (1989) *Virology*, 173 (1), 317–322.
10 Yu, H., Huang, S., Chokhawala, H., Sun, M., Zheng, H., and Chen, X. (2006) *Angew. Chem. Int. Ed.*, 45 (24), 3938–3944.
11 Boons, G.J. and Demchenko, A.V. (2000) *Chem. Rev.*, 100 (12), 4539–4565.
12 Chen, X. and Varki, A. (2010) *ACS Chem. Biol.*, 5 (2), 163–176.
13 Kiefel, M.J. and von Itzstein, M. (2002) *Chem. Rev.*, 102 (2), 471–490.
14 Ando, H. and Imanura, A. (2004) *Trends Glycosci. Glycotechnol.*, 16, 293–303.
15 Boons, G.J. and Demchenko, A.V. (2003) The Chemistry of Sialic Acid, in Carbohydrate-Based Drug Discovery (ed Wong, C.-H.), Wiley-VCH Verlag GmbH & Co. KGaA, Weinheim.
16 Halcomb, R.L. and Chappell, M.D. (2001) In: Glycochemistry: Principles, Synthesis, and Applications (ed Wang P.G. and Bertozzi C.R.), Marcel Dekker, New York, pp. 177–220.
17 Hasegawa, A. and Kiso, M. (1997) Chemical Synthesis of Sialyl Glycosides, in Preparative Carbohydrate Chemistry (ed Hanessian, S.), Marcel Dekker, New York.
18 Kiso, M., Ishida, H., and Ito, H. (2000) Special Problems in Glycosylation Reactions: Sialidations, in Carbohydrates in Chemistry and Biology (ed Ernst, B., Hart, G. W. and Sinay, P.), Wiley-VCH Verlag GmbH, Weinheim.
19 Ress, D.K. and Linhardt, R.J. (2004) *Curr. Org. Synth.*, 1 (1), 31–46.
20 Roy, R. (1997) Recent developments in the rational design of multivalent glycoconjugates, in Glycoscience Synthesis of Substrate Analogs and Mimetics (ed Driguez, H. and Thiem, J.), Springer Berlin Heidelberg.
21 Deninno, M.P. (1991) *Synthesis*, (8), 583–593.
22 Ogawa, T. and Sugimoto, M. (1985) *Carbohydr. Res.*, 135 (2), C5–C9.
23 Okamoto, K. and Goto, T. (1990) *Tetrahedron*, 46 (17), 5835–5857.
24 Numata, M., Sugimoto, M., Koike, K., and Ogawa, T. (1987) *Carbohydr. Res.*, 163 (2), 209–225.
25 Paulsen, H. and Tietz, H. (1984) *Carbohydr. Res.*, 125 (1), 47–64.
26 Kondo, H., Ichikawa, Y., and Wong, C.-H. (1992) *J. Am. Chem. Soc.*, 114 (22), 8748–8750.
27 Martin, T.J. and Schmidt, R.R. (1992) *Tetrahedron Lett.*, 33 (41), 6123–6126.
28 Sim, M.M., Kondo, H., and Wong, C.-H. (1993) *J. Am. Chem. Soc.*, 115 (6), 2260–2267.
29 Zheng, M. and Ye, X.-S. (2012) *Tetrahedron*, 68 (5), 1475–1482.
30 Martin, T.J., Brescello, R., Toepfer, A., and Schmidt, R.R. (1993) *Glycoconjugate J.*, 10 (1), 16–25.

31 De Meo, C. and Parker, O. (2005) *Tetrahedron: Asymmetry*, 16 (2), 303–307.
32 Zhang, Z., Ollmann, I.R., Ye, X.-S., Wischnat, R., Baasov, T., and Wong, C.-H. (1999) *J. Am. Chem. Soc.*, 121 (4), 734–753.
33 Hasegawa, A., Ohki, H., Nagahama, T., Ishida, H., and Kiso, M. (1991) *Carbohydr. Res.*, 212, 277–281.
34 Tanaka, H., Adachi, M., and Takahashi, T. (2005) *Chem. Eur. J.*, 11 (3), 849–862.
35 Hasegawa, A., Nagahama, T., Ohki, H., Hotta, K., Ishida, H., and Kiso, M. (1991) *J. Carbohydr. Chem.*, 10 (3), 493–498.
36 Dziadek, S., Brocke, C., and Kunz, H. (2004) *Chem. Eur. J.*, 10 (17), 4150–4162.
37 Marra, A. and Sinay, P. (1990) *Carbohydr. Res.*, 195 (2), 303–308.
38 Martichonok, V. and Whitesides, G.M. (1996) *J. Org. Chem.*, 61 (5), 1702–1706.
39 Qu, H., Liu, J.-M., Wdzieczak-Bakala, J., Lu, D., He, X., Sun, W., Sollogoub, M., and Zhang, Y. (2014) *Eur. J. Med. Chem.*, 75, 247–257.
40 Cai, S.T. and Yu, B. (2003) *Org. Lett.*, 5 (21), 3827–3830.
41 Liu, Y., Ruan, X., Li, X., and Li, Y. (2008) *J. Org. Chem.*, 73 (11), 4287–4290.
42 Tanaka, S.-i., Goi, T., Tanaka, K., and Fukase, K. (2007) *J. Carbohydr. Chem.*, 26 (7–9), 369–394.
43 Danishefsky, S.J., Deninno, M.P., and Chen, S. (1988) *J. Am. Chem. Soc.*, 110 (12), 3929–3940.
44 Haberman, J.M. and Gin, D.Y. (2001) *Org. Lett.*, 3 (11), 1665–1668.
45 Haberman, J.M. and Gin, D.Y. (2003) *Org. Lett.*, 5 (14), 2539–2541.
46 Hanashima, S., Akai, S., and Sato, K. (2008) *Tetrahedron Lett.*, 49 (34), 5111–5114.
47 Takahashi, T., Tsukamoto, H., and Yamada, H. (1997) *Tetrahedron Lett.*, 38 (47), 8223–8226.
48 Castro-Palomino, J.C., Tsvetkov, Y.E., and Schmidt, R.R. (1998) *J. Am. Chem. Soc.*, 120 (22), 5434–5440.
49 Ercegovic, T. and Magnusson, G. (1995) *J. Org. Chem.*, 60 (11), 3378–3384.
50 Ercegovic, T. and Magnusson, G. (1996) *J. Org. Chem.*, 61 (1), 179–184.
51 Hossain, N. and Magnusson, G. (1999) *Tetrahedron Lett.*, 40 (11), 2217–2220.
52 Ito, T., Couceiro, J.N.S.S., Kelm, S., Baum, L.G., Krauss, S., Castrucci, M.R., Donatelli, I., Kida, H., Paulson, J.C., Webster, R.G., and Kawaoka, Y. (1998) *J. Virol.*, 72 (9), 7367–7373.
53 Ito, Y. and Ogawa, T. (1990) *Tetrahedron*, 46 (1), 89–102.
54 Martichonok, V. and Whitesides, G.M. (1996) *J. Am. Chem. Soc.*, 118 (35), 8187–8191.
55 Okamoto, K., Kondo, T., and Goto, T. (1987) *Tetrahedron*, 43 (24), 5909–5918.
56 Ito, Y. and Ogawa, T. (1988) *Tetrahedron Lett.*, 29 (32), 3987–3990.
57 De Meo, C. and Priyadarshani, U. (2008) *Carbohydr. Res.*, 343 (10–11), 1540–1552.
58 Uchinashi, Y., Nagasaki, M., Zhou, J., Tanaka, K., and Fukase, K. (2011) *Org. Biomol. Chem.*, 9 (20), 7243–7248.
59 Crich, D. and Li, W. (2007) *J. Org. Chem.*, 72 (20), 7794–7797.
60 Crich, D. and Li, W. (2007) *J. Org. Chem.*, 72 (7), 2387–2391.
61 Farris, M.D. and De Meo, C. (2007) *Tetrahedron Lett.*, 48 (7), 1225–1227.
62 Harris, B.N., Patel, P.P., Gobble, C.P., Stark, M.J., and De Meo, C. (2011) *Eur. J. Org. Chem.*, 2011 (20–21), 4023–4027.
63 Hsu, C.-H., Chu, K.-C., Lin, Y.-S., Han, J.-L., Peng, Y.-S., Ren, C.-T., Wu, C.-Y., and Wong, C.-H. (2010) *Chem. Eur. J.*, 16 (6), 1754–1760.

64 Liang, F.-f., Chen, L., and Xing, G.-w. (2009) *Synlett*, (3), 425–428.
65 Lin, C.-C., Huang, K.-T., and Lin, C.-C. (2005) *Org. Lett.*, 7 (19), 4169–4172.
66 Sun, B., Srinivasan, B., and Huang, X.F. (2008) *Chem. Eur. J.*, 14 (23), 7072–7081.
67 Tanaka, H., Tateno, Y., Nishiura, Y., and Takahashi, T. (2008) *Org. Lett.*, 10 (24), 5597–5600.
68 Tanaka, K., Goi, T., and Fukase, K. (2005) *Synlett*, (19), 2958–2962.
69 Wang, Y.-J., Jia, J., Gu, Z.-Y., Liang, F.-F., Li, R.-C., Huang, M.-H., Xu, C.-S., Zhang, J.-X., Men, Y., and Xing, G.-W. (2011) *Carbohydr. Res.*, 346 (11), 1271–1276.
70 Yu, C.-S., Niikura, K., Lin, C.-C., and Wong, C.-H. (2001) *Angew. Chem. Int. Ed.*, 40 (15), 2900–2903.
71 Mandhapati, A.R., Rajender, S., Shaw, J., and Crich, D. (2015) *Angew. Chem. Int. Ed.*, 54 (4), 1275–1278.
72 Rao, J. and Zhu, X. (2015) *Tetrahedron Lett.*, 56 (37), 5168–5171.
73 Sun, B. and Jiang, H. (2012) *Tetrahedron Lett.*, 53 (42), 5711–5715.
74 Wang, Y., Xu, F.-F., and Ye, X.-S. (2012) *Tetrahedron Lett.*, 53 (28), 3658–3662.
75 Boltje, T.J., Heise, T., Rutjes, F.P.J.T., and van Delft, F.L. (2013) *Eur. J. Org. Chem.*, 2013 (24), 5257–5261.
76 Gong, J., Liu, H., Nicholls, J.M., and Li, X. (2012) *Carbohydr. Res.*, 361, 91–99.
77 Gu, Z.-y., Zhang, J.-x., and Xing, G.-w. (2012) *Chem. Asian J.*, 7 (7), 1524–1528.
78 Demchenko, A.V. and Boons, G.J. (1998) *Tetrahedron Lett.*, 39 (19), 3065–3068.
79 Ando, H., Koike, Y., Ishida, H., and Kiso, M. (2003) *Tetrahedron Lett.*, 44 (36), 6883–6886.
80 De Meo, C., Demchenko, A.V., and Boons, G.-J. (2002) *Aust. J. Chem.*, 55 (1–2), 131–134.
81 Tanaka, H., Nishiura, Y., and Takahashi, T. (2006) *J. Am. Chem. Soc.*, 128 (22), 7124–7125.
82 Tanaka, H., Nishiura, Y., and Takahashi, T. (2009) *J. Org. Chem.*, 74 (11), 4383–4386.
83 De Meo, C., Farris, M., Ginder, N., Gulley, B., Priyadarshani, U., and Woods, M. (2008) *Eur. J. Org. Chem.*, 2008 (21), 3673–3677.
84 Crich, D. and Wu, B. (2008) *Org. Lett.*, 10 (18), 4033–4035.
85 Xing, G.-w., Chen, L., and Liang, F.-f. (2009) *Eur. J. Org. Chem.*, 2009 (34), 5963–5970.
86 Chen, L., Liang, F., Xu, M., Xing, G., and Deng, Z. (2009) *Acta Chim. Sinica*, 67 (12), 1355–1362.
87 Navuluri, C. and Crich, D. (2013) *Angew. Chem. Int. Ed.*, 52 (43), 11339–11342.
88 Sun, B. and Jiang, H. (2011) *Tetrahedron Lett.*, 52 (45), 6035–6038.
89 Chu, K.-C., Ren, C.-T., Lu, C.-P., Hsu, C.-H., Sun, T.-H., Han, J.-L., Pal, B., Chao, T.-A., Lin, Y.-F., Wu, S.-H., Wong, C.-H., and Wu, C.-Y. (2011) *Angew. Chem. Int. Ed.*, 50 (40), 9391–9395.
90 Liao, G., Zhou, Z., and Guo, Z. (2015) *Chem. Commun.*, (47), 9647–9650.
91 Shivatare, S.S., Chang, S.-H., Tsai, T.-I., Tseng, S.Y., Shivatare, V.S., Lin, Y.-S., Cheng, Y.-Y., Ren, C.-T., Lee, C.-C.D., Pawar, S., Tsai, C.-S., Shih, H.-W., Zeng, Y.-F., Liang, C.-H., Kwong, P.D., Burton, D.R., Wu, C.-Y., and Wong, C.-H. (2016) *Nat. Chem.*, 8 (4), 338–346.
92 Noel, A., Delpech, B., and Crich, D. (2012) *Org. Lett.*, 14 (5), 1342–1345.
93 Noel, A., Delpech, B., and Crich, D. (2012) *Org. Lett.*, 14 (16), 4138–4141.
94 Liang, C.-F., Yan, M.-C., Chang, T.-C., and Lin, C.-C. (2009) *J. Am. Chem. Soc.*, 131 (9), 3138–3139.

95 Liang, C.-F., Kuan, T.-C., Chang, T.-C., and Lin, C.-C. (2012) *J. Am. Chem. Soc.*, 134 (38), 16074–16079.
96 Kancharla, P.K., Navuluri, C., and Crich, D. (2012) *Angew. Chem. Int. Ed.*, 51 (44), 11105–11109.
97 Ye, X.-S., Huang, X., and Wong, C.-H. (2001) *Chem. Commun.*, (11), 974–975.
98 Crich, D. and Wu, B. (2008) *Tetrahedron*, 64 (9), 2042–2047.
99 Wang, Y. and Ye, X.-S. (2009) *Tetrahedron Lett.*, 50 (27), 3823–3826.
100 Premathilake, H.D., Gobble, C.P., Pornsuriyasak, P., Hardimon, T., Demchenko, A.V., and De Meo, C. (2012) *Org. Lett.*, 14 (4), 1126–1129.
101 Tsvetkov, Y.E. and Schmidt, R.R. (1994) *Tetrahedron Lett.*, 35 (46), 8583–8586.

Index

Selective Glycosylations: Synthetic Methods and Catalysts, First Edition. Edited by Clay S. Bennett.